Intermediate Probability

Intermediate Probability

A Computational Approach

Marc S. Paolella

Swiss Banking Institute, University of Zurich, Switzerland

John Wiley & Sons, Ltd

Other Wiley Editorial Offices

John Wiley & Sons Inc., 111 River Street, Hoboken, NJ 07030, USA

Jossey-Bass, 989 Market Street, San Francisco, CA 94103-1741, USA

Wiley-VCH Verlag GmbH, Boschstr. 12, D-69469 Weinheim, Germany

John Wiley & Sons Australia Ltd, 42 McDougall Street, Milton, Queensland 4064, Australia

John Wiley & Sons (Asia) Pte Ltd, 2 Clementi Loop #02-01, Jin Xing Distripark, Singapore 129809

John Wiley & Sons Canada Ltd, 6045 Freemont Blvd, Mississauga, Ontario, L5R 4J3, Canada

Wiley also publishes its books in a variety of electronic formats. Some content that appears
in print may not be available in electronic books.

Anniversary Logo Design: Richard J. Pacifico

Library of Congress Cataloging-in-Publication Data

Paolella, Marc S.
Intermediate probability : a computational approach / Marc S. Paolella.
     p. cm.
ISBN 978-0-470-02637-3 (cloth)
1. Distribution (Probability theory)–Mathematical models. 2. Probabilities. I. Title.
QA273.6.P36 2007
519.2 – dc22

                                                          2007020127

British Library Cataloguing in Publication Data

A catalogue record for this book is available from the British Library

ISBN-13: 978-0-470-02637-3

Typeset in 10/12 Times by Laserwords Private Limited, Chennai, India

This book is printed on acid-free paper responsibly manufactured from sustainable forestry
in which at least two trees are planted for each one used for paper production.

Chapter Listing

Preface

Part I   Sums of Random Variables

1   **Generating functions**
2   **Sums and other functions of several random variables**
3   **The multivariate normal distribution**

Part II   Asymptotics and Other Approximations

4   **Convergence concepts**
5   **Saddlepoint approximations**
6   **Order statistics**

Part III   More Flexible and Advanced Random Variables

7   **Generalizing and mixing**
8   **The stable Paretian distribution**
9   **Generalized inverse Gaussian and generalized hyperbolic distributions**
10 **Noncentral distributions**

Appendix

A   **Notation and distribution tables**

References
Index

Contents

Preface   xi

Part I   Sums of Random Variables   **1**

1   Generating functions   **3**

1.1   The moment generating function   3
    1.1.1   Moments and the m.g.f.   4
    1.1.2   The cumulant generating function   7
    1.1.3   Uniqueness of the m.g.f.   11
    1.1.4   Vector m.g.f.   14
1.2   Characteristic functions   17
    1.2.1   Complex numbers   17
    1.2.2   Laplace transforms   22
    1.2.3   Basic properties of characteristic functions 23
    1.2.4   Relation between the m.g.f. and c.f.   25
    1.2.5   Inversion formulae for mass and density functions 27
    1.2.6   Inversion formulae for the c.d.f.   36
1.3   Use of the fast Fourier transform   40
    1.3.1   Fourier series   40
    1.3.2   Discrete and fast Fourier transforms   48
    1.3.3   Applying the FFT to c.f. inversion   50
1.4   Multivariate case   53
1.5   Problems   55

2   Sums and other functions of several random variables   **65**

2.1   Weighted sums of independent random variables   65
2.2   Exact integral expressions for functions of two continuous random
    variables   72

| | 2.3 | Approximating the mean and variance | 85 |
| | 2.4 | Problems | 90 |

3 | | **The multivariate normal distribution** | **97** |
	3.1	Vector expectation and variance	97
	3.2	Basic properties of the multivariate normal	100
	3.3	Density and moment generating function	106
	3.4	Simulation and c.d.f. calculation	108
	3.5	Marginal and conditional normal distributions	111
	3.6	Partial correlation	116
	3.7	Joint distribution of $\overline{X}$ and S^2 for i.i.d. normal samples	119
	3.8	Matrix algebra	122
	3.9	Problems	124

Part II Asymptotics and Other Approximations | | | **127** |

4 | | **Convergence concepts** | **129** |
	4.1	Inequalities for random variables	130
	4.2	Convergence of sequences of sets	136
	4.3	Convergence of sequences of random variables	142
		4.3.1 Convergence in probability	142
		4.3.2 Almost sure convergence	145
		4.3.3 Convergence in r-mean	150
		4.3.4 Convergence in distribution	153
	4.4	The central limit theorem	157
	4.5	Problems	163

5 | | **Saddlepoint approximations** | **169** |
	5.1	Univariate	170
		5.1.1 Density saddlepoint approximation	170
		5.1.2 Saddlepoint approximation to the c.d.f.	175
		5.1.3 Detailed illustration: the normal–Laplace sum	179
	5.2	Multivariate	184
		5.2.1 Conditional distributions	184
		5.2.2 Bivariate c.d.f. approximation	186
		5.2.3 Marginal distributions	189
	5.3	The hypergeometric functions $_1F_1$ and $_2F_1$	193
	5.4	Problems	198

6   Order statistics   **203**

   6.1   Distribution theory for i.i.d. samples   204
      6.1.1   Univariate   204
      6.1.2   Multivariate   210
      6.1.3   Sample range and midrange   215
   6.2   Further examples   219
   6.3   Distribution theory for dependent samples   230
   6.4   Problems   231

Part III   More Flexible and Advanced Random Variables   **237**

7   Generalizing and mixing   **239**

   7.1   Basic methods of extension   239
      7.1.1   Nesting and generalizing constants   240
      7.1.2   Asymmetric extensions   244
      7.1.3   Extension to the real line   247
      7.1.4   Transformations   249
      7.1.5   Invention of flexible forms   252
   7.2   Weighted sums of independent random variables   254
   7.3   Mixtures   254
      7.3.1   Countable mixtures   255
      7.3.2   Continuous mixtures   258
   7.4   Problems   269

8   The stable Paretian distribution   **277**

   8.1   Symmetric stable   277
   8.2   Asymmetric stable   281
   8.3   Moments   287
      8.3.1   Mean   287
      8.3.2   Fractional absolute moment proof I   288
      8.3.3   Fractional absolute moment proof II   293
   8.4   Simulation   296
   8.5   Generalized central limit theorem   297

9   Generalized inverse Gaussian and generalized hyperbolic distributions   **299**

   9.1   Introduction   299
   9.2   The modified Bessel function of the third kind   300
   9.3   Mixtures of normal distributions   303

	9.3.1	Mixture mechanics	303
	9.3.2	Moments and generating functions	304
9.4	The generalized inverse Gaussian distribution		306
	9.4.1	Definition and general formulae	306
	9.4.2	The subfamilies of the GIG distribution family	308
9.5	The generalized hyperbolic distribution		315
	9.5.1	Definition, parameters and general formulae	315
	9.5.2	The subfamilies of the GHyp distribution family	317
	9.5.3	Limiting cases of GHyp	327
9.6	Properties of the GHyp distribution family		328
	9.6.1	Location−scale behaviour of GHyp	328
	9.6.2	The parameters of GHyp	329
	9.6.3	Alternative parameterizations of GHyp	330
	9.6.4	The shape triangle	332
	9.6.5	Convolution and infinite divisibility	336
9.7	Problems		338

10 Noncentral distributions — **341**

10.1	Noncentral chi-square	341
	10.1.1 Derivation	341
	10.1.2 Moments	344
	10.1.3 Computation	346
	10.1.4 Weighted sums of independent central χ^2 random variables	347
	10.1.5 Weighted sums of independent $\chi^2(n_i, \theta_i)$ random variables	351
10.2	Singly and doubly noncentral F	357
	10.2.1 Derivation	357
	10.2.2 Moments	358
	10.2.3 Exact computation	360
	10.2.4 Approximate computation methods	363
10.3	Noncentral beta	369
10.4	Singly and doubly noncentral t	370
	10.4.1 Derivation	371
	10.4.2 Saddlepoint approximation	378
	10.4.3 Moments	381
10.5	Saddlepoint uniqueness for the doubly noncentral F	382
10.6	Problems	384

A   Notation and distribution tables — **389**

References — **401**

Index — **413**

Preface

This book is a sequel to Volume I, *Fundamental Probability: A Computational Approach* (2006), `http://www.wiley.com/WileyCDA/WileyTitle/productCd-04 70025948.html`, which covered the topics typically associated with a first course in probability at an undergraduate level. This volume is particularly suited to beginning graduate students in statistics, finance and econometrics, and can be used independently of Volume I, although references are made to it. For example, the third equation of Chapter 2 in Volume I is referred to as (I.2.3), whereas (2.3) means the third equation of Chapter 2 of the present book. Similarly, a reference to Section I.2.3 means Section 3 of Chapter 2 in Volume I.

The presentation style is the same as that in Volume I. In particular, computational aspects are incorporated throughout. Programs in Matlab are given for all computations in the text, and the book's website will provide these programs, as well as translations in the R language. Also, as in Volume I, emphasis is placed on solving more practical and challenging problems than is often done in such a course. As a case in point, Chapter 1 emphasizes the use of characteristic functions for calculating the density and distribution of random variables by way of (i) numerically computing the integrals involved in the inversion formulae, and (ii) the use of the fast Fourier transform. As many students may not be comfortable with the required mathematical machinery, a stand-alone introduction to complex numbers, Fourier series and the discrete Fourier transform are given as well.

The remaining chapters, in brief, are as follows.

Chapter 2 uses the tools developed in Chapter 1 to calculate the distribution of sums of random variables. I start with the usual, algebraically trivial examples using the moment generating function (m.g.f.) of independent and identically distributed (i.i.d) random variables (r.v.s), such as gamma and Bernoulli. More interesting and useful, but less commonly discussed, is the question of how to compute the distribution of a sum of independent r.v.s when the resulting m.g.f. is not 'recognizable', e.g., a sum of independent gamma r.v.s with different scale parameters, or the sum of binomial r.v.s with differing values of p, or the sum of independent normal and Laplace r.v.s.

Chapter 3 presents the multivariate normal distribution. Along with numerous examples and detailed coverage of the standard topics, computational methods for calculating the c.d.f. of the bivariate case are discussed, as well as partial correlation, which is required for understanding the partial autocorrelation function in time series analysis.

Chapter 4 is on asymptotics. As some of this material is mathematically more challenging, the emphasis is on providing careful and highly detailed proofs of basic results and as much intuition as possible.

Chapter 5 gives a basic introduction to univariate and multivariate saddlepoint approximations, which allow us to quickly and accurately invert the m.g.f. of sums of independent random variables without requiring the numerical integration (and occasional numeric problems) associated with the inversion formulae. The methods complement those developed in Chapters 1 and 2, and will be used extensively in Chapter 10. The beauty, simplicity, and accuracy of this method are reason enough to discuss it, but its applicability to such a wide range of topics is what should make this methodology as much of a standard topic as is the central limit theorem. Much of the section on multivariate saddlepoint methods was written by my graduate student and fellow researcher, Simon Broda.

Chapter 6 deals with order statistics. The presentation is quite detailed, with numerous examples, as well as some results which are not often seen in textbooks, including a brief discussion of order statistics in the non-i.i.d. case.

Chapter 7 is somewhat unique and provides an overview on how to help 'classify' some of the hundreds of distributions available. Of course, not all methods can be covered, but the ideas of nesting, generalizing, and asymmetric extensions are introduced. Mixture distributions are also discussed in detail, leading up to derivation of the variance–gamma distribution.

Chapter 8 is about the stable Paretian distribution, with emphasis on its computation, basic properties, and uses. With the unprecedented growth of it in applications (due primarily to its computational complexity having been overcome), this should prove to be a useful and timely topic well worth covering. Sections 8.3.2 and 8.3.3 were written together with my graduate student and fellow researcher, Yianna Tchopourian.

Chapter 9 is dedicated to the (generalized) inverse Gaussian and (generalized) hyperbolic distributions, and their connections. In addition to being mathematically intriguing, they are well suited for modelling a wide variety of phenomena. The author of this chapter, and all its problems and solutions, is my academic colleague Walther Paravicini.

Chapter 10 provides a quite detailed account of the singly and doubly noncentral F, t and beta distributions. For each, several methods for the exact calculation of the distribution are provided, as well as discussion of approximate methods, most notably the saddlepoint approximation.

The Appendix contains a list of tables, including those for abbreviations, special functions, general notation, generating functions and inversion formulae, distribution naming conventions, distributional subsets (e.g., $\chi^2 \subseteq \mathrm{Gam}$ and $\mathrm{N} \subseteq \mathrm{S}\alpha\mathrm{S}$), Student's t generalizations, noncentral distributions, relationships among major distributions, and mixture relationships.

As in Volume I, the examples are marked with symbols to designate their relative importance, with $\ominus$, $\odot$ and $\circledast$ indicating low, medium and high importance, respectively. Also as in Volume I, there are many exercises, and they are furnished with stars to indicate their difficulty and/or amount of time required for solution. Solutions to all exercises, in full detail, are available for instructors, as are lecture notes for beamer presentation. As discussed in the Preface to Volume I, *not everything in the text is supposed to be (or could be) covered in the classroom.* I prefer to use lecture time for discussing the major results and letting students work on some problems (algebraically and with a computer), leaving some derivations and examples for reading outside of the classroom.

The companion website for the book is `http://www.wiley.com/go/intermediate`.

ACKNOWLEDGEMENTS

I am indebted to Ronald Butler for teaching and working with me on several saddle-point approximation projects, including work on the doubly noncentral F distribution, the results of which appear in Chapter 10. The results on the saddlepoint approximation for the doubly noncentral t distribution represent joint work with Simon Broda. As mentioned above, Simon also contributed greatly to the section on multivariate saddlepoint methods. He has also devised some advanced exercises in Chapters 1 and 10, programmed Pan's (1968) method for calculating the distribution of a weighted sum of independent, central χ^2 r.v.s (see Section 10.1.4), and has proofread numerous sections of the book. Besides helping to write the technical sections in Chapter 8, Yianna Tchopourian has proofread Chapter 4 and singlehandedly tracked down the sources of all the quotes I used in this book. This book project has been significantly improved because of their input and I am extremely greatful for their help.

It is through my time as a student of, and my later joint work and common research ideas with, Stefan Mittnik and Svetlozar (Zari) Rachev that I became aware of the usefulness and numeric tractability via the fast Fourier transform of the stable Paretian distribution (and numerous other fields of knowledge in finance and statistics). I wish to thank them for their generosity, friendship and guidance over the last decade.

As already mentioned, Chapter 9 was written by Walther Paravicini, and he deserves all the credit for the well-organized presentation of this interesting and nontrivial subject matter. Furthermore, Walther has proofread the entire book and made substantial suggestions and corrections for Chapter 1, as well as several hundred comments and corrections in the remaining chapters. I am highly indebted to Walther for his substantial contribution to this book project.

One of my goals with this project was to extend the computing platform from Matlab to the R language, so that students and instructors have the choice of which to use. I wish to thank Sergey Goriatchev, who has admirably done the job of translating all the Matlab programs appearing in Volume I into the R language; those for the present volume are in the works. The Matlab and R code for both books will appear on the books' web pages.

Finally, I thank the editorial team Susan Barclay, Kelly Board, Richard Leigh, Simon Lightfoot, and Kathryn Sharples at John Wiley & Sons, Ltd for making this project go as smoothly and pleasantly as possible. A special thank-you goes to my copy editor, Richard Leigh, for his in-depth proofreading and numerous suggestions for improvement, not to mention the masterful final appearance of the book.

PART I

SUMS OF RANDOM VARIABLES

1

Generating functions

The shortest path between two truths in the real domain passes through the complex domain.   (Jacques Hadamard)

There are various integrals of the form

$$\int_{-\infty}^{\infty} g(t, x) \, dF_X(x) = \mathbb{E}[g(t, X)] \tag{1.1}$$

which are often of great value for studying r.v.s. For example, taking $g(n, x) = x^n$ and $g(n, x) = |x|^n$, for $n \in \mathbb{N}$, give the algebraic and absolute moments, respectively, while $g(n, x) = x_{[n]} = x(x - 1) \cdots (x - n + 1)$ yields the factorial moments of X, which are of use for lattice r.v.s. Also important (if not essential) for working with lattice distributions with nonnegative support is the *probability generating function*, obtained by taking $g(t, x) = t^x$ in (1.1), i.e., $\mathbb{P}_X(t) := \sum_{x=0}^{\infty} t^x p_x$, where $p_x = \Pr(X = x)$.[1]

For our purposes, we will concentrate on the use of the two forms $g(t, x) = \exp(tx)$ and $g(t, x) = \exp(itx)$, which are not only applicable to both discrete and continuous r.v.s, but also, as we shall see, of enormous theoretical and practical use.

1.1 The moment generating function

The *moment generating function* (m.g.f.), of random variable X is the function $\mathbb{M}_X$: $\mathbb{R} \mapsto \mathbb{X}_{\geq 0}$ (where $\mathbb{X}$ is the extended real line) given by $t \mapsto \mathbb{E}\left[e^{tX}\right]$. The m.g.f. $\mathbb{M}_X$ is said to exist if it is finite on a neighbourhood of zero, i.e., if there is an $h > 0$ such that, $\forall t \in (-h, h)$, $\mathbb{M}_X(t) < \infty$. If $\mathbb{M}_X$ exists, then the largest (open) interval $\mathcal{I}$

---

[1] Probability generating functions arise ubiquitously in the study of stochastic processes (often the 'next course' after an introduction to probability such as this). There are numerous books, at various levels, on stochastic processes; three highly recommended 'entry-level' accounts which make generous use of probability generating functions are Kao (1996), Jones and Smith (2001), and Stirzaker (2003). See also Wilf (1994) for a general account of generating functions.

around zero such that $\mathbb{M}_X(t) < \infty$ for $t \in \mathcal{I}$ is referred to as the *convergence strip (of the m.g.f.) of* X.

1.1.1 Moments and the m.g.f.

A fundamental result is that, if $\mathbb{M}_X$ exists, then all positive moments of X exist. This is worth emphasizing:

$$\text{If } \mathbb{M}_X \text{ exists, then } \forall r \in \mathbb{R}_{>0}, \ \mathbb{E}\left[|X|^r\right] < \infty. \tag{1.2}$$

To prove (1.2), let r be an arbitrary positive (real) number, and recall that $\lim_{x \to \infty} x^r/e^x = 0$, as shown in (I.7.3) and (I.A.36). This implies that, $\forall t \in \mathbb{R} \setminus 0$, $\lim_{x \to \infty} x^r/e^{|tx|} = 0$. Choose an $h > 0$ such that $(-h, h)$ is in the convergence strip of X, and a value t such that $0 < t < h$ (so that $\mathbb{E}\left[e^{tX}\right]$ and $\mathbb{E}\left[e^{-tX}\right]$ are finite). Then there must exist an x_0 such that $|x|^r < e^{|tx|}$ for $|x| > x_0$. For $|x| \le x_0$, there exists a finite constant K_0 such that $|x|^r < K_0 e^{|tx|}$. Thus, there exists a K such that $|x|^r < K e^{|tx|}$ for all x, so that, from the inequality-preserving nature of expectation (see Section I.4.4.2), $\mathbb{E}\left[|X|^r\right] \le K \mathbb{E}\left[e^{|tX|}\right]$. Finally, from the trivial identity $e^{|tx|} \le e^{tx} + e^{-tx}$ and the linearity of the expectation operator, $\mathbb{E}\left[e^{|tX|}\right] \le \mathbb{E}\left[e^{tX}\right] + \mathbb{E}\left[e^{-tX}\right] < \infty$, showing that $\mathbb{E}\left[|X|^r\right]$ is finite.

Remark: This previous argument also shows that, if the m.g.f. of X is finite on the interval $(-h, h)$ for some $h > 0$, then so is the m.g.f. of r.v. $|X|$ on the same neighbourhood. Let $|t| < h$, so that $\mathbb{E}\left[e^{t|X|}\right]$ is finite, and let $k \in \mathbb{N} \cup 0$. From the Taylor series of e^x, it follows that $0 \le |tX|^k/k! \le e^{|tX|}$, implying $\mathbb{E}\left[|tX|^k\right] \le k! \mathbb{E}\left[e^{|tX|}\right] < \infty$. Moreover, for all $N \in \mathbb{N}$,

$$S(N) := \sum_{k=0}^{N} \left| \frac{\mathbb{E}\left[|tX|^k\right]}{k!} \right| = \sum_{k=0}^{N} \frac{\mathbb{E}\left[|tX|^k\right]}{k!} = \mathbb{E}\left(\sum_{k=0}^{N} \frac{|tX|^k}{k!} \right) \le \mathbb{E}\left[e^{|tX|}\right],$$

so that

$$\lim_{N \to \infty} S(N) = \sum_{k=0}^{\infty} \frac{\mathbb{E}\left[|tX|^k\right]}{k!} \le \mathbb{E}\left[e^{|tX|}\right]$$

and the infinite series converges absolutely. Now, as $\left|\mathbb{E}\left[(tX)^k\right]\right| \le \mathbb{E}\left[|tX|^k\right] < \infty$, it follows that the series $\sum_{k=0}^{\infty} \mathbb{E}\left[(tX)^k\right]/k!$ also converges. As $\sum_{k=0}^{\infty}(tX)^k/k!$ converges pointwise to e^{tX}, and $|e^{tX}| \le e^{|tX|}$, the dominated convergence theorem applied to the integral of the expectation operator implies

$$\lim_{N \to \infty} \mathbb{E}\left[\sum_{k=0}^{N} \frac{(tX)^k}{k!} \right] = \mathbb{E}\left[e^{tX}\right].$$

That is,

$$M_X(t) = \mathbb{E}\left[e^{tX}\right] = \mathbb{E}\left[\sum_{k=0}^{\infty} \frac{(tX)^k}{k!}\right] = \sum_{k=0}^{\infty} \frac{t^k}{k!}\mathbb{E}\left[X^k\right], \tag{1.3}$$

which is important for the next result. ■

It can be shown that termwise differentiation of (1.3) is valid, so that the jth derivative with respect to t is

$$M_X^{(j)}(t) = \sum_{i=j}^{\infty} \frac{t^{i-j}}{(i-j)!}\mathbb{E}\left[X^i\right] = \sum_{n=0}^{\infty} \frac{t^n}{n!}\mathbb{E}\left[X^{n+j}\right]$$

$$= \mathbb{E}\left[\sum_{n=0}^{\infty} \frac{(tX)^n X^j}{n!}\right] = \mathbb{E}\left[X^j \sum_{n=0}^{\infty} \frac{(tX)^n}{n!}\right] = \mathbb{E}\left[X^j e^{tX}\right], \tag{1.4}$$

or

$$\boxed{M_X^{(j)}(t)\Big|_{t=0} = \mathbb{E}\left[X^j\right].}$$

Similarly, it can be shown that we are justified in arriving at (1.4) by simply writing

$$M_X^{(j)}(t) = \frac{d^j}{dt^j}\mathbb{E}\left[e^{tX}\right] = \mathbb{E}\left[\frac{d^j}{dt^j}e^{tX}\right] = \mathbb{E}\left[X^j e^{tX}\right].$$

In general, if $M_Z(t)$ is the m.g.f. of r.v. Z and $X = \mu + \sigma Z$, then it is easy to show that

$$M_X(t) = \mathbb{E}\left[e^{tX}\right] = \mathbb{E}\left[e^{t(\mu+\sigma Z)}\right] = e^{t\mu}M_Z(t\sigma). \tag{1.5}$$

The next two examples illustrates the computation of the m.g.f. in a discrete and continuous case, respectively.

⊖ ***Example 1.1*** Let $X \sim \text{DUnif}(\theta)$ with p.m.f. $f_X(x; \theta) = \theta^{-1}\mathbb{I}_{\{1,2,\ldots,\theta\}}(x)$. The m.g.f. of X is

$$M_X(t) = \mathbb{E}\left[e^{tX}\right] = \frac{1}{\theta}\sum_{j=1}^{\theta} e^{tj},$$

so that

$$M_X'(t) = \frac{1}{\theta}\sum_{j=1}^{\theta} j e^{tj}, \qquad \mathbb{E}[X] = M_X'(0) = \frac{1}{\theta}\sum_{j=1}^{\theta} j = \frac{\theta+1}{2},$$

and

$$M_X''(t) = \frac{1}{\theta} \sum_{j=1}^{\theta} j^2 e^{tj}, \qquad \mathbb{E}[X^2] = M_X''(0) = \frac{1}{\theta} \sum_{j=1}^{\theta} j^2 = \frac{(\theta+1)(2\theta+1)}{6},$$

from which it follows that

$$\mathbb{V}(X) = \mu_2' - \mu^2 = \frac{(\theta+1)(2\theta+1)}{6} - \left(\frac{\theta+1}{2}\right)^2 = \frac{(\theta-1)(\theta+1)}{12},$$

recalling (I.4.40). More generally, letting $X \sim \text{DUnif}(\theta_1, \theta_2)$ with p.d.f. $f_X(x; \theta_1, \theta_2) = (\theta_2 - \theta_1 + 1)^{-1} \mathbb{I}_{\{\theta_1, \theta_1+1, \dots, \theta_2\}}(x)$,

$$\mathbb{E}[X] = \frac{1}{2}(\theta_1 + \theta_2) \quad \text{and} \quad \mathbb{V}(X) = \frac{1}{12}(\theta_2 - \theta_1)(\theta_2 - \theta_1 + 2),$$

which can be shown directly using the m.g.f., or by simple symmetry arguments. ∎

⊖ **Example 1.2**   Let $U \sim \text{Unif}(0, 1)$. Then,

$$M_U(t) = \mathbb{E}[e^{tU}] = \int_0^1 e^{tu} \, du = \frac{e^t - 1}{t}, \quad t \neq 0,$$

which is finite in any neighbourhood of zero, and continuous at zero, as, via l'Hôpital's rule,

$$\lim_{t \to 0} \frac{e^t - 1}{t} = \lim_{t \to 0} \frac{e^t}{1} = 1 = \int_0^1 e^{0u} \, du = M_U(0).$$

The Taylor series expansion of $M_U(t)$ around zero is

$$\frac{e^t - 1}{t} = \frac{1}{t}\left(t + \frac{t^2}{2} + \frac{t^3}{6} + \cdots\right) = 1 + \frac{t}{2} + \frac{t^2}{6} + \cdots = \sum_{j=0}^{\infty} \frac{1}{j+1} \frac{t^j}{j!}$$

so that, from (1.3),

$$\mathbb{E}[U^r] = (r+1)^{-1}, \quad r = 1, 2, \dots. \tag{1.6}$$

In particular,

$$\mathbb{E}[U] = \frac{1}{2}, \quad \mathbb{E}[U^2] = \frac{1}{3}, \quad \mathbb{V}(U) = \frac{1}{3} - \frac{1}{4} = \frac{1}{12}.$$

Of course, (1.6) could have been derived with much less work and in more generality, as

$$\mathbb{E}[U^r] = \int_0^1 u^r \, du = (r+1)^{-1}, \quad r \in \mathbb{R}_{>0}.$$

For $X \sim \text{Unif}(a, b)$, write $X = U(b - a) + a$ so that, from the binomial theorem and (1.6),

$$\mathbb{E}[X^r] = \sum_{j=0}^{r} \binom{r}{j} a^{r-j} (b-a)^j \frac{1}{j+1} = \frac{b^{r+1} - a^{r+1}}{(b-a)(r+1)}, \qquad (1.7)$$

where the last equality is given in (I.1.57). Alternatively, we can use the location–scale relationship (1.5) with $\mu = a$ and $\sigma = b - a$ to get

$$\mathbb{M}_X(t) = \frac{e^{tb} - e^{ta}}{t(b-a)}, \quad t \neq 0, \qquad \mathbb{M}_X(0) = 1.$$

Then, with $j = i - 1$ and $t \neq 0$,

$$\mathbb{M}_X(t) = \frac{1}{t(b-a)} \left(\sum_{i=0}^{\infty} \frac{(tb)^i}{i!} - \sum_{k=0}^{\infty} \frac{(ta)^k}{k!} \right) = \sum_{i=1}^{\infty} \frac{b^i - a^i}{i!(b-a)} t^{i-1}$$

$$= \sum_{j=0}^{\infty} \frac{b^{j+1} - a^{j+1}}{(j+1)!(b-a)} t^j = \sum_{j=0}^{\infty} \frac{b^{j+1} - a^{j+1}}{(j+1)(b-a)} \frac{t^j}{j!},$$

which, from (1.3), yields the result in (1.7). ∎

1.1.2 The cumulant generating function

The *cumulant generating function* (c.g.f.), is defined as

$$\boxed{\mathbb{K}_X(t) = \log \mathbb{M}_X(t).} \qquad (1.8)$$

The terms κ_i in the series expansion $\mathbb{K}_X(t) = \sum_{r=0}^{\infty} \kappa_r t^r / r!$ are referred to as the *cumulants* of X, so that the ith derivative of $\mathbb{K}_X(t)$ evaluated at $t = 0$ is κ_i, i.e.,

$$\boxed{\kappa_i = \mathbb{K}_X^{(i)}(t)\Big|_{t=0}.}$$

It is straightforward to show that

$$\boxed{\kappa_1 = \mu, \qquad \kappa_2 = \mu_2, \qquad \kappa_3 = \mu_3, \qquad \kappa_4 = \mu_4 - 3\mu_2^2} \qquad (1.9)$$

(see Problem 1.1), with higher-order terms given in Stuart and Ord (1994, Section 3.14).

⊛ ***Example 1.3*** From Problem I.7.17, the m.g.f. of $X \sim N(\mu, \sigma^2)$ is given by

$$\mathbb{M}_X(t) = \exp\left\{ \mu t + \frac{1}{2} \sigma^2 t^2 \right\}, \qquad \mathbb{K}_X(t) = \mu t + \frac{1}{2} \sigma^2 t^2. \qquad (1.10)$$

Thus,

$$\mathbb{K}'_X(t) = \mu + \sigma^2 t, \quad \mathbb{E}[X] = \mathbb{K}'_X(0) = \mu, \quad \mathbb{K}''_X(t) = \sigma^2, \quad \mathbb{V}(X) = \mathbb{K}''_X(0) = \sigma^2,$$

and $\mathbb{K}_X^{(i)}(t) = 0$, $i \geq 3$, so that $\mu_3 = 0$ and $\mu_4 = \kappa_4 + 3\mu_2^2 = 3\sigma^4$, as also determined directly in Example I.7.3. This also shows that X has skewness $\mu_3/\mu_2^{3/2} = 0$ and kurtosis $\mu_4/\mu_2^2 = 3$. ∎

◎ **Example 1.4** For $X \sim \text{Poi}(\lambda)$,

$$\mathbb{M}_X(t) = \mathbb{E}[e^{tX}] = \sum_{x=0}^{\infty} \frac{e^{tx}e^{-\lambda}\lambda^x}{x!} = e^{-\lambda} \sum_{x=0}^{\infty} \frac{(\lambda e^t)^x}{x!}$$

$$= \exp\left(-\lambda + \lambda e^t\right). \tag{1.11}$$

As $\mathbb{K}_X^{(r)}(t) = \lambda e^t$ for $r \geq 1$, it follows that $\mathbb{E}[X] = \mathbb{K}'_X(t)\big|_{t=0} = \lambda$ and $\mathbb{V}(X) = \mathbb{K}''_X(t)\big|_{t=0} = \lambda$. This calculation should be compared with that in (I.4.34). Once the m.g.f. is available, higher moments are easily obtained, in particular,

$$\text{skew}(X) = \mu_3/\mu_2^{3/2} = \lambda/\lambda^{3/2} = \lambda^{-1/2} \to 0$$

and

$$\text{kurt}(X) = \mu_4/\mu_2^2 = \left(\kappa_4 + 3\mu_2^2\right)/\mu_2^2 = \left(\lambda + 3\lambda^2\right)/\lambda^2 \to 3,$$

as $\lambda \to \infty$. That is, as λ increases, the skewness and kurtosis of a Poisson random variable tend towards the skewness and kurtosis of a normal random variable. ∎

◎ **Example 1.5** For $X \sim \text{Gam}(a, b)$, the m.g.f. is, with $y = x(b - t)$,

$$\mathbb{M}_X(t) = \mathbb{E}[e^{tX}]$$

$$= \frac{b^a}{\Gamma(a)} \int_0^{\infty} x^{a-1}e^{-x(b-t)}\,dx = (b - t)^{-a}b^a \int_0^{\infty} \frac{1}{\Gamma(a)}y^{a-1}e^{-y}\,dy$$

$$= \left(\frac{b}{b-t}\right)^a, \quad t < b.$$

From this,

$$\mathbb{E}[X] = \frac{d\mathbb{M}_X(t)}{dt}\bigg|_{t=0} = a\left(\frac{b}{b-t}\right)^{a-1}b(b-t)^{-2}\bigg|_{t=0} = \frac{a}{b}$$

or, more easily, with $\mathbb{K}_X(t) = a(\ln b - \ln(b - t))$, (1.9) implies

$$\kappa_1 = \mathbb{E}[X] = \frac{d\mathbb{K}_X(t)}{dt}\bigg|_{t=0} = \frac{a}{b-t}\bigg|_{t=0} = \frac{a}{b} \tag{1.12}$$

and

$$\kappa_2 = \mu_2 = \mathbb{V}(X) = \left.\frac{d\mathbb{K}_X^2(t)}{dt^2}\right|_{t=0} = \left.\frac{a}{(b-t)^2}\right|_{t=0} = \frac{a}{b^2}.$$

Similarly,

$$\mu_3 = \frac{2a}{b^3} \quad \text{and} \quad \kappa_4 = \frac{6a}{b^4},$$

i.e., $\mu_4 = \kappa_4 + 3\mu_2^2 = 3a(2+a)/b^4$, so that the skewness and kurtosis are

$$\frac{\mu_3}{\mu_2^{3/2}} = \frac{2a/b^3}{\left(a/b^2\right)^{3/2}} = \frac{2}{\sqrt{a}} \quad \text{and} \quad \frac{\mu_4}{\mu_2^2} = \frac{3a(2+a)/b^4}{\left(a/b^2\right)^2} = \frac{3(2+a)}{a}. \tag{1.13}$$

These converge to 0 and 3, respectively, as a increases.   ∎

⊖ **Example 1.6**   From density (I.7.51), the m.g.f. of a location-zero, scale-one logistic random variable is (with $y = \left(1 + e^{-x}\right)^{-1}$), for $|t| < 1$,

$$\mathbb{M}_X(t) = \mathbb{E}\left[e^{tX}\right] = \int_{-\infty}^{\infty} \left(e^{-x}\right)^{1-t}\left(1+e^{-x}\right)^{-2} dx$$

$$= \int_0^1 \left(\frac{1-y}{y}\right)^{1-t} y^2 y^{-1}(1-y)^{-1}\, dy = \int_0^1 (1-y)^{-t} y^t\, dy$$

$$= B(1-t, 1+t) = \Gamma(1-t)\,\Gamma(1+t).$$

If, in addition, $t \neq 0$, the m.g.f. can also be expressed as

$$\mathbb{M}_X(t) = t\Gamma(t)\,\Gamma(1-t) = t\frac{\pi}{\sin \pi t}, \tag{1.14}$$

where the second identity is *Euler's reflection formula.*[2]   ∎

For certain problems, the m.g.f. can be expressed recursively, as the next example shows.

⊖ **Example 1.7**   Let $N_m \sim \text{Consec}(m, p)$, i.e., N_m is the random number of Bernoulli trials, each with success probability p, which need to be conducted until m successes in a row occur. The mean of N_m was computed in Example I.8.13 and the variance

---

[2] Andrews, Askey and Roy (1999, pp. 9–10) provide four different methods for proving Euler's reflection formula; see also Jones (2001, pp. 217–18), Havil (2003, p. 59), or Schiff (1999, p. 174). As an aside, from (1.14) with $t = 1/2$, it follows that $\Gamma(1/2) = \sqrt{\pi}$.

and m.g.f. in Problem I.8.13. In particular, from (I.8.52), with $\mathbb{M}_m(t) := \mathbb{M}_{N_m}(t)$ and $q = 1 - p$,

$$\mathbb{M}_m(t) = \frac{pe^t \mathbb{M}_{m-1}(t)}{1 - q\mathbb{M}_{m-1}(t)\,e^t}. \tag{1.15}$$

This can be recursively evaluated with $\mathbb{M}_1(t) = pe^t / \left(1 - qe^t\right)$ for $t \neq -\ln(1 - p)$, from the geometric distribution. Example 1.20 below illustrates how to use (1.15) to obtain the p.m.f. Problem 1.10 uses (1.15) to compute $\mathbb{E}[N_m]$. ∎

Calculation of the m.g.f. can also be useful for determining the expected value of particular functions of random variables, as illustrated next.

⊖ ***Example 1.8***   To determine $\mathbb{E}[\ln X]$ when $X \sim \chi_v^2$, we could try to directly integrate, i.e.,

$$\mathbb{E}[\ln X] = \frac{1}{2^{v/2}\Gamma(v/2)} \int_0^\infty (\ln x)\, x^{v/2-1} e^{-x/2}\, dx, \tag{1.16}$$

but this seems to lead nowhere. Note instead that the m.g.f. of $Z = \ln X$ is

$$\mathbb{M}_Z(t) = \mathbb{E}\left[e^{tZ}\right] = \mathbb{E}\left[X^t\right] = \frac{1}{2^{v/2}\Gamma(v/2)} \int_0^\infty x^{t+v/2-1} e^{-x/2}\, dx$$

or, with $y = x/2$,

$$\mathbb{M}_Z(t) = \frac{2^{t+v/2-1+1}}{2^{v/2}\Gamma(v/2)} \int_0^\infty y^{t+v/2-1} e^{-y}\, dy = 2^t \frac{\Gamma(t + v/2)}{\Gamma(v/2)}.$$

Then, with $d2^t/dt = 2^t \ln 2$,

$$\frac{d}{dt}\mathbb{M}_Z(t) = \frac{1}{\Gamma(v/2)} \left(2^t\, \Gamma'(t + v/2) + 2^t\, \ln 2\, \Gamma(t + v/2)\right)$$

and

$$\mathbb{E}[\ln X] = \frac{d}{dt}\mathbb{M}_Z(t)\bigg|_{t=0} = \frac{\Gamma'(v/2)}{\Gamma(v/2)} + \ln 2 = \psi(v/2) + \ln 2.$$

Having seen the answer, the integral (1.16) is easy; differentiating $\Gamma(v/2)$ with respect to $v/2$, using (I.A.43), and setting $y = 2x$,

$$\Gamma'\left(\frac{v}{2}\right) = \int_0^\infty \frac{d}{d(v/2)} x^{v/2-1} e^{-x}\, dx = \int_0^\infty x^{v/2-1} (\ln x)\, e^{-x}\, dx$$

$$= \int_0^\infty \left(\frac{y}{2}\right)^{v/2-1} \left(\ln \frac{y}{2}\right) e^{-y/2} \frac{dy}{2}$$

$$= \frac{1}{2^{v/2}} \int_0^\infty y^{v/2-1} (\ln y)\, e^{-y/2}\, dy - \frac{\ln 2}{2^{v/2}} \int_0^\infty y^{v/2-1} e^{-y/2}\, dy$$

$$= \Gamma(v/2)\, \mathbb{E}[\ln X] - (\ln 2)\, \Gamma(v/2),$$

giving $\mathbb{E}[\ln X] = \Gamma'(v/2)/\Gamma(v/2) + \ln 2$. ∎

1.1.3 Uniqueness of the m.g.f.

Under certain conditions, the m.g.f. uniquely determines or *characterizes* the distribution. To be more specific, we need the concept of equality in distribution: Let r.v.s X and Y be defined on the (induced) probability space $\{\mathbb{R}, \mathcal{B}, \Pr(\cdot)\}$, where $\mathcal{B}$ is the Borel σ-field generated by the collection of intervals $(a, b]$, $a, b \in \mathbb{R}$. Then X and Y are said to be *equal in distribution*, written $X \overset{d}{=} Y$, if

$$\Pr(X \in A) = \Pr(Y \in A) \quad \forall A \in \mathcal{B}. \tag{1.17}$$

The uniqueness result states that for r.v.s X and Y and some $h > 0$,

$$\mathbb{M}_X(t) = \mathbb{M}_Y(t) \; \forall \; |t| < h \quad \Rightarrow \quad X \overset{d}{=} Y. \tag{1.18}$$

See Section 1.2.4 below for some insight into why this result is true. As a concrete example, if the m.g.f. of an r.v. X is the same as, say, that of an exponential r.v., then one can conclude that X is exponentially distributed.

A similar notion applies to sequences of r.v.s, for which we need the concept of convergence in distribution. For a sequence of r.v.s X_n, $n = 1, 2, \ldots$, we say that X_n *converges in distribution* to X, written $X_n \overset{d}{\to} X$, if $F_{X_n}(x) \to F_X(x)$ as $n \to \infty$, for all points x such that $F_X(x)$ is continuous. Section 4.3.4 provides much more detail. It is important to note that if F_X is continuous, then it need not be the case that the F_{X_n} are continuous.

If X_n converges in distribution to a random variable which is, say, normally distributed, we will write $X_n \overset{d}{\to} N(\cdot, \cdot)$, where the mean and variance of the specified normal distribution are constants, and do not depend on n. Observe that $X_n \overset{d}{\to} N\left(\mu, \sigma^2\right)$ implies that, for n sufficiently large, the distribution of X_n can be adequately approximated by that of a $N\left(\mu, \sigma^2\right)$ random variable. We will denote this by writing $X_n \overset{\text{app}}{\sim} N\left(\mu, \sigma^2\right)$. This notation also allows the right-hand-side (r.h.s.) variable to depend on n; for example, we will write $S_n \overset{\text{app}}{\sim} N(n, n)$ to indicate that, as n increases, the distribution of S_n can be adequately approximated by a $N(n, n)$ random variable. In this case, we cannot write $S_n \overset{d}{\to} N(n, n)$, but it is true that $n^{-1/2}(S_n - n) \overset{d}{\to} N(0, 1)$.

We are now ready to state the convergence result for m.g.f.s. Let X_n be a sequence of r.v.s such that the corresponding m.g.f.s $\mathbb{M}_{X_n}(t)$ exist for $|t| < h$, for some $h > 0$, and all $n \in \mathbb{N}$. If X is a random variable whose m.g.f. $\mathbb{M}_X(t)$ exists for $|t| \leq h_1 < h$ for some $h_1 > 0$ and $\mathbb{M}_{X_n}(t) \to \mathbb{M}_X(t)$ as $n \to \infty$ for $|t| < h_1$, then $X_n \overset{d}{\to} X$. This convergence result also applies to the c.g.f. (1.8).

$\ominus$ **Example 1.9**
(a) Let X_n, $n = 1, 2, \ldots$, be a sequence of r.v.s such that $X_n \sim \text{Bin}(n, p_n)$, with $p_n = \lambda/n$, for some constant value $\lambda \in \mathbb{R}_{>0}$, so that $\mathbb{M}_{X_n}(t) = \left(p_n e^t + 1 - p_n\right)^n$ (see Problem 1.4), or

$$\mathbb{M}_{X_n}(t) = \left(\frac{\lambda}{n}e^t + 1 - \frac{\lambda}{n}\right)^n = \left(1 + \frac{\lambda}{n}\left(e^t - 1\right)\right)^n.$$

For all $h > 0$ and $|t| < h$, $\lim_{n \to \infty} \mathbb{M}_{X_n}(t) = \exp\left\{\lambda\left(e^t - 1\right)\right\} = \mathbb{M}_P(t)$, where $P \sim$ Poi (λ). That is, $X_n \xrightarrow{d}$ Poi (λ). Informally speaking, the binomial distribution with increasing n and decreasing p, such that np is a constant, approaches a Poisson distribution. This was also shown in Chapter I.4 by using the p.m.f. of a binomial random variable.

(b) Let $P_\lambda \sim$ Poi (λ), $\lambda \in \mathbb{R}_{>0}$, and $Y_\lambda = (P_\lambda - \lambda)/\sqrt{\lambda}$. From (1.5),

$$\mathbb{M}_{Y_\lambda}(t) = \exp\left\{\lambda\left(e^{t/\sqrt{\lambda}} - 1\right) - t\sqrt{\lambda}\right\}.$$

Writing

$$e^{t/\sqrt{\lambda}} = 1 + \frac{t}{\lambda^{1/2}} + \frac{t^2}{2\lambda} + \frac{t^3}{3!\lambda^{3/2}} + \cdots,$$

we see that

$$\lim_{\lambda \to \infty}\left[\lambda\left(e^{t/\sqrt{\lambda}} - 1\right) - t\sqrt{\lambda}\right] = \frac{t^2}{2},$$

or $\lim_{\lambda \to \infty} \mathbb{M}_{Y_\lambda}(t) = \exp\left(t^2/2\right)$, which is the m.g.f. of a standard normal random variable. That is, $Y_\lambda \xrightarrow{d} \text{N}(0, 1)$ as $\lambda \to \infty$. This should not be too surprising in light of the skewness and kurtosis results of Example 1.4.

(c) Let $P_\lambda \sim$ Poi (λ) with $\lambda \in \mathbb{N}$, and $Y_\lambda = (P_\lambda - \lambda)/\sqrt{\lambda}$. Then

$$p_{1,\lambda} := \frac{e^{-\lambda}\lambda^\lambda}{\lambda!} = \Pr(P_\lambda = \lambda) = \Pr(\lambda - 1 < P_\lambda \leq \lambda) = \Pr\left(\frac{-1}{\sqrt{\lambda}} < Y_\lambda \leq 0\right).$$

From the result in part **(b)** above, the limiting distribution of Y_λ is standard normal, motivating the conjecture that

$$\Pr\left(\frac{-1}{\sqrt{\lambda}} < Y_\lambda \leq 0\right) \approx \Phi(0) - \Phi\left(-\lambda^{-1/2}\right) =: p_{2,\lambda}, \tag{1.19}$$

where $\approx$ means that, as $\lambda \to \infty$, the ratio of the two sides approaches unity. To informally verify (1.19), Figure 1.1 plots the relative percentage error (RPE), $100(p_{2,\lambda} - p_{1,\lambda})/p_{1,\lambda}$, on a log scale, as a function of λ.

The mean value theorem (Section I.A.2.2.2) implies the existence of an $x_\lambda \in \left(-\lambda^{-1/2}, 0\right)$ such that

$$\frac{\Phi(0) - \Phi\left(-\lambda^{-1/2}\right)}{0 - \left(-\lambda^{-1/2}\right)} = \Phi'(x_\lambda) = \phi(x_\lambda).$$

Clearly, $x_\lambda \in \left(-\lambda^{-1/2}, 0\right) \to 0$ as $\lambda \to \infty$, so that

$$\Phi(0) - \Phi\left(-\lambda^{-1/2}\right) = \frac{\lambda^{-1/2}}{\sqrt{2\pi}} \exp\left\{-\frac{1}{2}x_\lambda^2\right\} \approx \frac{\lambda^{-1/2}}{\sqrt{2\pi}}.$$

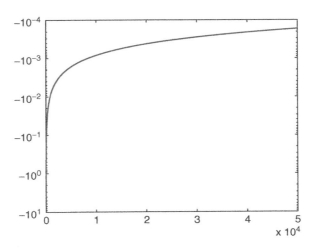

Figure 1.1   The relative percentage error of (1.19) as a function of λ

Combining these results yields

$$\frac{e^{-\lambda}\lambda^{\lambda}}{\lambda!} \approx \frac{\lambda^{-1/2}}{\sqrt{2\pi}},$$

or, rearranging, $\lambda! \approx \sqrt{2\pi}\,\lambda^{\lambda+1/2}e^{-\lambda}$. We understand this to mean that, for large λ, $\lambda!$ can be accurately approximated by the r.h.s. quantity, which is Stirling's approximation. ∎

⊖ *Example 1.10*
(a) Let $b > 0$ be a fixed value and, for any $a > 0$, let $X_a \sim \text{Gam}\,(a, b)$ and $Y_a = (X_a - a/b)\,/\sqrt{a/b^2}$. Then, for $t < a^{1/2}$,

$$\mathbb{M}_{Y_a}(t) = e^{-t\sqrt{a}}\mathbb{M}_{X_a}\left(\frac{b}{\sqrt{a}}t\right) = e^{-t\sqrt{a}}\left(\frac{1}{1 - a^{-1/2}t}\right)^a,$$

or $\mathbb{K}_{Y_a}(t) = -t\sqrt{a} - a\log\left(1 - a^{-1/2}t\right)$. From (I.A.114),

$$\log(1 + x) = \sum_{i=1}^{\infty} (-1)^{i+1} \frac{x^i}{i},$$

so that

$$\log\left(1 - a^{-1/2}t\right) = -\frac{t}{a^{1/2}} - \frac{t^2}{2a} - \frac{t^3}{3a^{3/2}} - \cdots$$

and $\lim_{a\to\infty}\mathbb{K}_{Y_a}(t) = t^2/2$. Thus, as $a \to \infty$, $Y_a \overset{d}{\to} \text{N}\,(0, 1)$, or, for large a, $X_a \overset{\text{app}}{\sim} \text{N}\left(a/b, a/b^2\right)$. Again, recall the skewness and kurtosis results of Example 1.5.

(b) Now let $S_n \sim \text{Gam}(n, 1)$ for $n \in \mathbb{N}$, so that, for large n, $S_n \overset{\text{app}}{\sim} \text{N}(n, n)$. The definition of convergence in distribution, and the continuity of the c.d.f. of S_n and that of its limiting distribution, informally suggest the limiting behaviour of the p.d.f. of S_n, i.e.,

$$f_{S_n}(s) = \frac{1}{\Gamma(n)} s^{n-1} \exp(-s) \approx \frac{1}{\sqrt{2\pi n}} \exp\left(-\frac{(s-n)^2}{2n^2}\right).$$

Choosing $s = n$ leads to $\Gamma(n+1) = n! \approx \sqrt{2\pi}(n+1)^{n+1/2} \exp(-n-1)$. From (I.A.46), $\lim_{n \to \infty} (1 + \lambda/n)^n = e^\lambda$, so

$$(n+1)^{n+1/2} = n^{n+1/2}\left(1 + \frac{1}{n}\right)^{n+1/2} \approx n^{n+1/2} e,$$

and substituting this into the previous expression for $n!$ yields Stirling's approximation $n! \approx \sqrt{2\pi}\, n^{n+1/2}\, e^{-n}$. ∎

1.1.4   Vector m.g.f.

Analogous to the univariate case, the (joint) m.g.f. of the vector $\mathbf{X} = (X_1, \ldots, X_n)$ is defined as

$$\boxed{\mathbb{M}_{\mathbf{X}}(\mathbf{t}) = \mathbb{E}\left[e^{\mathbf{t}'\mathbf{X}}\right], \quad \mathbf{t} = (t_1, \ldots, t_n),}$$

and exists if the expectation is finite on an open rectangle of $\mathbf{0}$ in $\mathbb{R}^n$, i.e., if there is a $\varepsilon > 0$ such that $\mathbb{E}[e^{\mathbf{t}'\mathbf{X}}]$ is finite for all $\mathbf{t}$ such that $|t_i| < \varepsilon$ for $i = 1, \ldots, n$.

As in the univariate case, if the joint m.g.f. exists, then it characterizes the distribution of $\mathbf{X}$ and, thus, all the marginals as well. In particular,

$$\mathbb{M}_{\mathbf{X}}((0, \ldots, 0, t_i, 0, \ldots, 0)) = \mathbb{E}\left[e^{t_i X_i}\right] = \mathbb{M}_{X_i}(t_i), \quad i = 1, \ldots, n.$$

Generalizing (1.4) and assuming the validity of exchanging derivative and integral,

$$\frac{\partial^k \mathbb{M}_{\mathbf{X}}(\mathbf{t})}{\partial t_1^{k_1} \partial t_2^{k_2} \cdots \partial t_n^{k_n}} = \int_{-\infty}^{\infty} \cdots \int_{-\infty}^{\infty} x_1^{k_1} x_2^{k_2} \cdots x_n^{k_n} \exp\{t_1 x_1 + t_2 x_2 + \cdots + t_n x_n\} f_{\mathbf{X}}(\mathbf{x})\, d\mathbf{x},$$

so that the integer product moments of $\mathbf{X}$, $\mathbb{E}\left[\prod_{i=1}^n X_i^{k_i}\right]$ for $k_i \in \mathbb{N}$, are given by

$$\left.\frac{\partial^k \mathbb{M}_{\mathbf{X}}(\mathbf{t})}{\partial t_1^{k_1} \partial t_2^{k_2} \cdots \partial t_n^{k_n}}\right|_{\mathbf{t}=\mathbf{0}} = \int_{-\infty}^{\infty} \cdots \int_{-\infty}^{\infty} x_1^{k_1} x_2^{k_2} \cdots x_n^{k_n} f_{\mathbf{X}}(\mathbf{x})\, dx_1\, dx_2 \cdots dx_n \qquad (1.20)$$

for $k = \sum_{i=1}^{n} k_i$ and such that $k_i = 0$ means that the derivative with respect to t_i is not taken. For example, if X and Y are r.v.s with m.g.f. $\mathbb{M}_{X,Y}(t_1, t_2)$, then

$$\mathbb{E}[XY] = \left. \frac{\partial^2 \mathbb{M}_{X,Y}(t_1, t_2)}{\partial t_1 \partial t_2} \right|_{t_1 = t_2 = 0}$$

and

$$\mathbb{E}[X^2] = \left. \frac{\partial^2 \mathbb{M}_{X,Y}(t_1, t_2)}{\partial t_1^2} \right|_{t_1 = t_2 = 0} = \left. \frac{\partial^2 \mathbb{M}_{X,Y}(t_1, 0)}{\partial t_1^2} \right|_{t_1 = 0}.$$

⊖ ***Example 1.11*** (Example I.8.12 cont.) Let $f_{X,Y}(x, y) = e^{-y} \mathbb{I}_{(0,\infty)}(x) \mathbb{I}_{(x,\infty)}(y)$ be the joint density of r.v.s X and Y. The m.g.f. is

$$\mathbb{M}_{X,Y}(t_1, t_2) = \int_0^\infty \int_x^\infty \exp\{t_1 x + t_2 y - y\} \, dy \, dx \qquad (1.21)$$

$$= \int_0^\infty \frac{1}{1 - t_2} \exp\{x(t_1 + t_2 - 1)\} \, dx$$

$$= \frac{1}{(1 - t_1 - t_2)(1 - t_2)}, \quad t_1 + t_2 < 1, \quad t_2 < 1, \qquad (1.22)$$

so that $\mathbb{M}_{X,Y}(t_1, 0) = (1 - t_1)^{-1}$, $t_1 < 1$, and $\mathbb{M}_{X,Y}(0, t_2) = (1 - t_2)^{-2}$, $t_2 < 1$. From Example 1.5, this implies that $X \sim \text{Exp}(1)$ and $Y \sim \text{Gam}(2, 1)$. Also,

$$\left. \frac{\partial \mathbb{M}_{X,Y}(t_1, 0)}{\partial t_1} \right|_{t_1 = 0} = (1 - t_1)^{-2} \big|_{t_1 = 0} = 1,$$

$$\left. \frac{\partial \mathbb{M}_{X,Y}(0, t_2)}{\partial t_2} \right|_{t_2 = 0} = 2(1 - t_2)^{-3} \big|_{t_2 = 0} = 2,$$

and

$$\frac{\partial^2 \mathbb{M}_{X,Y}(t_1, t_2)}{\partial t_1 \partial t_2} = \frac{3t_2 + t_1 - 3}{(t_1 + t_2 - 1)^3 (t_2 - 1)^2}, \quad \left. \frac{\partial^2 \mathbb{M}_{X,Y}(t_1, t_2)}{\partial t_1 \partial t_2} \right|_{t_1 = t_2 = 0} = 3,$$

so that $\mathbb{E}[X] = 1$, $\mathbb{E}[Y] = 2$ and $\text{Cov}(X, Y) = \mathbb{E}[XY] - \mathbb{E}[X]\mathbb{E}[Y] = 1$. ∎

The following result is due to Sawa (1972, p. 658), , and he used it for evaluating the moments of an estimator arising in an important class of econometric models; see also Sawa (1978). Let X_1 and X_2 be r.v.s such that $\Pr(X_1 > 0) = 1$, with joint m.g.f. $\mathbb{M}_{X_1, X_2}(t_1, t_2)$ which exists for $t_1 < \epsilon$ and $|t_2| < \epsilon$, for $\epsilon > 0$. Then, if it exists, the kth-order moment, $k \in \mathbb{N}$, of X_2/X_1 is given by

$$\mathbb{E}\left[\left(\frac{X_2}{X_1}\right)^k\right] = \frac{1}{\Gamma(k)} \int_{-\infty}^0 (-t_1)^{k-1} \left[\frac{\partial^k}{\partial t_2^k} \mathbb{M}_{X_1, X_2}(t_1, t_2) \right]_{t_2 = 0} dt_1. \qquad (1.23)$$

To informally verify this, assume we may reverse the order of the expectation with either the derivative or integral with respect to t_1 and t_2, so that the r.h.s. of (1.23) is

$$\frac{1}{\Gamma(k)} \int_{-\infty}^{0} (-t_1)^{k-1} \left[\frac{\partial^k}{\partial t_2^k} \mathbb{E}\left[e^{t_1 X_1} e^{t_2 X_2} \right] \right]_{t_2=0} dt_1$$

$$= \frac{1}{\Gamma(k)} \mathbb{E}\left[\left[\frac{\partial^k}{\partial t_2^k} e^{t_2 X_2} \right]_{t_2=0} \int_{-\infty}^{0} (-t_1)^{k-1} e^{t_1 X_1} dt_1 \right]$$

$$= \frac{1}{\Gamma(k)} \mathbb{E}\left[X_2^k \int_0^{\infty} u^{k-1} e^{-u X_1} du \right] = \mathbb{E}\left[\left(\frac{X_2}{X_1} \right)^k \right].$$

By working with $\mathbb{M}_{X_2,X_1}(t_2, t_1)$ instead of $\mathbb{M}_{X_1,X_2}(t_1, t_2)$, an expression for $\mathbb{E}[(X_1/X_2)^k]$ immediately results, though in terms of the more natural $\mathbb{M}_{X_1,X_2}(t_1, t_2)$, we get the following. Similar to (1.23), let X_1 and X_2 be r.v.s such that $\Pr(X_2 > 0) = 1$, with joint m.g.f. $\mathbb{M}_{X_1,X_2}(t_1, t_2)$ which exists for $|t_1| < \epsilon$ and $t_2 > -\epsilon$, for $\epsilon > 0$. Then the kth-order moment, $k \in \mathbb{N}$, of X_1/X_2 is given by

$$\mathbb{E}\left[\left(\frac{X_1}{X_2} \right)^k \right] = \frac{1}{\Gamma(k)} \int_0^{\infty} t_2^{k-1} \left[\frac{\partial^k}{\partial t_1^k} \mathbb{M}_{X_1,X_2}(t_1, -t_2) \right]_{t_1=0} dt_2, \qquad (1.24)$$

if it exists. To confirm this, the r.h.s. of (1.24) is (indulging in complete lack of rigour),

$$\frac{1}{\Gamma(k)} \int_0^{\infty} t_2^{k-1} \left[\frac{\partial^k}{\partial t_1^k} \mathbb{E}\left[e^{t_1 X_1} e^{-t_2 X_2} \right] \right]_{t_1=0} dt_2$$

$$= \frac{1}{\Gamma(k)} \mathbb{E}\left[\left[\frac{\partial^k}{\partial t_1^k} e^{t_1 X_1} \right]_{t_1=0} \int_0^{\infty} t_2^{k-1} e^{-t_2 X_2} dt_2 \right] = \mathbb{E}\left[\left(\frac{X_1}{X_2} \right)^k \right].$$

Remark: A rigorous derivation of (1.23) and (1.24) is more subtle than it might appear. A flaw in Sawa's derivation is noted by Mehta and Swamy (1978), who provide a more rigorous derivation of this result. However, even the latter authors did not correctly characterize Sawa's error, as pointed out by Meng (2005), who provides the (so far) definitive conditions and derivation of the result for the more general case of $\mathbb{E}[X_1^k/X_2^b]$, $k \in \mathbb{N}$, $b \in \mathbb{R}_{>0}$, and also references to related results and applications.[3] Meng also provides several interesting examples of the utility of working with the joint m.g.f., including relationships to earlier work by R. A. Fisher. An important use of (1.24) arises in the study of ratios of quadratic forms.

The inequality

$$\mathbb{E}\left[\left(\frac{X_2}{X_1} \right)^k \right] \geq \frac{\mathbb{E}\left[X_2^k \right]}{\mathbb{E}\left[X_1^k \right]} \qquad (1.25)$$

is shown in Mullen (1967).   ■

---

[3] Lange (2003, p. 39) also provides an expression for $\mathbb{E}[X_1^k/X_2^b]$.

⊖ **Example 1.12**   (Example 1.11 cont.) From (1.22) and (1.24),

$$\mathbb{E}\left[\frac{X}{Y}\right] = \int_0^\infty \left[\frac{\partial}{\partial t_1} \frac{1}{(1 - t_1 + t_2)(1 + t_2)}\right]_{t_1=0} dt_2 = \int_0^\infty (1 + t_2)^{-3} \, dt_2 = \frac{1}{2}$$

and

$$\mathbb{E}\left[\left(\frac{X}{Y}\right)^2\right] = 2\int_0^\infty t_2 (1 + t_2)^{-4} \, dt_2 = \frac{1}{3},$$

so that $\mathbb{V}(X/Y) = 1/12$. This is confirmed another way in Problem 2.6.   ■

1.2 Characteristic functions

Similar to the m.g.f., the *characteristic function* (c.f.) of r.v. X is defined as $\mathbb{E}\left[e^{itX}\right]$, where $i^2 = -1$, and is usually denoted as $\varphi_X(t)$. The c.f. is fundamental to probability theory and of much greater importance than the m.g.f. Its widespread use in intro- ductory expositions of probability theory, however, is hampered because its involves notions from complex analysis, with which not all students are familiar. This is reme- died to some extent via Section 1.2.1, which provides enough material for the reader to understand the rest of the chapter. More detailed treatments of c.f.s can be found in textbooks on advanced probability theory such as Wilks (1963), Billingsley (1995), Shiryaev (1996), Fristedt and Gray (1997), Gut (2005), or the book by Lukacs (1970), which is dedicated to the topic.

While it may not be too shocking that complex analysis arises in the theoretical underpinnings of probability theory, it might come as a surprise that it greatly assists *numerical* aspects by giving rise to expressions for real quantities which would other- wise not have been at all obvious. This, in fact, is true in general in mathematics (see the quote by Jacques Hadamard at the beginning of this chapter).

1.2.1 Complex numbers

Should I refuse a good dinner simply because I do not understand the process of digestion?
(Oliver Heaviside)

The *imaginary unit* i is defined to be a number having the property that

$$i^2 = -1. \tag{1.26}$$

One can use i in calculations as one does any ordinary real number such as $1, -1$ or $\sqrt{2}$, so expressions such as $1 + i$, i^5 or $3 - 5i$ can be interpreted naively. We define the set of all complex numbers to be $\mathbb{C} := \{a + bi \mid a, b \in \mathbb{R}\}$. If $z = a + bi$, then $\mathrm{Re}(z) := a$ and $\mathrm{Im}(z) := b$ are the real and imaginary parts of z.

The set of complex numbers is closed under addition and multiplication, i.e., sums and products of complex numbers are also complex numbers. In particular,

$$(a + bi) + (c + di) = (a + c) + (b + d)i$$
$$(a + bi) \cdot (c + di) = (ac - bd) + (bc + ad)i.$$

As a special case, note that $i^3 = -i$ and $i^4 = 1$. Therefore we have $i = i^5 = i^9 = \ldots$.
For each complex number $z = a + bi$, its *complex conjugate* is defined as $\bar{z} = a - bi$. The product $z \cdot \bar{z} = (a + bi)(a - bi) = a^2 - b^2i^2 = a^2 + b^2$ is always a non-negative real number. The sum is

$$z + \bar{z} = (a + bi) + (a - bi) = 2a = 2\,\mathrm{Re}(z). \tag{1.27}$$

The absolute value of z, or its *(complex) modulus*, is defined to be

$$|z| = |a + bi| = \sqrt{z\bar{z}} = \sqrt{a^2 + b^2}. \tag{1.28}$$

Simple calculations show that

$$|z_1 z_2| = |z_1||z_2|, \quad |z_1 + z_2| \le |z_1| + |z_2|, \quad \overline{z_1 z_2} = \overline{z_1}\,\overline{z_2}, \quad z_1, z_2 \in \mathbb{C}. \tag{1.29}$$

A sequence z_n of complex numbers is said to converge to some complex number $z \in \mathbb{C}$ iff the sequences $\mathrm{Re}\,z_n$ and $\mathrm{Im}\,z_n$ converge to $\mathrm{Re}\,z$ and $\mathrm{Im}\,z$, respectively. Hence, the series $\sum_{n=1}^{\infty} z_n$ converges if the series $\sum_{n=1}^{\infty} \mathrm{Re}\,z_n$ and $\sum_{n=1}^{\infty} \mathrm{Im}\,z_n$ converge separately. As in $\mathbb{R}$, define the exponential function by

$$\exp(z) = \sum_{k=0}^{\infty} \frac{z^k}{k!}, \quad z \in \mathbb{C}.$$

It can be shown that, as in $\mathbb{R}$, $\exp(z_1 + z_2) = \exp(z_1)\exp(z_2)$ for every $z_1, z_2 \in \mathbb{C}$. The definitions of the fundamental trigonometric functions in (I.A.28), i.e.,

$$\cos(z) = \sum_{k=0}^{\infty}(-1)^k \frac{z^{2k}}{(2k)!} \quad \text{and} \quad \sin(z) = \sum_{k=0}^{\infty}(-1)^k \frac{z^{2k+1}}{(2k+1)!},$$

also hold for complex numbers. In particular, if z takes the form $z = it$, where $t \in \mathbb{R}$, then $\exp(z)$ can be expressed as

$$\exp(it) = \sum_{k=0}^{\infty} \frac{(it)^k}{k!} = \sum_{k=0}^{\infty} \frac{i^k t^k}{k!} = \sum_{k=0}^{\infty}(-1)^k \frac{t^{2k}}{(2k)!} + i\sum_{k=0}^{\infty}(-1)^k \frac{t^{2k+1}}{(2k+1)!}, \tag{1.30}$$

i.e., from (I.A.28),

$$\boxed{\exp(it) = \cos(t) + i\sin(t).} \tag{1.31}$$

This relation is of fundamental importance, and is known as the *Euler formula*.[4]

---

[4] Named after the prolific Leonhard Euler (1707–1783), though (as often with naming conventions) it was actually discovered and published years before, in 1714, by Roger Cotes (1682–1716).

It easily follows from (1.31) that

$$\sin z = \frac{e^{iz} - e^{-iz}}{2i} \quad \text{and} \quad \cos z = \frac{e^{iz} + e^{-iz}}{2}. \tag{1.32}$$

Also, from (1.31) using $t = \pi$, we have $\cos \pi + i \sin \pi = -1$, or $e^{i\pi} + 1 = 0$, which is a simple but famous equation because it contains five of the most important quantities in mathematics. Similarly, $\exp(2\pi i) = 1$, so that, for $z \in \mathbb{C}$,

$$\exp(z + 2\pi i) = \exp(z) \exp(2\pi i) = \exp(z),$$

and one says that exp is a $2\pi i$-*cyclic* function. Lastly, with $z = a + ib \in \mathbb{C}$, (1.31) gives

$$\exp(\bar{z}) = \exp(a - bi) = \exp(a) \exp(-bi) = \exp(a)\big[\cos(-b) + i \sin(-b)\big]$$
$$= \exp(a)\big[\cos(b) - i \sin(b)\big] = \exp(a)\overline{\exp(ib)} = \overline{\exp(a + ib)} = \overline{\exp(z)}.$$

As a shorthand for $\cos(t) + i \sin(t)$, one sometimes sees $\mathrm{cis}(t) := \cos(t) + i \sin(t)$, i.e., $\mathrm{cis}(t) = \exp(it)$.

A complex-valued function can also be integrated: the Riemann integral of a complex-valued function is the sum of the Riemann integrals of its real and imaginary parts.

⊛ ***Example 1.13*** For $s \in \mathbb{R} \setminus 0$, we know that $\int e^{st} \, dt = s^{-1} e^{st}$, but what if $s \in \mathbb{C}$? Let $s = x + iy$, and use (1.31) and the integral results in Example I.A.24 to write

$$\int e^{(x+iy)t} \, dt = \int e^{xt} \cos(yt) \, dt + i \int e^{xt} \sin(yt) \, dt$$

$$= \frac{e^{xt}}{x^2 + y^2} (x \cos(yt) + y \sin(yt)) + i \frac{e^{xt}}{x^2 + y^2} (x \sin(yt) - y \cos(yt)).$$

This, however, is the same as $s^{-1} e^{st}$, as can be seen by writing

$$\frac{e^{st}}{s} = \frac{e^{xt}(\cos(yt) + i \sin(yt))}{x + iy} \frac{x - iy}{x - iy},$$

with $(x + iy)(x - iy) = x^2 + y^2$ and multiplying out the numerator. Thus,

$$\int e^{st} \, dt = s^{-1} e^{st}, \quad s \in \mathbb{C} \setminus 0, \tag{1.33}$$

a result which will be used below. ∎

A geometric approach to the complex numbers represents them as vectors in the plane, with the real term on the horizontal axis and the imaginary term on the vertical axis. Thus, the sum of two complex numbers can be interpreted as the sum of two vectors, and the modulus of $z \in \mathbb{C}$ is the length from 0 to z in the complex plane, recalling Pythagoras' theorem. The *unit circle* is the circle in the complex plane of radius 1 centred at 0, and includes all complex numbers of absolute value 1, i.e., such that $|z| = 1$; see Figure 1.2(a). If $t \in \mathbb{R}$, then the number $\exp(it)$ is contained in the unit circle, because

$$|\exp(it)| = \sqrt{\cos^2(t) + \sin^2(t)} = 1, \quad t \in \mathbb{R}. \tag{1.34}$$

For example, if $z = a + bi \in \mathbb{C}$, $a, b \in \mathbb{R}$, then (1.31) implies

$$\exp(z) = \exp(a + bi) = \exp(a)\exp(bi) = \exp(a)\big[\cos(b) + i\sin(b)\big],$$

and from (1.34),

$$|\exp(z)| = |\exp(a)||\exp(bi)| = \exp(a) = \exp(\mathrm{Re}(z)). \tag{1.35}$$

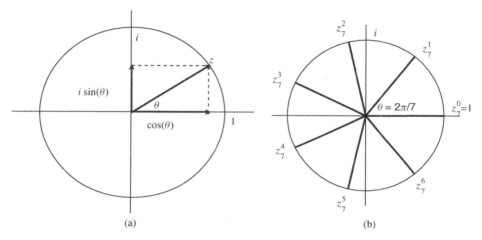

(a)   (b)

Figure 1.2 (a) Geometric representation of complex number $z = \cos(\theta) + i\sin(\theta)$ in the complex plane. (b) Plot of powers of $z_n = \exp(2\pi i/n)$ for $n = 7$, demonstrating that $\sum_{j=0}^{n-1} z_n^j = 0$

From the depiction of z as a vector in the complex plane, polar coordinates can also be used to represent z when $z \neq 0$. Let $r = |z| = \sqrt{a^2 + b^2}$ and define the *(complex) argument*, or *phase angle*, of z, denoted $\arg(z)$, to be the angle, say θ (in radians, measured counterclockwise from the positive real axis, modulo 2π), such that $a = r\cos(\theta)$ and $b = r\sin(\theta)$, i.e., for $a \neq 0$, $\arg(z) := \arctan(b/a)$. This is shown in Figure 1.2(a) for $r = 1$. From (1.31),

$$z = a + bi = r\cos(\theta) + ir\sin(\theta) = r\,\mathrm{cis}(\theta) = re^{i\theta},$$

and, as $r = |z|$ and $\theta = \arg(z)$, we can write

$$\mathrm{Re}(z) = a = |z|\cos(\arg z) \quad \text{and} \quad \mathrm{Im}(z) = b = |z|\sin(\arg z). \tag{1.36}$$

Now observe that, if $z_j = r_j \exp(i\theta_j) = r_j\,\mathrm{cis}(\theta_j)$, then

$$z_1 z_2 = r_1 r_2 \exp(i(\theta_1 + \theta_2)) = r_1 r_2\,\mathrm{cis}(\theta_1 + \theta_2), \tag{1.37}$$

so that

$$\arg(z_1 z_2 \cdots z_n) = \arg(z_1) + \arg(z_2) + \cdots + \arg(z_n) \quad \text{and} \quad \arg z^n = n \arg(z).$$

The following two examples illustrate very simple results which are used below in Example 1.25.

⊖ **Example 1.14**   Let $z = 1 - ik$ for some $k \in \mathbb{R}$. Set $z = re^{i\theta}$ so that $r = \sqrt{1 + k^2}$ and $\theta = \arctan(-k/1) = -\arctan(k)$. Then, with $z^m = r^m e^{i\theta m}$, $|z^m| = r^m = \left(1 + k^2\right)^{m/2}$ and $\arg(z^m) = \theta m = -m \arctan(k)$.   ∎

⊖ **Example 1.15**   Let $z = \exp\{ia/(1 - ib)\}$ for $a, b \in \mathbb{R}$. As

$$\frac{ia}{1 - ib} = -\frac{ab}{1 + b^2} + i\frac{a}{1 + b^2},$$

we can write

$$re^{i\theta} = z = \exp\left(-\frac{ab}{1 + b^2} + i\frac{a}{1 + b^2}\right) = \exp\left(-\frac{ab}{1 + b^2}\right)\exp\left(i\frac{a}{1 + b^2}\right),$$

for

$$r = \exp\left(-\frac{ab}{1 + b^2}\right) \quad \text{and} \quad \theta = \frac{a}{1 + b^2} \text{ modulo } 2\pi.$$

   ∎

The next example derives a simple but highly useful result which we will require when working with the discrete Fourier transform.

⊛ **Example 1.16**   Recall that the length of the unit circle is 2π, and let θ be the phase angle of the complex number z measured in radians (the arc length of the piece of the unit circle from $1 + 0i$ to z in Figure 1.2(a)). Then the quantity $z_n := \exp(2\pi i/n)$, $n \in \mathbb{N}$, plotted as a vector, will 'carve out' an nth of the unit circle, and, from (1.37), n equal pieces of the unit circle are obtained by plotting $z_n^0, z_n^1, \ldots, z_n^{n-1}$. This is shown in Figure 1.2(b) for $n = 7$. When seen as vectors emanating from the centre, it is clear that their sum is zero, i.e., for any $n \in \mathbb{N}$, $\sum_{j=0}^{n-1} z_n^j = 0$. More generally, for $k \in \{1, \ldots, n-1\}$, $\sum_{j=0}^{n-1}\left(z_n^k\right)^j = 0$, and clearly, $\sum_{j=0}^{n-1}\left(z_n^0\right)^j = n$. Because $z_n^n = 1 = z_n^{-n}$, this can be written as

$$\sum_{j=0}^{n-1}\left(z_n^k\right)^j = \begin{cases} n, & \text{if } k \in n\mathbb{Z} := \{0, n, -n, 2n, -2n, \ldots\}, \\ 0, & \text{if } k \in \mathbb{Z} \setminus n\mathbb{Z}. \end{cases} \tag{1.38}$$

The first part of (1.38) is trivial. For the second part, let $k \in \mathbb{Z} \setminus n\mathbb{Z}$ and $b := \sum_{j=0}^{n-1}(z_n^k)^j$. Note that $(z_n^k)^n = (z_n^n)^k = 1$. It follows that

$$z_n^k b = \sum_{j=0}^{n-1}(z_n^k)^{j+1} = \sum_{i=1}^{n-1}(z_n^k)^i + (z_n^k)^n = \sum_{i=0}^{n-1}(z_n^k)^i = b.$$

As $z_n^k \neq 1$ it follows that $b = 0$. See also Problem 1.23.   ∎

1.2.2 Laplace transforms

The *Laplace transform* of a real or complex function g of a real variable is denoted by $\mathcal{L}\{g\}$, and defined by

$$G(s) := \mathcal{L}\{g\}(s) := \int_0^\infty g(t)\,e^{-st}\,dt, \qquad (1.39)$$

for all real or complex numbers s, if the integral exists (see below). From the form of (1.39), there is clearly a relationship between the Laplace transform and the moment generating function, and indeed, the m.g.f. is sometimes referred to as a two-sided Laplace transform. We study it here instead of in Section 1.1 above because we allow s to be complex. The Laplace transform is also related to the Fourier transform, which is discussed below in Section 1.3 and Problem 1.19.

1.2.2.1 Existence of the Laplace transform

The integral (1.39) exists for $\mathrm{Re}(s) > \alpha$ if g is continuous on $[0, \infty)$ and g has *exponential order* α, i.e., $\exists\, \alpha \in \mathbb{R}$, $\exists\, M > 0$, $\exists\, t_0 \geq 0$ such that $|g(t)| \leq Me^{\alpha t}$ for $t \geq t_0$.[5] To see this, let g be of exponential order α and (piecewise) continuous. Then (as g is bounded on all subintervals on $\mathbb{R}_{\geq 0}$), $\exists M > 0$ such that $|g(t)| \leq Me^{\alpha t}$ for $t \geq 0$, and, with $s = x + iy$,

$$\int_0^u \left| g(t)\,e^{-st} \right| dt \leq M \int_0^u \left| e^{-(s-\alpha)t} \right| dt = M \int_0^u \left| e^{-(x-\alpha)t} \right| dt = M \int_0^u e^{-(x-\alpha)t}\,dt,$$

where the second to last equality follows from (1.35), i.e.,

$$\left| e^{-st} e^{\alpha t} \right| = \left| e^{-xt} \right| \left| e^{-iyt} \right| \left| e^{\alpha t} \right| = \left| e^{-xt} \right| \left| e^{\alpha t} \right|.$$

As $x = \mathrm{Re}(s) > \alpha$,

$$\lim_{u \to \infty} M \int_0^u e^{-(x-\alpha)t}\,dt = M \lim_{u \to \infty} \frac{1 - e^{-(x-\alpha)u}}{x - \alpha} = \frac{M}{x - \alpha} < \infty, \qquad (1.40)$$

showing that, under the stated conditions on g and s, the integral defining $\mathcal{L}\{g\}(s)$ converges absolutely, and thus exists.

1.2.2.2 Inverse Laplace transform

If G is a function defined on some part of the real line or the complex plane, and there exists a function g such that $\mathcal{L}\{g\}(s) = G(s)$ then, rather informally, this function g is referred to as the *inverse Laplace transform* of G, denoted by $\mathcal{L}^{-1}\{G\}$. Such an inverse Laplace transform need not exist, and if it exists, it will not be unique. If g is a function of a real variable such that $\mathcal{L}\{g\} = G$ and h is another function which is almost everywhere identical to g but differs on a finite set (or, more generally, on a set of measure zero), then, from properties of the Riemann (or Lebesgue) integral,

---

[5] Continuity of g on $[0, \infty)$ can be weakened to *piecewise continuity* on $[0, \infty)$. This means that $\lim_{t \downarrow 0} g(t)$ exists, and g is continuous on every finite interval $(0, b)$, except at a finite number of points in $(0, b)$ at which g has a *jump discontinuity*, i.e., g has a jump discontinuity at x if the limits $\lim_{t \uparrow x} g(t)$ and $\lim_{t \downarrow x} g(t)$ are finite, but differ. Notice that a piecewise continuous function is bounded on every bounded subinterval of $[0, \infty)$.

their Laplace transforms are identical, i.e., $\mathcal{L}\{g\} = G = \mathcal{L}\{h\}$. So both g and h could be regarded as versions of $\mathcal{L}^{-1}\{G\}$. If, however, functions g and h are continuous on $[0, \infty)$, such that $\mathcal{L}\{g\} = G = \mathcal{L}\{h\}$, then it can be proven that $g = h$, so in this case, there is a distinct choice of $\mathcal{L}^{-1}\{G\}$. See Beerends *et al.* (2003, p. 304) for a more rigorous discussion.

The linearity property of the Riemann integral implies the linearity property of Laplace transforms, i.e., for constants c_1 and c_2, and two functions $g_1(t)$ and $g_2(t)$ with Laplace transforms $\mathcal{L}\{g_1\}$ and $\mathcal{L}\{g_2\}$, respectively,

$$\mathcal{L}\{c_1 g_1 + c_2 g_2\} = c_1 \mathcal{L}\{g_1\} + c_2 \mathcal{L}\{g_2\}. \tag{1.41}$$

Also, by applying $\mathcal{L}^{-1}$ to both sides of (1.41),

$$c_1 g_1(t) + c_2 g_2(t) = \mathcal{L}^{-1}\{\mathcal{L}\{c_1 g_1 + c_2 g_2\}\} = \mathcal{L}^{-1}\{c_1 \mathcal{L}\{g_1\} + c_2 \mathcal{L}\{g_2\}\},$$

we see that $\mathcal{L}^{-1}$ is also a linear operator. Problem 1.17 proves a variety of further results involving Laplace transforms.

◎ **Example 1.17** Let $g : [0, \infty) \to \mathbb{C}$, $t \mapsto e^{it}$. Then its Laplace transform at $s \in \mathbb{C}$ with $\mathrm{Re}(s) > 0$ is, from (1.33),

$$\mathcal{L}\{g\}(s) = \int_0^\infty e^{it} e^{-st}\, dt = \int_0^\infty e^{t(i-s)} dt = \left. \frac{e^{t(i-s)}}{i-s} \right|_0^\infty = \frac{1}{i-s}\left(\lim_{t \to \infty} e^{-t(s-i)} - 1\right)$$

$$= \frac{1}{s-i} = \frac{s+i}{(s+i)(s-i)} = \frac{s}{s^2+1} + i\frac{1}{s^2+1}.$$

Now, (1.31) and (1.41) imply $\mathcal{L}\{g\} = \mathcal{L}\{\cos\} + i\mathcal{L}\{\sin\}$ or

$$\int_0^\infty \cos(t)\, e^{-st}\, dt = \frac{s}{s^2+1}, \qquad \int_0^\infty \sin(t)\, e^{-st}\, dt = \frac{1}{s^2+1}. \tag{1.42}$$

Relations (1.42) are derived directly in Example I.A.24. See Example 1.22 for their use. ∎

1.2.3 Basic properties of characteristic functions

For the c.f. of r.v. X, using (1.31) and the notation defined in (I.4.31),

$$\varphi_X(t) = \int_{-\infty}^\infty e^{itx}\, dF_X(x)$$

$$= \int_{-\infty}^\infty \cos(tx)\, dF_X(x) + i \int_{-\infty}^\infty \sin(tx)\, dF_X(x)$$

$$= \mathbb{E}[\cos(tX)] + i\,\mathbb{E}[\sin(tX)].$$

As

$$\cos(-\theta) = \cos(\theta) \quad \text{and} \quad \sin(-\theta) = -\sin(\theta), \tag{1.43}$$

it follows that

$$\varphi_X(-t) = \overline{\varphi}_X(t), \tag{1.44}$$

where $\overline{\varphi}_X$ is the complex conjugate of φ_X. Also, from (1.27) and (1.44),

$$\varphi_X(t) + \varphi_X(-t) = 2\,\mathrm{Re}(\varphi_X(t)). \tag{1.45}$$

Contrary to the m.g.f., the c.f. will always exist: from (1.34),

$$|\varphi_X(t)| = \left| \int_{-\infty}^{\infty} e^{itx}\, dF_X(x) \right| \leq \int_{-\infty}^{\infty} \left| e^{itx} \right|\, dF_X(x) = \int_{-\infty}^{\infty} dF_X(x) = 1. \tag{1.46}$$

Remark: A set of necessary and sufficient conditions for a function to be a c.f. is given by *Bochner's theorem*: A complex-valued function φ of a real variable t is a characteristic function iff (i) $\varphi(0) = 1$, (ii) φ is continuous, and (iii) for any positive integer n, real values $t_1, \ldots, t_n$, and complex values $\xi_1, \ldots, \xi_n$, the sum

$$S = \sum_{j=1}^{n} \sum_{k=1}^{n} \varphi(t_j - t_k) \xi_j \overline{\xi}_k \geq 0, \tag{1.47}$$

i.e., S is real and nonnegative. (The latter two conditions are equivalent to stating that f is a nonnegative definite function.) Note that, if φ_X is the c.f. of r.v. X, then, from (1.28) and (1.29),

$$S = \mathbb{E}\left[\sum_{j=1}^{n} \sum_{k=1}^{n} \left[\exp\left(i(t_j - t_k)X \right) \right] \xi_j \overline{\xi}_k \right] = \mathbb{E}\left[\left| \sum_{j=1}^{n} e^{it_j X} \xi_j \right|^2 \right] \geq 0,$$

which shows that if φ is a c.f., then it satisfies (1.47). See also Fristedt and Gray (1997, p. 227). The proof of the converse is more advanced; for it, and alternative criteria, see Lukacs (1970, Section 4.2) and Berger (1993, pp. 58–59), and the references therein. ∎

The *uniqueness theorem*, first proven in Lévy (1925), states that, for r.v.s X and Y,

$$\varphi_X = \varphi_Y \quad \Leftrightarrow \quad X \overset{d}{=} Y. \tag{1.48}$$

Proofs can be found in Lukacs (1970, Section 3.1) or Gut (2005, p. 160 and 250).

⊖ **Example 1.18**   Recall the probability integral transform at the end of Section I.7.3. If X is a continuous random variable with c.d.f. F_X, the c.f. of $Y = F_X(X)$ is

$$\phi_Y(s) = \mathbb{E}\left[e^{is F_X(X)} \right] = \int_{-\infty}^{\infty} e^{is F_X(t)} f_X(t)\, dt$$

so that, with $u = F_X(t)$ and $du = f_X(t)\, dt$,

$$\phi_Y(s) = \int_0^1 e^{isu}\, du = \frac{e^{is} - 1}{is},$$

which is the c.f. of a uniform random variable, implying $Y \sim \mathrm{Unif}(0,1)$. ∎

If φ_X is real, then $\varphi_X(t) = \int_{-\infty}^{\infty} \cos(tx) \, dF(x)$ and $\int_{-\infty}^{\infty} \sin(tx) \, dF_X(x) = 0$. This is the case, for example, if X is a continuous r.v. with p.d.f. $f_X(x)$ which is symmetric about zero because, with $u = -x$,

$$\int_0^{\infty} \sin(tx) f(x) \, dx = -\int_0^{-\infty} \sin(-tu) f(-u) \, du = -\int_{-\infty}^0 \sin(tu) f(u) \, du.$$

In fact, it can be proven that, for r.v. X with c.f. φ_X and p.m.f. or p.d.f. f_X, φ_X is real iff f_X is symmetric about zero (see Resnick, 1999, p. 297).

1.2.4   Relation between the m.g.f. and c.f.

1.2.4.1   If the m.g.f. exists on a neighbourhood of zero

Let X be an r.v. whose m.g.f. exists (i.e., $\mathbb{E}\left[e^{tX}\right]$ is finite on a neighbourhood of zero). At first glance, comparing the definitions of the m.g.f. and the c.f., it seems that we could just write

$$\varphi_X(t) = \mathbb{M}_X(it). \tag{1.49}$$

As an illustration of how to apply this formula, consider the case $X \sim N(0, 1)$. Then $\mathbb{M}_X(t) = e^{t^2/2}$. Plugging it into the right-hand side of this formula yields $\varphi_X(t) = e^{-t^2/2}$. This can be shown rigorously using complex analysis.

Note that, formally, (1.49) does not make sense because we are plugging a complex variable into a function that only admits real arguments. But in the vast majority of real applications, such as the calculation in the Gaussian case, the m.g.f. will usually have a functional form which allows for complex arguments in an obvious manner. In the remainder of the text, we shall use the relation $\varphi_X(t) = \mathbb{M}_X(it)$ if $\mathbb{M}_X$ is finite on a neighbourhood of zero.

This approach has a nontrivial mathematical background, an understanding of which is not required to be able to apply the formula. The following discussion sheds some light on this issue, and can be skipped.

If $\mathbb{M}_X$ is finite on the interval $(-h, h)$ for some $h > 0$, then the complex function $z \mapsto \mathbb{E}\left[e^{zX}\right]$ is well defined on the strip $\{z \in \mathbb{C} : |\mathrm{Re}(z)| < h\}$ (see Gut, 2005, p. 190). Note that, on the imaginary line, this function is the c.f. of X, while, on the real interval $(-h, h) \subseteq \mathbb{R}$, it is the m.g.f. It can be shown that the function $z \mapsto \mathbb{E}\left[e^{zX}\right]$ is what is called 'complex analytic',[6] which might be translated as 'nice and smooth'. The so-called *identity theorem* asserts in our case that a complex analytic function on the strip is uniquely determined by its values on the interval $(-h, h)$.

On the one hand, this shows that the m.g.f. completely determines the function $z \mapsto \mathbb{E}\left[e^{zX}\right]$ and, hence, the c.f. As a consequence, *if the m.g.f. of X exists on a*

---

[6] Complex analytic functions are also called 'holomorphic' and are the main objects of study in complex analysis; see Palka (1991) for an introduction.

neighbourhood of 0, then it uniquely determines the distribution of X. This was stated in (1.18).

On the other hand, this theorem can be used to actually *find* the c.f. As stated above, the m.g.f. will usually come in a functional form which suggests an extension to the complex plane (or at least to the strip). In our Gaussian example, the extension $z \mapsto e^{z^2/2}$ of $t \mapsto e^{t^2/2}$ immediately comes to mind. As a matter of fact, the extension thus constructed will normally be 'nice and smooth', so an application of the identity theorem tells us that it is the only 'nice and smooth' extension to the strip and thus coincides with the function $z \mapsto \mathbb{E}\left[e^{zX}\right]$. Hence plugging in the argument *it* really does result in $\varphi_X(t)$.

In the Gaussian example, $z \mapsto e^{z^2/2}$ is indeed a complex analytic function, so this shows that $\mathbb{E}\left[e^{zX}\right] = e^{z^2/2}$ and, in particular, that $\mathbb{E}\left[e^{itX}\right] = e^{-t^2/2}$.

1.2.4.2 The m.g.f. and c.f. for nonnegative X

Up to this point, only the case where the m.g.f. $\mathbb{M}_X$ is defined on a neighbourhood of zero has been considered. There is another important case where we can apply (1.49). If X is a nonnegative r.v., then $\mathbb{M}_X(t)$ is finite for all $t \in (-\infty, 0]$. It might well be defined for a larger set, but to give a meaning to (1.49), one only needs the fact that the m.g.f. is defined for all negative t.

If $X \geq 0$, then, similarly to the above digression on complex analysis, the function $z \mapsto \mathbb{E}\left[e^{zX}\right]$ is well defined on the left half-plane $\{z \in \mathbb{C} : \operatorname{Re}(z) < 0\}$. Again, this function is 'complex analytic' and hence uniquely determined by the m.g.f. (defined on $(-\infty, 0]$). It can be shown that $z \mapsto \mathbb{E}\left[e^{zX}\right]$ is at least continuous on $\{z \in \mathbb{C} : \operatorname{Re}(z) \leq 0\}$, which is the left half-plane plus the imaginary axis. So, a more involved application of the identity theorem tells us that the m.g.f. uniquely determines the c.f. in this case, too. As a consequence, *the m.g.f. of an r.v. $X \geq 0$ uniquely determines the distribution of X.* A rigorous proof of this result can be found in Fristedt and Gray (1997, p. 218, Theorem 6).

Equation (1.49) is still applicable as a rule of thumb, and the more complicated background that has just been outlined entails some inconveniences which are illustrated by the following example.

⊖ **Example 1.19** The Lévy distribution was introduced in Examples I.7.6 and I.7.15, and will be revisited in Section 8.2 in the context of the asymmetric stable Paretian distribution and Section 9.4.2 in the context of the generalized inverse Gaussian distribution. Its p.d.f. is, for $\chi > 0$,

$$f_X(x) = \left(\frac{\chi}{2\pi}\right)^{1/2} x^{-3/2} \exp\left[-\frac{\chi}{2x}\right] \mathbb{I}_{(0,\infty)}(x),$$

and its m.g.f. is shown in (9.34) to be, for $t \leq 0$ (and not on a neighbourhood of zero),

$$\mathbb{M}_X(t) = \exp\left[-\sqrt{\chi}\sqrt{-2t}\right], \qquad t \leq 0.$$

Evaluating $\mathbb{M}_X(it)$ requires knowing the square root of $-2it$. Consider the case $t > 0$. Then we have a choice because $\left((1-i)\sqrt{t}\right)^2 = (1-i)^2 t = -2it$ and also $\left(-(1-i)\sqrt{t}\right)^2 = -2it$. For negative t, we have a similar choice.

To see which is the right choice, we have to identify the extension $z \mapsto \mathbb{E}\left[e^{zX}\right]$ of the m.g.f. to the left half-plane: what is the nice and smooth extension of $t \mapsto \exp\left[-\sqrt{\chi}\sqrt{-2t}\right]$ for all $z \in \mathbb{C}$ such that $\mathrm{Re}(z) < 0$? If $\mathrm{Re}(z) < 0$, then $\mathrm{Re}(-2z) > 0$, in other words, $-2z$ lies in the right half-plane. Therefore $-2z$ has two square roots, one in the right half-plane and one in the left half-plane. Call the one in the right half-plane $\sqrt{-2z}$ (implying that the other is $-\sqrt{-2z}$). Then the function $z \mapsto \exp\left[-\sqrt{\chi}\sqrt{-2z}\right]$ is nice and smooth on the left half-plane and extends the m.g.f.

It can be continuously extended to $z \in \mathbb{C}$ with $\mathrm{Re}(z) \leq 0$ (and further) by always taking the proper root of $-2z$, which means that, for $t > 0$, the proper choice of $\sqrt{-2it}$ will be $(1 - i)\sqrt{t}$ (being in the right half-plane), while, for $t < 0$, the proper choice of $\sqrt{-2it}$ will be $(1 + i)\sqrt{-t}$ (also being the root in the right half-plane). This can be combined as

$$\sqrt{-2it} = \sqrt{|t|}\left(1 - i\,\mathrm{sgn}\,(t)\right), \qquad t \in \mathbb{R},$$

so that

$$\varphi_X(t) = \exp\left\{-\chi^{1/2}|t|^{1/2}\left[1 - i\,\mathrm{sgn}(t)\right]\right\}.$$

This result is also seen in the more general context of the c.f. of the stable Paretian distribution. In particular, taking $c = \chi$, $\mu = 0$, $\beta = 1$ and $\alpha = 1/2$ in (8.8) yields the same expression. ∎

1.2.5 Inversion formulae for mass and density functions

If you are going through hell, keep going. (Sir Winston Churchill)

By *inversion formula* we mean an integral expression whose integrand involves the c.f. of a random variable X and a number x, and whose result is the p.m.f., p.d.f. or c.d.f. of X at x.

Some of the proofs below make use of the following results for constants $k \in \mathbb{R} \setminus 0$ and $T \in \mathbb{R}_{\geq 0}$:

$$\int_{-T}^{T} \cos(kt)\,\mathrm{d}t = \frac{2\sin(kT)}{k} \quad \text{and} \quad \int_{-T}^{T} \sin(kt)\,\mathrm{d}t = 0. \tag{1.50}$$

The first follows from (I.A.27), i.e., that $\mathrm{d}\sin x/\mathrm{d}x = \cos x$, the fundamental theorem of calculus, a simple change of variable, and (1.43). The second can be derived similarly and/or confirmed quickly from a plot of sin.

We begin with expressions for the probability mass function. Let X be a discrete r.v. with support $\{x_j\}_{j=1}^{\infty}$, such that $x_j = x_k$ iff $j = k$, and mass function f_X. Then

$$f_X(x_j) = \lim_{T \to \infty} \frac{1}{2T} \int_{-T}^{T} e^{-itx_j} \varphi_X(t)\,\mathrm{d}t. \tag{1.51}$$

Proof (outline):[7] First observe that, from Euler's formula (1.31) and (1.50),

$$\frac{1}{2T} \int_{-T}^{T} e^{it(x_k - x_j)}\, dt = \frac{1}{2T} \int_{-T}^{T} \cos t(x_k - x_j)\, dt + \frac{i}{2T} \int_{-T}^{T} \sin t(x_k - x_j)\, dt$$

$$= \frac{1}{2T} \int_{-T}^{T} \cos t(x_k - x_j)\, dt \tag{1.52}$$

which, if $x_k = x_j$, is 1, and otherwise, using (1.50), is $\sin(A)/A$, where $A = T(x_k - x_j)$. Clearly, $\lim_{T \to \infty} \sin(A)/A = 0$, so that

$$\lim_{T \to \infty} \frac{1}{2T} \int_{-T}^{T} e^{it(x_k - x_j)}\, dt = \begin{cases} 1, & \text{if } x_k = x_j, \\ 0, & \text{if } x_k \neq x_j. \end{cases}$$

Next,

$$\frac{1}{2T} \int_{-T}^{T} e^{-itx_j} \varphi_X(t)\, dt = \frac{1}{2T} \int_{-T}^{T} e^{-itx_j} \sum_{k=1}^{\infty} e^{itx_k} f_X(x_k)\, dt$$

$$= \sum_{k=1}^{\infty} f_X(x_k) \frac{1}{2T} \int_{-T}^{T} e^{it(x_k - x_j)}\, dt$$

and

$$\lim_{T \to \infty} \frac{1}{2T} \int_{-T}^{T} e^{-itx_j} \varphi_X(t)\, dt = \sum_{k=1}^{\infty} f(x_k) \lim_{T \to \infty} \frac{1}{2T} \int_{-T}^{T} e^{it(x_k - x_j)}\, dt = f_X(x_j),$$

where the interchange of integral and summation, and limit and summation, can be shown to be valid. ∎

To illustrate (1.51), let $X \sim \mathrm{Bin}(10, 0.3)$ and consider calculating the p.m.f. at $x = 4$, the true value of which is 0.200121. The program in Listing 1.1 calculates the p.m.f. via (1.51), using the c.f. $\varphi_X(t) = \mathbb{M}_X(it)$. The binomial m.g.f. is shown in Problem 1.4 to be $\mathbb{M}_X(t) = (pe^t + 1 - p)^n$. The discrepancies of the computed values from the true p.m.f. value, as a function of T, are plotted in Figure 1.3.

```
function pmf = schlecht(x,n,p,T)
pmf=quadl(@ff,-T,T,1e-6,0,n,p,x) / 2 / T;

function I=ff(tvec,n,p,x);
q=1-p; I=zeros(size(tvec));
for loop=1:length(tvec)
  t=tvec(loop); cf=(p*exp(i*t)+q)^n; I(loop)= ( exp(-i*t*x) * cf );
end
```

Program Listing 1.1 Uses (1.51) to compute the p.m.f. of $X \sim \mathrm{Bin}(n, p)$ at x

---

[7] The proofs given in this section are not rigorous, but should nevertheless convince most readers of the validity of the results. Complete proofs are of course similar, and are available in advanced textbooks such as those mentioned at the beginning of Section 1.2. See, for example, Shiryaev (1996, p. 283) regarding this proof, and, for even more detail, Lukacs (1970, Section 3.2).

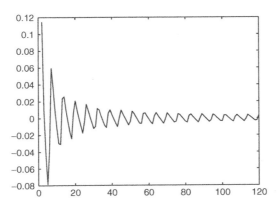

Figure 1.3 Discrepancy of the probability calculated via (1.51) (disregarding the complex part, which is nonzero for $T < \infty$) for $X \sim \text{Bin}(n, p)$ at $x = 4$ from the true p.m.f. value as a function of T, $T = 2, 3, \ldots, 120$

It does indeed appear that the magnitude of the discrepancy becomes smaller as T increases. More importantly, however, the discrepancy appears to be exactly zero on a regularly spaced set of T-values. These values occur for values of T which are integral multiples of π. This is verified in Figure 1.4, which is the same as Figure 1.3, but with T chosen as $\pi, 2\pi, \ldots$. This motivates the following claim, which is indeed a well-known fact and far more useful for numeric purposes:

$$f_X(x_j) = \frac{1}{2\pi} \int_{-\pi}^{\pi} e^{-itx_j} \varphi_X(t) \, dt. \tag{1.53}$$

See, for example, Feller (1971, p. 511) or Gut (2005, p. 164).

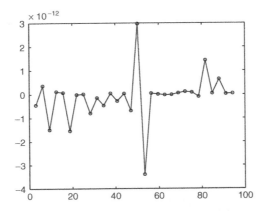

Figure 1.4 Same as Figure 1.3 but using values $T = \pi, 2\pi, \ldots, 30\pi$

⊖ ***Example 1.20*** (Example 1.7 cont.) The m.g.f. of N_m (number of i.i.d. Bernoulli trials with success probability p required until m consecutive successes occur) was found to be, with $q = 1 - p$,

$$\mathbb{M}_m(t) = \frac{pe^t \mathbb{M}_{m-1}(t)}{1 - q\mathbb{M}_{m-1}(t)\,e^t}, \qquad \mathbb{M}_1(t) = pe^t / \left(1 - qe^t\right), \qquad t \neq -\ln(1 - p),$$

and the c.f. is $\mathbb{M}_m(it)$. Using (1.53), the program in Listing 1.2 computes the p.m.f., which is plotted in Figure 1.5 for $m = 3$ and $p = 1/2$. This is indeed the same as in Figure I.2.1(a), which used a recursive expression for the p.m.f. ■

```
function f = consecpmf (xvec,m,p)
xl=length(xvec); f=zeros(xl,1); tol=1e-5;
for j=1:xl
  x=xvec(j); f(j)=quadl(@fun,-pi,pi,tol,0,x,m,p);
end
f=real(f); % roundoff error sometimes gives an imaginary
           % component of the order of 1e-16.

function I=fun(t,x,m,p), I=exp(-i*t*x) .* cf(t,m,p) / (2*pi);

function g = cf(t,m,p) % cf
q=1-p;
if m==1
  g = p*exp(i*t) ./ (1-q*exp(i*t));
else
  kk = exp(i*t) .* cf(t,m-1,p); g = (p*kk) ./ (1-q*kk);
end
```

Program Listing 1.2 Computes the p.m.f. via inversion formula (1.53) of the number of i.i.d. Ber(p) trials required until m consecutive successes occur

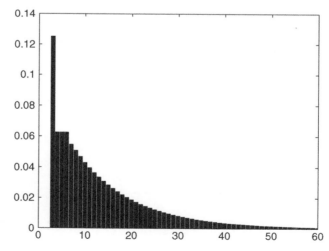

Figure 1.5 The p.m.f. of the number of children (assuming equal boy and girl probabilities) until one obtains three girls in a row, obtained by computing (1.53). Compare with Figure I.2.1(a)

Now consider the continuous case. If X is a continuous random variable with p.d.f. f_X and $\int_{-\infty}^{\infty} |\varphi_X(t)| \, dt < \infty$, then

$$f_X(x) = \frac{1}{2\pi} \int_{-\infty}^{\infty} e^{-itx} \varphi_X(t) \, dt. \tag{1.54}$$

Proof (outline): Take $T > 0$ so that, assuming we can change the order of integration and using (1.52) and (1.50) (and ignoring the 'measure zero' case of $x = y$),

$$\int_{-T}^{T} e^{-itx} \varphi_X(t) \, dt = \int_{-T}^{T} e^{-itx} \int_{-\infty}^{\infty} e^{ity} f_X(y) \, dy \, dt$$

$$= \int_{-\infty}^{\infty} f_X(y) \int_{-T}^{T} e^{it(y-x)} \, dt \, dy$$

$$= \int_{-\infty}^{\infty} f_X(y) \frac{2 \sin T(y-x)}{y-x} \, dy.$$

Then, with $A = T(y - x)$ and $dy = dA/T$,

$$\int_{-\infty}^{\infty} e^{-itx} \varphi_X(t) \, dt = 2 \lim_{T \to \infty} \int_{-\infty}^{\infty} f_X(y) \frac{\sin T(y-x)}{y-x} \, dy$$

$$= 2 \lim_{T \to \infty} \int_{-\infty}^{\infty} f_X \left(x + \frac{A}{T} \right) \frac{\sin A}{A} \, dA$$

$$= 2\pi f_X(x),$$

where the last equality makes the assumption (which can be justified) that interchanging the limit and integral is valid, and because

$$\int_{0}^{\infty} \frac{\sin x}{x} \, dx = \frac{\pi}{2}, \tag{1.55}$$

as shown in Example I.A.30. ■

Remark: If the cumulant generating function $\mathbb{K}_X(s) = \ln \mathbb{M}_X(s)$ of X exists, then (1.54) can be expressed as

$$f_X(x) = \frac{1}{2\pi i} \int_{-i\infty}^{i\infty} \exp\{\mathbb{K}_X(s) - sx\} \, ds, \tag{1.56}$$

as shown in Problem 1.12. ■

⊖ ***Example 1.21*** From (1.10), if $Z \sim \mathrm{N}(0, 1)$, then $\mathbb{M}_Z(t) = e^{t^2/2}$, and $\varphi_Z(t) = \mathbb{M}_Z(it) = e^{-t^2/2}$, which is real, as Z is symmetric about zero. Going the other way, from (1.54),

$$f_Z(x) = \frac{1}{2\pi} \int_{-\infty}^{\infty} e^{-itx} e^{-t^2/2} \, dt = \frac{1}{2\pi} \int_{-\infty}^{\infty} e^{-\frac{1}{2}(t^2 + 2itx)} \, dt$$

$$= \frac{1}{2\pi} \int_{-\infty}^{\infty} e^{-\frac{1}{2}\left(t^2 + 2t(ix) + (ix)^2\right)} e^{\frac{1}{2}(ix)^2} \, dt$$

$$= \frac{1}{\sqrt{2\pi}} e^{\frac{1}{2}(ix)^2} \frac{1}{\sqrt{2\pi}} \int_{-\infty}^{\infty} e^{-\frac{1}{2}(t+ix)^2} \, dt$$

$$= \frac{1}{\sqrt{2\pi}} e^{-x^2/2},$$

using $u = t + ix$ and $dt = du$ to resolve the integral. ∎

⊙ **Example 1.22** Consider deriving the density corresponding to $\varphi_X(t) = e^{-c|t|}$ for $c > 0$. Using the fact that $e^{-ixt} = \cos(tx) - i \sin(tx)$ and separating negative and positive t,

$$2\pi f_X(x) = \int_{-\infty}^{\infty} e^{-ixt} e^{-c|t|} \, dt$$

$$= \int_{-\infty}^{0} \cos(tx) e^{ct} \, dt - i \int_{-\infty}^{0} \sin(tx) e^{ct} \, dt$$

$$+ \int_{0}^{\infty} \cos(tx) e^{-ct} \, dt - i \int_{0}^{\infty} \sin(tx) e^{-ct} \, dt.$$

With the substitution $u = -ct$ in the first two integrals, $u = ct$ in the last two, and the basic facts (1.43), this simplifies to

$$f_X(x; c) = \frac{1}{c\pi} \int_0^{\infty} \cos(ux/c) e^{-u} \, du. \qquad (1.57)$$

To resolve this for $x \neq 0$, first set $t = ux/c$ and $s = 1/x$, and then use (1.42) to get

$$f_X(x; c) = \frac{s}{\pi} \int_0^{\infty} \cos(t) e^{-cst} \, dt = \frac{s}{\pi} \frac{cs}{(cs)^2 + 1}$$

$$= \frac{1}{\pi} \frac{cx^{-2}}{c^2 x^{-2} + 1} = \frac{1}{\pi} \frac{c}{c^2 + x^2} = \frac{1}{c\pi} \frac{1}{1 + (x/c)^2},$$

which is the Cauchy distribution with scale c. This integral also follows from (I.A.24) with $a = -1$ and $b = x/c$.

Observe that φ_X is real and f_X is symmetric about zero. ∎

We showed above that $\exp(\bar{z}) = \overline{\exp(z)}$, which implies $e^{-itx} = \overline{e^{itx}}$. We also saw above that $\overline{z_1 z_2} = \overline{z_1}\, \overline{z_2}$, $\varphi_X(-t) = \overline{\varphi}_X(t)$, and $z + \bar{z} = 2\operatorname{Re}(z)$. Then, with $v = -t$,

$$\int_{-\infty}^{0} e^{-itx} \varphi_X(t) \, dt = \int_0^{\infty} e^{ivx} \varphi_X(-v) \, dv = \int_0^{\infty} \overline{e^{-ivx}\overline{\varphi}_X(v)} \, dv$$

$$= \int_0^{\infty} \overline{e^{-ivx} \varphi_X(v)} \, dv$$

so that, for X continuous, (1.54) is

$$f_X(x) = \int_{-\infty}^{\infty} e^{-itx} \varphi_X(t) \, dt = \int_0^{\infty} e^{-itx} \varphi_X(t) \, dt + \int_{-\infty}^0 e^{-itx} \varphi_X(t) \, dt$$

$$= \int_0^{\infty} e^{-itx} \varphi_X(t) \, dt + \int_0^{\infty} \overline{e^{-ivx} \varphi_X(v)} \, dv$$

$$= 2 \int_0^{\infty} \text{Re} \left[e^{-itx} \varphi_X(t) \right] dt.$$

A similar calculation is done for the discrete case. Thus, the inversion formulae can also be written as

$$f_X(x_j) = \frac{1}{\pi} \int_0^{\pi} \text{Re} \left[e^{-itx_j} \varphi_X(t) \right] dt \qquad (1.58)$$

in the discrete case and

$$f_X(x) = \frac{1}{\pi} \int_0^{\infty} \text{Re} \left[e^{-itx} \varphi_X(t) \right] dt \qquad (1.59)$$

in the continuous case. These could be more suitable for numeric work than their counterparts (1.53) and (1.54) if a 'canned' integration routine is available but which does not support complex arithmetic.

Remark: In practice, the integral in (1.59) needs to be truncated at a point such that a specified degree of accuracy will be obtained. Ideally, one would accomplish this by algebraically determining an upper limit on the tail value of the integral. Less satisfactory, but still useful, is to verify that the absolute value of the integrand is, after some point, monotonically decreasing at a fast enough rate, and one then evaluates and accumulates the integral over contiguous pieces of the real line (possibly doubling in size each step) until the contribution to the integral is below a given threshold.

To avoid having to determine the upper bound in the integral, one can transform the integrand such that the range of integration is over a finite interval, say $(0, 1)$. One way of achieving this is via the substitution $u = 1/(1+t)$ (so that $t = (1-u)/u$ and $dt = -u^{-2} du$), which leads to

$$f_X(x) = \frac{1}{\pi} \int_0^1 h_x \left(\frac{1-u}{u} \right) u^{-2} \, du, \qquad h_x(t) = \text{Re} \left[e^{-itx} \varphi_X(t) \right]. \qquad (1.60)$$

Alternatively, use of $u = t/(t+1)$ in (1.59) leads to

$$f_X(x) = \frac{1}{\pi} \int_0^1 h_x \left(\frac{u}{1-u} \right) (1-u)^{-2} \, du. \qquad (1.61)$$

Use of (1.61) is actually equivalent to (1.60) and just results in the same integrand, flipped over the $u = 1/2$ axis. There are two caveats to this approach. First, as the integrand in (1.60) cannot be evaluated at zero, a similar problem remains in the sense of determining how close to zero the lower integration limit should be chosen (and similarly for (1.61) at the upper bound of one). Second, the transformation (1.60) or (1.61) could induce pathological oscillatory behaviour in the integrand, making it difficult, or impossible, to accurately integrate. This is the case in Example 1.23 and Problem 1.11. ∎

⊖ **Example 1.23**　Problem 1.13 shows that the c.f. of $L \sim \text{Lap}(0, 1)$ is given by $\varphi_L(t) = (1 + t^2)^{-1}$. Thus, $L/\lambda \sim \text{Lap}(0, \lambda)$ has c.f.

$$\varphi_{L/\lambda}(t) = \mathbb{E}\left[e^{itL/\lambda}\right] = \mathbb{E}\left[e^{i(t/\lambda)L}\right] = \varphi_L(t/\lambda) = \left(1 + t^2/\lambda^2\right)^{-1}.$$

From Example I.9.5, we know that $X - Y \sim \text{Lap}(0, \lambda)$, where $X, Y \overset{\text{i.i.d.}}{\sim} \text{Exp}(\lambda)$, so that $\varphi_{X-Y}(t) = \left(1 + t^2/\lambda^2\right)^{-1}$. This result is generalized in Example 2.10, which shows that if $X, Y \overset{\text{i.i.d.}}{\sim} \text{Gam}(a, b)$, then the c.f. of $D = X - Y$ is $\varphi_D(t) = \left(1 + t^2/b^2\right)^{-a}$. From (I.6.3) and (I.7.9), $\text{Var}(D) = 2a/b^2$, so that setting $b = \sqrt{2a}$ gives a unit variance. The p.d.f. is symmetric about zero, which is obvious from the definition of D, and also because φ_D is real.

For $a \neq 1$, $\varphi_D(t)$ is not recognized to be the c.f. of any common random variable, so the inversion formulae need to be applied to compute the p.d.f. of D. To use (1.59), a finite value for the upper limit of integration is required, and a plot of the integrand for $a = 1$ and a few x-values confirms that 200 is adequate for reasonable accuracy. The program in Listing 1.3 can be used to compute the p.d.f. over a grid of x-values. As a decreases, the tails of the distribution become fatter. The reader is encouraged to confirm that the use of transformation (1.60) is problematic in this case. See also Example 7.19 for another way of computing the p.d.f. ∎

```
function f = pdfinvgamdiff(xvec,a,b)
lo=1e-8; hi=200; tol=1e-8;
xl=length(xvec); f=zeros(xl,1);
for loop=1:xl
  x=xvec(loop); f(loop)=quadl(@gammdiffcf,lo,hi,tol,[],x,a,b) / pi;
end;

function I=gammdiffcf(tvec,x,a,b);
for ii=1:length(tvec)
  t=tvec(ii); cf=(1+(t/b).^2).^(-a); z =exp(-i*t*x).*cf; I(ii)=real(z);
end
```

Program Listing 1.3　Computes the p.d.f., based on (1.59) with an upper integral limit of 200, of the difference of two i.i.d. Gam(a, b) r.v.s at the values in the vector `xvec`

⊙ **Example 1.24**　Let X_1, X_2 be continuous r.v.s with joint p.d.f. f_{X_1, X_2}, joint c.f. φ_{X_1, X_2}, and such that $\Pr(X_2 > 0) = 1$ and $\mathbb{E}[X_2] = \mu_2 < \infty$. Geary (1944) showed that the density of $R = X_1/X_2$ can be written

$$f_R(r) = \frac{1}{2\pi i} \int_{-\infty}^{\infty} \left[\frac{\partial \varphi_{X_1, X_2}(s, t)}{\partial t} \right]_{t=-rs} ds. \tag{1.62}$$

To prove this, as in Daniels (1954), consider a new density

$$f_{Y_1, Y_2}(y_1, y_2) = \frac{y_2}{\mu_2} f_{X_1, X_2}(y_1, y_2),$$

and observe that

$$\int_0^{\infty} \int_{-\infty}^{\infty} f_{Y_1, Y_2}(y_1, y_2) \, dy_1 \, dy_2 = \frac{1}{\mu_2} \int_0^{\infty} y_2 \int_{-\infty}^{\infty} f_{X_1, X_2}(y_1, y_2) \, dy_1 \, dy_2$$

$$= \frac{1}{\mu_2} \int_0^{\infty} y_2 f_{X_2}(y_2) \, dy_2 = 1.$$

Next, define the r.v. $W = Y_1 - rY_2$, and use a result in Section I.8.2.6 (or a straightforward multivariate transformation) to get

$$f_W(w) = \int_0^{\infty} f_{Y_1, Y_2}(w + ry_2, y_2) \, dy_2.$$

Then

$$f_W(0) = \int_0^{\infty} f_{Y_1, Y_2}(ry_2, y_2) \, dy_2 = \frac{1}{\mu_2} \int_0^{\infty} y_2 f_{X_1, X_2}(ry_2, y_2) \, dy_2 = \frac{1}{\mu_2} f_R(r),$$

where the last equality follows from (2.10), as derived later in Section 2.2. Hence, from (1.54),

$$f_R(r) = \mu_2 f_W(0) = \frac{\mu_2}{2\pi} \int_{-\infty}^{\infty} \varphi_W(s) \, ds,$$

where

$$\varphi_W(s) = \mathbb{E}\left[e^{isW}\right] = \mathbb{E}\left[e^{isY_1 - isrY_2}\right] = \varphi_{Y_1, Y_2}(s, t)\big|_{t=-rs}.$$

It remains to show that

$$\varphi_{Y_1, Y_2}(s, t) = \frac{1}{i\mu_2} \frac{\partial}{\partial t} \varphi_{X_1, X_2}(s, t),$$

which follows because

$$\varphi_{Y_1, Y_2}(s, t) = \mathbb{E}\left[e^{isY_1 + itY_2}\right] = \int_0^{\infty} \int_{-\infty}^{\infty} e^{isy_1 + ity_2} f_{Y_1, Y_2}(y_1, y_2) \, dy_1 \, dy_2$$

$$= \int_0^{\infty} \int_{-\infty}^{\infty} e^{isy_1 + ity_2} \frac{y_2}{\mu_2} f_{X_1, X_2}(y_1, y_2) \, dy_1 \, dy_2$$

$$= \frac{1}{i\mu_2} \int_0^{\infty} \int_{-\infty}^{\infty} \frac{\partial}{\partial t} \left[e^{isy_1 + ity_2} f_{X_1, X_2}(y_1, y_2)\right] dy_1 \, dy_2$$

$$= \frac{1}{i\mu_2} \frac{\partial}{\partial t} \int_0^\infty \int_{-\infty}^\infty \left[e^{isy_1 + ity_2} \right] f_{X_1, X_2}(y_1, y_2) \, dy_1 \, dy_2$$

$$= \frac{1}{i\mu_2} \frac{\partial}{\partial t} \varphi_{X_1, X_2}(s, t),$$

and the exchange of differentiation and integration can be justified by the results given in Section I.A.3.4 and (1.46).

A similar derivation shows that the m.g.f. of W, if it exists, is

$$\mathbb{M}_W(s) = \frac{1}{\mu_2} \frac{\partial}{\partial t} \mathbb{M}_{X_1, X_2}(s, t)\Big|_{t=-rs}, \tag{1.63}$$

so that

$$f_R(r) = \mu_2 f_W(0), \tag{1.64}$$

where W has m.g.f. (1.63). ∎

1.2.6   Inversion formulae for the c.d.f.

We begin with a result which has limited computational value, but is useful for establishing the uniqueness theorem and also can be used to establish (1.54).

For the c.f. φ_X of r.v. X with c.d.f. F_X, Lévy (1925) showed that, if F_X is continuous at the two points $a \pm h$, $h > 0$, then

$$\boxed{F_X(a+h) - F_X(a-h) = \lim_{T \to \infty} \frac{1}{\pi} \int_{-T}^T \frac{\sin(ht)}{t} e^{-ita} \varphi_X(t) \, dt.} \tag{1.65}$$

Proof (outline): This follows Wilks (1963, p. 116). We will require the identity

$$\sin(b) - \sin(c) = 2 \sin\left(\frac{b-c}{2}\right) \cos\left(\frac{b+c}{2}\right) \tag{1.66}$$

from (I.A.30), and the results from Example I.A.21, namely

$$\int_{-T}^T t^{-1} \sin(bt) \sin(ct) \, dt = 0 \tag{1.67a}$$

and

$$\int_{-T}^T t^{-1} \sin(bt) \cos(ct) \, dt = 2 \int_0^T t^{-1} \sin(bt) \cos(ct) \, dt. \tag{1.67b}$$

Let

$$G(a, T, h) = \frac{1}{\pi} \int_{-T}^T \frac{\sin(ht)}{t} e^{-ita} \varphi_X(t) \, dt$$

$$= \frac{1}{\pi} \int_{-T}^T \frac{\sin(ht)}{t} e^{-ita} \int_{-\infty}^\infty e^{itx} \, dF_X(x) \, dt$$

$$= \int_{-\infty}^{\infty} \frac{1}{\pi} \int_{-T}^{T} \frac{\sin(ht)}{t} e^{-ita} e^{itx} \, dt \, dF_X(x)$$

$$= \int_{-\infty}^{\infty} m(x, a, T, h) \, dF_X(x),$$

where the exchange of integrals can be justified, and

$$m(x, a, T, h) = \frac{1}{\pi} \int_{-T}^{T} \frac{\sin(ht)}{t} e^{it(x-a)} \, dt$$

$$\overset{(1.31)}{=} \frac{1}{\pi} \int_{-T}^{T} \frac{\sin(ht)}{t} \left[\cos\left(t\,(x-a)\right) + i \sin\left(t\,(x-a)\right)\right] \, dt$$

$$\overset{(1.67)}{=} \frac{2}{\pi} \int_{0}^{T} \frac{\sin(ht)}{t} \cos\left(t\,(x-a)\right) \, dt.$$

Now, with $b = t\,(x - a + h)$ and $c = t\,(x - a - h)$,

$$\frac{b - c}{2} = ht, \qquad \frac{b + c}{2} = t\,(x - a),$$

and (1.66) implies

$$m(x, a, T, h) = \frac{1}{\pi} \int_{0}^{T} \frac{\sin\left(x - a + h\right)t}{t} \, dt - \frac{1}{\pi} \int_{0}^{T} \frac{\sin\left(x - a - h\right)t}{t} \, dt.$$

It is easy to show (see Problem 1.14) that

$$\lim_{T \to \infty} m(x, a, T, h) = \begin{cases} 0, & \text{if } x \notin [a - h, a + h], \\ 1/2, & \text{if } x = a - h \text{ or } x = a + h, \\ 1, & \text{if } x \in (a - h, a + h), \end{cases} \tag{1.68}$$

so that, assuming the exchange of limit and integral and using the continuity of F_X at the two points $a \pm h$,

$$\lim_{T \to \infty} G(a, T, h) = \int_{-\infty}^{\infty} \lim_{T \to \infty} m(x, a, T, h) \, dF_X(x)$$

$$= \int_{a-h}^{a+h} dF_X(x) = F_X(a + h) - F_X(a - h),$$

which is (1.65).   ∎

Remarks:
(a) If two r.v.s X and Y have the same c.f., then, from (1.65),

$$F_X(a + h) - F_X(a - h) = F_Y(a + h) - F_Y(a - h)$$

for all pairs of points $(a \pm h)$ for which F_X and F_Y are continuous. This, together with the fact that F_X and F_Y are right-continuous c.d.f.s, implies that $F_X(x) = F_Y(x)$, i.e., the uniqueness theorem.

(b) It is also straightforward to derive the p.d.f. inversion formula (1.54) from (1.65); see Problem 1.15. ■

We now turn to expressions for the c.d.f. which lend themselves to computation. If the cumulant generating function exists, as is often the case, a useful integral expression for the c.d.f. is derived in Problem 1.12. More generally, Gil-Peleaz (1951) derived the inversion formula for continuous r.v. X with c.f. φ_X,

$$F_X(x) = \frac{1}{2} + \frac{1}{2\pi} \int_0^\infty \frac{e^{itx}\varphi_X(-t) - e^{-itx}\varphi_X(t)}{it} \, dt, \qquad (1.69)$$

which can be proved as follows.

Proof (outline): From the definition of φ_X and expression (1.32) for $\sin z$, let

$$I(a, b) = \int_a^b \frac{e^{itx}\varphi_X(-t) - e^{-itx}\varphi_X(t)}{it} \, dt$$

$$= \int_a^b \int_{-\infty}^\infty \frac{e^{-it(y-x)} - e^{it(y-x)}}{it} f_X(y) \, dy \, dt$$

$$= -2 \int_a^b \int_{-\infty}^\infty \frac{\sin(t(y-x))}{t} f_X(y) \, dy \, dt$$

for $0 < a < b < \infty$. Also let

$$s(z) = \frac{2}{\pi} \int_0^\infty \frac{\sin tz}{t} \, dt = \begin{cases} -1, & \text{if } z < 0, \\ 0, & \text{if } z = 0, \\ 1, & \text{if } z > 0 \end{cases}$$

(which easily follows from (1.55) after substituting $y = tz$), so that

$$\int_{-\infty}^\infty s(y-x) f_X(y) \, dy = \int_{-\infty}^x s(y-x) f_X(y) \, dy + \int_x^\infty s(y-x) f_X(y) \, dy$$

$$= -F_X(x) + (1 - F_X(x)) = 1 - 2F_X(x).$$

Assuming the order of integration can be changed,[8]

$$\frac{1}{\pi} I(a, b) = -\int_{-\infty}^\infty \frac{2}{\pi} \int_a^b \frac{\sin t(y-x)}{t} \, dt f_X(y) \, dy$$

and

$$\lim_{\substack{a \to 0 \\ b \to \infty}} \frac{1}{\pi} I(a, b) = -\int_{-\infty}^\infty s(y-x) f_X(y) \, dy = 2F_X(x) - 1,$$

from which (1.69) follows. ■

---

[8] To verify that the order can be changed, one needs

$$\left| \int_a^b \frac{\sin t(y-x)}{t} \, dt \right| \leq \int_a^b \left| \frac{\sin t(y-x)}{t} \right| \, dt \leq \int_a^b \frac{1}{t} \, dt \leq \int_a^b \frac{1}{a} \, dt \leq \frac{b-a}{a}.$$

For numeric inversion purposes, an expression related to (1.69) is even more useful. As $\sin(-\theta) = -\sin(\theta)$,

$$
e^{-itx}\varphi_X(t) = \int_{-\infty}^{\infty} e^{-it(x-y)} f_X(y)\,dy
$$

$$
= \int_{-\infty}^{\infty} \cos(t(x-y)) f_X(y)\,dy - i \int_{-\infty}^{\infty} \sin(t(x-y)) f_X(y)\,dy
$$

and

$$
I(0,\infty) = +2 \int_0^{\infty} \frac{1}{t} \int_{-\infty}^{\infty} \sin(t(x-y)) f_X(y)\,dy\,dt
$$

$$
= -2 \int_0^{\infty} \frac{1}{t} \operatorname{Im}\left(e^{-itx}\varphi_X(t)\right) dt,
$$

recalling $\operatorname{Im}(a+bi) = b$. That is,

$$
F_X(x) = \frac{1}{2} - \frac{1}{\pi}\int_0^{\infty} g_x(t)\,dt, \tag{1.70}
$$

where, using (1.36),

$$
g_x(t) = \frac{\operatorname{Im} z_x(t)}{t} = \frac{|z_x(t)|\sin(\arg z_x(t))}{t} \quad \text{and} \quad z_x(t) = e^{-itx}\varphi_X(t).
$$

As in (1.60) and (1.61), (1.70) can sometimes be more effectively computed as either

$$
F_X(x) = \frac{1}{2} - \frac{1}{\pi}\int_0^1 g_x\left(\frac{1-u}{u}\right) u^{-2}\,du \tag{1.71}
$$

or

$$
F_X(x) = \frac{1}{2} - \frac{1}{\pi}\int_0^1 g_x\left(\frac{u}{1-u}\right)(1-u)^{-2}\,du. \tag{1.72}
$$

⊖ ***Example 1.25*** For $n, \theta \in \mathbb{R}_{>0}$, let X be the r.v. with

$$
\varphi_X(t) = (1-2it)^{-n/2}\exp\left\{\frac{it\theta}{1-2it}\right\}, \tag{1.73}
$$

which is the c.f. of a so-called *noncentral* χ^2 *distribution* with n degrees of freedom and noncentrality parameter θ. Let $z_x(t) = e^{-itx}\varphi_X(t)$. From (1.37) and the results in Examples 1.14 and 1.15,

$$
|z_x(t)| = \left|e^{-itx}\right|\,|\varphi_X(t)| = |\varphi_X(t)| = \left|(1-2it)^{-n/2}\right|\left|\exp\left\{\frac{it\theta}{1-2it}\right\}\right|
$$

$$
= (1+4t^2)^{-n/4}\exp\left(-\frac{2\theta t^2}{1+4t^2}\right),
$$

and

$$\arg z_x(t) = \arg e^{-itx} + \arg \varphi_X(t) = -tx + \frac{n}{2} \arctan(2t) + \frac{t\theta}{1 + 4t^2},$$

so that (1.70) yields the impressive looking

$$F_X(x) = \frac{1}{2} - \frac{1}{\pi} \int_0^\infty \frac{\sin\left(-tx + \frac{n}{2}\arctan(2t) + t\theta/(1 + 4t^2)\right)}{t\left(1 + 4t^2\right)^{n/4} \exp\left(2\theta t^2/(1 + 4t^2)\right)} \, dt. \qquad (1.74)$$

(Other, more preferable, methods to compute F_X will be discussed in Chapter 10.) The next section discusses a method of computing the p.d.f. ∎

1.3   Use of the fast Fourier transform

The fast Fourier transform (FFT) is a numerical procedure which vastly speeds up the calculation of the discrete Fourier transform (DFT), to be discussed below. The FFT is rightfully considered to be one of the most important scientific developments in the twentieth century because of its applicability to a vast variety of fields, from medicine to financial option pricing (see, for example, Carr and Madan, 1999), not to mention signal processing and electrical engineering. For our purposes, the application of interest is the fast numeric inversion of the characteristic function of a random variable for a (potentially large) grid of points. For a small number of points, the numeric integration methods of Sections 1.2.5 and 1.2.6 are certainly adequate, but if, say, a plot of the p.d.f. is required, or, more importantly, a likelihood calculation (which involves the evaluation of the p.d.f. at all the observed data points), then such methods are far too slow for practical use. For example, in financial applications, one often has thousands of observations.

A full appreciation of the FFT requires a book-length treatment, as well as a deeper understanding of the mathematics of Fourier analysis. Our goal here is far more modest; we wish to gain a heuristic, basic understanding of Fourier series, Fourier transforms, and the FFT, and concentrate on their use for inverting the c.f.

1.3.1   Fourier series

We begin with a rudimentary discussion of Fourier series, based on the presentations in Dyke (2001) and Schiff (1999), which are both highly recommended starting points for students who wish to pursue the subject further. More advanced but still accessible treatments are provided in Bachman, Narici and Beckenstein (2000) and Beerends *et al.* (2003).

Consider the *periodic function* shown in Figure 1.6, as a function of time t in, say, seconds. The *period* is the time (in seconds) it takes for the function to repeat, and its *frequency* is the number of times the wave repeats in a given time unit, say one second. Because the function in Figure 1.6 (apparently) repeats every one second, its *fundamental period* is $T = 1$, and its *fundamental frequency* is $f = 1/T$. The trigonometric functions sin and cos are natural period functions, and form the basis into which many periodic functions can be decomposed.

Let θ denote the angle in the usual depiction of the unit circle (similar to that in Figure 1.2), measured in radians (the arc length), of which the total length around the circle is 2π. Let θ be the function of time t given by $\theta(t) = \omega t$, so that ω is the *angular velocity*. If we take $\omega = 2\pi/T$ and let t move from 0 to T, then $\theta(t) = \omega t$ will move from zero to 2π, and thus $\cos(\omega t)$ will make one complete cycle, and similarly for sin. Now with $n \in \mathbb{N}$, $\cos(n\omega t)$ will make n complete cycles, where the angular velocity is $n\omega$. This (hopefully) motivates one to consider using $\cos(n\omega t)$ and $\sin(n\omega t)$, for $n = 0, 1, \ldots$, as a 'basis', in some sense, for decomposing periodic functions such as that shown in Figure 1.6. That is, we wish to express a periodic function g with fundamental period T as

$$g(t) = \sum_{n=0}^{\infty} \left[a_n \cos(n\omega t) + b_n \sin(n\omega t) \right], \quad \omega = \frac{2\pi}{T}, \tag{1.75}$$

where the a_i and b_i represent the *amplitudes* of the components. Notice that, for $n = 0$, $\cos(n\omega t) = 1$ and $\sin(n\omega t) = 0$, so that a_0 is the constant term of the function, and there is no b_0 term. The function shown in Figure 1.6 is the weighted sum of four sin terms. Notice that if n is not an integer, then $\cos(n\omega t)$ will not coincide at $t = 0$ and $t = T$, and so will not be of value for decomposing functions with fundamental period T.

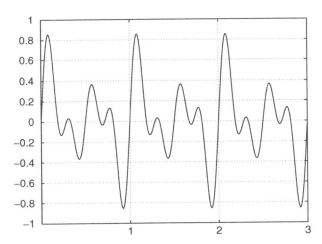

Figure 1.6   The periodic function (1.75) as a function of t, with all $a_i = 0$, $b_1 = 0.1$, $b_2 = 0.4$, $b_3 = 0.2$, $b_4 = 0.3$, and fundamental period $T = 1$

Recall from linear algebra the concepts of basis, span, dimension, orthogonality, linear independence, and orthonormal basis. In particular, for a given set S of vectors in $\mathbb{R}^n$, an orthonormal basis B can be constructed (e.g., via the Gramm–Schmidt process), such that any vector v which belongs to the span of S can be uniquely represented by the elements in B. The natural orthonormal basis for $\mathbb{R}^n$ is the set of unit vectors

$\mathbf{e}_1 = (1, 0, \ldots, 0)'$, $\mathbf{e}_2 = (0, 1, 0, \ldots, 0)'$, $\ldots$, $\mathbf{e}_n = (0, \ldots, 0, 1)'$, which are orthonormal because the inner product (in this case, the dot product) of $\mathbf{e}_i$ with itself is one, i.e., $\langle \mathbf{e}_i, \mathbf{e}_i \rangle = 1$, and that of $\mathbf{e}_i$ and $\mathbf{e}_j$, $i \neq j$, is zero. If $\mathbf{v} \in \mathbb{R}^n$, then it is trivial to see that, with the $\mathbf{e}_i$ so defined,

$$\mathbf{v} = \sum_{i=1}^{n} \langle \mathbf{v}, \mathbf{e}_i \rangle \mathbf{e}_i. \tag{1.76}$$

However, (1.76) holds for any orthonormal basis of $\mathbb{R}^n$. To illustrate, take $n = 3$ and the nonorthogonal basis $\mathbf{a}_1 = (1, 0, 0)'$, $\mathbf{a}_2 = (1, 1, 0)'$, $\mathbf{a}_3 = (1, 1, 1)'$, and use Gramm–Schmidt to orthonormalize this; in Matlab, this is done with

```
A=[1,0,0; 1,1,0; 1,1,1]', oA=orth(A)
```

Then type

```
v=[1,5,-3]'; a=zeros(3,1);
for i=1:3, a = a + ( v' * oA(:,i) ) * oA(:,i) ; end, a
```

to verify the claim.

In the Fourier series setting, it is helpful and common to start with taking $T = 2\pi$, so that $\omega = 1$, and assuming that f is defined over $[-\pi, \pi]$, and periodic, such that $g(x) = g(x + T) = g(x + 2\pi)$, $x \in \mathbb{R}$. (This range is easily extended to $[-r, r]$ for any $r > 0$, or to $[0, T]$, for any $T > 0$, as seen below.) Then we are interested in the sequence of functions

$$\{2^{-1/2}, \cos(n\cdot), \sin(n\cdot), \ n \in \mathbb{N}\},$$

because it forms an *infinite orthonormal basis*[9] for the set of (piecewise) continuous functions on $[-\pi, \pi]$ with respect to the inner product

$$\langle f, g \rangle := \frac{1}{\pi} \int_{-\pi}^{\pi} f(t)\overline{g(t)} \, dt, \tag{1.77}$$

where $\overline{g}$ is the complex conjugate of function g. It is straightforward to verify orthonormality of these functions, the easiest of which is

$$\left\langle \frac{1}{\sqrt{2}}, \frac{1}{\sqrt{2}} \right\rangle = \frac{1}{\pi} \int_{-\pi}^{\pi} \frac{1}{2} \, dt = 1.$$

The others require the use of basic trigonometric identities. For example, as $\cos^2(x) + \sin^2(x) = 1$ and, from (I.A.23b), $\cos(x + y) = \cos x \cos y - \sin x \sin y$, we have $\cos(2x) = \cos^2(x) - \sin^2(x) = 1 - 2\sin^2(x)$, implying

$$\left\langle \sin(nt), \sin(nt) \right\rangle = \frac{1}{\pi} \int_{-\pi}^{\pi} \sin^2(nt) \, dt = \frac{1}{2\pi} \int_{-\pi}^{\pi} (1 - \cos(2nt)) \, dt = 1.$$

---

[9] The concept of an infinite orthonormal basis is the cornerstone of the theory of so-called Hilbert spaces, and usually discussed in the framework of functional analysis. This is a fundamentally important branch of mathematics, with numerous real applications, including many in probability and statistics. Very accessible introductions are given by Light (1990) and Ha (2006).

Again using (I.A.23b), and the fact that $\cos(t) = \cos(-t)$,

$$\cos(m+n) = \cos(m)\cos(n) - \sin(m)\sin(n),$$

$$\cos(m-n) = \cos(m)\cos(n) + \sin(m)\sin(n),$$

and summing,

$$\cos(m+n) + \cos(m-n) = 2\cos(m)\cos(n),$$

we see, for $m \neq n$, that

$$\langle \cos(mt), \cos(nt) \rangle = \frac{1}{2\pi} \int_{-\pi}^{\pi} \left[\cos((m+n)t) + \cos((m-n)t) \right] dt$$

$$= \frac{1}{2\pi} \left. \frac{\sin((m+n)t)}{m+n} + \frac{\sin((m-n)t)}{m-n} \right|_{-\pi}^{\pi} = 0.$$

The reader is encouraged to verify similarly that $\langle \cos(nt), \cos(nt) \rangle = 1$, $\langle 1, \cos(nt) \rangle = \langle 1, \sin(nt) \rangle = 0$, $\langle \sin(mt), \cos(nt) \rangle = 0$, and $\langle \cos(mt), \sin(nt) \rangle = 0$. Thus, based on (1.76), it would appear that

$$a_0 = \left\langle g, \frac{1}{\sqrt{2}} \right\rangle = \frac{1}{\pi} \int_{-\pi}^{\pi} \frac{1}{\sqrt{2}} g(x)\,dx,$$

$$a_n = \frac{1}{\pi} \int_{-\pi}^{\pi} g(t)\cos(nt)\,dt, \quad \text{and} \quad b_n = \frac{1}{\pi} \int_{-\pi}^{\pi} g(t)\sin(nt)\,dt, \quad n = 1, 2, \ldots,$$

and the Fourier series of g is $g(t) = a_0/\sqrt{2} + \sum_{n=1}^{\infty} \left[a_n \cos(nt) + b_n \sin(nt) \right]$. Equivalently, and often seen, one writes

$$g(t) = \frac{a_0}{2} + \sum_{n=1}^{\infty} \left[a_n \cos(nt) + b_n \sin(nt) \right], \tag{1.78}$$

with

$$a_n = \frac{1}{\pi} \int_{-\pi}^{\pi} g(t)\cos(nt)\,dt \quad \text{and} \quad b_n = \frac{1}{\pi} \int_{-\pi}^{\pi} g(t)\sin(nt)\,dt, \quad n = 0, 1, 2, \ldots, \tag{1.79}$$

so that the expression for a_n includes $n = 0$ (and, as $\sin(0) = 0$, $b_0 = 0$). The values of the a_i and b_i can be informally verified by multiplying the expression in (1.78) by, say, $\cos(nt)$, and integrating termwise.

Remarks:
(a) The rigorous justification of termwise integration in (1.78) requires more mathematical sophistication, one requirement of which is that $\lim_{i \to \infty} \langle \mathbf{v}, \mathbf{e}_i \rangle = 0$, where the

e_i form an infinite orthonormal basis for some space V, and $\mathbf{v} \in V$. As a practical matter, the decomposition (1.75) will involve only a finite number of terms, with all the higher-frequency components being implicitly set to zero.

(b) Recall that an even function is symmetric about zero, i.e., $g : D \mapsto \mathbb{R}$ is even if $g(-x) = g(x)$ for all $x \in D$, and g is odd if $g(-x) = -g(x)$ (see the remarks after Example I.A.27. Thus, cos is even, and sin is odd, which implies that, if g is an even function and it has a Fourier expansion (1.75), then it will have no sin terms, i.e., all the $b_i = 0$, and likewise, if g is odd, then $a_i = 0$, $i \geq 1$. ∎

⊖ **Example 1.26** Let g be the periodic function with $g(t) = t^2$ for $-r \leq t \leq r, r > 0$, and such that $g(t) = g(t + 2r)$. To deal with the fact that the range is relaxed such that r need not be π, observe that if $-r \leq t \leq r$, then $-\pi < \pi t/r < \pi$, so that

$$s(t) := \sin\left(\frac{\pi}{r}t\right) = \sin\left(\frac{\pi}{r}t + 2\pi\right) = \sin\left(\frac{\pi}{r}(t + 2r)\right) = s(t + 2r),$$

i.e., the function s has period $2r$, and likewise for $c(t) := \cos(\pi t/r)$. Thus,

$$g(t) = \frac{a_0}{2} + \sum_{n=1}^{\infty} [a_n s(nt) + b_n c(nt)]$$

$$= \frac{a_0}{2} + \sum_{n=1}^{\infty} \left[a_n \sin\left(\frac{n\pi t}{r}\right) + b_n \cos\left(\frac{n\pi t}{r}\right)\right], \quad (1.80)$$

where

$$a_n = \frac{1}{r} \int_{-r}^{r} g(t) \cos\left(\frac{n\pi t}{r}\right) dt, \quad b_n = \frac{1}{r} \int_{-r}^{r} g(t) \sin\left(\frac{n\pi t}{r}\right) dt, \quad (1.81)$$

which, in this case, are

$$a_0 = \frac{1}{r} \int_{-r}^{r} t^2 \, dt = \frac{2}{3}r^2,$$

$$a_n = \frac{1}{r} \int_{-r}^{r} t^2 \cos\left(\frac{n\pi t}{r}\right) dt = \frac{4r^2 \cos(\pi n)}{\pi^2 n^2} = \frac{4r^2 (-1)^n}{\pi^2 n^2}, \quad n = 1, 2, \ldots,$$

$$b_n = 0,$$

having used repeated integration by parts for a_n, and where the b_n are zero because f is an even function. Thus,

$$g(t) = \frac{r^2}{3} + \frac{4r^2}{\pi^2} \sum_{n=1}^{\infty} \frac{(-1)^n}{n^2} \cos\left(\frac{n\pi t}{r}\right). \quad (1.82)$$

Figure 1.7 plots (1.82) for $r = 1$, truncating the infinite sum to have N terms. Series (1.82) with $t = 0$ and $r = \pi$ implies $0^2 = \pi^2/3 + 4\sum_{n=1}^{\infty} (-1)^n n^{-2}$, i.e.,

$$\frac{\pi^2}{12} = \frac{1}{1^2} - \frac{1}{2^2} + \frac{1}{3^2} - \cdots,$$

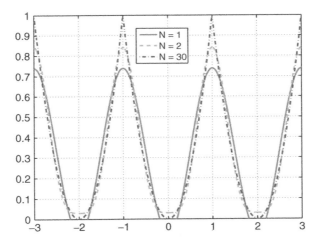

Figure 1.7 Fourier series (1.82) for $r = 1$, truncating the infinite sum to have N terms

and, with $t = \pi = r$, $\pi^2 = \pi^2/3 + 4\sum_{n=1}^{\infty} n^{-2}$, i.e.,

$$\boxed{\frac{\pi^2}{6} = \sum_{n=1}^{\infty} \frac{1}{n^2},}$$

which was stated without proof in (I.A.88).

Example 1.27 Assume that $g(t) = t^2 - t$ for $t \in [0, T]$, and $g(t) = g(t + T)$. This is plotted in Figure 1.8 for $T = 2$, using the Matlab code

```
g=[]; T=2;
for t=0:0.01:4*T
  u=t; while u>T, u=u-T; end, g=[g u^2-u];
end
h=plot(0:0.01:4*T,g,'r.'), set(h,'MarkerSize',3)
```

Similar to Example 1.26, it is straightforward to verify that, for a periodic function defined on $t \in [0, T]$ and such that $g(t) = g(t + T)$,

$$g(t) = \frac{a_0}{2} + \sum_{n=1}^{\infty} \left[a_n \cos(n\omega t) + b_n \sin(n\omega t) \right], \quad \omega = \frac{2\pi}{T}, \qquad (1.83)$$

with, for $n = 0, 1, 2, \ldots,$

$$a_n = \frac{2}{T} \int_0^T g(t) \cos(n\omega t)\, dt, \quad b_n = \frac{2}{T} \int_0^T g(t) \sin(n\omega t)\, dt. \qquad (1.84)$$

In this case, a bit of straightforward work yields

$$a_0 = \frac{T(2T - 3)}{3}, \quad a_n = \frac{T^2}{\pi^2 n^2}, \quad b_n = \frac{(1 - T)T}{n\pi}, \quad n = 1, 2, \ldots, \qquad (1.85)$$

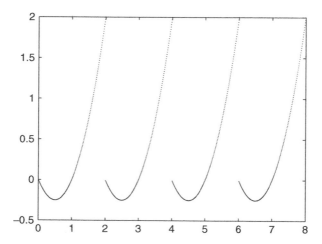

Figure 1.8 The periodic function $g(t) = t^2 - t$ for $t \in [0, T]$, $T = 2$, with $g(t) = f(t+T)$

and Figure 1.9 plots (1.83) with (1.85), for several values of N, the truncation value of the infinite sum. We see that, compared to the function in Example 1.26, N has to be quite large to obtain reasonable accuracy. ∎

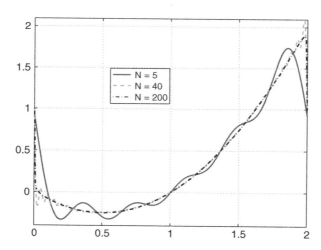

Figure 1.9 Approximation (1.83) and (1.85) to the function shown in Figure 1.8, truncating the infinite sum to have N terms

From (1.32) and (1.83),

$$g(t) = \frac{a_0}{2} + \sum_{n=1}^{\infty} \left[\frac{a_n}{2} \left(e^{in\omega t} + e^{-in\omega t} \right) + \frac{b_n}{2i} \left(e^{in\omega t} - e^{-in\omega t} \right) \right]$$

$$= \frac{a_0}{2} + \frac{1}{2} \sum_{n=1}^{\infty} \left[(a_n - ib_n) e^{in\omega t} + (a_n + ib_n) e^{-in\omega t} \right],$$

and from (1.84), for $n \geq 1$,

$$a_n = \frac{2}{T} \int_0^T g(t) \cos(n\omega t) \, dt = \frac{1}{T} \int_0^T g(t) e^{in\omega t} \, dt + \frac{1}{T} \int_0^T g(t) e^{-in\omega t} \, dt,$$

$$ib_n = \frac{2i}{T} \int_0^T g(t) \sin(n\omega t) \, dt = \frac{1}{T} \int_0^T g(t) e^{in\omega t} \, dt - \frac{1}{T} \int_0^T g(t) e^{-in\omega t} \, dt,$$

so that

$$a_n - ib_n = \frac{2}{T} \int_0^T g(t) e^{-in\omega t} \, dt, \quad a_n + ib_n = \frac{2}{T} \int_0^T g(t) e^{in\omega t} \, dt,$$

yielding

$$g(t) = \frac{a_0}{2} + \sum_{n=1}^{\infty} \left[A_n e^{in\omega t} + B_n e^{-in\omega t} \right],$$

where

$$A_n = \frac{a_n - ib_n}{2} = \frac{1}{T} \int_0^T g(t) e^{-in\omega t} \, dt, \quad B_n = \frac{a_n + ib_n}{2} = \frac{1}{T} \int_0^T g(t) e^{in\omega t} \, dt.$$

The two terms in the sum are easily combined to give the *complex Fourier series expansion*

$$g(t) = \sum_{n=-\infty}^{\infty} C_n e^{in\omega t}, \quad C_n = \frac{1}{T} \int_0^T g(t) e^{-in\omega t} \, dt, \quad \omega = \frac{2\pi}{T}. \qquad (1.86)$$

(Note that the term with $n = 0$ is correct.) As a check, note that, for $k = (n - m) \neq 0$,

$$\int_0^T e^{i(n-m)\omega t} \, dt = \int_0^T \exp\left(\frac{2\pi i k t}{T}\right) \, dt = \int_0^T \cos\left(\frac{2\pi k t}{T}\right) \, dt + i \int_0^T \sin\left(\frac{2\pi k t}{T}\right) \, dt$$

$$= \frac{T}{2\pi k} \sin(2\pi k) + i \frac{T}{2\pi k} (1 - \cos 2\pi k) = 0,$$

so that

$$\int_0^T e^{i(n-m)\omega t} \, dt = \begin{cases} T, & \text{if } n = m, \\ 0, & \text{if } n \neq m, \end{cases}$$

i.e., the expression for C_m from (1.86) is (assuming the validity of exchanging sum and integral)

$$\frac{1}{T} \int_0^T g(t) e^{-im\omega t} \, dt = \frac{1}{T} \int_0^T \left(\sum_{n=-\infty}^{\infty} C_n e^{in\omega t} \right) e^{-im\omega t} \, dt$$

$$= \frac{1}{T} \sum_{n=-\infty}^{\infty} C_n \int_0^T e^{i(n-m)\omega t} \, dt = C_m,$$

as expected. It should be clear from the form of (1.86) that, if we started with g defined on $[-r, r]$ for $r > 0$, periodic with period $T = 2r$ such that $g(x) = g(x + 2r)$, then

$$g(t) = \sum_{n=-\infty}^{\infty} C_n e^{in\omega t}, \quad C_n = \frac{1}{2r} \int_{-r}^{r} g(t) e^{-in\omega t} \, dt, \quad \omega = \frac{2\pi}{T} = \frac{\pi}{r}, \quad (1.87)$$

which the reader is encouraged to check using (1.80) and (1.81).

1.3.2   Discrete and fast Fourier transforms

Based on (1.86), it would (informally) seem natural to construct a discrete version of the C_n by taking

$$G_n = \frac{1}{T} \sum_{t=0}^{T-1} g_t e^{-2\pi i n t/T}, \quad n = 0, \ldots, T - 1, \quad (1.88)$$

where $g_t = g(t)$, with the subscript serving as a reminder that g is evaluated at discrete points. For vector $\mathbf{g} = (g_0, \ldots, g_{T-1})$, we will write $\mathbf{G} = \mathcal{F}(\mathbf{g})$, where $\mathbf{G} = (G_0, \ldots, G_{T-1})$. Expression (1.88) is the definition of the *(forward, complex) discrete Fourier transform*, or DFT. The T values of g_t can be recovered from the *inverse discrete Fourier transform*, defined as

$$g_t = \sum_{n=0}^{T-1} G_n e^{2\pi i n t/T}, \quad t = 0, \ldots, T - 1, \quad (1.89)$$

and we write $\mathbf{g} = \mathcal{F}^{-1}(\mathbf{G})$. The programs in Listings 1.4 and 1.5 compute (1.88) and (1.89), respectively.

```
function G=dft(g)
  T=length(g); z=exp(-2*pi*i/T); G=zeros(T,1); t=0:(T-1);
  for n=0:(T-1),  w=z.^(t*n); G(n+1)=sum(g.*w'); end, G=G/T;
```

Program Listing 1.4   The forward discrete Fourier transform (1.88)

```
function g=idft(G)
  T=length(G); z=exp(2*pi*i/T); g=zeros(T,1); t=0:(T-1);
  for n=0:(T-1),  w=z.^(t*n); g(n+1)=sum(G.*w'); end
```

Program Listing 1.5   The inverse discrete Fourier transform (1.89)

Function $\mathcal{F}^{-1}$ is the inverse of $\mathcal{F}$ in the sense that, for a set of complex numbers $\mathbf{g} = (g_0, \ldots, g_{T-1})$, $\mathbf{g} = \mathcal{F}^{-1}(\mathcal{F}(\mathbf{g}))$. To see this, let $\mathbf{G} = \mathcal{F}(\mathbf{g})$ and define $z_T := e^{2\pi i/T}$. From (1.89) and (1.88), the tth element of $\mathcal{F}^{-1}(\mathbf{G})$, for $t \in \{0, 1, \ldots, T - 1\}$, is

$$\sum_{n=0}^{T-1} G_n e^{2\pi i n t/T} = \sum_{n=0}^{T-1} G_n z_T^{nt} = \sum_{n=0}^{T-1} \left(\frac{1}{T} \sum_{s=0}^{T-1} g_s z_T^{-ns} \right) z_T^{nt} = \frac{1}{T} \sum_{s=0}^{T-1} g_s \left(\sum_{n=0}^{T-1} (z_T^{t-s})^n \right).$$

Now recalling (1.38), we see that, if $s = t$, then the inner sum is T, and zero otherwise, so that the tth element of $\mathcal{F}^{-1}(\mathbf{G})$ is g_t. This holds for all $t \in \{0, 1, \ldots, T - 1\}$, so that $\mathcal{F}^{-1}(\mathcal{F}(\mathbf{g})) = \mathbf{g}$. This is numerically illustrated in the program in Listing 1.6.

```
T=2^8; g=randn(T,1)+i*randn(T,1); G=dft(g); gg=idft(G); max(abs(g-gg))
```

Program Listing 1.6 Illustrates numerically that $\mathbf{g} = \mathcal{F}^{-1}(\mathcal{F}(\mathbf{g}))$

The *fast Fourier transform*, or FFT, is the DFT, but calculated in a judicious manner which capitalizes on redundancies. Numerous algorithms exist which are optimized for various situations, the simplest being when $T = 2^p$ is a power of 2, which we illustrate. As in Černý (2004, Section 7.4.1), splitting the DFT (1.88) into even and odd indices gives, with $0 \le n \le T - 1$ and $z_T := e^{-2\pi i/T}$,

$$G_n = \frac{1}{T}\sum_{t=0}^{T-1} g_t \left(z_T^n\right)^t = \frac{1}{T}\sum_{t=0}^{(T/2)-1} g_{2t}\left(z_T^n\right)^{2t} + \frac{1}{T}\sum_{t=0}^{(T/2)-1} g_{2t+1}\left(z_T^n\right)^{2t+1}$$

$$= \frac{1}{T}\sum_{t=0}^{(T/2)-1} g_{2t}\left(z_T^{2n}\right)^{t} + \frac{1}{T}z_T^n\sum_{t=0}^{(T/2)-1} g_{2t+1}\left(z_T^{2n}\right)^{t}$$

$$=: S_1 + S_2.$$

As $z_T^{T/2} = \left(e^{-2\pi i/T}\right)^{T/2} = e^{-\pi i} = -1$ and $z_T^T = e^{-2\pi i} = 1$, the DFT of the $(n + T/2)$th element is

$$G_{n+T/2} = \frac{1}{T}\sum_{t=0}^{(T/2)-1} g_{2t}\left(z_T^{n+T/2}\right)^{2t} + \frac{1}{T}\sum_{t=0}^{(T/2)-1} g_{2t+1}\left(z_T^{n+T/2}\right)^{2t+1}$$

$$= \frac{1}{T}\sum_{t=0}^{(T/2)-1} g_{2t}\left(z_T^{2n}z_T^T\right)^{t} + \frac{1}{T}\left(z_T^n z_T^{T/2}\right)\sum_{t=0}^{(T/2)-1} g_{2t+1}\left(z_T^{2n}z_T^T\right)^{t}$$

$$= \frac{1}{T}\sum_{t=0}^{(T/2)-1} g_t\left(z_T^{2n}\right)^{t} - \frac{1}{T}z_T^n\sum_{t=0}^{(T/2)-1} g_t\left(z_T^{2n}\right)^{t}$$

$$= S_1 - S_2,$$

so that the sums S_1 and S_2 only need to be computed for $n = 0, 1, \ldots, (T/2) - 1$. Recalling that T is a power of 2, this procedure can be applied to S_1 and S_2, and so on, p times. It is easy to show that this recursive method will require, approximately, of the order of $T(\log_2(T) + 1)$ complex-number multiplications, while the DFT requires of the order of T^2. For $T = 2^6 = 64$ data points, the ratio $T/\log_2(T)$ is about 10, i.e., the FFT is 10 times faster, while with $T = 2^{10} = 1024$ data points, the FFT is about 100 times faster.

In Matlab (version 7.1), the command `fft` is implemented in such a way that it computes (what we, and most books, call) the inverse Fourier transform (1.89), while the command `ifft` computes the DFT (1.88).[10] This is illustrated in the code in Listing 1.7.

```
T=2^10; g=randn(T,1)+i*randn(T,1);
s=cputime; G1=dft(g);   t1=cputime-s % how long does the direct DFT take?
s=cputime; for r=1:1000, G2=ifft(g); end, t2=(cputime-s)/1000
timeratio=t1/t2   % compare to the FFT
thedifference = max(abs(G1-G2)) % is zero, up to roundoff error
```

Program Listing 1.7   The DFT (1.88) is computed via the FFT by calling Matlab's `ifft` command

1.3.3 Applying the FFT to c.f. inversion

Similar to Černý (2004, Section 7.5.3), let X be a continuous random variable with p.d.f. f_X and c.f. φ_X, so that, for $\ell, u \in \mathbb{R}$ and $T \in \mathbb{N}$,

$$\varphi_X(s) = \int_{-\infty}^{\infty} e^{isx} f_X(x)\,dx \approx \int_{\ell}^{u} e^{isx} f_X(x)\,dx \approx \sum_{n=0}^{T-1} e^{isx_n} P_n , \qquad (1.90)$$

with

$$P_n := f_X(x_n)\,\Delta x, \quad x_n := \ell + n\,(\Delta x), \quad \Delta x := \frac{u - \ell}{T},$$

whereby the second approximation is valid (in the sense that it can be made arbitrarily accurate by choosing T large enough) from the definition of the Riemann integral, and the first approximation is valid (in the sense that it can be made arbitrarily accurate by choosing $-\ell$ and u large enough) because, as an improper integral, φ_X exists, and we *assume* that f is such that $\lim_{x \to \infty} f_X(x) = \lim_{x \to -\infty} f_X(x) = 0$. (While virtually all p.d.f.s in practice will satisfy this assumption, it is not true that if f_X is a proper continuous p.d.f., then $\lim_{x \to \infty} f_X(x) = \lim_{x \to -\infty} f_X(x) = 0$. See Problem 1.20.)

We assume that $\varphi_X(s)$ is numerically available for any s, and it is the P_n which we wish to recover. This will be done using the DFT (via the FFT). With the inverse DFT (1.89) in mind, dividing both sides of (1.90) by $e^{is\ell}$ and equating $sn\,(\Delta x)$ with $2\pi nt/T$ gives

$$\varphi_X(s)\,e^{-is\ell} \approx \sum_{n=0}^{T-1} e^{isn(\Delta x)} P_n = \sum_{n=0}^{T-1} e^{2\pi int/T} P_n =: g_t, \qquad (1.91)$$

yielding $\mathbf{g} = \mathcal{F}^{-1}(\mathbf{P})$, with $\mathbf{g} = (g_0, \ldots, g_{T-1})$ and $\mathbf{P} = (P_0, \ldots, P_{T-1})$. As $s\,(\Delta x) = 2\pi t/T$, it makes sense to take a T-length grid of s-values as, say,

$$s_t = \frac{2\pi t}{T\,(\Delta x)}, \quad t = -\frac{T}{2}, -\frac{T}{2} + 1, \ldots, \frac{T}{2} - 1,$$

---

[10] To make matters more confusing, according to the Matlab's help file on its FFT, it defines the DFT and its inverse as we do (except for where the factor $1/T$ is placed), but this is not what they compute.

so $g_t \approx \varphi_X(s_t) e^{-i s_t \ell}$ from (1.91). The relation between $\mathcal{F}$ and $\mathcal{F}^{-1}$ implies $\mathbf{P} = \mathcal{F}(\mathbf{g})$, so applying the DFT (via the FFT) to $\mathbf{g}$ yields $\mathbf{P}$, the nth element of which is $P_n = f_X(x_n) \Delta x$. Finally, dividing this $\mathbf{P}$ by Δx yields the p.d.f. of X at the T-length grid of x-values ℓ, $\ell + (\Delta x)$, $\ldots$, $\ell + (T-1)(\Delta x) \approx u$. Clearly, choosing (Δx) small and T large (and ℓ and u appropriately) will yield greater accuracy, though the higher T is, the longer the FFT calculation. Typically, T is chosen as a power of 2, so that the most efficient FFT algorithms are possible. The code in Listing 1.8 implements this for inverting the normal and χ^2 characteristic functions, and plots the RPE of the computed p.d.f. versus the true p.d.f.

```
T=2^15; dx=0.002; % tuning parameters which influence quality

% ell is the lower limit in finite integral approximation to the c.f.
ell=-3; % use this when inverting the standard normal.
ell=0; % use 0 for the chi^2.

t=-(T/2):(T/2)-1; s = 2*pi*t/T/dx;

%%%%%%%%%%%%% Choose the c.f.  %%%%%%%%%%%%%%%%%%%%%%%%%%%%%%
phi = exp(-(s.^2)/2); % the standard normal c.f.
% df=4; phi = (1-2*i*s).^(-df/2); % the chi^2(df) c.f.
%%%%%%%%%%%%%%%%%%%%%%%%%%%%%%%%%%%%%%%%%%%%%%%%%%%%%%%%%%%%%

g = phi.*exp(-i*s*ell);
P=ifft(g); % Remember: in Matlab, ifft computes the DFT
pdf = P/dx;

%%%%%%%%%%%%%%%%%%%%%%%%%%%%%%%%%%%%%%%%%%%%%%%%%%%%%%%%%%%%%
if max(abs(imag(pdf)))>1e-7, 'problem!', end
%% There should be no imaginary part, so check for this
%%%%%%%%%%%%%%%%%%%%%%%%%%%%%%%%%%%%%%%%%%%%%%%%%%%%%%%%%%%%%

%%%%%%%%%%%%%%%%%%%%%%%%%%%%%%%%%%%%%%%%%%%%%%%%%%%%%%%%%%%%%
f=abs(pdf); f=f(end:-1:1); % this is necessary.
%%%%%%%%%%%%%%%%%%%%%%%%%%%%%%%%%%%%%%%%%%%%%%%%%%%%%%%%%%%%%

x=ell+(dx/1):dx:T*dx; % next 2 lines in case x is off by one
if length(x)<length(f), x=[x (T+1)*dx]; end
if length(x)>length(f), x=x(1:end-1); end

xup=4; % set to around 20 for the chi^2
loc=find(diff(x>xup)); % just up to x=xup
x=x(10:loc); f=f(10:loc);

true = normpdf(x); plot(x,100*(f-true)./true)
%true = chi2pdf(x,df); plot(x,100*(f-true)./true)
```

Program Listing 1.8 Illustrates the use of the FFT for inverting the standard normal c.f., or the χ^2, at an equally spaced grid of points. The program in Listing 1.9 is similar, but is more advanced and is recommended in practice

The implementation in Listing 1.8 can be improved in several ways. One is to dynamically choose the step size Δx and the exponent p in the grid size $T = 2^p$ as a function of where the density is to be calculated. This requires having an upper bound on p based on available computer memory required for the FFT. In addition, for given values of p and Δx (i.e., for the same computational cost), higher accuracy can be obtained by using a variation on the above method, as detailed in Mittnik, Doganoglu and Chenyao (1999) and the references therein. We implement this in the program in Listing 1.9; it should be used in practice instead of the simpler code in Listing 1.8.

Moreover, the program in Listing 1.8 only delivers the p.d.f. at an equally spaced grid of points. This is fine for plotting purposes, but if the likelihood is being calculated, the data points are almost certainly not equally spaced. If the density of X is desired at a particular set of points $\mathbf{z} = (z_1, \ldots, z_k)$, then, provided that $\ell \leq z_1$ and $z_k \leq u$, some form of interpolation can be used to get $f_X(\mathbf{z})$. This could easily be accomplished by using the built-in Matlab function `interp1`, but this is problematic when a massive number of data points are involved (such as $T = 2^{14}$ or more). As such, the program in Listing 1.9 calls the function `wintp1`, given in Listing 1.10, which sets up a moving window for the (linear) interpolation, resulting in a sizeable speed increase.

```
function pdf=fftnoncentralchi2(z,df,noncen)

% set the tuning parameters for the FFT
pmax=18; step=0.01; p=14;
maxz=round(max(abs(z)))+5;
while((maxz/step+1)>2^(p-1))
  p=p+1;
end;
if p > pmax, p=pmax; end
if maxz/step+1 >2^(p-1)
  step=(maxz+1)*1.001/(2^(p-1));
end;

% compute the pdf on a grid of points
zs=sort(z); [xgrd,bigpdf]=runthefft(p,step, df,noncen);

% Now use linear interpolation to get the pdf at the desired values.
pdf=wintp1(xgrd,bigpdf,zs);


function [x,p]=runthefft(n,h, df,noncen)
n=2^n; x=(0:n-1)'*h-n*h/2; s=1/(h*n);
t=2*pi*s*((0:n-1)'-n/2);
sgn=ones(n,1); sgn(2:2:n)=-1*ones(n/2,1);
%%%%%%%%% This is now the c.f. %%%%%%%%%
CF = (1-2*i*t).^(-df/2) .* exp((i*t*noncen)./(1-2*i*t));
%%%%%%%%%%%%%%%%%%%%%%%%%%%%%%%%%%%%%%%%%
phi=sgn.*CF; phi(n/2+1)=sgn(n/2+1); p=s.*abs(fft(phi));
```

Program Listing 1.9 Uses the FFT and linear interpolation to invert the specified c.f. and compute the p.d.f. at the vector of points z. In this case, the c.f. corresponds to a noncentral χ^2(df,noncen) random variable, as in (1.73). The code for function `wintp1` is in Listing 1.10

```
function r=wintpl(x,y,z,w)
% interpolate with a window of size w on z.
if nargin<4, w=length(z); end;
m=length(z); n=floor(m/w); r=[];
for i=1:n
 dz=z((i-1)*w+1:i*w); k=interval(x,dz);
 t=intpol(x(k),y(k),dz); r=[r;t];
end;
if (rem(m,w)~=0),
    e=rem(m,w); g=[n*w+1:n*w+e]; dz=z(g);k=interval(x,dz);
    t=intpol(x(k),y(k),dz); r=[r;t];
end;

function i=interval(x,y)
% find the x: min(x)<min(y) and max(x)> max(y)
i1=find(x<min(y)); i2=find(x>max(y)); i=i1(size(i1,1)):i2(1);
j=i(1); while (x(j)>y(1)), j=j-1; end;
k=max(i); while(x(k)<max(y)),   k=k+1; end;
i=[(j:i(1)-1)';i;(max(i)+1:k)'];

function r=intpol(x,y,z);
% linear interp of z to f(x,y), needs min(x)< z < max(x)
n=length(z); r=[]; j=1; i=1;
while i<=n, flag=i;
  while (flag==i)
    if (z(i)>=x(j) & z(i)<x(j+1))
      r=[r;y(j+1)-(x(j+1)-z(i))*(y(j+1)-y(j))/(x(j+1)-x(j))];
        i=i+1;
    else, j=j+1; end;
  end;
end;
```

Program Listing 1.10 Interpolation program called by the function in Listing 1.9. This was written by Toker Doganoglu and used in the implementation detailed in Mittnik, Doganoglu and Chenyao (1999).

To illustrate, we use the c.f. in (1.73), corresponding to the noncentral χ^2 distribution. For $\theta = 0$, this reduces to the c.f. of the usual χ^2 r.v., so that we can easily check the results, while methods of computing the p.d.f. of the noncentral χ^2 distribution to machine precision are discussed in detail in Chapter 10. Figure 1.10 shows the p.d.f. as computed via program fftnoncentralchi2 using 5 degrees of freedom and several values of the noncentrality parameter θ. What is remarkable is the massive cut in computation time compared to the use of the p.d.f. inversion formula in Section 1.2.5, which would have to be evaluated for each ordinate in Figure 1.10.

1.4 Multivariate case

This section extends some of the ideas for univariate continuous r.v.s to the multivariate case. Paralleling the vector m.g.f., the c.f. of the d-dimensional continuous random vector $\mathbf{X} = (X_1, \ldots, X_d)$ is given by

$$\varphi_{\mathbf{X}}(\mathbf{t}) = \mathbb{E}\left[e^{i\mathbf{t}'\mathbf{X}}\right], \tag{1.92}$$

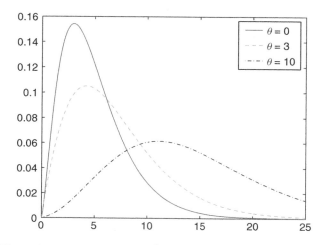

Figure 1.10 The noncentral $\chi^2(5, \theta)$ p.d.f. computed via the FFT

where $\mathbf{t} = (t_1, \ldots, t_d)$. The p.d.f. of $\mathbf{X}$ can be recovered by

$$f_{\mathbf{X}}(\mathbf{x}) = \frac{1}{(2\pi)^d} \int_{-\infty}^{\infty} \cdots \int_{-\infty}^{\infty} e^{-i\mathbf{t}'\mathbf{X}} \varphi_{\mathbf{X}}(\mathbf{t}) \, dt_1 \cdots dt_d,$$

which reduces to (1.54) when $d = 1$.

Consider the case when $\mathbb{M}_X$ exists, and $\varphi_X(t) = \mathbb{M}_X(it)$. In terms of the c.g.f. $\mathbb{K}_{\mathbf{X}}$, the multivariate survivor function

$$\overline{F}_{\mathbf{X}}(\mathbf{x}) = \Pr(X_1 > x_1, \ldots, X_d > x_d)$$

can be computed as

$$\overline{F}_{\mathbf{X}}(\mathbf{x}) = \frac{1}{(2\pi i)^d} \int_{c_1-i\infty}^{c_1+i\infty} \cdots \int_{c_d-i\infty}^{c_d+i\infty} \exp\left(\mathbb{K}_{\mathbf{X}}(\mathbf{t}) - \mathbf{t}'\mathbf{x}\right) \frac{d\mathbf{t}}{\prod_{j=1}^{d} t_j}, \qquad (1.93)$$

where $c_k \in \mathbb{R}_{>0}$, $k = 1, 2, \ldots, d$, such that they are in the convergence region of $\mathbb{K}_{\mathbf{X}}$. See Kolassa (2003, p. 276) for proof. Expression (1.93) reduces to (1.100), given below in Problem 1.12 for $d = 1$.

In the case where the c.g.f. is not available, or, more likely, the multivariate c.d.f. $F_{\mathbf{X}}$ is desired (instead of the survivor function), Shephard (1991) has generalized Gil-Peleaz's univariate result to the d-dimensional case when the mean of $\mathbf{X}$ exists. He shows that, for $d \geq 2$,

$$u(\mathbf{x}) := 2^d F_{\mathbf{X}}(x_1, \ldots, x_d)$$

$$- 2^{d-1}[F_{\mathbf{X}}(x_2, x_3, \ldots, x_d) + \cdots + F_{\mathbf{X}}(x_1, \ldots, x_{d-2}, x_{d-1})]$$

$$+ 2^{d-2}[F_{\mathbf{X}}(x_3, x_4, \ldots, x_d) + \cdots + F_{\mathbf{X}}(x_1, \ldots, x_{d-3}, x_{d-2})]$$

$$+ \cdots + (-1)^d \tag{1.94}$$

is given by

$$
u(\mathbf{x}) = \begin{cases}
\dfrac{2(-2)^d i^{d-1}}{(2\pi)^d} \displaystyle\int_0^\infty \cdots \int_0^\infty \underset{t_2}{\Delta} \cdots \underset{t_d}{\Delta} \, \mathrm{Im} \left[\dfrac{e^{-i\mathbf{t}'\mathbf{x}} \varphi_{\mathbf{X}}(\mathbf{t})}{\prod_{j=1}^d t_j} \right] d\mathbf{t}, & \text{if } d \text{ is odd,} \\[4mm]
\dfrac{2(-2)^d i^d}{(2\pi)^d} \displaystyle\int_0^\infty \cdots \int_0^\infty \underset{t_2}{\Delta} \cdots \underset{t_d}{\Delta} \, \mathrm{Re} \left[\dfrac{e^{-i\mathbf{t}'\mathbf{x}} \varphi_{\mathbf{X}}(\mathbf{t})}{\prod_{j=1}^d t_j} \right] d\mathbf{t}, & \text{if } d \text{ is even,}
\end{cases}
$$

$$\tag{1.95}$$

where $\underset{t}{\Delta}\, \eta(t) = \eta(t) + \eta(-t)$. (A similar, but slightly more complicated expression which is also valid for $d = 1$ is also given by Shephard).

For the bivariate case $d = 2$, use of (1.94) and (1.95) yields the expression

$$F_{\mathbf{X}}(x_1, x_2) = -\frac{1}{4} + \frac{1}{2}\big[F_X(x_1) + F_X(x_2)\big] - \frac{1}{2\pi^2} \int_0^\infty \int_0^\infty g(x_1, x_2, t_1, t_2)\, dt_1\, dt_2, \tag{1.96}$$

where

$$g(x_1, x_2, t_1, t_2) = \mathrm{Re}\left[\frac{\exp\{-it_1 x_1 - it_2 x_2\}\varphi(t_1, t_2)}{t_1 t_2} - \frac{\exp\{-it_1 x_1 + it_2 x_2\}\varphi(t_1, -t_2)}{t_1 t_2} \right].$$

Note that, given only $\varphi_{\mathbf{X}}(t_1, t_2)$, the univariate c.d.f.s $F_{X_1}(x_1)$ and $F_{X_2}(x_2)$ can be evaluated by univariate inversion of $\varphi_{X_1}(t_1) \equiv \varphi_{\mathbf{X}}(t_1, 0)$ and $\varphi_{X_2}(t_2) \equiv \varphi_{\mathbf{X}}(0, t_2)$, respectively. See Problem 1.21 for an illustration.

1.5 Problems

Opportunity is missed by most people because it is dressed in overalls and looks like work. (Thomas Edison)

It is common sense to take a method and try it; if it fails, admit it frankly and try another. But above all, try something. (Franklin D. Roosevelt)

1.1. ★ Derive the results in (1.9).

1.2. ★ Let $Y \sim \mathrm{Lap}\,(0, 1)$. The raw integer moments of Y were shown in Chapter I.7 to be $\mathbb{E}[Y^r] = r!$, $r = 0, 2, 4, \ldots$, and $\mathbb{E}[Y^r] = 0$, for odd r. Verify these for the first six integer moments of Y by using the m.g.f. Also show that the variance is 2 and the kurtosis is 6.

1.3. ★ Let $X \sim \mathrm{Gam}\,(\alpha, \beta)$. Derive the m.g.f., expected value and variance of $Z = \ln X$. Also compute $\mathrm{Cov}\,(Z, X)$.

1.4. ★ Calculate the cumulant generating function and, from this, the mean, variance, skewness, and kurtosis of the binomial distribution with parameters n and p. What happens to the skewness and kurtosis as $n \to \infty$?

1.5. ★ Let $X \sim \text{NBin}(r, p)$, with p.m.f. (I.4.19), i.e.,

$$f_X(x; r, p) = \binom{r + x - 1}{x} p^r (1 - p)^x \mathbb{I}_{\{0,1,\ldots\}}(x).$$

(a) Show that the m.g.f. of X is

$$\mathbb{M}_X(t) = \left(\frac{p}{1 - (1 - p)e^t} \right)^r, \quad t < -\log(1 - p). \tag{1.97}$$

(b) Similar to Problem 1.4, use the c.g.f. to calculate the mean, variance, skewness, and kurtosis of X. What happens to the skewness and kurtosis as $r \to \infty$?

(c) Using (1.97), calculate the c.g.f. of a geometric random variable Y with p.m.f. (I.4.18), i.e., when the (single) success is counted. Use it to calculate the mean and variance of Y. Extend this to the negative binomial case.

1.6. For $X \sim \text{Poi}(\lambda)$, calculate the mean and variance using just the m.g.f. (i.e., not the c.g.f.).

1.7. ★ Let $X_1, \ldots, X_{k+1}$ be independent random variables, each from a Poisson distribution with parameters $\lambda_1, \ldots, \lambda_{k+1}$.

(a) Derive the m.g.f. of $Y = \sum_{i=1}^{k+1} X_i$ and determine its distribution.

(b) Show that the conditional distribution of $X_1, \ldots, X_k$ given $X_1 + \cdots + X_{k+1} = n$ is multinomial with the parameters n, λ_1/λ, λ_2/λ, $\ldots$, λ_k/λ, where $\lambda = \lambda_1 + \cdots + \lambda_{k+1}$. Observe that this generalizes the result in (I.8.6) of Example I.8.4.

1.8. ★ For continuous positive random variable X with m.g.f. $\mathbb{M}_X(t)$,

$$\mathbb{E}\left[X^{-1}\right] = \int_0^\infty \mathbb{M}_X(-t) \, dt \tag{1.98}$$

and, more generally,

$$\mathbb{E}\left[X^{-c}\right] = \frac{1}{\Gamma(c)} \int_0^\infty t^{c-1} \mathbb{M}_X(-t) \, dt, \tag{1.99}$$

as shown by Cressie *et al.* (1981).

(a) Let $Z \sim \text{Gam}(\alpha, \beta)$. Calculate $\mathbb{E}\left[Z^{-c}\right]$ directly and using (1.99).

(b) Prove (1.99). Hint: First verify that $\int_0^\infty e^{-tx} t^{c-1} \, dt = \Gamma(c) x^{-c}$.

1.9. ★ ★ Show that the m.g.f. $\mathbb{M}_K(t) =: \mathbb{M}_N(t)$ for the r.v. K in Banach's matchbox problem (see Problem I.4.13) with N matches satisfies

$$\mathbb{M}_N(t) = \frac{e^{2t}}{2e^t - 1} \mathbb{M}_{N-1}(t) + \frac{e^t - 1}{2e^t - 1} \binom{N - \frac{1}{2}}{N}$$

Using this, compute $\mathbb{M}_N'(t)$ and $\mathbb{E}[K]$. (Contributed by Markus Haas)

1.10. Use (1.15) to compute $\mathbb{E}[N_m]$. (Contributed by Markus Haas)

1.11. ★ Let $X \sim \text{Unif}(0, 1)$, for which we certainly do not need inversion formulae to compute the p.d.f. and c.d.f., but their use is instructive.

(a) Compute the c.f. $\varphi_X(t)$.

(b) For $F_X = 0.5$, the integrand in (1.70) has to be zero. Verify this directly.

(c) Consider what happens to the integrand at the extremes of the range of integration in (1.70), i.e., with $z(t)$ and $g(t)$ as defined in (1.70), for $x \in (0, 1)$, compute $\lim_{t \to \infty} z(t)$, $\lim_{t \to \infty} g(t)$, $\lim_{t \to 0} z(t)$, and $\lim_{t \to 0} g(t)$.

(d) For this c.f., decide if the use of (1.70) or (1.71) is preferred. Using $x = 0.2$, plot the integrands.

1.12. ★ Recall inversion formula (1.54), i.e.,

$$f_X(x) = \frac{1}{2\pi} \int_{-\infty}^{\infty} e^{-itx} \varphi_X(t) \, dt,$$

for X continuous. (Contributed by Simon Broda)

(a) Derive the expression in (1.56), i.e.,

$$f_X(x) = \frac{1}{2\pi i} \int_{-i\infty}^{i\infty} \exp\{\mathbb{K}_X(s) - sx\} \, ds,$$

from (1.54).

(b) The integral in (1.56) can also be expressed as

$$f_X(x) = \frac{1}{2\pi i} \int_{c-i\infty}^{c+i\infty} \exp\{\mathbb{K}_X(s) - sx\} \, ds,$$

where c is an arbitrary real constant in the convergence strip of $\mathbb{K}_X$ This is called the Fourier–Mellin integral and is a standard result in the theory of Laplace transforms; see Schiff (1999, Ch. 4). Write out the double integral for the survivor function

$$\overline{F}_X(x) = 1 - F_X(x) = \int_x^{\infty} f_X(y) \, dy,$$

exchange the order of the integrals, and confirm that c needs to be positive for this to produce the valid integral expression

$$\overline{F}_X(x) = \frac{1}{2\pi i} \int_{c-i\infty}^{c+i\infty} \exp\{\mathbb{K}_X(s) - sx\} \, \frac{ds}{s}, \quad c > 0. \tag{1.100}$$

Similarly, derive the expression for the c.d.f. given by

$$F_X(x) = -\frac{1}{2\pi i} \int_{c-i\infty}^{c+i\infty} \exp\{\mathbb{K}_X(s) - sx\} \, \frac{ds}{s}, \quad c < 0. \tag{1.101}$$

(c) Write a program to compute the c.d.f. based on the previous result, and compute it using the c.f. for the uniform distribution.

1.13. ★ As in Gut (2005, p. 442), a random variable X is said to have an *infinitely divisible distribution* iff, for each $n \in \mathbb{N}$, there exist i.i.d. r.v.s $X_{n,k}$, $1 \le k \le n$, such that

$$X \stackrel{d}{=} \sum_{k=1}^{n} X_{n,k},$$

recalling (1.17). Equivalently, X is has an infinitely divisible distribution iff

$$\varphi_X(t) = \left[\varphi_{X_{n,1}}(t)\right]^n, \quad \forall n \in \mathbb{N}, \tag{1.102}$$

For example, if $\varphi_X(t) = e^{-|t|}$, then taking $\varphi_{X_{n,1}}(t)$ to be $e^{-|t|/n}$ shows that

$$\varphi_X(t) = \left[e^{-|t|/n}\right]^n = e^{-|t|}.$$

(See also Example 1.22). This demonstrates that a Cauchy random variable is infinitely divisible and, if $X_i \sim \mathrm{Cau}(c_i)$, then $X = \sum_{i=1}^{n} X_i \sim \mathrm{Cau}\left(\sum_{i=1}^{n} c_i\right)$.
   Let $X \sim \mathrm{Lap}(0, 1)$. First show that the characteristic function of X is given by

$$\varphi_X(t) = \left(1 + t^2\right)^{-1}.$$

Then show that a Laplace r.v. is infinitely divisible.

1.14. ★ Verify (1.68).

1.15. ★ Derive (1.54) from (1.65).

1.16. Use Euler's formula to show that $i^i = e^{-\pi/2}$.

1.17. ★★ This problem proves some useful properties of the Laplace transform and is based on Schiff (1999).

(a) Let f be (piecewise) continuous on $[0, \infty)$ and let $F(s) = \int_0^\infty e^{-st} f(t)\, dt$ converge uniformly for all $s \in D \subset \mathbb{C}$, i.e., for all $s \in D$,

$$\text{for any } \epsilon > 0, \ \exists\, T > 0 \text{ s.t., for } \tau \geq T, \quad \left| \int_\tau^\infty e^{-st} f(t)\, dt \right| < \epsilon. \quad (1.103)$$

Prove that

$$\lim_{s \to s_0} \int_0^\infty e^{-st} f(t)\, dt = \int_0^\infty \lim_{s \to s_0} e^{-st} f(t)\, dt, \quad (1.104)$$

i.e., that $F(s_0) = \lim_{s \to s_0} F(s)$. Hint: This amounts to showing that

$$\left| \int_0^\infty e^{-st} f(t)\, dt - \int_0^\infty e^{-s_0 t} f(t)\, dt \right|$$

is arbitrarily small for s sufficiently close to s_0, and consider that

$$\left| \int_0^\infty e^{-st} f(t)\, dt - \int_0^\infty e^{-s_0 t} f(t)\, dt \right| \leq \int_0^{t_0} \left| e^{-st} - e^{-s_0 t} \right| |f(t)|\, dt$$

$$+ \left| \int_{t_0}^\infty \left(e^{-st} - e^{-s_0 t} \right) |f(t)|\, dt \right|.$$

(b) Recall Fubini's theorem (I.A.159), which states that if $f : A \subset \mathbb{R}^2 \to \mathbb{R}$ is continuous and A is closed and bounded, then

$$\int_a^b \int_0^\tau f(x, t)\, dt\, dx = \int_0^\tau \int_a^b f(x, t)\, dx\, dt. \quad (1.105)$$

To extend this to

$$\int_a^b \int_0^\infty f(x, t)\, dt\, dx = \int_0^\infty \int_a^b f(t, x)\, dx\, dt, \quad (1.106)$$

assume $f : A \subset \mathbb{R}^2 \to \mathbb{R}$ is continuous on each rectangle $a \leq x \leq b$, $0 \leq t \leq T$, $T > 0$, and $\int_0^\infty f(x, t)\, dt$ converges uniformly for all $x \in [a, b]$, i.e.,

$$\text{for any } \epsilon > 0, \ \exists\, T > 0 \text{ s.t., for } \tau \geq T, \quad \left| \int_\tau^\infty f(x, t)\, dt \right| < \frac{\epsilon}{b - a}. \quad (1.107)$$

Then taking $\lim_{\tau \to \infty}$ of both sides of (1.105) yields

$$\lim_{\tau \to \infty} \int_a^b \int_0^\tau f(x, t)\, dt\, dx = \int_0^\infty \int_a^b f(t, x)\, dx\, dt,$$

and the l.h.s. converges if the 'tail area' $\lim_{\tau \to \infty} \int_a^b \int_\tau^\infty f(x, t)\, dt\, dx = 0$. But this is true because, for $\tau \geq T$, (1.107) implies $\int_a^b \int_\tau^\infty f(x, t)\, dt\, dx < \epsilon$.

Let $f(x, t)$ and $D_1 f = (\partial/\partial x) f$ be continuous on each rectangle $a \le x \le b$, $0 \le t \le T, T > 0$, assume $F(x) = \int_0^\infty f(x, t)\, dt$ exists and $\int_0^\infty D_1 f(x, t)\, dt$ converges uniformly. Use (1.104) and (1.106) to show that

$$\frac{d}{dx} F(x) = \int_0^\infty D_1 f(x, t)\, dt, \quad a < x < b. \tag{1.108}$$

(c) Let f be (piecewise) continuous on $[0, \infty)$ of exponential order α, and let $F(s) = \mathcal{L}\{f\}$. Show that $(-1)^n t^n f(t)$ is of exponential order, and then use (1.108) to show that, for real s,

$$\frac{d}{ds} F(s) = \mathcal{L}\{-tf(t)\}, \quad s > \alpha.$$

Induction then shows that

$$\frac{d^n}{ds^n} F(s) = \mathcal{L}\{(-1)^n t^n f(t)\}, \quad n \in \mathbb{N}, \quad s > \alpha. \tag{1.109}$$

(d) Let f be (piecewise) continuous on $[0, \infty)$ of exponential order α, and let $F(s) = \mathcal{L}\{f\}$, such that $\lim_{t \downarrow 0} f(t)/t$ exists. Use (1.106) to show that, for real s,

$$\int_s^\infty F(x)\, dx = \mathcal{L}\left\{\frac{f(t)}{t}\right\}, \quad s > \alpha. \tag{1.110}$$

Use this to simplify the integral $\int_0^\infty e^{-st} t^{-1} \sin(t)\, dt$.

(e) Let f and f' be continuous functions on $[0, \infty)$, and f is of exponential order α. Show that

$$\mathcal{L}\{f'(t)\} = s\mathcal{L}\{f(t)\} - f(0). \tag{1.111}$$

(f) The *convolution* of two functions, f and g, is given by

$$(f * g)(t) = \int_0^t f(x) g(t - x)\, dx, \tag{1.112}$$

if the integral exists, and is a central concept when studying the sum of random variables (Chapter 2). Let f and g be (piecewise) continuous on $[0, \infty)$ and of exponential order α. Show that

$$\mathcal{L}\{f * g\} = \mathcal{L}\{f\}\mathcal{L}\{g\}, \quad \text{Re}(s) > \alpha. \tag{1.113}$$

Hint: It will be necessary to reverse the order of two improper integrals, the justification for which can be shown as follows. As in Schiff (1999, p. 93), let $h(x, y)$ be a function such that

$$\int_0^\infty \int_0^\infty |h(x, y)|\, dx\, dy < \infty,$$

i.e., h is absolutely convergent. Define

$$a_{mn} = \int_n^{n+1} \int_m^{m+1} |h(x, y)| \, dx \, dy \quad \text{and} \quad b_{mn} = \int_n^{n+1} \int_m^{m+1} h(x, y) \, dx \, dy,$$

and use (I.A.97) to show that

$$\int_0^\infty \int_0^\infty h(x, y) \, dx \, dy = \int_0^\infty \int_0^\infty h(x, y) \, dy \, dx.$$

1.18. ★ Similar to (1.112), for $\mathbf{g} = (g_0, \ldots, g_{T-1})$ and $\mathbf{h} = (h_0, \ldots, h_{T-1})$ vectors of complex numbers, their (circular) convolution is $\mathbf{c} = (c_0, \ldots, c_{T-1}) = \mathbf{g} * \mathbf{h}$, where

$$c_t = \sum_{j=0}^{T-1} g_{t-j} h_j, \quad \text{with } g_{t-j} = g_{T+t-j}. \tag{1.114}$$

Show algebraically that $\mathcal{F}(\mathbf{g} * \mathbf{h}) = T \mathcal{F}(\mathbf{g}) \, \mathcal{F}(\mathbf{h})$, where the product of the two DFTs means the elementwise product of their components. Also write a Matlab program to numerically verify this.

1.19. ★ ★ The *Fourier transform* is the result of considering what happens to the Fourier series (1.87) when letting the fundamental period T of function g tend towards infinity, i.e., letting the fundamental frequency $\Delta f := 1/T$ go to zero. The C_n coefficients measure the contribution (amplitude) of particular discrete frequencies, and we are interested in the limiting case for which the C_n becomes a continuum, so that the expression for $g(t)$ can be replaced by an integral. To this end, let $f_n = n\omega/(2\pi) = n/T$, so that, with $r = T/2$, taking the limit of the expressions in (1.87) in such a way that, as n and T approach ∞, $f_n = n/T$ stays finite gives

$$g(t) = \lim_{T \to \infty} \sum_{n=-\infty}^{\infty} C_n e^{2\pi i f_n t} = \lim_{T \to \infty} \sum_{n=-\infty}^{\infty} \left[\int_{-T/2}^{T/2} g(t) \, e^{-2\pi i f_n t} \, dt \right] e^{2\pi i f_n t} \, \Delta f$$

$$= \int_{-\infty}^{\infty} \left[\int_{-\infty}^{\infty} g(t) \, e^{-2\pi i f t} \, dt \right] e^{2\pi i f t} \, df,$$

where f now denotes the continuous limiting function of f_n. Let $\omega = 2\pi f$, with $d\omega = 2\pi \, df$, and set $G(\omega)$ to be the inner integral, so that we obtain the pair

$$\boxed{G(\omega) = \int_{-\infty}^{\infty} g(t) \, e^{-i\omega t} \, dt \qquad g(t) = \frac{1}{2\pi} \int_{-\infty}^{\infty} G(\omega) \, e^{i\omega t} \, d\omega,} \tag{1.115}$$

where the former is the *Fourier transform*, and the latter is the *inverse Fourier transform*.

For $T > 0$ and $a \in \mathbb{R}$, let $g(t) = a\mathbb{I}(|t| \leq T)$. Compute the Fourier transform $G(\omega)$ of $g(t)$ and verify it via the inverse Fourier transform. See the footnote for a hint (and remove one star from the difficulty level).[11]

1.20. ★ Try the following question on the mathematician of your choice. If f is a probability density function, is it true that $f(x) \to 0$ when $x \to \pm\infty$? The answer is no. We can construct a counter-example as follows (leaving aside some technical details concerning convergence).

(Contributed by Walther Paravicini)

(a) Assume that f_n is a p.d.f. for each natural number n. Provided that we can give a meaning to the weighted sum $f(x) := \sum_{n=1}^{\infty} 2^{-n} f_n(x)$, $x \in \mathbb{R}$, convince yourself that f is again a p.d.f.

(b) For all natural numbers n, let $\sigma_n > 0$ and $\mu_n \in \mathbb{R}$. Let f_n be the p.d.f. of an r.v. distributed as $N(\mu_n, \sigma_n^2)$. For a given n, calculate the maximum of $f_n(x)$, $x \in \mathbb{R}$, in terms of σ_n and μ_n.

(c) Find a choice of σ_n and μ_n such that the maximum of $2^{-n} f_n(x)$, $x \in \mathbb{R}$, does not converge to 0 if $n \to \infty$ and such that the argument where the maximum is attained converges to ∞.

(d) Write a Matlab program which calculates the density of the so-constructed p.d.f. $f(x) = \sum_{n=1}^{\infty} 2^{-n} f_n(x)$, $x \in \mathbb{R}$. Produce a plot of the density which shows the behaviour of $f(x)$ for large x. For a suitable choice of σ_n and μ_n, the plot should show a series of ever-increasing spikes which get steeper and steeper.

1.21. ★ This problem illustrates the numeric implementation of (1.94) and (1.95) in a nontrivial case, but one for which we can straightforwardly calculate the correct answer to an arbitrary precision for comparison purposes, viz., the bivariate normal distribution, which was first seen in Example I.8.5 and will be more formally introduced and detailed in Chapter 3. At this point, interest centres purely on the mechanics of inverting a bivariate c.f.   (Contributed by Simon Broda)

Let $\mathbf{X} = (X_1, X_2)'$ follow a bivariate normal distribution, i.e.,

$$\begin{bmatrix} X_1 \\ X_2 \end{bmatrix} \sim N\left(\begin{bmatrix} \mu_1 \\ \mu_2 \end{bmatrix}, \begin{bmatrix} \sigma_1^2 & \rho\sigma_1\sigma_2 \\ \rho\sigma_1\sigma_2 & \sigma_2^2 \end{bmatrix} \right), \qquad (1.116)$$

with density

$$f_{X_1,X_2}(x, y) = K \exp\left\{ -\frac{X^2 - 2\rho XY + Y^2}{2(1 - \rho^2)} \right\},$$

where

$$K^{-1} = 2\pi\sigma_1\sigma_2 (1 - \rho^2)^{1/2}, \qquad X = \frac{x - \mu_1}{\sigma_1}, \qquad Y = \frac{y - \mu_2}{\sigma_2},$$

for $\mu_1, \mu_2 \in \mathbb{R}$, $\sigma_1, \sigma_2 \in \mathbb{R}_{>0}$, and $|\rho| < 1$.

---

[11] Hint: For the Fourier transform $G(\omega)$ of $g(t)$, use (1.32), and for the inverse Fourier transform, first show that $\sin(a - b) + \sin(a + b) = 2\sin(a)\sin(b)$, and use this with the results in Example I.A.21 and a result mentioned in Example I.A.30.

Preferable methods for evaluation of $F_{X_1, X_2}(x_1, x_2)$ are discussed in Section 3.4. The joint c.f. of (X_1, X_2) is given by $\varphi(t_1, t_2) = \mathbb{M}_X(it)$ from (3.16), i.e.,

$$\varphi(t_1, t_2) = \exp\left[i(t_1\mu_1 + t_2\mu_2) - \frac{1}{2}(\sigma_1^2 t_1^2 + 2\rho\sigma_1\sigma_2 t_1 t_2 + \sigma_2^2 t_2^2) \right]. \quad (1.117)$$

Write a Matlab program to accomplish this.

1.22. ★ ★ This problem informally derives inversion formulae for the ratio of two random variables. (Contributed by Simon Broda)

(a) Let X_1 and X_2 be continuous r.v.s with c.f. φ_{X_1, X_2} and such that $\mathbb{E}[X_1]$ and $\mathbb{E}[X_2]$ exist. Show that an inversion formula for the density of $R := X_1/X_2$ is given by

$$f_R(r) = \frac{1}{\pi^2} \int_0^\infty \int_{-\infty}^\infty \operatorname{Re}\left[\frac{\varphi_2(t_1, -t_2 - rt_1)}{t_2} \right] dt_1 \, dt_2, \quad (1.118)$$

where

$$\varphi_2(t_1, t_2) = \frac{\partial}{\partial t_2} \varphi_{X_1, X_2}(t_1, t_2).$$

Hint: First show that, with $X_3 := X_1 - rX_2$,

$$\Pr\left(\frac{X_1}{X_2} < r \right) = \Pr(X_3 < 0) + \Pr(X_2 < 0) - 2\Pr(X_3 < 0 \wedge X_2 < 0), \quad (1.119)$$

and use this in conjunction with (1.96) and the fact that

$$\varphi_{X_3, X_2}(t_1, t_2) = \varphi_{X_1, X_2}(t_1, t_2 - rt_1)$$

to derive an inversion formula for the c.d.f. of R in terms of the joint c.f. of (X_1, X_2). Differentiate with respect to r and use the fact that $\varphi_{X_1, X_2}(t_1, t_2) = \overline{\varphi}_{X_1, X_2}(-t_1, -t_2)$.

(b) For the case where $F_{X_2}(0) = 0$, Geary's result (1.62) can be used, i.e.,

$$f_R(r) = \frac{1}{2\pi i} \int_{-\infty}^\infty \left[\frac{\partial \varphi_{X_1, X_2}(s, t)}{\partial t} \right]_{t=-rs} ds.$$

In the event that $F_{X_2}(0)$ is positive but 'sufficiently small', (1.62) can be used to approximate (1.118). Consider $R_1 = X_1/X_2$, where (X_1, X_2) has a bivariate normal distribution as in (1.116), with joint c.f. given in (1.117). Write a Matlab program that evaluates the density of R by means of (1.62) and (1.118). Use your program to evaluate the density of R and $R_2 = X_2/X_1$, where $\mu_1 = 2$, $\mu_2 = 0.1$, $\sigma_1 = \sigma_2 = 1$, $\rho = 0$, and demonstrate that the relative error incurred by using (1.62) over (1.118) is larger for R. See also Example 2.18 on how to compute the p.d.f. of R_1 and R_2.

1.23. Prove the second part of (1.38) by using the closed-form solution of a finite geometric sum for complex numbers. It is interesting to see this optically: for $n = 6$, write a program which plots the $(z_n^k)^j$ on a circle for $j = 0, \ldots, n - 1$, but each with a slight perturbation so that duplicates can be recognized.

2

Sums and other functions of several random variables

Everything should be made as simple as possible, but not simpler.
(Albert Einstein)

Results for sums of r.v.s in the discrete case and some basic examples were given in Chapter I.6. The importance and ubiquitousness of sums also holds in the continuous case, which is the main subject of this chapter. Some relationships involving sums and ratios of fundamental random variables are summarized in Table A.11.

2.1 Weighted sums of independent random variables

Let X_i, $i = 1, \ldots, n$, be independent r.v.s, each of which possesses an m.g.f., and define $Y = \sum_{i=1}^{n} a_i X_i$. Then

$$\mathbb{M}_Y(s) = \mathbb{E}\left[e^{sY}\right] = \mathbb{E}\left[e^{sa_1 X_1}\right] \cdots \mathbb{E}\left[e^{sa_n X_n}\right] = \prod_{i=1}^{n} \mathbb{M}_{X_i}(sa_i), \qquad (2.1)$$

from (I.5.20). As an important special case, for independent r.v.s X and Y,

$$\boxed{\mathbb{M}_{X+Y}(s) = \mathbb{M}_X(s)\, \mathbb{M}_Y(s)} \qquad (2.2)$$

and, using the c.f.,

$$\boxed{\varphi_{X+Y}(t) = \varphi_X(t)\, \varphi_Y(t).} \qquad (2.3)$$

Intermediate Probability: A Computational Approach   M. Paolella
© 2007 John Wiley & Sons, Ltd

⊛ **Example 2.1**   If X and Y are independent Poisson distributed with parameters λ_1 and λ_2, then, from (2.2),

$$\mathbb{M}_{X+Y}(t) = \mathbb{M}_X(t)\,\mathbb{M}_Y(t) = \exp\left\{\lambda_1\left(e^t - 1\right)\right\}\exp\left\{\lambda_2\left(e^t - 1\right)\right\}$$
$$= \exp\left\{(\lambda_1 + \lambda_2)\left(e^t - 1\right)\right\}$$

is the m.g.f. of a Poisson random variable with parameter $\lambda_1 + \lambda_2$, which determines the distribution of $X + Y$. This derivation can be compared with the method of convolution as used in Example I.6.10; see also Problem 1.7(**a**).   ■

⊛ **Example 2.2**   Let $X_i \overset{\text{ind}}{\sim} \text{Bin}(n_i, p)$ with $\mathbb{M}_{X_i}(t) = \left(pe^t + q\right)^{n_i}$ from Example 1.9. From (2.2), the m.g.f. of $S = \sum_{i=1}^k X_i$ is $\mathbb{M}_S(t) = \left(pe^t + q\right)^{n.}$, where $n. = \sum_{i=1}^k n_i$. From the uniqueness of m.g.f.s, it follows that $S \sim \text{Bin}(n., p)$. Compare this to the derivation of S for $k = 2$ in Problem I.6.1.   ■

The m.g.f. of $X \sim \text{NBin}(r, p)$ was shown in Problem 1.5 to be

$$\mathbb{M}_X(t) = p^r\left(1 - (1-p)e^t\right)^{-r}, \quad t < -\log(1-p).$$

Thus, similar to the previous example, it is clear that a sum of independent negative binomial r.v.s (each with the same value of p) is also negative binomially distributed. Random variables which satisfy this property, i.e., their sum belongs to the same distribution family as that of the components, are said to be *closed under addition*. In the binomial and negative binomial cases, if the success probability p differs among the components, then they are no longer closed under addition. In the continuous case, this property holds for normal random variables (with no constraints on their means and variances, except that they are finite), as is shown in Section 3.3, and also for gamma r.v.s when they have the same scale parameter, as is shown next.

⊛ **Example 2.3**   If $X_i \overset{\text{ind}}{\sim} \text{Gam}(\alpha_i, \beta)$, then $\sum_i X_i \sim \text{Gam}\left(\sum_i \alpha_i, \beta\right)$; in words, a sum of independent gamma random variables each with the same *scale* parameter also follows a gamma distribution. This follows from (2.2) and Example 1.5; it was also proven directly in Example I.9.11. Two special cases are of great interest:

(i)
$$\boxed{\text{If } X_i \overset{\text{i.i.d.}}{\sim} \text{Exp}(\lambda)\,,\text{ then } \sum_i^n X_i \sim \text{Gam}(n, \lambda)}$$

(see Example I.9.3 for a direct proof), and

(ii)
$$\boxed{\text{If } Y_i \overset{\text{ind}}{\sim} \chi^2(v_i)\,,\text{ then } \sum_i^n Y_i \sim \chi^2(v), \ v = \sum_i^n v_i.}$$

Special case (ii) and Example I.7.2 together show that, if $Z_i \overset{\text{i.i.d.}}{\sim} \text{N}(0, 1)$, $i = 1, \ldots, n$, then $\sum_{i=1}^{n} Z_i^2 \sim \chi^2(n)$. More generally,

$$
\text{if } X_i \overset{\text{ind}}{\sim} \text{N}\left(\mu_i, \sigma_i^2\right), \quad \text{then} \quad \sum_{i=1}^{n} \left(\frac{X_i - \mu_i}{\sigma_i}\right)^2 \sim \chi^2(n), \tag{2.4}
$$

a simple result of enormous importance in statistics. ∎

⊖ ***Example 2.4***   (Difference between two Gumbel r.v.s) Let $X_i \overset{\text{i.i.d.}}{\sim} \text{Gum}(0, 1)$, $i = 1, 2$. The m.g.f. of $X = X_1$ is, from (I.7.53) with $y = \exp(-x)$ (and $x = -\ln y$, $dx = -y^{-1}\, dy$),

$$
\begin{aligned}
\mathbb{M}_X(t) &= \mathbb{E}\left[e^{tX}\right] \\
&= \int_{-\infty}^{\infty} \exp\left(-x(1-t) - e^{-x}\right) dx = \int_{-\infty}^{\infty} \left[\exp(-x)\right]^{1-t} \exp\left(-e^{-x}\right) dx \\
&= -\int_{\infty}^{0} y^{-t} \exp(-y)\, dy = \int_{0}^{\infty} y^{-t+1-1} \exp(-y)\, dy \\
&= \Gamma(1-t), \quad t < 1,
\end{aligned} \tag{2.5}
$$

and $\mathbb{K}_X(t) = \ln \mathbb{M}_X(t)$. With $\psi(z) = d \ln \Gamma(z)/dz$ defined to be the *digamma* function (see Problem 2.19 below), we know from (1.9) that

$$
\mathbb{E}[X] = \kappa_1 = \mathbb{K}_X^{(1)}(t)\Big|_{t=0} = -\psi(1-t)|_{t=0} = \gamma,
$$

where $\gamma \approx 0.5772156649$ is Euler's constant (see Examples I.A.38 and I.A.53). Similarly,

$$
\mathbb{V}(X) = \kappa_2 = \frac{d}{dt}\left(-\psi(1-t)\right)\Big|_{t=0} = \psi'(1)
$$

which, from (2.42), is $\pi^2/6$.

The m.g.f. of $D = X_1 - X_2$ is then, from the independence of X_1 and X_2,

$$
\begin{aligned}
\mathbb{E}\left[e^{tD}\right] &= \mathbb{E}\left[e^{t(X_1 - X_2)}\right] \\
&= \mathbb{E}\left[e^{tX_1}\right] \mathbb{E}\left[e^{-tX_2}\right] = \mathbb{M}_{X_1}(t)\, \mathbb{M}_{X_2}(-t) \\
&= \Gamma(1-t)\, \Gamma(1+t),
\end{aligned}
$$

so that, from Example 1.6, $D \sim \text{Log}(0, 1)$, i.e., the difference between two i.i.d. Gumbel r.v.s follows a logistic distribution. ∎

⊙ ***Example 2.5***   Let $X_i \overset{\text{ind}}{\sim} \text{N}\left(\mu_i, \sigma_i^2\right)$ and define $S_n = \sum_{i=1}^{n} X_i$ and $D = X_1 - X_2$. The m.g.f. of S_n is, from (2.2) and the independence of the X_i, given by

$$M_{S_n}(t) = \prod_{i=1}^{n} M_{X_i}(t) = \prod_{i=1}^{n} \exp\left\{\mu_i t + \frac{1}{2}\sigma_i^2 t^2\right\} = \exp\left\{t\sum_{i=1}^{n}\mu_i + \frac{1}{2}t^2\sum_{i=1}^{n}\sigma_i^2\right\}$$

which, from the uniqueness of m.g.f.s, implies that $S_n \sim N\left(\sum_{i=1}^{n}\mu_i, \sum_{i=1}^{n}\sigma_i^2\right)$. Similarly,

$$M_D(t) = M_{X_1}(t)\,M_{X_2}(-t) = \exp\left\{\mu_1 t + \frac{1}{2}\sigma_1^2 t^2\right\}\exp\left\{-\mu_2 t + \frac{1}{2}\sigma_i^2 t^2\right\}$$

$$= \exp\left\{(\mu_1 - \mu_2)\,t + \frac{1}{2}t^2\left(\sigma_1^2 + \sigma_2^2\right)\right\},$$

showing that $D \sim N\left(\mu_1 - \mu_2, \sigma_1^2 + \sigma_2^2\right)$. The joint m.g.f. of S_2 and D is

$$M_{S_2, D}(s, d) = \mathbb{E}\left[e^{s(X_1 + X_2) + d(X_1 - X_2)}\right] = \mathbb{E}\left[e^{(s+d)X_1 + (s-d)X_2}\right]$$

$$= M_{X_1}(s + d)\,M_{X_2}(s - d)$$

$$= \exp\left\{\mu_1\,(s + d) + \frac{1}{2}\sigma_1^2\,(s + d)^2\right\}\exp\left\{\mu_2\,(s - d) + \frac{1}{2}\sigma_2^2\,(s - d)^2\right\}$$

$$= \exp\left\{(\mu_1 + \mu_2)\,s + \frac{1}{2}s^2\left(\sigma_1^2 + \sigma_2^2\right)\right\}$$

$$\times \exp\left\{(\mu_1 - \mu_2)\,d + \frac{1}{2}d^2\left(\sigma_2^2 + \sigma_1^2\right)\right\}\exp\left\{sd\left(\sigma_1^2 - \sigma_2^2\right)\right\}$$

$$= M_{S_2}(s)\,M_D(d)\exp\left\{sd\left(\sigma_1^2 - \sigma_2^2\right)\right\},$$

which factors only for $\sigma_1^2 = \sigma_2^2$. We conclude that, if $\sigma_1^2 = \sigma_2^2$, then S_2 and D are independent. See also Example 3.6. ∎

In the previous examples, the m.g.f. of a weighted sum of r.v.s was recognized as being that of a 'common' r.v., so that the p.d.f. and c.d.f. could be easily elicited. Matters in many settings of interest will, unfortunately, not always be so simple. In such cases, to evaluate the p.d.f. and c.d.f., the inversion formulae in Sections 1.2.5 and 1.2.6 need to be applied to the c.f.

⊖ ***Example 2.6***   The m.g.f. of $X \sim \text{Exp}(\lambda)$, with $f_X(x) = \lambda e^{-\lambda x}\mathbb{I}_{(0,\infty)}(x)$, is, from Example 1.5, $M_X(t) = \lambda/(\lambda - t)$, $t < \lambda$, so $\varphi_X(t) = M_X(it) = \lambda/(\lambda - it)$. Now let $X_j \overset{\text{ind}}{\sim} \text{Exp}(\lambda_j)$, $j = 1, \ldots, n$, and $S = \sum_{j=1}^{n} X_j$. Then

$$\varphi_S(t) = \prod_{j=1}^{n}\varphi_{X_j}(t) = \prod_{j=1}^{n}\frac{\lambda_j}{\lambda_j - it},$$

and the inversion formula (1.60) can be used to compute the density of S. Using weights $\lambda_j = j$, $i = 1, \ldots, 4$, the density of S is shown in Figure 2.1. It should be mentioned that, for this random variable with small n, the integrand in (1.60) can be pathologically oscillatory and jeopardize the accuracy of the integration. Another way of computing this density is given in Example 2.11 below. ∎

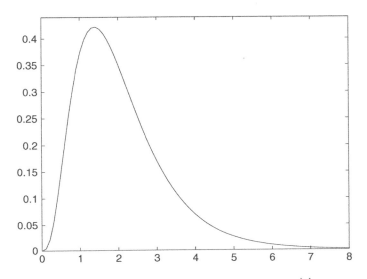

Figure 2.1 Density of $X_1 + X_2 + X_3 + X_4$, where $X_i \overset{\text{ind}}{\sim} \text{Exp}(i)$

⊚ ***Example 2.7*** (Example 2.2 cont.) Let $X_j \overset{\text{ind}}{\sim} \text{Bin}(n_j, p_j)$ and $S = \sum_{j=1}^{k} X_j$. For $k = 2$, the p.m.f. is given in (I.6.30) using the discrete convolution formula and is straightforwardly generalized for $k > 2$. However, the 'curse of dimensionality' quickly rears its ugly head as k grows, making exact calculations prohibitive in terms of both programming and run time.

Instead, from $\mathbb{M}_S(t) = \prod_{j=1}^{k}(p_j e^t + q_j)^{n_j}$ and $\varphi_S(t) = \mathbb{M}_S(it)$, (1.58) can be calculated to obtain the p.m.f. (Notice that k could be replaced by $\sum_{j=1}^{k} n_j$ and the n_j set to unity, without loss of generality.) This is accomplished by the program in Listing 2.1.

Calculating the p.m.f. for S for various parameter constellations reveals that the integrand in (1.58) is very well behaved, so that the inversion is very fast and accurate. Figure 2.2 shows the mass function for $k = 3$ with $n_j = 5j$ and $p_j = 0.8/2^{j-1}$, $j = 1, 2, 3$. In fact, as (1.58) (apparently) works for noninteger values of x, it was evaluated over a fine grid and shown as the dotted line in the figure.

Remark: It can be shown that the p.m.f. of S is unimodal: see Wang (1993), who provides a unified approach to studying various properties of S based on combinatoric arguments. ∎

⊚ ***Example 2.8*** Recall the discussion of occupancy distributions in Section I.4.2.6, in which Y_k denotes the number of cereal boxes one needs to purchase in order to get at least one of k different prizes, $2 \leq k \leq r$, where r is the total number of different prizes available. The exact mass function is easily calculated by using expression (I.4.28) for the survivor function.

```
function f = sumbincf(xvec,n,p)
tol=1e-7; xl=length(xvec); f=zeros(xl,1);
for loop=1:xl
  x=xvec(loop); f(loop)= quadl(@ff,0,pi,tol,0,n,p,x) / pi;
end

function I=ff(tvec,n,p,x);
q=1-p; I=zeros(size(tvec));
for loop=1:length(tvec)
  t=tvec(loop);
  cf=exp( sum( n.*log(p.*exp(i*t) + q) ) ); % ** HERE IS THE C.F.\ **
  I(loop)=real( exp(-i*t*x) * cf );
end
```

Program Listing 2.1 Computes the p.m.f. via (1.58) at each ordinate in xvec of a sum of k independent Bin(n_j, p_j) r.v.s, where $n = (n_1, \ldots, n_k)$ and $p = (p_1, \ldots, p_k)$. Observe how the c.f. is calculated by using logs

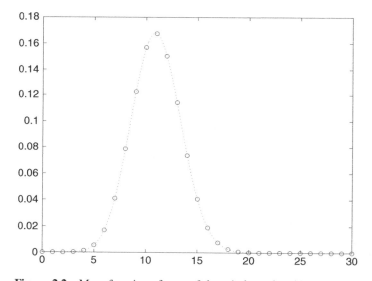

Figure 2.2 Mass function of sum of three independent binomial r.v.s

As in Example 6.8, Y_k can be expressed as $Y_k = \sum_{j=0}^{k-1} G_j$, where $G_j \overset{\text{ind}}{\sim} \text{Geo}(p_j)$ with p.m.f. (I..4.18), i.e., with $q_j = 1 - p_j$,

$$f_{G_j}(g; p_j) = p_j q_j^{g-1} \mathbb{I}_{\{1,2,\ldots\}}(g), \qquad p_j = \frac{r-j}{r}.$$

(Note that G_0 is degenerate.) From the results in Problem 1.5(**c**) and the independence of the G_j,

$$\mathbb{M}_{Y_k}(t) = \prod_{j=0}^{k-1} \mathbb{M}_{G_j}(t) = e^{kt} \prod_{j=0}^{k-1} \frac{p_j}{1 - q_j e^t}, \tag{2.6}$$

and $\varphi_{Y_k}(t) = \mathbb{M}_{Y_k}(it)$. The program in Listing 2.2 computes the p.m.f. of a sum of independent geometric r.v.s with arbitrary p_j (and is very similar to that in Listing 2.1). The p.m.f. of Y_k is then computed by calling this with the specific values of p_j associated with the occupancy problem, as done with the program in Listing 2.3. ∎

```
function f=sumgeocf(xvec,p)
tol=1e-7; xl=length(xvec); f=zeros(xl,1);
for loop=1:xl
  x=xvec(loop); f(loop)= quadl(@ff,0,pi,tol,0,x,p) / pi;
end

function I=ff(tvec,x,p);
q=1-p; I=zeros(size(tvec)); k=length(p);
for loop=1:length(tvec)
  t=tvec(loop);
  cf=exp( k*i*t + sum( log( p./(1-q.*exp(i*t)) ) ) ); % THE C.F.\
  I(loop)=real( exp(-i*t*x) * cf );
end
```

Program Listing 2.2 Computes the p.m.f. via (1.58) at each ordinate in xvec of a sum of k independent Geo(p_j) r.v.s with p.m.f. (I.4.18), where $p = (p_1, \ldots, p_k)$. Adjust the integration tolerance tol as needed

```
function f = occpmf(xvec,r,k)
i=0:(k-1); p=(r-i)/r; f = sumgeocf(xvec,p);
```

Program Listing 2.3 Computes the p.m.f. of the occupancy r.v. Y_k by calling the general program sumgeocf in Listing 2.2

⊖ **Example 2.9** Similar to Example 1.2 and Problem 1.11, for $X \sim \text{Unif}(a, b)$, $a < b$, the c.f. is

$$\varphi_X(t) = \mathbb{E}\left[e^{itX}\right] = \frac{1}{b-a} \int_a^b e^{itx} \, dx = \frac{e^{itb} - e^{ita}}{it(b-a)}.$$

Let $X_j \overset{\text{ind}}{\sim} \text{Unif}(a_j, b_j)$ and define $S = \sum_{j=1}^n X_j$ with support $\left(\sum_{j=1}^n a_j, \sum_{j=1}^n b_j\right)$. The c.f. is $\varphi_S(t) = \prod_{j=1}^n \varphi_{X_j}(t)$ and the inversion formulae can be straightforwardly applied. Problem 2.3 shows that S is symmetric about its mean

$$e[S] = \sum_{j=1}^n \mathbb{E}\left[X_j\right] = \frac{1}{2}\left(\sum_{j=1}^n a_j + \sum_{j=1}^n b_j\right),$$

which implies (from the discussion in Section 1.2.3) that the c.f. of $S - \mathbb{E}[S]$ is real. Indeed, for $n = 1$,

$$\mathbb{E}\left[e^{it(X-(a+b)/2)}\right] = e^{-\frac{1}{2}(a+b)it}\mathbb{E}\left[e^{itX}\right]$$

$$= e^{-\frac{1}{2}(a+b)it}\frac{e^{itb} - e^{ita}}{it\,(b-a)} = \frac{2}{t\,(a-b)}\sin\frac{1}{2}t\,(a-b)\,,$$

with the general case following easily. ∎

⊖ **Example 2.10** Consider the random variable $Z = X_1 - X_2$, where X_1 and X_2 are i.i.d. gamma r.v.s with density $f_X(x; a) = \Gamma(a)^{-1}e^{-x}x^{a-1}$. As $\mathbb{M}_X(t) = (1 - t)^{-a}$ and

$$\mathbb{M}_{(-X)}(t) = \mathbb{E}\left[e^{t(-X)}\right] = \mathbb{M}_X(-t)\,,$$

it follows that the m.g.f. of Z is given by

$$\mathbb{M}_Z(t) = \mathbb{E}\left[e^{tZ}\right] = \mathbb{E}\left[e^{t(X_1-X_2)}\right]$$

$$= \mathbb{M}_X(t)\,\mathbb{M}_X(-t) = (1 - t)^{-a}(1 + t)^{-a} = \left(1 - t^2\right)^{-a}\,.$$

With $\varphi_Z(t) = \mathbb{M}_Z(it) = \left(1 + t^2\right)^{-a}$, Example 1.23 used the inversion formulae to get the p.d.f. of Z. As φ_Z is real, we know that f_Z is symmetric about zero (see Section 1.2.3).

Now let $X_1, X_2 \overset{\text{i.i.d.}}{\sim} \text{Gam}(a, b)$ and $S = X_1 - X_2$. Then $\mathbb{M}_X(t) = b^a(b - t)^{-a}$, and a similar calculation gives $\mathbb{M}_S(t) = \left(1 - t^2/b^2\right)^{-a}$. This could also have been arrived at directly by observing that $S = Z/b$, and, thus,

$$\mathbb{M}_S(t) = \mathbb{M}_{Z/b}(t) = \mathbb{E}\left[e^{tZ/b}\right] = \mathbb{E}\left[e^{Z(t/b)}\right] = \mathbb{M}_Z(t/b) = \left(1 - t^2/b^2\right)^{-a}\,.$$

See also Example 7.19. ∎

2.2 Exact integral expressions for functions of two continuous random variables

While of great practical and theoretical importance, the use of the m.g.f. in the previous section was limited to determining just sums and differences of independent r.v.s. This section derives integral formulae for the density of sums, differences, products, and quotients of two, possibly dependent, r.v.s. These are all special cases of the multivariate transformation method detailed in Chapter I.9, but arise so frequently in applications that it is worth emphasizing them.

Moreover, while obtaining the m.g.f. or c.f. of a sum of independent r.v.s is easy, if its form is not recognizable, then calculation of the p.m.f. or p.d.f. of the sum will involve numeric integration (via the inversion formulae). The integral expressions developed in this chapter are far more straightforward to motivate and derive than those for the inversion formulae, which require use of complex numbers.

In the case of sums, the extension of the formulae presented below to more than two r.v.s is straightforward, but they are computationally of little use (see the discussion in Section I.6.5, so that approximations would be required. To this end, approximate expressions, based on Taylor series expansions, for the mean and variance of a general function of two or more r.v.s is developed in Section 2.3. These could be used, for example, in conjunction with a normal approximation.

If X and Y are jointly distributed continuous random variables with density $f_{X,Y}(x, y)$, then the densities of $S = X + Y$, $D = X - Y$, $P = XY$ and $R = X/Y$ can be respectively expressed as

$$f_S(s) = \int_{-\infty}^{\infty} f_{X,Y}(x, s - x)\, dx = \int_{-\infty}^{\infty} f_{X,Y}(s - y, y)\, dy, \qquad (2.7)$$

$$f_D(d) = \int_{-\infty}^{\infty} f_{X,Y}(x, x - d)\, dx = \int_{-\infty}^{\infty} f_{X,Y}(d + y, y)\, dy, \qquad (2.8)$$

$$f_P(p) = \int_{-\infty}^{\infty} \frac{1}{|x|} f_{X,Y}\left(x, \frac{p}{x}\right) dx = \int_{-\infty}^{\infty} \frac{1}{|y|} f_{X,Y}\left(\frac{p}{y}, y\right) dy, \qquad (2.9)$$

$$f_R(r) = \int_{-\infty}^{\infty} \frac{|x|}{r^2} f_{X,Y}\left(x, \frac{x}{r}\right) dx = \int_{-\infty}^{\infty} |y|\, f_{X,Y}(ry, y)\, dy. \qquad (2.10)$$

The first of these is just the convolution of X and Y; it was given in (I.8.42) and (I.9.3), having been derived in two different ways (conditioning and bivariate transformation, respectively). The product formula (2.9) was also derived using a bivariate transformation in Example I.9.2. The expressions for f_D and f_R could also be straightforwardly derived using these methods, but we show yet another way below. For each of the four cases S, D, P, and R given above, the arguments in $f_{X,Y}(\cdot, \cdot)$ are easy to remember by recalling the derivation of the discrete (and independent) counterparts in Chapter I.4. For example, in (2.7), $s = x + (s - x) = (s - y) + y$; and in (2.8), $d = x - (x - d) = (d + y) - y$.

To prove (2.8), using $Z = D$ and with $u = x - y$,

$$F_Z(z) = \Pr(Z \leq z) = \int\!\!\int_{x - y \leq z} f_{X,Y}(x, y)\, dy\, dx = \int_{-\infty}^{\infty} \int_{x - z}^{\infty} f_{X,Y}(x, y)\, dy\, dx$$

$$= \int_{-\infty}^{\infty} \int_{z}^{-\infty} f_{X,Y}(x, x - u)\,(-du)\, dx = \int_{-\infty}^{z} \int_{-\infty}^{\infty} f_{X,Y}(x, x - u)\, dx\, du$$

so that, from (I.4.1),

$$f_Z(z) = \frac{d}{dz} F_Z(z) = \int_{-\infty}^{\infty} f_{X,Y}(x, x - z)\, dx.$$

The second formula follows similarly: with $u = x - y$,

$$\iint\limits_{x-y\leq z} f_{X,Y}(x,y)\,dx\,dy = \int_{-\infty}^{\infty}\int_{-\infty}^{y+z} f_{X,Y}(x,y)\,dx\,dy$$

$$= \int_{-\infty}^{\infty}\int_{-\infty}^{z} f_{X,Y}(u+y,y)\,du\,dy$$

$$= \int_{-\infty}^{z}\int_{-\infty}^{\infty} f_{X,Y}(u+y,y)\,dy\,du$$

and, from (I.4.1),

$$f_Z(z) = \frac{d}{dz}F_Z(z) = \int_{-\infty}^{\infty} f_{X,Y}(z+y,y)\,dy.$$

If the difference $D' = Y - X$ is desired, simply replacing d with $-d$ in (2.8) gives

$$f_{D'}(d) = \int_{-\infty}^{\infty} f_{X,Y}(x, x+d)\,dx = \int_{-\infty}^{\infty} f_{X,Y}(y-d, y)\,dy.$$

The derivation of (2.7), (2.9), and (2.10) are similar. In addition, for $R' = Y/X$,

$$f_{R'}(r) = \int_{-\infty}^{\infty} |x|\, f_{X,Y}(x, rx)\,dx. \tag{2.11}$$

The mechanical application of the calculus in the above derivations (or use of a bivariate Jacobian transformation, introduced in Chapter I.9) somewhat robs these formulae of their intuitive appeal: Figure 2.3 graphically illustrates what is being computed (and informally also why the limiting operations of calculus work). First recall that, if X is a continuous r.v. with p.d.f. f_X and support S_X, then, for all $x \in S_X$, $\Pr(X = x) = 0$, while, for small $\epsilon > 0$, $\Pr(x < X < x + \epsilon) \approx f_X(x)\epsilon$. A similar statement holds for bivariate r.v.s via $\Pr(x < X < x + \epsilon_1, y < Y < y + \epsilon_2) \approx f_{X,Y}(x, y)\epsilon_1\epsilon_2$. Figure 2.3(a) corresponds to the sum $S = X + Y$ depicted for a constant value $S = s$. In particular, for any x, the probability of the event $(x < X < x + dx, s < S < s + ds)$ is approximated by the area of the shaded parallelogram times the density $f_{X,Y}$ evaluated at x and $s - x$, i.e., $f_{X,Y}(x, s-x)\,dx\,ds$. The probability that $S = s$ is then obtained by 'summing' over all possible x, i.e.,

$$\Pr(s < S < s + ds) \approx \left[\int f_{X,Y}(x, s-x)\,dx\right]ds,$$

from which the first integral expression in (2.7) follows. The expression for the difference $D = X - Y$ is similarly obtained from Figure 2.3(b).

Figure 2.3(c) corresponds to the product $P = XY$, with the area of the shaded regions, when treated as rectangles, given by dx times the absolute value of $(p + dp)/x - p/x$ or $dx\,dp/|x|$. (Observe that the vertical distance between the two curves,

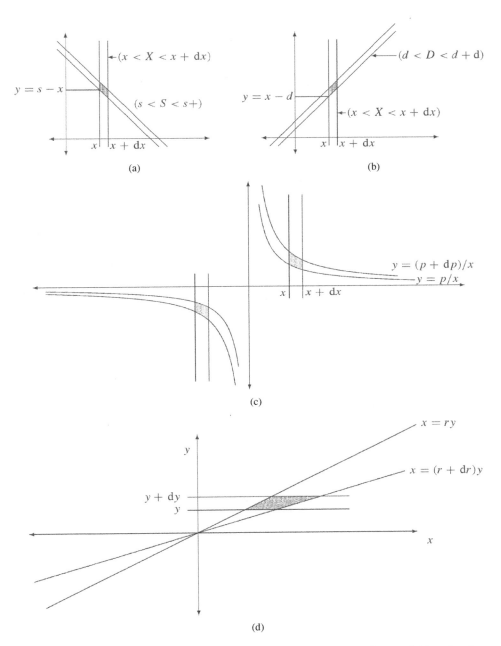

Figure 2.3 Graphical depiction of (a) the sum, (b) difference, (c) product, and (d) ratio of r.v.s X and Y

$\mathrm{d}p/|x|$, decreases as $|x|$ grows.) Integrating out x from the expression

$$\Pr(x < X < x + \mathrm{d}x, \ p < P < p + \mathrm{d}p) \approx f_{X,Y}(x, p/x)|x|^{-1}\, \mathrm{d}x\, \mathrm{d}p$$

yields the middle equation in (2.9). Finally, Figure 2.3(d) corresponds to the ratio $R = X/Y$, with the area of the shaded region, when treated as a parallelogram, given by $\mathrm{d}y$ times $|y|\, \mathrm{d}r$.

⊙ **Example 2.11**   (Example 2.6 cont.) Let $X_i \overset{\text{ind}}{\sim} \mathrm{Exp}(\lambda_i)$, $i = 1, \dots, n$, and $S = \sum_{i=1}^{n} X_i$. First let $n = 2$. From (2.7),

$$f_S(s) = \int_{-\infty}^{\infty} f_{X_1}(x) f_{X_2}(s - x)\, \mathrm{d}x = \int_0^s \lambda_1 e^{-\lambda_1 x} \lambda_2 e^{-\lambda_2(s-x)}\, \mathrm{d}x$$

$$= \lambda_1 \lambda_2 e^{-\lambda_2 s} \int_0^s e^{-(\lambda_1 - \lambda_2)x}\, \mathrm{d}x = \lambda_1 \lambda_2 e^{-\lambda_2 s} \frac{1}{\lambda_1 - \lambda_2}\left(1 - e^{-(\lambda_1 - \lambda_2)s}\right)$$

$$= \frac{\lambda_1}{\lambda_1 - \lambda_2} \lambda_2 e^{-\lambda_2 s} + \frac{\lambda_2}{\lambda_2 - \lambda_1} \lambda_1 e^{-\lambda_1 s}. \tag{2.12}$$

This nice structure is indeed preserved in the general case. It can be proven by induction (see, for example, Ross, 1997, p. 246) that

$$f_S(s) = \sum_{i=1}^{n} C_{i,n} \lambda_i e^{-\lambda_i s}, \qquad C_{i,n} = \prod_{j \neq i} \frac{\lambda_j}{\lambda_j - \lambda_i}. \tag{2.13}$$

From this, the c.d.f. is clearly

$$F_S(s) = \int_0^s f_S(t)\, \mathrm{d}t = \sum_{i=1}^{n} C_{i,n} \int_0^s \lambda_i e^{-\lambda_i t}\, \mathrm{d}t = \sum_{i=1}^{n} C_{i,n}\left(1 - e^{-\lambda_i s}\right). \tag{2.14}$$

The program in Listing 2.4 computes the p.d.f. and c.d.f. at a grid of points for a given vector of λ-values.   ■

```
function [f,F] = expsum (xvec,lamvec)
n=length(lamvec); lamvec=reshape(lamvec,1,n);
for i=1:n
  C(i)=1;
  for j=1:n
    if i ~= j
      C(i) = C(i) * lamvec(j)/(lamvec(j)-lamvec(i));
    end
  end
end
f=zeros(length(xvec),1);
for i=1:length(xvec)
  x=xvec(i);
  f(i) = sum(C.*lamvec.*exp(-lamvec*x));
  F(i) = sum(C.*(1 - exp(-lamvec*x)));
end
```

Program Listing 2.4   Computes (2.13) and (2.14)

⊙ **Example 2.12** (Example 2.11 cont.) Coelho (1998) derived a tractable expression for the p.d.f. of $Q = \sum_{i=1}^{n} X_i$, where $X \stackrel{\text{ind}}{\sim} \text{Gam}(a_i, b_i)$ and $a_i \in \mathbb{N}$. In particular, for $n = 2$, with $A = a_1 + a_2$, $B = b_1 + b_2$ and $b_1 \neq b_2$, density $f_Q(q; a_1, b_1, a_2, b_2)$ is given by

$$b_1^{a_1} b_2^{a_2} \sum_{i=1}^{2} (-1)^{A-a_i} \left(\sum_{j=1}^{a_i} \frac{\binom{A-j-1}{a_i-j}}{(j-1)!} q^{j-1} (2b_i - B)^{j-A} \right) e^{-b_i q} \mathbb{I}_{(0,\infty)}(q). \tag{2.15}$$

It is straightforward to check that, if $a_1 = a_2 = 1$, then (2.15) reduces to

$$f_Q(q; 1, b_1, 1, b_2) = \left(\frac{b_1}{b_1 - b_2} b_2 e^{-b_2 q} + \frac{b_2}{b_2 - b_1} b_1 e^{-b_1 q} \right) \mathbb{I}_{(0,\infty)}(q),$$

as given in (2.12). ∎

⊖ **Example 2.13** Let $X_i \stackrel{\text{i.i.d.}}{\sim} \text{Unif}(0, 1)$, $i = 1, 2$, and consider computing the probability that their sum exceeds one. Given that the density of X_i is symmetric with expected value $1/2$, we might assume from a symmetry argument that $\Pr(X_1 + X_2 > 1) = 1/2$. To check,

$$\Pr(X_1 + X_2 > 1) = \int\int_{x_1+x_2>1} f_{X_1}(x_1) f_{X_2}(x_2) \, dx_1 \, dx_2 = \int_0^1 \int_{1-x_2}^1 dx_1 \, dx_2 = \frac{1}{2}.$$

A method which implicitly avoids the double integral is to write

$$\Pr(X_2 > 1 - X_1) = \int_0^1 \Pr(X_2 > 1 - X_1 \mid X_1 = x_1) f_{X_1}(x_1) \, dx_1$$

$$= \int_0^1 x_1 \, dx_1 = \frac{1}{2},$$

because, from the uniform distribution function,

$$\Pr(X_2 > 1 - X_1 \mid X_1 = x_1) = \Pr(X_2 > 1 - x_1) = 1 - (1 - x_1) = x_1.$$

Of course, this is ultimately the same as the previous derivation, i.e.,

$$\Pr(X_2 > 1 - X_1) = \int_0^1 \Pr(X_2 > 1 - X_1 \mid X_1 = x_1) f_{X_1}(x_1) \, dx_1$$

$$= \int_0^1 \int_{1-x_1}^1 \mathbb{I}_{(0,1)}(x_2) \, dx_2 \, \mathbb{I}_{(0,1)}(x_1) \, dx_1 = \frac{1}{2}.$$

An alternative way uses the convolution formula (2.7) to first obtain the density of $S = X_1 + X_2$, which is interesting in its own right. As the joint density factors (via independence),

$$f_S(s) = \int_{-\infty}^{\infty} f_{X_1}(x_1) f_{X_2}(s - x_1) \, dx_1 = \int_{-\infty}^{\infty} \mathbb{I}_{(0,1)}(x_1) \mathbb{I}_{(0,1)}(s - x_1) \, dx_1.$$

The latter indicator function is $0 < s - x_1 < 1 \Leftrightarrow s - 1 < x_1 < s$, so that both conditions (a) $0 < x_1 < 1$ and (b) $s - 1 < x_1 < s$ must be satisfied. If $0 < s < 1$, then the lower bound of (a) and the upper bound of (b) are binding, while, for $1 \leq s < 2$, the opposite is true. As there is no overlap between the two cases,

$$f_S(s) = \int_0^s \mathbb{I}_{(0,1)}(s)\, dx_1 + \int_{s-1}^1 \mathbb{I}_{[1,2)}(s)\, dx_1 = s\mathbb{I}_{(0,1)}(s) + (2-s)\,\mathbb{I}_{[1,2)}(s), \quad (2.16)$$

which is a triangle when plotted. Thus,

$$\Pr(X_1 + X_2 > 1) = \int_1^2 (2-s)\, ds = \frac{1}{2}.$$

The expression for the density of S may also be obtained graphically. Similar to the illustration in Figure 2.3(a), and using the fact that f_{X_1,X_2} is just $\mathbb{I}_{(0,1)}(x_1)\,\mathbb{I}_{(0,1)}(x_2)$, we see that $\Pr(s < S < s + ds)$ for $0 \leq s < 1$ is given by the area of the shaded region in Figure 2.4(a). This is the difference in the area of the two triangles, or

$$\frac{1}{2}(s + ds)^2 - \frac{1}{2}s^2 = s\, ds + \frac{1}{2}(ds)^2 \approx s\, ds.$$

Similarly, for $s \geq 1$ (Figure 2.4(b)), the area of the shaded region is

$$\frac{1}{2}(2 - s)^2 - \frac{1}{2}(2 - s - ds)^2 = (2 - s)\, ds - \frac{1}{2}(ds)^2 \approx (2 - s)\, ds.$$

Thus, f_S is as given in (2.16). ∎

⊖ **Example 2.14** Let $X_i \overset{\text{i.i.d.}}{\sim} \text{Par I}(b, 1)$, $i = 1, 2$, and define $Z = X_1 X_2$. From Section I.7.1(8) or Table I.C.4, $f_X(x; b, 1) = bx^{-(b+1)}\mathbb{I}_{(1,\infty)}(x)$, so that (2.9) yields

$$f_Z(z) = \int_{-\infty}^{\infty} \frac{1}{|x_1|} bx_1^{-(b+1)}\, b\left(\frac{z}{x_1}\right)^{-(b+1)} \mathbb{I}_{(1,\infty)}(x_1)\,\mathbb{I}_{(1,\infty)}(z/x_1)\, dx_1$$

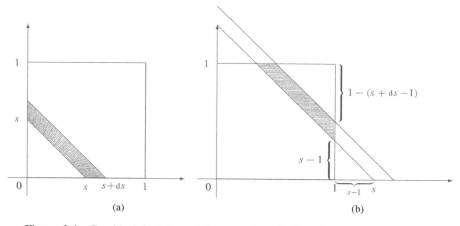

(a)   (b)

Figure 2.4 Graphical depiction of the sum of two independent standard uniform r.v.s

and, using the indicator functions to help us avoid silly mistakes,

$$\mathbb{I}_{(1,\infty)}\left(x_1\right)\mathbb{I}_{(1,\infty)}\left(z/x_1\right) \neq 0 \Leftrightarrow 1 < x_1 < \infty, \ 1 < z/x_1 < \infty$$

$$\Leftrightarrow 1 < x_1 < z < \infty$$

$$\Leftrightarrow \mathbb{I}_{(1,z)}\left(x_1\right)\mathbb{I}_{(1,\infty)}\left(z\right)$$

which gives

$$f_Z\left(z\right) = \mathbb{I}_{(1,\infty)}\left(z\right)b^2 z^{-(b+1)} \int_1^z \frac{1}{x_1}\,\mathrm{d}x_1 = b^2\left(\ln z\right)z^{-(b+1)}\mathbb{I}_{(1,\infty)}\left(z\right).$$

Alternatively, the c.d.f. of the product is given by

$$F_Z\left(z\right) = \Pr\left(Z \leq z\right) = \Pr\left(X_1 X_2 \leq z\right) = \Pr\left(\ln X_1 + \ln X_2 \leq \ln z\right),$$

and, from (I.7.65), the distribution of $Y = \ln X_1$ is

$$f_Y\left(y\right) = f_{X_1}\left(x\right)\left|\frac{\mathrm{d}x}{\mathrm{d}y}\right| = b\left(e^y\right)^{-(b+1)}\mathbb{I}_{(1,\infty)}\left(e^y\right)e^y = be^{-by}\mathbb{I}_{(0,\infty)}\left(y\right),$$

an exponential. From Example 2.3, $S = \ln X_1 + \ln X_2 \sim \mathrm{Gam}\left(2, b\right)$, with density $f_S(s) = b^2 s e^{-bs}\mathbb{I}_{(0,\infty)}(s)$. Thus, $F_Z\left(z\right) = F_S(\ln z)$ and we want

$$f_Z(z) = \frac{\mathrm{d}}{\mathrm{d}z}F_Z(z) = \frac{\mathrm{d}}{\mathrm{d}z}\int_0^{\ln z} f_S\left(s\right)\mathrm{d}s.$$

Recall Leibniz' rule (see Section I.A.3.4.3), which states that, if $I = \int_{\ell(z)}^{h(z)} f\left(s, z\right)\mathrm{d}s$, then

$$\frac{\partial I}{\partial z} = \int_{\ell(z)}^{h(z)} \frac{\partial f}{\partial z}\,\mathrm{d}s + f\left(h\left(z\right), z\right)\frac{\mathrm{d}h}{\mathrm{d}z} - f\left(\ell\left(z\right), z\right)\frac{\mathrm{d}\ell}{\mathrm{d}z}.$$

In our case, $f\left(s, z\right) = f_S\left(s\right) = b^2 s e^{-bs}\mathbb{I}_{(0,\infty)}(s)$, and

$$f_Z(z) = \frac{\mathrm{d}}{\mathrm{d}z}\int_0^{\ln z} f_S\left(s\right)\mathrm{d}s$$

$$= \int_0^{\ln z} 0\,\mathrm{d}s + \left(b^2\left(\ln z\right)e^{-b(\ln z)}\right)\mathbb{I}_{(0,\infty)}\left(\ln z\right)\frac{1}{z} - f\left(0\right)\cdot 0$$

$$= b^2\left(\ln z\right)z^{-(b+1)}\mathbb{I}_{(1,\infty)}\left(z\right),$$

which agrees with the previous derivation. ∎

⊖ ***Example 2.15***   Example I.9.7 illustrated via a bivariate transformation that the ratio of a standard normal r.v. to the square root of an independent chi-square r.v. divided by its degrees of freedom follows a Student's t distribution. This is also easily shown using (2.10). Let $Z \sim N(0, 1)$ independent of $C \sim \chi_v^2$ and define $X = \sqrt{C/v}$ with density

$$f_X(x) = \frac{2^{-v/2+1} v^{v/2}}{\Gamma(v/2)} x^{v-1} e^{-\left(vx^2\right)/2} \mathbb{I}_{(0,\infty)}(x)$$

from Problem I.7.5. Then

$$f_T(t) = \int_{-\infty}^{\infty} |x| f_Z(tx) f_X(x) \, dx = K \int_0^\infty x \exp\left[-\frac{1}{2}(tx)^2\right] x^{v-1} e^{-\left(vx^2\right)/2} \, dx$$

$$= K \int_0^\infty x^v \exp\left(-\frac{1}{2}x^2(t^2 + v)\right) dx, \tag{2.17}$$

where $K := \left[2^{-v/2+1} v^{v/2}\right] / \left[\sqrt{2\pi}\,\Gamma(v/2)\right]$. Substituting $y = x^2$, $dx/dy = y^{-1/2}/2$,

$$f_T(t) = K \int_0^\infty x^{2\frac{v}{2}} e^{-\left(x^2/2\right)\left(v+t^2\right)} \, dx = K \int_0^\infty y^{\frac{v}{2}} e^{-y\frac{1}{2}\left(v+t^2\right)} \frac{1}{2} y^{-1/2} \, dy$$

$$= \frac{K}{2} \int_0^\infty y^{\frac{v-1}{2}} e^{-y\frac{1}{2}\left(v+t^2\right)} \, dy = \frac{K}{2} \int_0^\infty y^{\frac{v+1}{2}-1} e^{-y\frac{1}{2}\left(v+t^2\right)} \, dy$$

$$= \frac{K}{2} \Gamma\left(\frac{v+1}{2}\right) \left[\frac{1}{2}(v+t^2)\right]^{-(v+1)/2},$$

as $\int_0^\infty e^{-zx} x^{t-1} \, dx = \Gamma(t) z^{-t}$ (see the gamma p.d.f.). Simplifying, all the powers of 2 cancel, so that

$$f_T(t) = \frac{\Gamma\left(\frac{v+1}{2}\right) v^{v/2}}{\sqrt{\pi}\,\Gamma(v/2)} \left[v + t^2\right]^{-(v+1)/2} = \frac{v^{-1/2}}{B\left(\frac{v}{2}, \frac{1}{2}\right)} \left[1 + t^2/v\right]^{-(v+1)/2},$$

i.e., $T \sim t(v)$.

Another way of deriving the t is by using the conditioning result (I.8.40), which states that, for some event E,

$$\Pr(E) = \int \Pr(E \mid X = x) f_X(x) \, dx.$$

For $T = Z/X$,

$$f_T(t) = F_T'(t) = \frac{d}{dt} \Pr\left(Z < t\sqrt{\chi_v^2/v}\right) = \frac{d}{dt} \int \Pr(Z < tX \mid X = x) f_X(x) \, dx$$

$$= \frac{d}{dt} \int \Pr(Z < tx) f_X(x) \, dx = \frac{d}{dt} \int \int_{-\infty}^{tx} f_Z(z) \, dz \cdot f_X(x) \, dx$$

and, substituting,

$$
\begin{aligned}
f_T(t) &= K \frac{d}{dt} \int_0^\infty \int_{-\infty}^{tx} e^{-\frac{1}{2}z^2} \, dz \cdot x^{\nu-1} e^{-(\nu x^2)/2} \, dx \\
&= K \int_0^\infty x^{\nu-1} e^{-(\nu x^2)/2} \frac{d}{dt} \int_{-\infty}^{tx} e^{-\frac{1}{2}z^2} \, dz \, dx \\
&= K \int_0^\infty x^{\nu-1} e^{-(\nu x^2)/2} \cdot \left[e^{-\frac{1}{2}(tx)^2} x - e^{-\infty} \cdot 0 \right] dx \\
&= K \int_0^\infty x^\nu e^{-(x^2/2)(\nu+t^2)} \, dx
\end{aligned}
$$

from Leibniz' rule. This is the same integral as in the previous derivation. See also Problem 2.18. ∎

⊖ *Example 2.16* Let $X_0 \sim N(0,1)$ independent of $Y_0 \sim \text{Lap}(0,1)$. For a known value of c, $0 \le c \le 1$, define

$$
Z = c X_0 + (1-c) Y_0 = X + Y,
$$

where $X = c X_0 \sim N(0, c^2)$ and, with $k = 1-c$, $Y = k Y_0 \sim \text{Lap}(0, k)$. (Note that $X_0 + Y_0$ is just a scale transformation of Z with $c = 1/2$.) We will refer to this as the normal–Laplace convolution distribution, and write $Z \sim \text{NormLap}(c)$. It easily follows that Z is symmetric about $\mathbb{E}[Z] = 0$ and $\mathbb{V}(Z) = c^2 + 2(1-c)^2$.

Because the tails of the Laplace are heavier than those of the normal, the weighted sum Z allows for a smooth continuum of tail behaviour (constrained, of course, to lie between the two extremes offered by the normal and Laplace). This property is useful when modelling data which exhibit tails which are at least somewhat larger than the normal – this being a quite likely occurrence in practice, a fact observed already at the end of the nineteenth century and notoriously prominent in recent times with the enormous availability and variety of data.[1]

The exact density of Z for $c \in (0,1)$ is obtained from the convolution formula (2.7) as

$$
f_Z(z) = \int_{-\infty}^\infty f_X(z-y) f_Y(y) \, dy = \frac{A+B}{c(1-c) 2\sqrt{2\pi}},
$$

where

$$
A = \int_{-\infty}^0 \exp\left(-\frac{1}{2} \left(\frac{z-y}{c} \right)^2 + \frac{y}{1-c} \right) dy \tag{2.18}
$$

---

[1] See the book by Kotz, Podgorski and Kozubowski (2001) for a detailed account of the theory and applications of the Laplace distribution. Several distributions exist besides the NormLap which nest the normal and Laplace, the most common of which being the generalized exponential, popularized in the book of Box and Tiao (1973, 1992); see Problem 7.16. Haas, Mittnik and Paolella (2006) demonstrate the value of the NormLap for modelling financial returns data and show it to be superior to the GED and other distributional models which nest the normal and Laplace. Chapters 7 and 9 will examine various other alternatives to the normal and Laplace distributions which are fatter-tailed and, possibly, skewed.

and

$$B = \int_0^\infty \exp\left(-\frac{1}{2}\left(\frac{z-y}{c}\right)^2 - \frac{y}{1-c}\right) dy, \qquad (2.19)$$

which would need to be numerically evaluated (see Problem 2.8). Alternatively, note that A can be written as

$$A = \exp\left\{-\frac{z^2}{2c^2}\right\} \int_{-\infty}^0 \exp\left\{-\frac{1}{2c^2}\left(y^2 - 2zy - \frac{2c^2 y}{1-c}\right)\right\} dy$$

$$= \exp\left\{-\frac{z^2}{2c^2}\right\} \int_{-\infty}^0 \exp\left\{-\frac{1}{2c^2}\left(y^2 - 2ys + s^2 - s^2\right)\right\} dy, \quad s = z + \frac{c^2}{1-c},$$

$$= \exp\left\{-\frac{z^2}{2c^2}\right\} \exp\left\{\frac{s^2}{2c^2}\right\} \int_{-\infty}^0 \exp\left\{-\frac{1}{2}\left(\frac{y-s}{c}\right)^2\right\} dy$$

$$= \exp\left\{\frac{s^2 - z^2}{2c^2}\right\} \sqrt{2\pi} c \, \Phi\left(-\frac{s}{c}\right),$$

where $\Phi(x)$ is the standard normal c.d.f. at x. Similarly, with $\overline{\Phi}(x) = 1 - \Phi(x)$,

$$B = \exp\left\{\frac{d^2 - z^2}{2c^2}\right\} \sqrt{2\pi} c \, \overline{\Phi}\left(-\frac{d}{c}\right), \qquad d = z - \frac{c^2}{1-c},$$

so that f_Z can be written as

$$f_Z(z) = \frac{1}{2(1-c)} \left(\exp\left\{\frac{s^2 - z^2}{2c^2}\right\} \Phi\left(-\frac{s}{c}\right) + \exp\left\{\frac{d^2 - z^2}{2c^2}\right\} \overline{\Phi}\left(-\frac{d}{c}\right)\right). \quad (2.20)$$

As function Φ is efficiently implemented in computing platforms, (2.20) is essentially closed-form and can be evaluated extremely fast over a grid of ordinates (particularly so when using the 'vectorized' capabilities of Matlab and related software). It is important, however, that $\overline{\Phi}(x) = 1 - \Phi(x)$ be calculated as $\Phi(-x)$ for large x (Problem 2.8).

Plots of f_Z for $c = 0.1, 0.3, 0.5, 0.7$, and 0.9 are shown in Figure 2.5(a) using a log scale to help distinguish them. As c moves from 0 to 1 (from Laplace to normal), the log p.d.f. varies from a triangle to a parabola. The different tail behaviour is made clear in Figure 2.5(b).

Finally, the c.d.f. of Z can be expressed as

$$F_Z(z) = \Phi\left(\frac{z}{c}\right) + \frac{1}{2} \exp\left\{\frac{1}{2}r^2\right\} \left[e^s \Phi\left(-\frac{z}{c} - r\right) - e^{-s} \Phi\left(\frac{z}{c} - r\right)\right], \quad (2.21)$$

where $s = z/(1-c)$ and $r = c/(1-c)$; and the raw moments are given by

$$\mathbb{E}[Z^m] = (1-c)^m m! \sum_{i=0}^{m/2} \frac{1}{i!} \left(\frac{r^2}{2}\right)^i, \qquad (2.22)$$

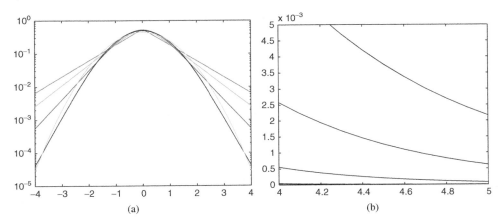

Figure 2.5 (a) The p.d.f. of the convolution of independent normal and Laplace random variables using $c = 0.1, 0.3, 0.5, 0.7$, and 0.9, plotted on a log scale. (b) The same p.d.f.s shown in a segment of the right tail; the top line corresponds to $c = 0.1$ (90 % Laplace, 10 % normal), the next line from the top corresponds to $c = 0.3$, etc.

as derived in Problem 2.8. For example, from (2.22), the kurtosis of Z is given by

$$\frac{\mu_4}{\mu_2^2} = 3 + \frac{3}{1 + r^2 + \frac{1}{4}r^4}, \quad r = \frac{c}{1-c}, \tag{2.23}$$

which is larger than three (the kurtosis of the normal) for $c < 1$. ∎

Example 2.17 (Example I.8.18 cont.) Let $X, Y \overset{\text{i.i.d.}}{\sim} N(0, 1)$ and again consider the distribution of ratio $R = X/Y$. From the first equation in (2.10),

$$f_R(r) = \frac{1}{2\pi} \int_{-\infty}^{\infty} \frac{|x|}{r^2} \exp\left\{-\frac{1}{2}\left(x^2 + \left(\frac{x}{r}\right)^2\right)\right\} dx$$

$$= \frac{1}{\pi r^2} \int_0^{\infty} x \exp\left\{-\frac{1}{2}\left(\frac{x^2}{k}\right)\right\} dx,$$

where $k = r^2 / (1 + r^2)$ and, with $u = \frac{1}{2}\left(\frac{x^2}{k}\right)$, $x = \sqrt{2ku}$, $dx = k (2ku)^{-1/2} du$,

$$\int_0^{\infty} x \exp\left\{-\frac{1}{2}\left(\frac{x^2}{k}\right)\right\} dx = k \int_0^{\infty} \exp\{-u\} du = k,$$

so that

$$f_R(r) = \frac{1}{\pi r^2} k = \frac{1}{\pi} \frac{1}{1 + r^2},$$

i.e., R is a Cauchy random variable. An alternative way of resolving the integral can be seen as follows. For $S \sim N(0, k)$ and $Z \sim N(0, 1)$,

$$\int_0^\infty x \exp\left\{-\frac{1}{2}\left(\frac{x^2}{k}\right)\right\} dx = \frac{2}{2}\frac{\sqrt{2\pi}\sqrt{k}}{\sqrt{2\pi}\sqrt{k}}\int_0^\infty x \exp\left\{-\frac{1}{2}\left(\frac{x^2}{k}\right)\right\} dx$$

$$= \frac{\sqrt{2\pi}\sqrt{k}}{2}\mathbb{E}\,|S|\,,$$

where $\mathbb{E}\,|S| = \mathbb{E}\left|\sqrt{k}Z\right| = \sqrt{k}\sqrt{2/\pi}$ from Example I.7.3, so that

$$f_R\,(r) = \frac{1}{\pi r^2}\frac{\sqrt{2\pi}\sqrt{k}}{2}\sqrt{k}\sqrt{2/\pi} = \frac{k}{\pi r^2} = \frac{1}{\pi}\frac{1}{1+r^2}$$

as before.   ■

⊚ **Example 2.18**   (Example 2.17 cont.) From the symmetry of the problem, it should be clear that R and $R' = Y/X$ have the same distribution. This will not be the case in the more general situation in which $\mathbf{X} = (X_1\ X_2)' \sim \text{N}\left(\binom{\mu_1}{\mu_2}, \Sigma\right)$. The derivation of the density of R or R' is straightforward but somewhat tedious; see Hinkley (1969), Greenberg and Webster (1983, pp. 325–328) and the references therein. Consider the special case for which $\Sigma = \mathbf{I}_2$ and $R' = X_2/X_1$. Problem 2.7 shows that the density $f_{R'}\,(r)$ is

$$\exp\left(\frac{b^2/a - c}{2}\right)\frac{1}{2\pi}\left(\frac{b}{a}\sqrt{\frac{2\pi}{a}}\left(1 - 2\Phi\left(-\frac{b}{\sqrt{a}}\right)\right) + 2a^{-1}\exp\left(-\frac{b^2}{2a}\right)\right), \quad (2.24)$$

where $a = 1 + r^2$, $b = \mu_1 + r\mu_2$, $c = \mu_1^2 + \mu_2^2$ and Φ is the standard normal c.d.f. Figure 2.6 shows the density for $\mu_1 = 0.1$ and $\mu_2 = 2$. If desired, the c.d.f. could be numerically computed as $F_{R'}\,(r) = \int_{-\infty}^r f_{R'}\,(t)\,dt$.   ■

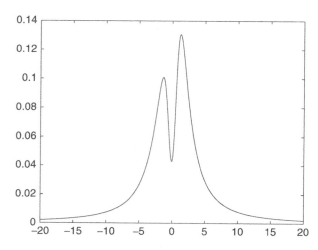

Figure 2.6   Density of X_2/X_1, where $\mathbf{X} \sim \text{N}\left([0.1,\ 2]',\ \mathbf{I}_2\right)$

In the previous examples, the integrals were analytically resolved; this will, unfortunately, not be the case for many cases of interest, so that numerical integration methods will be necessary.

⊖ **Example 2.19** Let $Z \sim N(0, \sigma^2)$ and $G \sim \text{Gam}(a, c)$ be independent random variables and $f_{\text{Gam}}(g; a, c) = c^a \Gamma(a)^{-1} e^{-cg} g^{a-1} \mathbb{I}_{(0,\infty)}(g)$. The r.v. $S = Z + G$ arises in the econometric analysis of frontier functions (see Greene, 1990). From the convolution formula (2.7) and their independence,

$$f_S(s) = \int_{-\infty}^{\infty} f_{Z,G}(s - g, g) \, dg$$

$$= \frac{c^a}{\sigma \sqrt{2\pi} \, \Gamma(a)} \int_0^{\infty} \exp\left\{-\frac{1}{2}\left(\frac{s-g}{\sigma}\right)^2\right\} \exp\{-cg\} g^{a-1} \, dg, \quad (2.25)$$

which needs to be evaluated numerically at each s. The inversion methods in Sections 1.2.5 and 1.3 can straightforwardly be applied to the c.f. of S,

$$\varphi_S(t; \sigma, a, c) = \varphi_G(t; a, c) \, \varphi_Z(t; \sigma)$$

$$= \left(\frac{c}{c - it}\right)^a \exp\left(-\frac{1}{2}\sigma^2 t^2\right)$$

from (2.3). Figure 2.7 plots (2.25) for $a = 4$ and three different values of σ; in Figure 2.7(a) $c = 1$, while in Figure 2.7(b) $c = 3$. ■

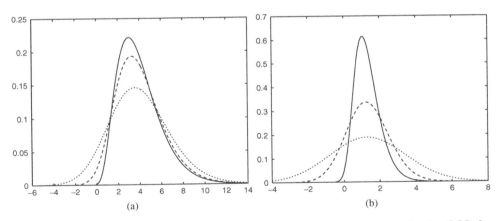

Figure 2.7 Convolution of independent normal and gamma random variables: density (2.25) for (a) $a = 4, c = 1$, (b) $a = 4, c = 3$ with $\sigma = 0.25$ (solid), 1 (dashed) and 2 (dotted)

2.3 Approximating the mean and variance

Let $g(\mathbf{x}) := g(x_1, x_2)$ be a bivariate continuous function which is defined on an open neighbourhood of $\mathbf{x}^0 = (x_1^0, x_2^0) \in \mathbb{R}^2$ and whose first and second derivatives exist. For convenience, let $\dot{g}_i(\mathbf{x}) = (\partial g / \partial x_i)(\mathbf{x})$, $\ddot{g}_{ij}(\mathbf{x}) = (\partial^2 g / \partial x_i \partial x_j)(\mathbf{x})$ and $\Delta_i = (x_i - x_i^0)$, $i = 1, 2$.

From (I.A.153), the two-term Taylor series approximation for $g(\mathbf{x})$ is given by

$$g(\mathbf{x}) \approx g\left(\mathbf{x}^0\right) + \Delta_1 \dot{g}_1\left(\mathbf{x}^0\right) + \Delta_2 \dot{g}_2\left(\mathbf{x}^0\right)$$

$$+ \frac{1}{2}\Delta_1^2 \ddot{g}_{11}\left(\mathbf{x}^0\right) + \frac{1}{2}\Delta_2^2 \ddot{g}_{22}\left(\mathbf{x}^0\right) + \Delta_1 \Delta_2 \ddot{g}_{12}\left(\mathbf{x}^0\right). \tag{2.26}$$

Now let $\mathbf{X} = (X_1, X_2)$ be a bivariate r.v. with mean $\mathbf{x}^0 = \left(\mu_{X_1}, \mu_{X_2}\right)$. Taking expectations of $g(\mathbf{X})$ gives

$$\mathbb{E}\left[\Delta_i \dot{g}_i\left(\mathbf{x}^0\right)\right] = \dot{g}_i\left(\mathbf{x}^0\right)\mathbb{E}\left[\left(X_i - \mu_{X_i}\right)\right] = 0, \quad i = 1, 2,$$

$$\mathbb{E}\left[\Delta_i^2 \ddot{g}_{ii}\left(\mathbf{x}^0\right)\right] = \ddot{g}_{ii}\left(\mathbf{x}^0\right)\mathbb{E}\left[\left(X_i - \mu_{X_i}\right)^2\right] = \ddot{g}_{ii}\left(\mathbf{x}^0\right)\mathbb{V}(X_i), \quad i = 1, 2,$$

$$\mathbb{E}\left[\Delta_i \Delta_j \ddot{g}_{ij}\left(\mathbf{x}^0\right)\right] = \ddot{g}_{ij}\left(\mathbf{x}^0\right)\mathbb{E}\left[\left(X_i - \mu_{X_i}\right)\left(X_j - \mu_{X_j}\right)\right] = \ddot{g}_{ij}\left(\mathbf{x}^0\right)\text{Cov}\left(X_i, X_j\right),$$

or

$$\mathbb{E}[g(\mathbf{X})] \approx g\left(\mathbf{x}^0\right) + \frac{1}{2}\ddot{g}_{11}\left(\mathbf{x}^0\right)\mathbb{V}(X_1) + \frac{1}{2}\ddot{g}_{22}\left(\mathbf{x}^0\right)\mathbb{V}(X_2)$$
$$+ \ddot{g}_{12}\left(\mathbf{x}^0\right)\text{Cov}(X_1, X_2). \tag{2.27}$$

We refer to (2.27) as the second-order Taylor series approximation of $\mathbb{E}[g(\mathbf{X})]$. As $\mathbb{E}[\Delta_i \dot{g}_i(\mathbf{x}^0)] = 0$, the first-order approximation just uses the zero-order term, i.e., $g(\mathbf{x}^0)$. For example, let Y be a univariate random variable. While $\mathbb{E}[1/Y] \neq 1/\mathbb{E}[Y]$, we see that it is a zero-order approximation; the second-order term is given below in (2.34).

To approximate the variance, use of only the zero-order term in (2.27) gives

$$\mathbb{V}(g(\mathbf{X})) = \mathbb{E}\left[(g(\mathbf{X}) - \mathbb{E}[g(\mathbf{X})])^2\right] \approx \mathbb{E}\left[(g(\mathbf{X}) - g(\mathbf{x}^0))^2\right]. \tag{2.28}$$

From (2.26), $g(\mathbf{X}) \approx g\left(\mathbf{x}^0\right) + \Delta_1 \dot{g}_1\left(\mathbf{x}^0\right) + \Delta_2 \dot{g}_2\left(\mathbf{x}^0\right)$ is the first-order approximation, and applying this to the r.h.s. of (2.28) gives

$$\mathbb{E}\left[(g(\mathbf{X}) - g(\mathbf{x}^0))^2\right] \approx \mathbb{E}\left[(\Delta_1 \dot{g}_1(\mathbf{x}^0) + \Delta_2 \dot{g}_2(\mathbf{x}^0))^2\right]$$

$$= \dot{g}_1^2(\mathbf{x}^0)\mathbb{E}[\Delta_1^2] + \dot{g}_2^2(\mathbf{x}^0)\mathbb{E}[\Delta_2^2] + 2\dot{g}_1(\mathbf{x}^0)\dot{g}_2(\mathbf{x}^0)\mathbb{E}[\Delta_1\Delta_2],$$

i.e.,

$$\mathbb{V}(g(\mathbf{X})) \approx \dot{g}_1^2\left(\mathbf{x}^0\right)\mathbb{V}(X_1) + \dot{g}_2^2\left(\mathbf{x}^0\right)\mathbb{V}(X_2)$$
$$+ 2\dot{g}_1\left(\mathbf{x}^0\right)\dot{g}_2\left(\mathbf{x}^0\right)\text{Cov}(X_1, X_2). \tag{2.29}$$

Because the approximation for $\mathbb{E}[g(\mathbf{X})]$ in (2.27) makes use of the second-order terms in the Taylor series, we would expect it to be more accurate than the approximate expression for $\mathbb{V}(g(\mathbf{X}))$, which uses only first-order terms.

The approximations (2.27) and (2.29) are easily extended to the more general case with $\mathbf{X} := (X_1, \ldots, X_n)$; the reader should verify that, with $\mathbf{x}^0 = \mathbb{E}[\mathbf{X}] = \left(\mu_{X_1}, \ldots, \mu_{X_n}\right)$,

$$\mathbb{E}\left[g\left(\mathbf{X}\right)\right] \approx g\left(\mathbf{x}^0\right) + \frac{1}{2}\sum_{i=1}^{n} \ddot{g}_{ii}\left(\mathbf{x}^0\right)\mathbb{V}\left(X_i\right) + \sum_{i=1}^{n}\sum_{j=i+1}^{n} \ddot{g}_{ij}\left(\mathbf{x}^0\right)\mathrm{Cov}\left(X_i, X_j\right)$$

(2.30)

and

$$\mathbb{V}\left(g\left(\mathbf{X}\right)\right) \approx \sum_{i=1}^{n} \dot{g}_i^2\left(\mathbf{x}^0\right)\mathbb{V}\left(X_i\right) + 2\sum_{i=1}^{n}\sum_{j=i+1}^{n} \dot{g}_i\left(\mathbf{x}^0\right)\dot{g}_j\left(\mathbf{x}^0\right)\mathrm{Cov}\left(X_i, X_j\right).$$

(2.31)

Perhaps the most prominent application of the above approximations is for $g\left(x, y\right) = x/y$. With

$$\dot{g}_1\left(x, y\right) = y^{-1} \qquad \ddot{g}_{11}\left(x, y\right) = 0 \qquad \ddot{g}_{12}\left(x, y\right) = -y^{-2}$$

$$\dot{g}_2\left(x, y\right) = -xy^{-2} \qquad \ddot{g}_{22}\left(x, y\right) = 2xy^{-3}$$

and letting $\sigma_X^2 = \mathbb{V}\left(X\right)$, $\sigma_Y^2 = \mathbb{V}\left(Y\right)$ and $\sigma_{X,Y} = \mathrm{Cov}\left(X, Y\right)$, (2.27) and (2.29) reduce to

$$\mathbb{E}\left[\frac{X}{Y}\right] \approx \frac{\mu_X}{\mu_Y}\left(1 + \frac{\sigma_Y^2}{\mu_Y^2} - \frac{\sigma_{X,Y}}{\mu_X\mu_Y}\right)$$

(2.32)

and $\mathbb{V}\left(X/Y\right) \approx \mu_Y^{-2}\sigma_X^2 + \mu_X^2\mu_Y^{-4}\sigma_Y^2 - 2\mu_X\mu_Y^{-3}\sigma_{X,Y}$, or

$$\mathbb{V}\left(\frac{X}{Y}\right) \approx \left(\frac{\mu_X}{\mu_Y}\right)^2\left(\frac{\sigma_X^2}{\mu_X^2} + \frac{\sigma_Y^2}{\mu_Y^2} - \frac{2\sigma_{X,Y}}{\mu_X\mu_Y}\right).$$

(2.33)

From (2.32) and (2.33), it is easy to derive moment approximations of $1/Y$ by setting $\mu_X = 1$ and $\sigma_X^2 = 0$, i.e., X is the degenerate r.v. $X = 1$ and $\sigma_{X,Y} = 0$. In particular,

$$\mathbb{E}\left[\frac{1}{Y}\right] \approx \frac{1}{\mu_Y} + \frac{\sigma_Y^2}{\mu_Y^3} \quad \text{and} \quad \mathbb{V}\left(\frac{1}{Y}\right) \approx \frac{\sigma_Y^2}{\mu_Y^4}.$$

(2.34)

Another popular application is with $g\left(x, y\right) = xy$, in which case

$$\mathbb{E}\left[XY\right] \approx \mu_X\mu_Y + \sigma_{X,Y} \quad \text{and} \quad \mathbb{V}\left(XY\right) \approx \mu_Y^2\sigma_X^2 + \mu_X^2\sigma_Y^2 + 2\mu_Y\mu_X\sigma_{X,Y}.$$

(2.35)

In stark comparison to the ratio case above, exact results are easy to obtain for the product, the trick being to use the identity

$$XY = \mu_X \mu_Y + (X - \mu_X) \mu_Y + (Y - \mu_Y) \mu_X + (X - \mu_X)(Y - \mu_Y).$$

Taking expected values of these terms, it follows immediately that

$$\mathbb{E}[XY] = \mu_X \mu_Y + \text{Cov}(X, Y).$$

For the variance, define for convenience $c_{ij} := \mathbb{E}[(X - \mu_X)^i (Y - \mu_Y)^j]$, so that, after a bit of simplifying,

$$\boxed{\mathbb{V}(XY) = \mu_Y^2 \sigma_X^2 + \mu_X^2 \sigma_Y^2 + 2\mu_X \mu_Y c_{11} - c_{11}^2 + c_{22} + 2\mu_Y c_{21} + 2\mu_X c_{12},}$$

provided, of course, that $\mathbb{V}(XY)$ exists. Thus, in (2.35), the approximation to the expected value is exact, while the approximate variance is missing several higher-order terms. As a special case, if $\mu_X = \mu_Y = 0$, then

$$\mathbb{V}(XY) = c_{22} - c_{11}^2 = \mathbb{E}[X^2 Y^2] - (\mathbb{E}[XY])^2.$$

If X and Y are independent, then $c_{11} = c_{12} = c_{21} = 0$ and $c_{22} = \mathbb{V}(X)\mathbb{V}(Y)$, so that

$$\boxed{X \perp Y \quad \Rightarrow \quad \mathbb{E}[XY] = \mu_X \mu_Y, \ \mathbb{V}(XY) = \mu_Y^2 \sigma_X^2 + \mu_X^2 \sigma_Y^2 + \sigma_X^2 \sigma_Y^2.} \quad (2.36)$$

This implies further that, if X and Y have the same mean μ and same variance σ^2, then

$$\mathbb{E}[XY] = \mu^2 \quad \text{and} \quad \mathbb{V}(XY) = \sigma^2 (2\mu^2 + \sigma^2). \quad (2.37)$$

⊖ **Example 2.20**   Let $C \sim \chi_k^2$ and $Y = k^{1/2} C^{-1/2}$. Then $\mathbb{E}[Y^2] = k\mathbb{E}[C^{-1}] = k/(k-2)$ from (I.7.44). Now, with $X \sim N(0, 1)$ independent of C, we know from Example I.9.7 that

$$T = \frac{X}{\sqrt{C/k}} = XY$$

is Student's t with k degrees of freedom. From (2.36) with $\mu_X = 0$ and $\sigma_X^2 = 1$, we see that $\mathbb{E}[T] = 0$ and

$$\mathbb{V}(T) = \mathbb{V}(XY) = \mu_Y^2 + 0 + \sigma_Y^2 = \mathbb{E}[Y^2] = \frac{k}{k-2},$$

showing (I.7.46).   ∎

⊖ **Example 2.21**   Let $X, Y \overset{\text{i.i.d.}}{\sim} N(\mu, \sigma^2)$ represent (or approximate) the lengths of pieces of plywood cut by an employee at a hardware store. Let $A = XY$ be the area of the rectangle constructed by joining the two pieces at right angles. Compute $\mathbb{E}[A]$ and $\mathbb{V}(A)$. Compare these to the mean and variance of $W = X^2$.

From (I.7.67), $\mathbb{E}[W] = \sigma^2 + \mu^2$ and $\mathbb{V}(W) = 2\sigma^2(2\mu^2 + \sigma^2)$, while from (2.37), $\mathbb{E}[A] = \mu^2$ and $\mathbb{V}(A) = \sigma^2(2\mu^2 + \sigma^2)$, i.e., the means are close if σ^2 is small relative to μ, but the variances differ by a factor of 2.[2]

The next example gives us a good reason to consider the general expressions (2.30) and (2.31), and shows an exception to the general rule that the expectation of a ratio is not the ratio of expectations.

◎ **Example 2.22** Let X_i be i.i.d. r.v.s with finite mean, and define $S := \sum_{i=1}^{n} X_i$ and $R_i := X_i/S$, $i = 1, \ldots, n$. It follows that

$$1 = \sum_{i=1}^{n} R_i = \mathbb{E}\left[\sum_{i=1}^{n} R_i\right] = n\mathbb{E}[R_1],$$

i.e., $\mathbb{E}[R_i] = n^{-1}$. Note that, if the X_i are positive r.v.s (or negative r.v.s), then $0 < R_i < 1$, and the expectation exists. Now let the X_i be i.i.d. positive r.v.s, and let λ_i, $i = 1, \ldots, n$, be a set of known constants. The expectation of

$$R := \frac{\sum_{i=1}^{n} \lambda_i X_i}{\sum_{i=1}^{n} X_i} = \frac{\sum_{i=1}^{n} \lambda_i X_i}{S} = \sum_{i=1}^{n} \lambda_i R_i$$

is

$$\mathbb{E}[R] = \sum_{i=1}^{n} \lambda_i \mathbb{E}[R_i] = n^{-1} \sum_{i=1}^{n} \lambda_i =: \bar{\lambda}.$$

A very useful special case is when $X_i \overset{\text{i.i.d.}}{\sim} \chi_1^2$, which arises in the study of ratios of quadratic forms. From (I.7.42), $\mathbb{E}[X_i] = 1$, so that

$$\mathbb{E}[R] = \frac{\sum_{i=1}^{n} \lambda_i}{n} = \frac{\mathbb{E}\left[\sum_{i=1}^{n} \lambda_i X_i\right]}{\mathbb{E}\left[\sum_{i=1}^{n} X_i\right]},$$

i.e., the expectation of the ratio R is, exceptionally, the ratio of expectations!

Consider the approximations (2.30) and (2.31) applied to $g(\mathbf{X}) = R$, with $X_i \overset{\text{i.i.d.}}{\sim} \chi_1^2$. It is easy to verify that

$$\dot{g}_i(\mathbf{x}) = \frac{S\lambda_i - \sum_{i=1}^{n} \lambda_i x_i}{S^2}, \quad \ddot{g}_{ii}(\mathbf{x}) = 2\frac{\left(\sum_{i=1}^{n} \lambda_i x_i\right) - S\lambda_i}{S^3},$$

---

[2] It is important to realize that A and W have different distributions, even though X and Y are i.i.d. This distinction was, for example, overlooked in a recent textbook, and was caught by a reviewer of that book, Faddy (2002), who (along with an otherwise positive review) said the mistake 'can only be described as a howler'.

which (noting that S evaluated at $\mathbf{x}_0 = (1, \ldots, 1)$ is n) yield $\dot{g}_i(\mathbf{x}_0) = (\lambda_i - \bar{\lambda})/n$ and $\ddot{g}_{ii}(\mathbf{x}_0) = 2(\bar{\lambda} - \lambda_i)/n^2$. Thus, (2.30) with $\mathbb{V}(X_i) = 2$ gives

$$\mathbb{E}[R] \approx \bar{\lambda} + \frac{1}{n^2} \sum_{i=1}^{n} (\bar{\lambda} - \lambda_i) = \bar{\lambda} = \mathbb{E}[R],$$

showing that (2.30) in this case is exact. For the variance, (2.31) gives $\mathbb{V}(R) \approx 2n^{-2} \sum_{i=1}^{n} (\lambda_i - \bar{\lambda})^2$. This turns out to be quite close to the exact value, which just replaces n^2 by $n(n+2)$ in the expression, as will be verified in a later chapter on quadratic forms. ∎

Complementing the preceding example, if $R = U/V$, where $U(\mathbf{X})$ and $V(\mathbf{X})$ are univariate functions of r.v.s $X_1, \ldots, X_n$, then $\mathrm{Cov}(R, V) = 0 \Leftrightarrow \mathbb{E}[R] = \mathbb{E}[U]/\mathbb{E}[V]$. This follows simply because

$$\mathrm{Cov}(R, V) = \mathbb{E}[RV] - \mathbb{E}[R]\mathbb{E}[V] = \mathbb{E}[U] - \mathbb{E}[R]\mathbb{E}[V],$$

as noted by Heijmans (1999). This would rule out the independence of U and V: if U and V are independent with $\mathbb{E}[U] \neq 0$ and $\Pr[V > 0] = 1$, then $\mathbb{E}[U/V] = \mathbb{E}[U]\mathbb{E}[V^{-1}]$, and Jensen's inequality, in particular Example I.4.27, implies that $\mathbb{E}[V^{-1}] > 1/\mathbb{E}[V]$, assuming the expectations exist.

2.4 Problems

When you make the finding yourself – even if you're the last person on Earth to see the light – you'll never forget it. (Carl Sagan)

The scholar who cherishes the love of comfort is not fit to be deemed a scholar. (Confucius)

2.1. Let $X_i \overset{\text{i.i.d.}}{\sim} N(0, 1)$.

    (a) Give the distribution of $R = X_1/X_2$.

    (b) Give the distribution of $S = R^2$.

    (c) Give the distribution of $E = s(\sum_{i=1}^{n} X_i) + c$.

    (d) Give the distribution of $Z = s(\sum_{i=1}^{n} X_i^2) + c$.

    (e) Give the distribution of $M = X_1/|X_2|$.

    (f) Give the distribution of $Y = |M|$.

    (g) Give the distribution of $Q = X_1^2/\sum_{i=2}^{n} X_i^2$.

(h) Let $U = \sum_{i=1}^{m} X_i^2$, $D = \sum_{i=m+1}^{n} X_i^2$ and $Z = U/D$. Give the distribution of

$$G = \frac{n-m}{m} Z \quad \text{and} \quad B = \frac{Z}{1+Z}.$$

2.2. ★ Let $\overline{X}$ and S_n^2 be the sample mean and variance of $X_i \overset{\text{i.i.d.}}{\sim} N(0,1)$, $i = 1, \ldots, n$.

(a) Give the distribution of $n\overline{X}^2$ and the p.d.f. of $(n-1)S_n^2$.

(b) Give the distribution of $A = n\overline{X}^2/S_n^2$.

(c) Give the distribution of

$$V = \frac{(n-1)^{-1}A}{1+(n-1)^{-1}A}.$$

(d) Show that

$$V = \frac{(\sum_{i=1}^{n} X_i)^2}{n\sum_{i=1}^{n} X_i^2}. \tag{2.38}$$

(e) Give the distribution of

$$H = nV = \frac{(\sum_{i=1}^{n} X_i)^2}{\sum_{i=1}^{n} X_i^2}.$$

(f) Make a plot comparing f_H to a kernel density estimate (see Section I.7.4.2) obtained by simulating H using, say, $n = 5$ and so confirm that f_H is correct. More interestingly, also simulate H but using Cauchy and Student's t data with 1 and 2 degrees of freedom instead, and overlay their kernel densities onto the plot.

2.3. ★ Recall Example 2.9.

(a) Algebraically prove that S is symmetric.

(b) Construct a computer program which computes the p.d.f. and c.d.f. of S. Plot the p.d.f. and c.d.f. of $S = X_1 + X_2 + X_3$, where $X_1 \sim \text{Unif}(-3,2)$, $X_2 \sim \text{Unif}(0,6)$ and $X_3 \sim \text{Unif}(8,9)$ (and they are independent).

2.4. ★ Let $S = X/\sqrt{3}$, where $X \sim t(3)$. Compute the density of S via transformation and also by analytically inverting its c.f., which is $\varphi_S(t) = (1+t)e^{-t}$.

2.5. ★ Derive (2.10).

2.6. ★ Derive (2.11) both analytically and using a graph similar to those in Figure 2.3. Use this to compute $f_{R'}$ and $\mathbb{E}[R']$ for the density $f_{X,Y}$ in Example 1.11. Then compute the density of X/Y and its expected value and variance.

2.7. ★ ★ Derive (2.24).

2.8. Recall Example 2.16 on the normal–Laplace distribution, i.e., the convolution of a normal and a Laplace random variable.

(a) ★ Show why $\overline{\Phi}(x) = 1 - \Phi(x)$ needs to be replaced by $\Phi(-x)$ for large x.

(b) ★ Examine what happens to the terms in (2.20) as $c \to 1$. As a particular case, take $c = 0.975$ and $x = 1$, for which (2.20) can no longer be used for calculation with software using double precision arithmetic.

(c) ★ One way to avoid this numerical problem is to numerically integrate the expressions in (2.18) and (2.19). Construct a program to do this as efficiently as possible.

(d) ★ ★ Derive (2.21) and (2.22).

2.9. ★ Let $X_i \overset{\text{i.i.d.}}{\sim} \text{Par I}(b, x_0)$ where $f_{\text{Par I}}(x; b, x_0) = b x_0^b x^{-(b+1)} \mathbb{I}_{[x_0, \infty)}(x)$ and $S_n = \sum_{i=1}^{n} \ln X_i$.

(a) Derive the distribution of $S_1 = \ln X_1$.

(b) Using (2.9), derive $f_Z(z)$, where $Z = X_1 X_2$.

(c) From this, derive $f_{S_2}(s)$, where $S_2 = \ln Z$.

(d) Derive the distribution of S_3 and guess that of S_n.

(e) Finally, derive the distribution of $Z_n = \prod_{i=1}^{n} X_i$.

2.10. ★ Let $X, Y \overset{\text{i.i.d.}}{\sim} \text{Unif}(0, 1)$.

(a) Compute $f_Z(z)$, where $Z := X - Y$.

(b) Compute $f_W(w)$, where $W := |Z| = |X - Y|$. To which family does $f_W(w)$ belong?

(c) Compute $f_A(a)$, where $A = (X + Y)/2$.

2.11. ★ ★ Let $X_i \overset{\text{i.i.d.}}{\sim} \text{Unif}(0, 1)$ and $S_n = \sum_{i=1}^{n} X_i$. The p.d.f. of S_n is given by

$$f_{S_n}(s) = \sum_{k=0}^{n} \binom{n}{k} (-1)^k \frac{(s-k)^{n-1} \mathbb{I}(s \geq k)}{(n-1)!}, \tag{2.39}$$

which is most easily determined using a Laplace transform (see Feller, 1971, p. 436; Resnick, 1992; Kao, 1996, pp. 33–34; Lange, 2003, pp. 33–34).[3] For $n = 1$, this reduces to

$$f_{S_1}(s) = \mathbb{I}(s \geq 0) - \mathbb{I}(s \geq 1) = \mathbb{I}(0 \leq s < 1),$$

---

[3] Note that Feller's equation (3.2) contains a mistake. And yes, the relevant page numbers from Kao (1996) and Lange (2003) do coincide!

while for $n = 2$,

$$f_{S_2}(s) = \binom{2}{0}(-1)^0 (s-0) \, \mathbb{I}(s \geq 0) + \binom{2}{1}(-1)^1 (s-1) \, \mathbb{I}(s \geq 1)$$

$$+ \binom{2}{2}(-1)^2 (s-2) \, \mathbb{I}(s \geq 2)$$

$$= s\mathbb{I}(s \geq 0) - 2(s-1)\mathbb{I}(s \geq 1) + (s-2)\mathbb{I}(s \geq 2)$$

$$= \begin{cases} s, & \text{if } 0 \leq s < 1, \\ s - 2(s-1), & \text{if } 1 \leq x < 2, \\ s - 2(s-1) + (s-2), & \text{if } \quad s = 2. \end{cases}$$

$$= \begin{cases} s, & \text{if } 0 \leq s < 1, \\ 2 - s, & \text{if } 1 \leq x < 2, \end{cases}$$

which is a kind of *symmetric triangular distribution*; see Problem 2.12 below.

(a) Show that (2.39) for $n = 3$ reduces to

$$f_{S_3}(s) = \frac{1}{2} s^2 \mathbb{I}_{(0,1)}(s) + \left(3s - s^2 - \frac{3}{2}\right) \mathbb{I}_{[1,2)}(s) + \frac{(s-3)^2}{2} \mathbb{I}_{[2,3)}(s).$$

$$\text{(2.40)}$$

(b) Derive (2.40) directly by convoluting f_{S_2} and f_{X_3}.

(c) Compute f_A, where $A = S_3/3$.

(d) Program (2.39) and, for several values of n, plot the location–scale transform of S_n such that it has mean zero and variance one. Compare the density (via overlaid plots and the RPE) to the standard normal density. It appears that, for $n \geq 12$, they coincide considerably. Indeed, the standardized sum of independent uniform r.v.s was used in the 'old days' as a cheap way of generating normal r.v.s. Explain why, for any n, the approximation will break down in the tails.

2.12. ★ ★ Consider the so-called *symmetric triangular distribution*, which we will denote as Symtri (a, b), and is given by the isosceles triangle with base on the x axis extending from a to b, with midpoint at $c = (a + b)/2$.

(a) What is h, the value of $f_X(x)$ at the mode of the distribution?

(b) Show that

$$f_X(x) = \begin{cases} h^2(x-a), & \text{if } a \leq x \leq c, \\ h - h^2(x-c), & \text{if } c \leq x \leq b. \end{cases}$$

(c) Derive $F_X(x)$, the c.d.f. of X in terms of x, a and b. Hint: Perhaps one integral is enough.

(d) Write the density $f_Y(y)$ using only one expression, where $Y \sim$ Symtri $(-d, d)$.

(e) Calculate $\mathbb{V}(Y)$.

(f) Calculate $\mathbb{V}(X)$, where $X \sim \text{Symtri}(a, b)$.

(g) Let $U_i \overset{\text{i.i.d.}}{\sim} \text{Unif}(k, k + a)$, $i = 1, 2$, and $Y \sim \text{Symtri}(-a, a)$, $k \in \mathbb{R}$, $a \in \mathbb{R}_{>0}$. Show that $\mathbb{M}_Y(s) = \mathbb{M}_{(U_1 - U_2)}(s)$, where $\mathbb{M}_X(s)$ is the m.g.f. of random variable X, i.e., the difference between two i.i.d. uniform r.v.s is symmetric triangular distributed. Hint: Recall that $\mathbb{M}_{(-Y)}(s) = \mathbb{M}_Y(-s)$.

(h) If $X_i \overset{\text{i.i.d.}}{\sim} \text{Symtri}(a, b)$, $i = 1, \ldots, n$, and $Y = \max(X_i)$, give the formula for $\Pr(Y < y)$ and evaluate for $\Pr\left(Y < \frac{a+b}{2}\right)$. Hint: See equation (6.4) for the c.d.f. of the maximum of independent random variables.

2.13. ★ Let $S = \sum_{i=1}^{N} X_i$ where N is a geometric random variable with parameter p, independent of $X_i \overset{\text{i.i.d.}}{\sim} \exp(\lambda)$.

(a) Derive the m.g.f., density and expected value of S.

(b) Derive $\mathbb{E}[S]$ directly using iterated expectations.

2.14. ★ Let $f_{X,Y}(x, y) = f_{X,Y}(x, y; a, b) = cx^a y^b \mathbb{I}_{(0,1)}(x) \mathbb{I}_{(0,x)}(y)$ for known a, $b > 0$.

(a) Calculate c.

(b) Compute the marginals f_X and f_Y and their expectations.

(c) Compute the conditionals $f_{X|Y}$ and $f_{Y|X}$, their expectations and the iterated expectations.

(d) Compute the density of $S = X + Y$ and specialize to the cases $a = b = 1$ and $a = 2$, $b = 3$.

2.15. ★ Assume $Y_1, \ldots, Y_n \overset{\text{i.i.d.}}{\sim} (\mu, \sigma_Y^2)$ independent of $Z \sim (0, \sigma_Z^2)$ and let $X_i = Y_i + Z$, $i = 1, \ldots, n$.

(a) Compare $\mathbb{E}[\overline{X}_n]$ with $\mathbb{E}[X_1]$.

(b) Compare $\mathbb{V}(\overline{X}_n)$ with $\mathbb{V}(X_1)/n$.

(c) Compare $\mathbb{E}[S_n^2(X)]$ with $\mathbb{V}(X_1)$.

2.16. From the joint density

$$f_{X,Y}(x, y) = \frac{1}{\Gamma(a) b^{a+1}} x^{a-1} \exp\left(-\frac{x+y}{b}\right) \mathbb{I}_{(0,\infty)}(x) \mathbb{I}_{(0,\infty)}(y)$$

with parameters $a, b \in \mathbb{R}_{>0}$, show that the p.d.f. of $R = X/Y$ is

$$f_R(r) = \frac{ar^{a-1}}{(r+1)^{a+1}} \mathbb{I}_{(0,\infty)}(r).$$

(Rohatgi, 1984, p. 472 (14))

Some properties of this density were examined in Example I.7.2.

2.17. Let S^2 be the sample variance of an i.i.d. normal sample. Compare the exact value of $\mathbb{E}\left[S^{-2}\right]$ to the approximation of it based on (2.34).

2.18. ★ Let $(X, Y)' \sim \mathrm{N}_2(\mathbf{0}, \Sigma)$, $\Sigma = \begin{bmatrix} 1 & \rho \\ \rho & 1 \end{bmatrix}$, define $U = X$, $V = Y/(X^2/1)^{1/2} = Y/|X|$. If $\rho = 0$, then, from Example 2.15, $V \sim t_1$. Derive the p.d.f. of V for $0 \le \rho < 1$, and show that $V \sim t_1$ iff $\rho = 0$, i.e., that it is also necessary that, in the ratio which constructs a Student's t r.v., the r.v.s in the numerator and denominator (Z and C, respectively, in Example 2.15) are independent in the case of one degree of freedom. As mentioned in Chen and Adatia (1997), it can be shown that this is true in the general case of n degrees of freedom, $n \in \mathbb{N}$.

2.19. ★ The digamma function is given by

$$\psi(s) = \frac{\mathrm{d}}{\mathrm{d}s} \ln \Gamma(s) = \int_0^\infty \left[\frac{e^{-t}}{t} - \frac{e^{-st}}{1 - e^{-t}} \right] \mathrm{d}t \qquad (2.41)$$

(Andrews, Askey and Roy, 1999, p. 26) with higher-order derivatives denoted as

$$\psi^{(n)}(s) = \frac{\mathrm{d}^n}{\mathrm{d}s^n} \psi(s) = \frac{\mathrm{d}^{n+1}}{\mathrm{d}s^{n+1}} \ln \Gamma(s) = (-1)^{n+1} \int_0^\infty \frac{t^n e^{-st}}{1 - e^{-t}} \mathrm{d}t,$$

$n = 1, 2, \ldots$, also known as the *polygamma function*. Numeric methods exist for their evaluation; see Abramowitz and Stegun (1972, Section 6.3). Matlab version 7 includes a function, `psi`, to compute $\psi^{(n)}(s)$, and to call the Maple engine from Matlab version 7 (via the Symbolic Math Toolbox), use `mfun('Psi',n,x)`.

A result of general interest (and required when we compute the variance of a Gumbel r.v.) is that

$$\psi'(1) = \int_0^\infty \frac{t e^{-t}}{1 - e^{-t}} \mathrm{d}t = \frac{\pi^2}{6}; \qquad (2.42)$$

see Andrews, Askey and Roy (1999, p. 51 and 55) for proof.

As convenient numeric methods were not always available for the computation of the polygamma function, it is of interest to know how one could compute them. We need only consider $s \in [1, 2]$, because, for s outside $[1, 2]$, the recursion (Abramowitz and Stegun, 1972, eqs 6.3.5 and 6.4.6)

$$\psi^{(n)}(s + 1) = \psi^{(n)}(s) + (-1)^n n! s^{-(n+1)} \qquad (2.43)$$

can be used.

This method, when applicable, is also of value for computing functions for which canned numeric procedures are not available. We consider truncating the expansions

$$\psi(s) = -\gamma + (s-1) \sum_{k=1}^{\infty} \frac{1}{k(k+s-1)}, \quad s \neq 0, -1, -2, \ldots \quad (2.44)$$

$$\psi^{(n)}(s) = (-1)^{n+1} n! \sum_{k=0}^{\infty} (s+k)^{-(n+1)}, \quad s \neq 0, -1, -2, \ldots, \quad n = 1, 2, \ldots$$

$$(2.45)$$

(Abramowitz and Stegun, 1972, eqs 6.3.16 and 6.4.10), where γ is Euler's constant, and then approximating the tail sum by its continuity corrected integral, e.g., for (2.45),

$$\psi^{(n)}(s) \simeq (-1)^{n+1} n! \left[\sum_{k=0}^{N_n} (s+k)^{-n-1} + \int_{N_n+\frac{1}{2}}^{\infty} \frac{dt}{(s+t)^{n+1}} \right].$$

This yields, for $1 \leq s \leq 2$,

$$\psi^{(n)}(s) \approx (-1)^{n+1} \left[n! \sum_{k=0}^{N_n} (s+k)^{-n-1} + \frac{1}{n} \left(s + N_n + \tfrac{1}{2} \right)^{-n} \right], \quad n \geq 1.$$

$$(2.46)$$

For (2.44), we get

$$\psi(s) \approx -\gamma + (s-1) \sum_{k=1}^{N_0} k^{-1} (k+s-1)^{-1} + \ln \left| \frac{N_0 + s - 0.5}{N_0 + 0.5} \right|. \quad (2.47)$$

Derive (2.46) and (2.47).

3

The multivariate normal distribution

Generally, a meaningful multivariate distribution is produced by extending some special features of a univariate probabilistic model. However, there are several potential multivariate extensions of a given univariate distribution, due to the impossibility of obtaining a standard set of criteria that can always be applied to produce a unique distribution which could unequivocally be called *the* multivariate version.

The well-known exception is the multivariate normal.

(Papageorgiou, 1997)

This chapter serves to introduce one of the most important multivariate distributions. Section 3.1 is concerned with the first two moments of vectors of random variables in general. Then, Section 3.2 introduces the multivariate normal and states its main properties without proof. These are developed in subsequent sections, along with other useful aspects of the multivariate normal. Section 3.8 is a review of matrix algebra.

3.1 Vector expectation and variance

Let $\mathbf{X} = (X_1, \ldots, X_n)'$ be an r.v. such that $\mathbb{E}[X_i] = \mu_i$, $\mathbb{V}(X_i) = \sigma_i^2$ and $\mathrm{Cov}(X_i, X_j) = \sigma_{ij}$. Then, for function $g : \mathbb{R}^n \to \mathbb{R}^m$ expressible as $\mathbf{g}(\mathbf{X}) = (g_1(\mathbf{X}), \ldots, g_m(\mathbf{X}))'$, the expected value of vector $\mathbf{g}(\mathbf{X})$ is defined by

$$\mathbb{E}[\mathbf{g}(\mathbf{X})] = (\mathbb{E}[g_1(\mathbf{X})], \ldots, \mathbb{E}[g_m(\mathbf{X})])',$$

Intermediate Probability: A Computational Approach M. Paolella
© 2007 John Wiley & Sons, Ltd

with the most prominent case being the elementwise identity function, i.e.,

$$\mathbf{g}(\mathbf{X}) = (g_1(\mathbf{X}), \ldots, g_n(\mathbf{X}))' = (X_1, \ldots, X_n)' = \mathbf{X},$$

so that

$$\boxed{\mathbb{E}[\mathbf{X}] := \mathbb{E}\left[(X_1, \ldots, X_n)'\right] = (\mu_1, \ldots, \mu_n)',} \tag{3.1}$$

usually denoted $\boldsymbol{\mu}_\mathbf{X}$ or just $\boldsymbol{\mu}$. Although one could analogously define the vector of variances as, say, $\mathbf{d}(\mathbf{X}) = \left(\sigma_1^2, \ldots, \sigma_n^2\right)$, it is more common to denote by $\mathbb{V}(\mathbf{X})$ the matrix of covariances,

$$\mathbb{V}(\mathbf{X}) := \mathbb{E}\left[(\mathbf{X} - \boldsymbol{\mu}_\mathbf{X})(\mathbf{X} - \boldsymbol{\mu}_\mathbf{X})'\right] = \begin{bmatrix} \sigma_1^2 & \sigma_{12} & \cdots & \sigma_{1n} \\ \sigma_{21} & \sigma_2^2 & \cdots & \sigma_{2n} \\ \vdots & \vdots & \ddots & \vdots \\ \sigma_{n1} & \sigma_{n2} & \cdots & \sigma_n^2 \end{bmatrix}, \tag{3.2}$$

which is symmetric and often designated by $\boldsymbol{\Sigma}_\mathbf{X}$ or just $\boldsymbol{\Sigma}$. Note that a particular element of $\boldsymbol{\Sigma}$ is given by

$$\sigma_{ij} = \mathbb{E}\left[(X_i - \mu_i)(X_j - \mu_j)\right]. \tag{3.3}$$

⊖ ***Example 3.1*** In Example I.6.7, the mean, variance, and covariance corresponding to the elements of a multinomial r.v., $\mathbf{X} = (X_1, \ldots, X_k) \sim \text{Multinom}(n, \mathbf{p}), \mathbf{p} = (p_1, \ldots, p_k)$, were shown to be $\mathbb{E}[X_i] = np_i$, $\mathbb{V}(X_i) = np_i q_i$, and $\text{Cov}(X_i, X_j) = -np_i p_j$, where $q_i = 1 - p_i$. Thus, $\mathbb{E}[\mathbf{X}] = n\mathbf{p}'$ and

$$\mathbb{V}(\mathbf{X}) = n \begin{bmatrix} p_1 q_1 & -p_1 p_2 & \cdots & -p_1 p_k \\ -p_2 p_1 & p_2 q_2 & \cdots & -p_2 p_k \\ \vdots & \vdots & \ddots & \vdots \\ -p_k p_1 & -p_k p_2 & \cdots & p_k q_k \end{bmatrix},$$

from (3.1) and (3.2), respectively. ∎

From (3.2), the vector of variances $\mathbf{d}$ is seen to be $\text{diag}(\boldsymbol{\Sigma})$. For any matrix $\mathbf{A} \in \mathbb{R}^{m \times n}$ and vector $\mathbf{b} \in \mathbb{R}^m$, it follows from the properties of univariate expected value that $\mathbb{E}[\mathbf{AX} + \mathbf{b}] = \mathbf{A}\boldsymbol{\mu}_\mathbf{X} + \mathbf{b}$, while

$$\begin{aligned} \mathbb{V}(\mathbf{AX} + \mathbf{b}) &= \mathbb{E}\left[((\mathbf{AX} + \mathbf{b}) - (\mathbf{A}\boldsymbol{\mu}_\mathbf{X} + \mathbf{b}))((\mathbf{AX} + \mathbf{b}) - (\mathbf{A}\boldsymbol{\mu}_\mathbf{X} + \mathbf{b}))'\right] \\ &= \mathbb{E}\left[(\mathbf{A}(\mathbf{X} - \boldsymbol{\mu}_\mathbf{X}))(\mathbf{A}(\mathbf{X} - \boldsymbol{\mu}_\mathbf{X}))'\right] \\ &= \mathbf{A}\mathbb{V}(\mathbf{X})\mathbf{A}', \end{aligned}$$

i.e.,

$$\mathbb{V}\left(\mathbf{AX} + \mathbf{b}\right) = \mathbf{A}\Sigma\mathbf{A}'. \tag{3.4}$$

An important special case is $m = 1$; if $\mathbf{a} = (a_1, a_2, \ldots, a_n)' \in \mathbb{R}^n$ is a column vector, then

$$\mathbb{V}\left(\mathbf{a}'\mathbf{X}\right) = \mathbf{a}'\Sigma\mathbf{a} = \sum_{i=1}^{n} a_i^2 \mathbb{V}\left(X_i\right) + \sum\sum_{i \neq j} a_i a_j \operatorname{Cov}\left(X_i, X_j\right), \tag{3.5}$$

as given in (I.6.4). From (3.5), we easily verify that $\Sigma = \mathbb{V}\left(\mathbf{X}\right)$ is positive semi-definite. Let $Y_{\mathbf{a}} = \mathbf{a}'\mathbf{X}$ for any vector $\mathbf{a} \in \mathbb{R}^n$ and note that

$$0 \leq \mathbb{V}\left(Y_{\mathbf{a}}\right) = \mathbb{V}\left(\mathbf{a}'\mathbf{X}\right) = \mathbf{a}'\mathbb{V}\left(\mathbf{X}\right)\mathbf{a}.$$

If $0 < \mathbb{V}\left(Y_{\mathbf{a}}\right)$ for all nonzero $\mathbf{a}$, i.e., $\mathbf{a} \in \mathbb{R}^n \setminus \mathbf{0}$, then $\Sigma > 0$, i.e., is positive definite.

As in the univariate case, it also makes sense to speak of the covariance of two r.v.s $\mathbf{X} = (X_1, \ldots, X_n)'$ and $\mathbf{Y} = (Y_1, \ldots, Y_m)'$, given by

$$\Sigma_{\mathbf{X},\mathbf{Y}} := \operatorname{Cov}\left(\mathbf{X}, \mathbf{Y}\right) := \mathbb{E}\left[(\mathbf{X} - \boldsymbol{\mu}_{\mathbf{X}})(\mathbf{Y} - \boldsymbol{\mu}_{\mathbf{Y}})'\right] = \begin{bmatrix} \sigma_{X_1, Y_1} & \sigma_{X_1, Y_2} & \cdots & \sigma_{X_1, Y_m} \\ \sigma_{X_2, Y_1} & \sigma_{X_2, Y_2} & \cdots & \sigma_{X_2, Y_m} \\ \vdots & \vdots & \ddots & \vdots \\ \sigma_{X_n, Y_1} & \sigma_{X_n, Y_2} & \cdots & \sigma_{X_n, Y_m} \end{bmatrix},$$

an $n \times m$ matrix with (i, j)th element $\sigma_{X_i, Y_j} = \operatorname{Cov}\left(X_i, Y_j\right)$. From symmetry, Cov $\left(\mathbf{X}, \mathbf{Y}\right) = \operatorname{Cov}\left(\mathbf{Y}, \mathbf{X}\right)'$. More generally,

$$\operatorname{Cov}\left(\mathbf{AX}, \mathbf{BY}\right) = \mathbb{E}\left[\mathbf{A}\left(\mathbf{X} - \boldsymbol{\mu}_{\mathbf{X}}\right)\left(\mathbf{Y} - \boldsymbol{\mu}_{\mathbf{Y}}\right)'\mathbf{B}'\right] = \mathbf{A}\Sigma_{\mathbf{X},\mathbf{Y}}\mathbf{B}',$$

with important special case

$$\operatorname{Cov}\left(\mathbf{a}'\mathbf{X}, \mathbf{b}'\mathbf{Y}\right) = \operatorname{Cov}\left(\sum_{i=1}^{n} a_i X_i, \sum_{j=1}^{m} b_j Y_j\right) = \mathbf{a}'\Sigma_{\mathbf{X},\mathbf{Y}}\mathbf{b} = \sum_{i=1}^{n}\sum_{j=1}^{m} a_i b_j \operatorname{Cov}\left(X_i, Y_j\right) \tag{3.6}$$

for vectors $\mathbf{a} = (a_1, a_2, \ldots, a_n) \in \mathbb{R}^n$ and $\mathbf{b} = (b_1, b_2, \ldots, b_m) \in \mathbb{R}^m$, as in (I.6.5). One could also express the variance of the weighted sum of multivariate r.v.s, $\mathbb{V}\left(\sum_{i=1}^{k} \mathbf{A}_i \mathbf{X}_i\right)$, in terms of weighted covariances of the $\mathbf{X}_i$. For two r.v.s, this is

$$\mathbb{V}\left(\mathbf{AX} + \mathbf{BY} + \mathbf{b}\right) = \mathbf{A}\mathbb{V}\left(\mathbf{X}\right)\mathbf{A}' + \mathbf{B}\mathbb{V}\left(\mathbf{Y}\right)\mathbf{B}' + \mathbf{A}\operatorname{Cov}\left(\mathbf{X}, \mathbf{Y}\right)\mathbf{B}' + \mathbf{B}\operatorname{Cov}\left(\mathbf{Y}, \mathbf{X}\right)\mathbf{A}',$$

assuming, of course, that the sizes of $\mathbf{AX}$, $\mathbf{BY}$ and $\mathbf{b}$ are the same.

3.2   Basic properties of the multivariate normal

The joint density of vector $\mathbf{Z} = (Z_1, \ldots Z_n)'$, where $Z_i \overset{\text{i.i.d.}}{\sim} N(0, 1)$ is

$$f_{\mathbf{Z}}(z) = \prod_{i=1}^{n} f_{Z_i}(z_i) = (2\pi)^{-n/2} \exp\left\{-\frac{1}{2} \sum_{i=1}^{n} z_i^2\right\} = (2\pi)^{-n/2} e^{-\mathbf{z}'\mathbf{z}/2}, \qquad (3.7)$$

and is referred to as the *standard (n-variate) multivariate normal* and denoted $\mathbf{Z} \sim N(\mathbf{0}, \mathbf{I}_n)$. As the components of $\mathbf{Z}$ are i.i.d., it follows that $\mathbb{E}[\mathbf{Z}] = \mathbf{0}$ and, recalling (3.3) and (I.5.20), $\mathbb{V}(\mathbf{Z}) = \mathbf{I}_n$. If instead, $X_1, \ldots, X_n \overset{\text{ind}}{\sim} N(\mu_i, \sigma_i^2)$, then their joint density is

$$f_{\mathbf{X}}(\mathbf{x}) = \prod_{i=1}^{n} f_N(x_i; \mu_i, \sigma_i^2) = \prod_{i=1}^{n} \frac{1}{\sigma_i \sqrt{2\pi}} \exp\left\{-\frac{1}{2}\left(\frac{x_i - \mu_i}{\sigma_i}\right)^2\right\}$$

$$= \frac{1}{\sqrt{(2\pi)^n \prod_{i=1}^{n} \sigma_i^2}} \exp\left\{-\frac{1}{2} \sum_{i=1}^{n}\left(\frac{x_i - \mu_i}{\sigma_i}\right)^2\right\}. \qquad (3.8)$$

⊖ ***Example 3.2***   (Example I.8.3 cont.) Recalling Example I.8.3, we see that normality of the two marginals does not imply joint normality. In fact, as in Pierce and Dykstra (1969), extending that density to the n-variate case as

$$f_{\mathbf{X}}(\mathbf{x}) = \frac{1}{(2\pi)^{n/2}} \exp\left\{-\frac{1}{2}\sum_{i=1}^{n} x_i^2\right\}\left[1 + \prod_{i=1}^{n}\left(x_i \exp\left\{-\frac{1}{2}x_i^2\right\}\right)\right],$$

a similar calculation shows that the joint density of the $n - 1$ r.v.s $\mathbf{Y}_j = \{\mathbf{X} \setminus X_j\}$ is $f_{\mathbf{Y}_j}(\mathbf{y}_j) = \int_{-\infty}^{\infty} f_{\mathbf{X}}(\mathbf{x})\, dx_j = I_1 + I_2$, where $I_2 = 0$ and

$$I_1 = \frac{1}{(2\pi)^{(n-1)/2}} \exp\left\{-\frac{1}{2}\sum_{i \neq j}^{n} x_i^2\right\},$$

showing that $\mathbf{Y}_j \sim N_{n-1}(\mathbf{0}, \mathbf{I})$, the joint density of $n - 1$ independent $N(0, 1)$ r.v.s. This implies that *any* k-size subset of $\mathbf{X}$ has a $N_k(\mathbf{0}, \mathbf{I})$ distribution, i.e., *all* $2^n - 2$ marginals are normally distributed.   ∎

The joint density (3.8) is still restricted to having independent components. Far more generally, $\mathbf{Y}$ is an (n-variate) multivariate normal r.v. if its density is given by

$$f_{\mathbf{Y}}(\mathbf{y}) = \frac{1}{|\Sigma|^{1/2} (2\pi)^{n/2}} \exp\left\{-\frac{1}{2}\left((\mathbf{y} - \boldsymbol{\mu})' \Sigma^{-1} (\mathbf{y} - \boldsymbol{\mu})\right)\right\}, \qquad (3.9)$$

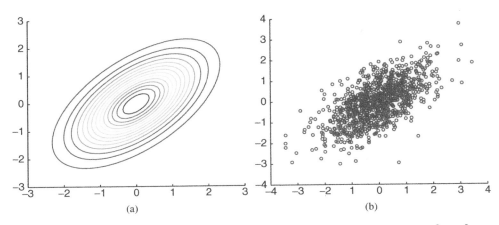

Figure 3.1 (a) Contour plot of the bivariate normal density (3.9) with $\mu_1 = \mu_2 = 0$, $\sigma_1^2 = \sigma_2^2 = 1$ and $\sigma_{12} = 0.6$. (b) Scatterplot of 1000 simulated values from that distribution

denoted $\mathbf{Y} \sim \mathrm{N}\,(\boldsymbol{\mu}, \boldsymbol{\Sigma})$, where $\boldsymbol{\mu} = (\mu_1, \ldots, \mu_n)' \in \mathbb{R}^n$ and $\boldsymbol{\Sigma} > 0$ with (i, j)th element σ_{ij}, $\sigma_i^2 := \sigma_{ii}$. If the number of components needs to be emphasized, we write $\mathbf{Y} \sim \mathrm{N}_n\,(\boldsymbol{\mu}, \boldsymbol{\Sigma})$. The form of (3.9) will be justified in Section 3.3 below.

⊖ **Example 3.3** The *standard bivariate normal* distribution has density (3.9), $n = 2$, $\mu_1 = \mu_2 = 0$, and $\sigma_1^2 = \sigma_2^2 = 1$, with σ_{12} such that $\boldsymbol{\Sigma}$ is positive definite, or $|\sigma_{12}| < 1$. Figure 3.1(a) illustrates (3.9) for $\sigma_{12} = 0.6$ with a contour plot of the density, while Figure 3.1(b) shows a scatterplot of 1000 simulated values of $\mathbf{Y}$ using the method of simulation discussed below in Section 3.4. The contour plot can be compared to Figure I.8.3, which shows a similar, but nonnormal, bivariate density.

Sungur (1990) derived the simple first-order approximation to the standard bivariate normal p.d.f.

$$f_{X,Y}(x, y) \approx \phi(x)\phi(y)(1 + \sigma_{12}xy), \qquad (3.10)$$

where ϕ is the univariate standard normal p.d.f. It is clearly exact when X and Y are independent ($\sigma_{12} = 0$), and worsens as the absolute correlation between X and Y increases. Figure 3.2(a) shows approximation (3.10) for $\sigma_{12} = 0.6$, which can be compared with the exact density in Figure 3.1. Figure 3.2(b) uses $\sigma_{12} = -0.2$, and shows that the approximation is reasonable for small $|\sigma_{12}|$. ∎

Important facts involving the multivariate normal distribution (3.9) are:

1. The first two moments are

$$\boxed{\mathbb{E}\,[\mathbf{Y}] = \boldsymbol{\mu} \quad \text{and} \quad \mathbb{V}\,(\mathbf{Y}) = \boldsymbol{\Sigma}}$$

and parameters $\boldsymbol{\mu}$ and $\boldsymbol{\Sigma}$ completely determine, or characterize, the distribution. Thus, if $\mathbf{X}$ and $\mathbf{Y}$ are both multivariate normal with the same mean and variance, then they have the same distribution.

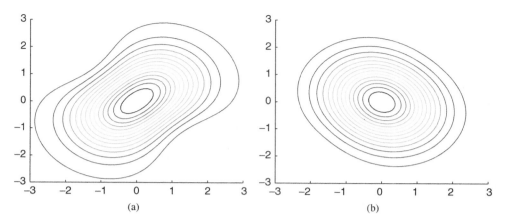

Figure 3.2 Contour plot of approximation (3.10) to the standard bivariate normal density for (a) $\sigma_{12} = 0.6$ and (b) $\sigma_{12} = -0.2$

2. All $2^n - 2$ marginals are normally distributed with mean and variance given appropriately from $\boldsymbol{\mu}$ and $\boldsymbol{\Sigma}$. For example, $Y_i \sim N\left(\mu_i, \sigma_i^2\right)$ and, for $i \neq j$,

$$
\begin{pmatrix} Y_i \\ Y_j \end{pmatrix} \sim N\left(\begin{bmatrix} \mu_i \\ \mu_j \end{bmatrix}, \begin{bmatrix} \sigma_i^2 & \sigma_{ij} \\ \sigma_{ij} & \sigma_j^2 \end{bmatrix} \right).
$$

3. An important special case is the bivariate normal,

$$
\begin{pmatrix} Y_1 \\ Y_2 \end{pmatrix} \sim N\left(\begin{bmatrix} \mu_1 \\ \mu_2 \end{bmatrix}, \begin{bmatrix} \sigma_1^2 & \rho\sigma_1\sigma_2 \\ \rho\sigma_1\sigma_2 & \sigma_2^2 \end{bmatrix} \right),
\tag{3.11}
$$

where, recalling (I.5.26), Corr $(Y_1, Y_2) = \rho$, as shown in Problem 3.9. The density in the bivariate case is not too unsightly without matrices:

$$
f_{Y_1, Y_2}(x, y) = K \exp\left\{ -\frac{\tilde{x}^2 - 2\rho\tilde{x}\tilde{y} + \tilde{y}^2}{2\left(1 - \rho^2\right)} \right\},
\tag{3.12}
$$

where

$$
K = \frac{1}{2\pi\sigma_1\sigma_2\left(1 - \rho^2\right)^{1/2}}, \qquad \tilde{x} = \frac{x - \mu_1}{\sigma_1}, \qquad \tilde{y} = \frac{y - \mu_2}{\sigma_2}.
$$

The marginal distributions are $Y_i \sim N\left(\mu_i, \sigma_i^2\right)$, $i = 1, 2$.

4. If Y_i and Y_j are jointly normally distributed, then they are independent iff Cov $\left(Y_i, Y_j\right) = 0$. (This will be proven below.) For the bivariate normal above, Y_1 and Y_2 are independent iff $\rho = 0$. For the general multivariate case, this extends to

nonoverlapping subsets, say $\mathbf{Y}_I$ and $\mathbf{Y}_J$, where I is an indexing set:[1]

$$\mathbf{Y}_I \perp \mathbf{Y}_J \quad \text{iff} \quad \text{Cov}\,(\mathbf{Y}_I, \mathbf{Y}_J) =: \Sigma_{IJ} = \mathbf{0}.$$

Remark: It is essential to emphasize that this result assumes that $\mathbf{Y}_{(i)}$ and $\mathbf{Y}_{(j)}$ are jointly normally distributed. It is **not** necessarily true that univariate normal r.v.s whose correlation is zero are independent. To take an example from Melnick and Tenenbein (1982), let $X \sim \mathrm{N}(0, 1)$ and set $Y = X$ if $|X| \le c$, and $Y = -X$ otherwise, where c is such that $\mathbb{E}[XY] = 0$. This leads to the equation

$$\Phi(c) = \frac{3}{4} + c\phi(c),$$

the solution of which is 1.53817225445505 to machine precision found using the Matlab code

```
a=optimset('tolx',1e-12); format long
fzero(inline('normcdf(c)-0.75 - c*normpdf(c)'),[1.5 1.6],a)
```

From symmetry, $Y \sim \mathrm{N}(0, 1)$. Clearly, X and Y are not jointly normally distributed, but they *are* uncorrelated! Figure 3.3 shows a scatterplot of 200 simulated (X, Y) points, confirming their massive nonlinear dependence structure, while simulating 1 million pairs and computing their sample covariance yields a value under $1/1000$. Other examples of uncorrelated marginally normal r.v.s such that the joint distribution is not bivariate normal are given in Kowalski (1973). ∎

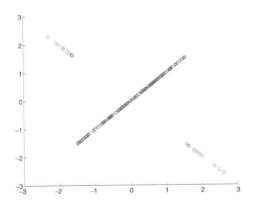

Figure 3.3 Scatterplot of marginally normal but not bivariate normal r.v.s

5. For nonoverlapping subsets $\mathbf{Y}_{(i)}$ and $\mathbf{Y}_{(j)}$ of $\mathbf{Y}$, the conditional distribution of $\mathbf{Y}_{(i)} \mid \mathbf{Y}_{(j)}$ is also normally distributed. The general case is given in (3.22) below;

---

[1] That is, $I = \{i_1, \ldots, i_k\}$, $1 \le k < n$, $i_j \in \mathcal{N}_n := \{1, 2, \ldots, n\}$ for $j = 1, \ldots, k$, such that $i_\ell \ne i_m$ for $\ell \ne m$, and $J = \mathcal{N}_n \setminus I$.

in the bivariate normal case, the conditionals are

$$
\begin{aligned}
Y_1 \mid Y_2 &\sim N\left(\mu_1 + \rho\sigma_1\sigma_2^{-1}(y_2 - \mu_2),\ \sigma_1^2(1-\rho^2)\right), \\
Y_2 \mid Y_1 &\sim N\left(\mu_2 + \rho\sigma_2\sigma_1^{-1}(y_1 - \mu_1),\ \sigma_2^2(1-\rho^2)\right).
\end{aligned}
\tag{3.13}
$$

6. Let $\mathbf{a} \in \mathbb{R}^n \setminus \mathbf{0}$ be column vector. The linear combination $L = \mathbf{a}'\mathbf{Y} = \sum_{i=1}^n a_i Y_i$ is normally distributed with mean $\mathbb{E}[L] = \sum_{i=1}^n a_i \mu_i = \mathbf{a}'\boldsymbol{\mu}$ and variance $\mathbb{V}(L)$ from (3.5). More generally, with $\mathbf{A} \in \mathbb{R}^{m \times n}$ a full rank matrix with $m \leq n$, the set of linear combinations

$$
\mathbf{L} = (L_1, \ldots, L_m)' = \mathbf{AY} \sim N\left(\mathbf{A}\boldsymbol{\mu},\ \mathbf{A}\Sigma\mathbf{A}'\right),
\tag{3.14}
$$

using (3.4).

⊛ ***Example 3.4***　Let $X_i \overset{\text{i.i.d.}}{\sim} N(0,1)$, $i = 1, 2$, and consider the joint density of their sum and difference, $\mathbf{Y} = (S, D)$, where $S = X_1 + X_2$ and $D = X_1 - X_2$. Let $\mathbf{A} = \left(\begin{smallmatrix} 1 & 1 \\ 1 & -1 \end{smallmatrix}\right)$ and $\mathbf{X} = (X_1, X_2) \sim N(\mathbf{0}, \mathbf{I}_2)$, so that $\mathbf{Y} = \mathbf{AX}$. From property 6 above, it follows that $\mathbf{Y} \sim N\left(\mathbf{A0},\ \mathbf{AIA}'\right)$ or $\mathbf{Y} \sim N(\mathbf{0}, 2\mathbf{I}_2)$, i.e., $S \sim N(0,2)$, $D \sim N(0,2)$ and S and D are independent. ∎

Remark: There is actually more to the previous example than meets the eye. It can be shown that the converse holds, namely that, if X_1 and X_2 are independent r.v.s such that their sum and difference are independent, then X_1 and X_2 are normally distributed. The result characterizes the normal distribution and was first shown by Sergei Natanovich Bernstein in 1941. See Seneta (1982) for a biography of Bernstein and a list of his publications. The theorem is also detailed in Bryc (1995, Section 5.1), where other characterizations of the normal distribution are presented. See also Section 3.7 and Kotz, Balakrishnan and Johnson (2000, Sections 45.7 and 46.5) for further detail on characterization of the normal distribution. ∎

⊖ ***Example 3.5***　The bivariate density

$$
f_{X,Y}(x,y) = C \exp\left\{-\left[x^2 + y^2 + 2xy(x + y + xy)\right]\right\},
\tag{3.15}
$$

given by Castillo and Galambos (1989) has the interesting (but not unique) property that both conditional distributions are normal but X, Y are obviously not jointly normal. The code in Listing 3.1 produces the plots in Figure 3.4. ∎

⊛ ***Example 3.6***　Let X and Y be bivariate normal with $\mu_1 = \mu_2 = 0$, $\sigma_1 = \sigma_2 = 1$ and correlation coefficient $\rho = 0.5$. To evaluate $\Pr(X > Y + 1)$, observe that, from fact 6 above and (I.6.4), $X - Y \sim N(0 - 0, 1 + 1 - 2(0.5)) = N(0,1)$, so that $\Pr(X > Y + 1) = 1 - \Phi(1) \approx 0.16$. To evaluate $\text{Corr}(X - Y + 1, X + Y - 2)$, note that

$$
\text{Cov}(X - Y + 1, X + Y - 2) = \mathbb{E}[(X - Y)(X + Y)] = \mathbb{E}[X^2] - \mathbb{E}[Y^2] = 0
$$

```
xvec=-6:0.1:6; yvec=xvec;
lx=length(xvec); ly=length(yvec); f=zeros(lx,ly);
for xloop=1:lx,    x=xvec(xloop);
  for yloop=1:ly, y=yvec(yloop);
    f(xloop,yloop)=exp(-(x^2+y^2+2*x*y*(x+y+x*y)));
  end
end
contour(xvec,yvec,f), grid
x1=find(xvec==-1); x2=find(xvec==0); x3=find(xvec==1);
figure, plot(yvec,f(x1,:),'r-',yvec,f(x2,:),'g--',yvec,f(x3,:),'b-.')
```

Program Listing 3.1   Generates the plots in Figure 3.4

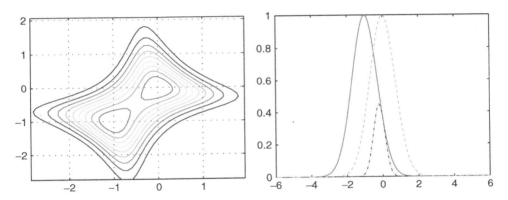

Figure 3.4   Contour plot of the bivariate density (3.15) (left) and three conditionals (right)

and thus the correlation is zero. From facts 4 and 6, we see that $X + Y$ and $X - Y$ are independent. See also Example 2.5 and Problem I.8.1.   ∎

Example 3.7   We wish to calculate the density of $X_1 \mid S$, where $X_i \overset{\text{i.i.d.}}{\sim} N\left(\mu, \sigma^2\right)$, $i = 1, \ldots, n$, and $S = \sum_{i=1}^{n} X_i$. Let $L_1 = X_1 = \mathbf{a}_1' \mathbf{X}$ and $L_2 = \sum_{i=1}^{n} X_i = \mathbf{a}_2' \mathbf{X}$ for $\mathbf{a}_1 = (1, 0, \ldots, 0)'$ and $\mathbf{a}_2 = (1, 1, \ldots, 1)'$, so that, from property 6,

$$\begin{bmatrix} L_1 \\ L_2 \end{bmatrix} \sim N\left(\begin{bmatrix} \mu \\ \sum_{i=1}^{n} \mu \end{bmatrix}, \begin{bmatrix} 1 & 0 & \cdots & 0 \\ 1 & 1 & \cdots & 1 \end{bmatrix} \Sigma \begin{bmatrix} 1 & 1 \\ 0 & 1 \\ \vdots & \vdots \\ 0 & 1 \end{bmatrix} \right)$$

$$= N\left(\begin{bmatrix} \mu \\ n\mu \end{bmatrix}, \begin{bmatrix} \sigma^2 & \sigma^2 \\ \sigma^2 & n\sigma^2 \end{bmatrix} \right),$$

with $\rho = n^{-1/2}$. From property 5, $(X_1 \mid S = s) \sim N\left(s/n, \; \sigma^2\left(1 - n^{-1}\right)\right)$.

A direct approach which uses only the fact that sums of normal are normal is as follows. We know $S \sim N(n\mu, n\sigma^2)$ and also that $S \mid X_1 = x \sim N(x + (n-1)\mu, (n-1)\sigma^2)$. Letting $\phi(x; \mu, \sigma^2)$ denote the $N(\mu, \sigma^2)$ density evaluated at x, Bayes' rule implies that $f_{X_1 \mid S}(x \mid s)$ is

$$\frac{f_{S \mid X_1}(s \mid x) f_{X_1}(x)}{f_S(s)} = \frac{\phi\left(s; x + (n-1)\mu, (n-1)\sigma^2\right) \phi\left(x; \mu, \sigma^2\right)}{\phi\left(s; n\mu, n\sigma^2\right)}$$

$$= \frac{\frac{1}{\sqrt{2\pi(n-1)\sigma^2}} \exp\left(-\frac{1}{2(n-1)\sigma^2}(s - (x + (n-1)\mu))^2\right) \frac{1}{\sqrt{2\pi\sigma^2}} \exp\left(-\frac{1}{2\sigma^2}(x - \mu)^2\right)}{\frac{1}{\sqrt{2\pi n\sigma^2}} \exp\left(-\frac{1}{2n\sigma^2}(s - n\mu)^2\right)}$$

$$= \frac{1}{\sqrt{2\pi v}} \exp\left(-\frac{1}{2v}\left((x - k)^2\right)\right),$$

after some simplification, where $v = (n-1)\sigma^2/n$ and $k = sn^{-1}$. Thus, $X_1 \mid S \sim N(k, v)$ as before. ∎

While the p.d.f. (3.9) is straightforward to evaluate numerically, the c.d.f. of the multivariate normal distribution is, in general, not, but simulation can be effectively used. This is discussed in Section 3.4 below.

3.3 Density and moment generating function

In addition to using the tools of vector m.g.f.s and their characterizing properties discussed earlier, this section will require some matrix concepts, such as rank, spectral decomposition and square root. These are briefly outlined in Section 3.8.

Let $\mathbf{Z} \sim N(\mathbf{0}, \mathbf{I}_n)$ so that $f_{\mathbf{Z}}(z) = (2\pi)^{-n/2} e^{-\mathbf{z}'\mathbf{z}/2}$. The m.g.f. of $\mathbf{Z}$ is

$$\mathbb{M}_{\mathbf{Z}}(\mathbf{t}) = \mathbb{E}\left[e^{\mathbf{t}'\mathbf{z}}\right] = (2\pi)^{-n/2} \int_{\mathbb{R}^n} \exp\left\{\mathbf{t}'\mathbf{z} - \mathbf{z}'\mathbf{z}/2\right\} d\mathbf{z}$$

$$= \prod_{i=1}^{n} \int_{-\infty}^{\infty} \frac{1}{\sqrt{2\pi}} \exp\left\{t_i z_i - \frac{1}{2} z_i^2\right\} dz_i = \prod_{i=1}^{n} \mathbb{M}_{Z_i}(t_i) = e^{\mathbf{t}'\mathbf{t}/2}.$$

Of interest are the m.g.f. and density of the generalized location–scale family $\mathbf{Y} = \mathbf{a} + \mathbf{BZ}$, $\mathbf{a} \in \mathbb{R}^m$, $\mathbf{B} \in \mathbb{R}^{m \times n}$, such that rank $(\mathbf{B}) = m \leq n$. Observe that $\boldsymbol{\mu} := \mathbb{E}[\mathbf{Y}] = \mathbf{a}$ and, from (3.4), $\boldsymbol{\Sigma} := \mathbb{V}(\mathbf{Y}) = \mathbf{BB}' \in \mathbb{R}^{m \times m}$. Furthermore,

$$\mathbb{M}_{\mathbf{Y}}(\mathbf{t}) = \mathbb{E}\left[e^{\mathbf{t}'\mathbf{Y}}\right] = \mathbb{E}\left[\exp\left\{\mathbf{t}'\mathbf{a} + \mathbf{t}'\mathbf{BZ}\right\}\right] = e^{\mathbf{t}'\boldsymbol{\mu}} \mathbb{M}_{\mathbf{Z}}(\mathbf{B}'\mathbf{t})$$

$$= \exp\left\{\mathbf{t}'\boldsymbol{\mu} + \mathbf{t}' \boldsymbol{\Sigma} \mathbf{t}/2\right\}. \tag{3.16}$$

To derive the density of $\mathbf{Y}$, first observe that $m = \text{rank}\,(\mathbf{B}) = \text{rank}\,(\mathbf{BB}') = \text{rank}\,(\Sigma)$, so that $\Sigma > 0$ and $\Sigma^{1/2} > 0$, where $\Sigma^{1/2}\,\Sigma^{1/2} = \Sigma$ (see Section 3.8 for details on the construction of $\Sigma^{1/2}$). Defining $\mathbf{L} = \Sigma^{-1/2}\,(\mathbf{Y} - \mu)$, we have, from (3.16),

$$\mathbb{M}_{\mathbf{L}}(\mathbf{t}) = \mathbb{E}\left[\exp\left\{\mathbf{t}'\,\Sigma^{-1/2}\,(\mathbf{Y} - \mu)\right\}\right] = \exp\left\{-\mathbf{t}'\,\Sigma^{-1/2}\mu\right\}\mathbb{M}_{\mathbf{Y}}(\Sigma^{-1/2}\,\mathbf{t})$$

$$= \exp\left\{-\mathbf{t}'\,\Sigma^{-1/2}\,\mu\right\}\exp\left\{\mathbf{t}'\,\Sigma^{-1/2}\,\mu + \tfrac{1}{2}\mathbf{t}'\,\Sigma^{-1/2}\,\Sigma\,\Sigma^{-1/2}\,\mathbf{t}\right\} = e^{\mathbf{t}'\mathbf{t}/2},$$

so that, from the characterization property of m.g.f.s, $\mathbf{L}$ is standard multivariate normal (of size m). As $\mathbf{Y} = \mu + \Sigma^{1/2}\mathbf{L}$ and $\partial\,\Sigma^{-1/2}\,(\mathbf{Y} - \mu)\,/\partial\mathbf{Y} = \Sigma^{-1/2}$, transforming and using a basic property of determinants yields

$$f_{\mathbf{Y}}\,(\mathbf{y}) = \left|\Sigma^{-1/2}\right| f_{\mathbf{L}}\left(\Sigma^{-1/2}\,(\mathbf{y} - \mu)\right) \tag{3.17}$$

$$= \frac{1}{\left|\Sigma^{1/2}\right|(2\pi)^{n/2}}\exp\left\{-\frac{1}{2}\,(\mathbf{y} - \mu)'\,\Sigma^{-1}\,(\mathbf{y} - \mu)\right\},$$

which is the multivariate normal density, as in (3.9), $\mathbf{Y} \sim \text{N}\,(\mu, \Sigma)$. From $\mathbb{M}_{\mathbf{Y}}$, it is clear that the distribution is completely determined by μ and Σ, i.e., its mean and variance–covariance matrix.

Furthermore, given any $\mu \in \mathbb{R}^n$ and any $n \times n$ positive definite matrix Σ, a random variable $\mathbf{Y}$ can be constructed such that $\mathbf{Y} \sim \text{N}\,(\mu, \Sigma)$. This is easily accomplished by letting $\mathbf{Z} \sim \text{N}\,(\mathbf{0}, \mathbf{I}_n)$ and taking $\mathbf{Y} = \mu + \Sigma^{1/2}\mathbf{Z}$. Then $\mathbb{E}\,[\mathbf{Y}] = \mu$ and, recalling (3.4), $\mathbb{V}\,(\mathbf{Y}) = \Sigma$. As the normal is characterized by its mean and variance, it follows that $\mathbf{Y} \sim \text{N}\,(\mu, \Sigma)$.

More generally, Problem 1.6 shows that, if $\mathbf{Y} \sim \text{N}\,(\mu, \Sigma)$, then, for appropriately sized $\mathbf{a}$ and $\mathbf{B}$,

$$\mathbf{X} = \mathbf{a} + \mathbf{BY} \sim \text{N}\,(\nu, \Omega), \quad \nu = \mathbf{a} + \mathbf{B}\mu, \quad \Omega = \mathbf{B}\,\Sigma\,\mathbf{B}'. \tag{3.18}$$

If $\mathbf{B}$ is $m \times n$ and either (i) $m \leq n$ and $\text{rank}\,(\mathbf{B}) < m$ or (ii) $m > n$, then Ω is only positive semi-definite with $\det\,(\Omega) = 0$ and, consequently, $\mathbf{X}$ does not have an m-dimensional p.d.f.

⊚ **Example 3.8**   Let $\mathbf{Y} = (Y_1, Y_2) \sim \text{N}_2\,(\mu, \Sigma)$ with $\text{Corr}\,(Y_1, Y_2) = \rho = \sigma_{12}/\sigma_1\sigma_2$, so that, with $Z_i = (Y_i - \mu_i)\,/\sigma_i$ and $z_i = (y_i - \mu_i)\,/\sigma_i$, $i = 1, 2$,

$$F_{\mathbf{Y}}\,(\mathbf{y}) = \text{Pr}\,(Y_1 \leq y_1,\ Y_2 \leq y_2) = \text{Pr}\,(Z_1 \leq z_1,\ Z_2 \leq z_2) = F_{\mathbf{Z}}\,(\mathbf{z}). \tag{3.19}$$

To see that $\mathbf{Z} = (Z_1, Z_2)$ is standard bivariate normal, take $\mathbf{a}$ and $\mathbf{B}$ as

$$\mathbf{Z} = \begin{bmatrix} Z_1 \\ Z_2 \end{bmatrix} = \begin{bmatrix} (Y_1 - \mu_1)\,/\sigma_1 \\ (Y_2 - \mu_2)\,/\sigma_2 \end{bmatrix} = \mathbf{a} + \mathbf{BY} = \begin{bmatrix} -\mu_1/\sigma_1 \\ -\mu_2/\sigma_2 \end{bmatrix}$$

$$+ \begin{bmatrix} \sigma_1^{-1} & 0 \\ 0 & \sigma_2^{-1} \end{bmatrix}\begin{bmatrix} Y_1 \\ Y_2 \end{bmatrix},$$

so that (3.18) implies

$$\mathbf{Z} \sim N(\mathbf{v}, \mathbf{\Omega}), \quad \mathbf{v} = \mathbf{a} + \mathbf{B}\boldsymbol{\mu} = \mathbf{0}, \quad \mathbf{\Omega} = \mathbf{B}\boldsymbol{\Sigma}\mathbf{B}' = \begin{bmatrix} 1 & \rho \\ \rho & 1 \end{bmatrix}.$$

Observe also that $\mathrm{Corr}(Y_1, Y_2) = \rho = \mathrm{Corr}(Z_1, Z_2)$.　■

An important special case of (3.18) is when $\mathbf{a} = \mathbf{0}$ and $\mathbf{B} = \mathbf{b} = (b_1, \ldots, b_n)$, so that

$$\mathbf{Y} = \sum_{i=1}^{n} b_i Z_i \sim N(\mathbf{b}'\boldsymbol{\mu}, \sigma^2), \quad \sigma^2 = \mathbf{b}\boldsymbol{\Sigma}\mathbf{b}',$$

and $\mathbf{b}'\boldsymbol{\mu}$ is a scalar. Moreover, if $\boldsymbol{\Sigma} = \mathrm{diag}(\sigma_1^2, \ldots, \sigma_n^2)$, then $\sigma^2 = \sum_{i=1}^{n} b_i^2 \sigma_i^2$, giving

$$\boxed{\text{if } X_i \stackrel{\text{ind}}{\sim} N(\mu_i, \sigma_i^2), \quad \text{then } \sum_{i=1}^{n} b_i X_i \sim N\left(\sum_{i=1}^{n} b_i \mu_i, \sum_{i=1}^{n} b_i^2 \sigma_i^2\right).}$$

3.4　Simulation and c.d.f. calculation

Generating a sample of observations from r.v. $\mathbf{Z} \sim N(\mathbf{0}, \mathbf{I}_n)$ is obviously quite easy: each of the n components of vector $\mathbf{Z}$ is i.i.d. standard univariate normal, for which simulation methods are well known. As $\mathbf{Y} = \boldsymbol{\mu} + \boldsymbol{\Sigma}^{1/2}\mathbf{Z}$ follows a $N(\boldsymbol{\mu}, \boldsymbol{\Sigma})$ distribution, realizations of $\mathbf{Y}$ can be obtained from the computed samples $\mathbf{z}$ as $\boldsymbol{\mu} + \boldsymbol{\Sigma}^{1/2}\mathbf{z}$, where $\boldsymbol{\Sigma}^{1/2}$ can be computed via the Cholesky decomposition (see Section 3.8).

For example, to simulate a pair of mean-zero bivariate normal r.v.s with covariance matrix

$$\boldsymbol{\Sigma} = \begin{bmatrix} 1 & \rho \\ \rho & 1 \end{bmatrix},$$

i.e., with unit variance and correlation ρ, first generate two i.i.d. standard normal r.v.s, say $\mathbf{z} = (z_1, z_2)'$, and then set $\mathbf{y} = \boldsymbol{\Sigma}^{1/2}\mathbf{z}$. The following code accomplishes this in Matlab for $\rho = 0.6$:

```
rho=0.6; [V,D]=eig([1 rho; rho 1]); C=V*sqrt(D)*V';
z=randn(2,1); y=C*z;
```

To generate T realizations of $\mathbf{y}$, the last line above could be repeated T times, or done 'all at once' as

```
T=100; z=randn(2,T); y=C*z;
```

which is easier, more elegant, and far faster in Matlab.

Similarly, to generate $T = 10^6$ pairs of bivariate normal r.v.s with

$$\mu = \begin{bmatrix} \mu_1 \\ \mu_2 \end{bmatrix} = \begin{bmatrix} 1 \\ 2 \end{bmatrix}, \quad \Sigma = \begin{bmatrix} \sigma_1^2 & \sigma_{12} \\ \sigma_{12} & \sigma_2^2 \end{bmatrix} = \begin{bmatrix} 3 & \rho\sigma_1\sigma_2 \\ \rho\sigma_1\sigma_2 & 4 \end{bmatrix}, \quad \rho = 0.5,$$

use

```
s1sqr=3; s2sqr=4; rho=0.5; sigma12=rho*sqrt(s1sqr*s2sqr)
[V,D]=eig([s1sqr sigma12; sigma12 s2sqr]); C=V*sqrt(D)*V';
T=1000000; z=randn(2,T); mu=[1 ; 2]; y= kron(mu,ones(1,T)) + C*z;
mean(y'), cov(y'), corr(y')
```

where the last line serves as verification.

With the ability to straightforwardly simulate r.v.s, the c.d.f. $\Pr(\mathbf{Y} \le \mathbf{y})$ can be calculated by generating a large sample of observations and computing the fraction which satisfy $\mathbf{Y} \le \mathbf{y}$. Continuing the previous calculation, the c.d.f. of $\mathbf{Y}$ at $(1.5, 2.5)$ can be approximated by

```
length( find(y(1,:)<1.5 \& y(2,:)<2.5) ) / length(y)
```

which, using $T = 10^6$, yields about three- to four-digit accuracy. The exact answer, to seven digits, is 0.4459117.

More generally, the probability of any region can be computed. For example, in the zero-mean, unit-variance bivariate case with correlation ρ, the probability that both Y_1 and Y_2 are positive (referred to as an *orthant* probability) can be computed with the following code over a grid of ρ-values:

```
rhovec=-0.99:0.01:0.99;  n=1000000;  emp1=zeros(1,length(rhovec));
for loop=1:length(rhovec)
  rho=rhovec(loop); [V,D]=eig([1 rho; rho 1]);
  C=V*sqrt(D)*V'; z=randn(2,n); y=C*z;
  emp1(loop)=length( find(y(1,:)>0 & y(2,:)>0) ) / length(y);
end
plot(rhovec,emp1)
```

The resulting plot is shown in Figure 3.5. Note that use of 1 million replications ensures a very smooth plot. In this case, a closed-form solution for the orthant

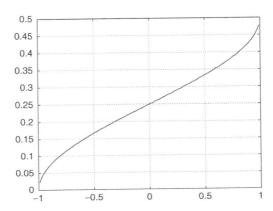

Figure 3.5 Orthant probability $\Pr(Y_1 > 0, Y_2 > 0)$ for $\mathbf{Y}$ bivariate normal versus correlation ρ

probability is $1/4 + \arcsin(\rho)/2\pi$, as first published by W. Sheppard in 1899. See Kotz, Balakrishnan and Johnson (2000, p. 265), Stirzaker (2003, p. 387), or the references in Fang, Kotz and Ng (1990, p. 53) for details on the derivation. Kotz, Balakrishnan and Johnson (2000, p. 263) give similar formulae for several other regions involving integers, e.g., $\Pr(Y_1 > 2, Y_2 > 1) = \sqrt{1/(8\pi)}(1 + \rho)^2$. Notice also that, from (I.5.3),

$$\Pr(Y_1 > 0, Y_2 > 0) = 1 - F_{Y_1}(0) - F_{Y_2}(0)$$

$$+ \Pr(Y_1 \leq 0, Y_2 \leq 0) = \Pr(Y_1 \leq 0, Y_2 \leq 0),$$

which is also intuitive from the symmetry of the distribution.

The above simulation technique can be used with more complicated regions. To calculate, say, the region given by $Y_2 > 0$ and $Y_1 < Y_2$, i.e.,

$$\int_0^\infty \int_0^y f_{Y_1, Y_2}(x, y)\, dx\, dy, \tag{3.20}$$

the previous code could be used, but with the fifth line replaced by

```
emp1(loop)=length( find(y(2,:)>0 & y(1,:)>y(2,:)) ) / length(y);
```

Computing this reveals that the probability is precisely half of the orthant probability just computed, which follows because this density is symmetric over the line $y_1 = y_2$.

In the bivariate case, numeric integration of the density (3.9) is still a feasible alternative to simulation, though hardly any faster. Calling `bvnrectangle([0 0]',[1 rho; rho 1], 0,8,0,8)` using the program in Listing 3.2 will compute (3.20).

```
function area=bvnrectangle(mu,Sig,xmin,xmax,ymin,ymax)

Siginv=inv(Sig); detSig=det(Sig);
area=dblquad(@bvnpdf,xmin,xmax,ymin,ymax,1e-6,@quadl,mu,Siginv,detSig);

function f=bvnpdf(xvec,y,mu,Siginv,detSig)
% multivariate normal pdf at x,y with mean mu and inverse var
      cov Siginv
% optionally pass detSig, the determinant of varcov matrix Sig
if nargin<4, detSig=1/det(Siginv); end
for xloop=1:length(xvec)
  x=xvec(xloop);
  if x>y
    f(xloop)=0;
  else
    v=[x;y]; e=(v-mu)' * Siginv * (v-mu);
    f(xloop)=exp(-e/2) / (2*pi) / sqrt(detSig);
  end
end
end
```

Program Listing 3.2 Integrates a region of the bivariate normal density with mean `mu` (pass as column vector) and variance–covariance matrix `Sig`

It works by integrating over a rectangle, as required by the Matlab function `dblquad`, and setting the density to zero over regions which should not be included.

The evaluation of the bivariate normal c.d.f. arises frequently in practice; notice from (3.19) that it is sufficient to construct an algorithm for the c.d.f. of the standard bivariate normal. This could be computed with bivariate integration as in program `bvnrectangle` with `xmin` and `ymin` taken to be large negative values, e.g., -10. This is far less efficient than summing enough terms of a convergent series, the most popular of which is the so-called *tetrachoric series*, developed by Pearson in 1901 (see Gupta, 1963). The convergence becomes slower as $|\rho| \rightarrow 1$, prompting the work of Vasicek (1998), who gives a different series expression which converges fast for large ρ. The derivation of this method, as well as the formulae for the tetrachoric series expansion, can be found there. The program in Listing 3.3 implements both methods based on the recommendation of Vasicek (1998) to use the tetrachoric series for $\rho^2 < 0.5$ and the new series otherwise.

```
function F = bvncdf(xvec,yvec,rho)
F=zeros(length(xvec),length(yvec));
for xloop=1:length(xvec), for yloop=1:length(yvec)
  x=xvec(xloop); y=yvec(yloop);
  if x==0 && y==0
    F(xloop,yloop) = 1/4 + asin(rho)/(2*pi);
  elseif y==0
    F(xloop,yloop)=mvncdf0(x,rho);
  elseif x==0
    F(xloop,yloop)=mvncdf0(y,rho);
  else
    F(xloop,yloop)=mvncdf0(x,(rho*x-y)/sqrt(x^2-2*rho*x*y+y^2))...
        *sign(x))+mvncdf0(y,(rho*y-x)/sqrt(x^2-2*rho*x*y+y^2)...
        *sign(y))-0.5*double(x*y<0);
  end
end, end
```

Program Listing 3.3 Computes the standard bivariate normal c.d.f. based on the results and recommendations in Vasicek (1998), in which all the formulae can be found; continued in Listing 3.4

Remark: Boys (1989) discusses other methods for computing bivariate normal probabilities, while Genz (2004) considers the bivariate and trivariate case for both the normal and Student's t. For higher dimensions, see Genz (1992), Hajivassiliou, McFadden and Ruud (1996), Vijverberg (2000) and the references therein. Gassmann, Deák and Szántai (2002) provide a short survey of competing methods and argue that there is no single superior method, but rather make recommendations for computation based on the actual probability, the dimension and the correlation structure. ∎

3.5 Marginal and conditional normal distributions

Let $\mathbf{X} \sim N_n (\boldsymbol{\mu}, \boldsymbol{\Sigma})$. For $1 \leq k \leq n$, the m.g.f. of the marginal distribution of the k-size subset of $\mathbf{X}$, $\{X_{i_1}, \ldots, X_{i_k}\}$, is given by $M_{\mathbf{X}}(\mathbf{t})$ with the $n - k$ elements $(1, \ldots, n) \setminus$

```
function F=mvncdf0(x,rho)
if rho^2 < 0.5 % Use tetrachoric series
  S=0;
  if rho~=0
    dS=1; k=0;
    while abs(dS)>1e-12
      k=k+1;
      dS=1/prod(1:k-1)/(2*k-1)*(-1)^(k-1)*2^(-k+1)...
          *Hermite(2*k-2,x)*rho^(2*k-1);
      S=S+dS;
    end
  end
  F=0.5*normcdf(x) + 1/(sqrt(2*pi))*normpdf(x)*S;
else % Use Vasicek (1998) series
  B=1/(2*pi)*sqrt(1-rho^2)*exp(-.5*(x^2/(1-rho^2)));
  A=-1/sqrt(2*pi)*abs(x)*normcdf(-abs(x)/sqrt(1-rho^2))+B;
  Q=A; k=0;
  while abs(A/Q)>1e-12
      k=k+1; B=(2*k-1)^2/(2*k*(2*k+1))*(1-rho^2)*B;
      A=-(2*k-1)/(2*k*(2*k+1))*x^2*A+B; Q=Q+A;
  end
  if rho>0,  F=min(normcdf(x),.5)-Q;
  else,      F=max(normcdf(x)-.5,0)+Q;
  end
end

function H=Hermite(k,x)
H=0;
for i=1:floor(k/2)+1
  H=H+prod(1:k)/prod(1:i-1)/prod(1:k-2*(i-1))...
      *(-1)^(i-1)*2^(-i+1)*x^(k-2*i+2);
end

function I=int(xvec,y,rho)
F=zeros(size(xvec)); mu=[0 0]'; si=[1 rho;rho 1];
for loop=1:length(xvec)
  x=xvec(loop); I(loop)=mvnpdf([x y]',mu,si);
end
```

Program Listing 3.4   Continued from Listing 3.3

$(i_1, \ldots, i_k)$ of $\mathbf{t}$ set to zero. From the form of $\mathbb{M}_{\mathbf{X}}(\mathbf{t})$, namely $\exp\left\{\mathbf{t}'\boldsymbol{\mu} + \frac{1}{2}\mathbf{t}'\boldsymbol{\Sigma}\mathbf{t}\right\}$, it is apparent that the resulting m.g.f. is that of a k-dimensional (multivariate) normal random variable with the appropriate elements of $\boldsymbol{\mu}$ and the appropriate rows and columns of $\boldsymbol{\Sigma}$ deleted. For instance, if $n = 3$ and $\mathbf{X} = (X_1, X_2, X_3)'$, then

$$\mathbb{M}_{\mathbf{X}}(\mathbf{t}) = \exp\left\{ \sum_{i=1}^{3} t_i \mu_i + \frac{1}{2} \begin{pmatrix} t_1 & t_2 & t_3 \end{pmatrix} \begin{pmatrix} \sigma_1^2 & \sigma_{12} & \sigma_{13} \\ \sigma_{12} & \sigma_2^2 & \sigma_{23} \\ \sigma_{13} & \sigma_{23} & \sigma_3^2 \end{pmatrix} \begin{pmatrix} t_1 \\ t_2 \\ t_3 \end{pmatrix} \right\}$$

and, for the three univariate marginals, $\mathbb{M}_{X_j}(t) = \exp\left\{\mu_j t + \sigma_j^2 t^2/2\right\}$, $j = 1, 2, 3$. Similarly,

$$\mathbb{M}_{(X_2, X_3)}(t_2, t_3) = \exp\left\{\sum_{i=2}^{3} t_i \mu_i + \frac{1}{2} \left(t_2 \quad t_3 \right) \begin{pmatrix} \sigma_2^2 & \sigma_{23} \\ \sigma_{23} & \sigma_3^2 \end{pmatrix} \begin{pmatrix} t_2 \\ t_3 \end{pmatrix} \right\}$$

$$= \exp\left\{\mu_2 t_2 + \mu_3 t_3 + \frac{1}{2}\left(t_2^2 \sigma_2^2 + 2 t_2 t_3 \sigma_{23} + t_3^2 \sigma_3^2\right)\right\},$$

with similar expressions for $\mathbb{M}_{(X_1, X_2)}$ and $\mathbb{M}_{(X_1, X_3)}$.

Now suppose that $\mathbf{Y} = (Y_1, \ldots, Y_n)' \sim N(\boldsymbol{\mu}, \boldsymbol{\Sigma})$ is partitioned into two subvectors $\mathbf{Y} = (\mathbf{Y}'_{(1)}, \mathbf{Y}'_{(2)})'$, where $\mathbf{Y}_{(1)} = (Y_1, \ldots, Y_p)'$ and $\mathbf{Y}_{(2)} = (Y_{p+1}, \ldots, Y_n)'$ with $\boldsymbol{\mu}$ and $\boldsymbol{\Sigma}$ partitioned accordingly such that $\mathbb{E}\left[\mathbf{Y}_{(i)}\right] = \boldsymbol{\mu}_{(i)}$, $\mathbb{V}\left(\mathbf{Y}_{(i)}\right) = \boldsymbol{\Sigma}_{ii}$, $i = 1, 2$, and $\mathrm{Cov}\left(\mathbf{Y}_{(1)}, \mathbf{Y}_{(2)}\right) = \boldsymbol{\Sigma}_{12}$, i.e., $\boldsymbol{\mu} = \left(\boldsymbol{\mu}'_{(1)}, \boldsymbol{\mu}'_{(2)}\right)'$ and

$$\boldsymbol{\Sigma} = \begin{bmatrix} \boldsymbol{\Sigma}_{11} & \vdots & \boldsymbol{\Sigma}_{12} \\ \cdots & \cdots & \cdots \\ \boldsymbol{\Sigma}_{21} & \vdots & \boldsymbol{\Sigma}_{22} \end{bmatrix}, \quad \boldsymbol{\Sigma}_{21} = \boldsymbol{\Sigma}'_{12}.$$

One of the most useful properties of the multivariate normal is that zero correlation implies independence, i.e., $\mathbf{Y}_{(1)}$ and $\mathbf{Y}_{(2)}$ are independent iff $\boldsymbol{\Sigma}_{12} = \mathbf{0}$. Recall that the covariance of two independent r.v.s is always zero, but the opposite need not be true. In the normal case, however, the form of the m.g.f. shows that it factors iff $\boldsymbol{\Sigma}_{12} = \mathbf{0}$, i.e., with $\mathbf{t} = \left(\mathbf{t}'_{(1)}, \mathbf{t}'_{(2)}\right)'$,

$$\mathbb{M}_{\mathbf{Y}}(\mathbf{t}) = \exp\left\{\mathbf{t}'\boldsymbol{\mu} + \frac{1}{2}\left(\mathbf{t}'_{(1)} \quad \mathbf{t}'_{(2)} \right) \begin{pmatrix} \boldsymbol{\Sigma}_{11} & \mathbf{0} \\ \mathbf{0} & \boldsymbol{\Sigma}_{22} \end{pmatrix} \begin{pmatrix} \mathbf{t}_{(1)} \\ \mathbf{t}_{(2)} \end{pmatrix} \right\}$$

$$= \exp\left\{\mathbf{t}'_{(1)}\boldsymbol{\mu}_{(1)} + \mathbf{t}'_{(2)}\boldsymbol{\mu}_{(2)} + \frac{1}{2}\left(\mathbf{t}'_{(1)}\boldsymbol{\Sigma}_{11}\mathbf{t}_{(1)} + \mathbf{t}'_{(2)}\boldsymbol{\Sigma}_{22}\mathbf{t}_{(2)}\right)\right\}$$

$$= \exp\left\{\mathbf{t}'_{(1)}\boldsymbol{\mu}_{(1)} + \frac{1}{2}\mathbf{t}'_{(1)}\boldsymbol{\Sigma}_{11}\mathbf{t}_{(1)}\right\} \exp\left\{\mathbf{t}'_{(2)}\boldsymbol{\mu}_{(2)} + \frac{1}{2}\mathbf{t}'_{(2)}\boldsymbol{\Sigma}_{22}\mathbf{t}_{(2)}\right\}$$

$$= \mathbb{M}_{\mathbf{Y}_{(1)}}\left(\mathbf{t}_{(1)}\right) \mathbb{M}_{\mathbf{Y}_{(2)}}\left(\mathbf{t}_{(2)}\right).$$

If $\boldsymbol{\Sigma}_{12} \neq \mathbf{0}$, then such a factorization is clearly not possible. In (I.5.20) we show that, if X_1 and X_2 are independent random variables, then for functions g_1 and g_2,

$$\mathbb{E}\left[g_1(X_1) g_2(X_2)\right] = \mathbb{E}\left[g_1(X_1)\right] \mathbb{E}\left[g_2(X_2)\right]. \tag{3.21}$$

The converse of this can also be shown (see, e.g., Gut, 2005, p. 70, and the references therein). In particular, r.v.s X_1 and X_2 are independent iff (3.21) holds for all bounded, continuous functions g_1, g_2. In particular, if the m.g.f. of X_1 and X_2 factors into their respective m.g.f.s, then X_1 and X_2 are independent. Thus,

$$\mathbb{M}_{\mathbf{Y}}(\mathbf{t}) = \mathbb{M}_{\mathbf{Y}_{(1)}}\left(\mathbf{t}_{(1)}\right) \mathbb{M}_{\mathbf{Y}_{(2)}}\left(\mathbf{t}_{(2)}\right) \quad \text{iff} \quad \mathbf{Y}_{(1)} \perp \mathbf{Y}_{(2)}.$$

The conditional distribution of $\mathbf{Y}_{(1)}$ given $\mathbf{Y}_{(2)}$ is considered next. If $\Sigma_{22} > 0$ (which is true if $\Sigma > 0$), then

$$\left(\mathbf{Y}_{(1)} \mid \mathbf{Y}_{(2)} = \mathbf{y}_{(2)}\right) \sim N\left(\boldsymbol{\mu}_{(1)} + \Sigma_{12}\Sigma_{22}^{-1}\left(\mathbf{y}_{(2)} - \boldsymbol{\mu}_{(2)}\right), \ \Sigma_{11} - \Sigma_{12}\Sigma_{22}^{-1}\Sigma_{21}\right). \quad (3.22)$$

To verify this, define the $p \times 1$ vector $\mathbf{X} = \mathbf{Y}_{(1)} - \boldsymbol{\mu}_{(1)} - \Sigma_{12}\Sigma_{22}^{-1}\left(\mathbf{Y}_{(2)} - \boldsymbol{\mu}_{(2)}\right)$ and note that, as a linear combination of the elements of $\mathbf{Y}$, r.v. $\mathbf{X}$ is also normally distributed with mean zero and variance $\mathbb{V}(\mathbf{X}) = \mathbb{E}[\mathbf{X}\mathbf{X}']$ given by

$$\mathbb{E}\left[\left\{\left(\mathbf{Y}_{(1)} - \boldsymbol{\mu}_{(1)}\right) - \Sigma_{12}\Sigma_{22}^{-1}\left(\mathbf{Y}_{(2)} - \boldsymbol{\mu}_{(2)}\right)\right\}\right.$$
$$\left. \times \left\{\left(\mathbf{Y}_{(1)} - \boldsymbol{\mu}_{(1)}\right)' - \left(\mathbf{Y}_{(2)} - \boldsymbol{\mu}_{(2)}\right)'\Sigma_{22}^{-1}\Sigma_{21}\right\}\right]$$
$$= \Sigma_{11} - \Sigma_{12}\Sigma_{22}^{-1}\Sigma_{21} - \Sigma_{12}\Sigma_{22}^{-1}\Sigma_{21} + \Sigma_{12}\Sigma_{22}^{-1}\Sigma_{22}\Sigma_{22}^{-1}\Sigma_{21}$$
$$= \Sigma_{11} - \Sigma_{12}\Sigma_{22}^{-1}\Sigma_{21}.$$

Similarly, $\left(\mathbf{X}', \mathbf{Y}'_{(2)}\right)'$ is also multivariate normal with covariance term

$$\mathbb{E}\left[\left(\mathbf{X} - \mathbf{0}\right)\left(\mathbf{Y}_{(2)} - \boldsymbol{\mu}_{(2)}\right)'\right] = \mathbb{E}\left[\left(\mathbf{Y}_{(1)} - \boldsymbol{\mu}_{(1)} - \Sigma_{12}\Sigma_{22}^{-1}\left(\mathbf{Y}_{(2)} - \boldsymbol{\mu}_{(2)}\right)\right)\left(\mathbf{Y}_{(2)} - \boldsymbol{\mu}_{(2)}\right)'\right]$$
$$= \Sigma_{12} - \Sigma_{12}\Sigma_{22}^{-1}\Sigma_{22} = \mathbf{0},$$

i.e., $\mathbf{X}$ and $\mathbf{Y}_{(2)}$ are independent and

$$\begin{bmatrix} \mathbf{X} \\ \mathbf{Y}_{(2)} \end{bmatrix} \sim N\left(\begin{bmatrix} \mathbf{0} \\ \boldsymbol{\mu}_{(2)} \end{bmatrix}, \begin{bmatrix} \mathbf{C} & \mathbf{0} \\ \mathbf{0} & \Sigma_{22} \end{bmatrix}\right), \quad \mathbf{C} = \Sigma_{11} - \Sigma_{12}\Sigma_{22}^{-1}\Sigma_{21}.$$

As

$$\mathbf{Y}_{(1)} = \mathbf{X} + \boldsymbol{\mu}_{(1)} + \Sigma_{12}\Sigma_{22}^{-1}\left(\mathbf{Y}_{(2)} - \boldsymbol{\mu}_{(2)}\right),$$

for fixed $\mathbf{Y}_{(2)} = \mathbf{y}_{(2)}$, (i.e., conditional on $\mathbf{Y}_{(2)} = \mathbf{y}_{(2)}$), $\mathbf{Y}_{(1)}$ is normally distributed with mean $\mathbb{E}[\mathbf{X}] + \boldsymbol{\mu}_{(1)} + \Sigma_{12}\Sigma_{22}^{-1}\left(\mathbf{y}_{(2)} - \boldsymbol{\mu}_{(2)}\right)$ and variance $\mathbb{V}(\mathbf{X}) = \Sigma_{11} - \Sigma_{12}\Sigma_{22}^{-1}\Sigma_{21}$, which is precisely (3.22).

The reader should verify that (3.22) indeed reduces to (3.13) in the bivariate normal case.

A special case of great importance is with $p = 1$ and, in what has become a standard notation, $Y = Y_{(1)}$ and $\mathbf{X} = \mathbf{Y}_{(2)}$ so that

$$\left(Y \mid \mathbf{X} = \mathbf{x}\right) \sim N\left(\mu_Y + \mathbf{b}\left(\mathbf{x} - \boldsymbol{\mu}_{\mathbf{X}}\right), \ \sigma^2\right),$$

where $\mu_Y := \mathbb{E}[Y]$, $\boldsymbol{\mu}_{\mathbf{X}} := \mathbb{E}[\mathbf{X}]$, $\mathbf{b} = \Sigma_{12}\Sigma_{22}^{-1}$, and $\sigma^2 := \mathbb{V}(Y) = \Sigma_{11} - \Sigma_{12}\Sigma_{22}^{-1}\Sigma_{21}$. The conditional mean $\mu_Y + \mathbf{b}\left(\mathbf{x} - \boldsymbol{\mu}_{\mathbf{X}}\right)$ is referred to as the *regression function* of Y on $\mathbf{X}$, with $\mathbf{b}$ the *regression coefficient*. Notice that the regression function is linear in $\mathbf{x}$ and that σ^2 does not depend on $\mathbf{x}$. This latter property is referred to as *homoscedasticity*.

⊖ **Example 3.9** Let

$$\mathbf{Y} = \begin{bmatrix} Y_1 \\ Y_2 \\ Y_3 \end{bmatrix} \sim N(\boldsymbol{\mu}, \boldsymbol{\Sigma}), \quad \boldsymbol{\mu} = \begin{bmatrix} 2 \\ 1 \\ 0 \end{bmatrix}, \quad \boldsymbol{\Sigma} = \begin{bmatrix} 2 & 1 & 1 \\ 1 & 3 & 0 \\ 1 & 0 & 1 \end{bmatrix}.$$

Because $\det(\boldsymbol{\Sigma}) = 2 \neq 0$, $\mathbf{Y}$ has a three-dimensional p.d.f. and is not degenerate.

1. The six marginal distributions are given by

$$Y_1 \sim N(2, 2), \quad Y_2 \sim N(1, 3), \quad Y_3 \sim N(0, 1),$$

$$\begin{bmatrix} Y_1 \\ Y_2 \end{bmatrix} \sim N\left(\begin{bmatrix} 2 \\ 1 \end{bmatrix}, \begin{bmatrix} 2 & 1 \\ 1 & 3 \end{bmatrix} \right), \quad \begin{bmatrix} Y_1 \\ Y_3 \end{bmatrix} \sim N\left(\begin{bmatrix} 2 \\ 0 \end{bmatrix}, \begin{bmatrix} 2 & 1 \\ 1 & 1 \end{bmatrix} \right),$$

and

$$\begin{bmatrix} Y_2 \\ Y_3 \end{bmatrix} \sim N\left(\begin{bmatrix} 1 \\ 0 \end{bmatrix}, \begin{bmatrix} 3 & 0 \\ 0 & 1 \end{bmatrix} \right).$$

2. To derive the distribution of $Y_2 \mid (Y_1, Y_3)$, first rewrite the density as

$$\begin{bmatrix} Y_2 \\ Y_1 \\ Y_3 \end{bmatrix} \sim N\left(\begin{bmatrix} \mu_{(1)} \\ \mu_{(2)} \end{bmatrix}, \begin{bmatrix} \Sigma_{11} & \Sigma_{12} \\ \Sigma_{21} & \Sigma_{22} \end{bmatrix} \right),$$

where $\mu_{(1)}$ and Σ_{11} are scalars, with

$$\begin{bmatrix} \mu_{(1)} \\ \mu_{(2)} \end{bmatrix} = \begin{bmatrix} 1 \\ \cdots \\ 2 \\ 0 \end{bmatrix}, \quad \begin{bmatrix} \Sigma_{11} & \Sigma_{12} \\ \Sigma_{21} & \Sigma_{22} \end{bmatrix} = \begin{bmatrix} 3 & 1 & 0 \\ 1 & 2 & 1 \\ 0 & 1 & 1 \end{bmatrix}.$$

Then, from (3.22),

$$Y_2 \mid (Y_1, Y_3) \sim N\left(\mu_{(1)} + \Sigma_{12}\Sigma_{22}^{-1}(\mathbf{y}_{(2)} - \mu_{(2)}), \ \Sigma_{11} - \Sigma_{12}\Sigma_{22}^{-1}\Sigma_{21} \right),$$

i.e., substituting and simplifying,

$$\mathbb{E}[Y_2 \mid (Y_1, Y_3)] = \mu_{(1)} + \Sigma_{12}\Sigma_{22}^{-1}(\mathbf{y}_{(2)} - \mu_{(2)})$$

$$= 1 + \begin{bmatrix} 1 & 0 \end{bmatrix} \begin{bmatrix} 2 & 1 \\ 1 & 1 \end{bmatrix}^{-1} \left(\begin{bmatrix} y_1 \\ y_3 \end{bmatrix} - \begin{bmatrix} 2 \\ 0 \end{bmatrix} \right)$$

$$= y_1 - y_3 - 1$$

and

$$\mathbb{V}(Y_2 \mid (Y_1, Y_3)) = \Sigma_{11} - \Sigma_{12}\Sigma_{22}^{-1}\Sigma_{21}$$

$$= 3 - \begin{bmatrix} 1 & 0 \end{bmatrix} \begin{bmatrix} 2 & 1 \\ 1 & 1 \end{bmatrix}^{-1} \begin{bmatrix} 1 \\ 0 \end{bmatrix} = 2,$$

so that

$$Y_2 \mid (Y_1, Y_3) \sim N(y_1 - y_3 - 1, \ 2).$$

3. The distribution of (X_1, X_2), where $X_1 = \sum_{i=1}^{3} Y_i$ and $X_2 = Y_1 - Y_3$, is determined by writing

$$\mathbf{X} = \begin{bmatrix} X_1 \\ X_2 \end{bmatrix} = \mathbf{KY}, \quad \text{where} \quad \mathbf{K} = \begin{bmatrix} 1 & 1 & 1 \\ 1 & 0 & -1 \end{bmatrix}, \quad \mathbf{Y} = \begin{bmatrix} Y_1 \\ Y_2 \\ Y_3 \end{bmatrix},$$

so that $\mathbf{X} \sim N(\mathbf{K}\boldsymbol{\mu}, \mathbf{K}\Sigma\mathbf{K}')$ or

$$\begin{bmatrix} X_1 \\ X_2 \end{bmatrix} \sim N\left(\begin{bmatrix} 3 \\ 2 \end{bmatrix}, \begin{bmatrix} 10 & 2 \\ 2 & 1 \end{bmatrix} \right).$$

∎

3.6 Partial correlation

Let $\mathbf{Y} \sim N_n(\boldsymbol{\mu}, \Sigma)$. As usual, the covariance between two of the univariate random variables in $\mathbf{Y}$, say Y_i and Y_j, is determined from the (i, j)th entry of Σ. In this section, we consider the covariance (actually the correlation) of Y_i and Y_j *when conditioning on a set of other variables in* $\mathbf{Y}$. This structure has various uses in statistical analysis involving multivariate normal r.v.s, one of which is the study of autoregressive models in time series analysis, which we will look at in detail in a later chapter.

Denote the (i, j)th element of $\mathbf{C} = \Sigma_{11} - \Sigma_{12}\Sigma_{22}^{-1}\Sigma_{21}$ by $\sigma_{ij|(p+1,\dots,n)}$. Motivated by the conditional variance in (3.22), the *partial correlation of* Y_i *and* Y_j, *given* $\mathbf{Y}_{(2)}$ is defined by

$$\rho_{ij|(p+1,\dots,n)} = \frac{\sigma_{ij|(p+1,\dots,n)}}{\sqrt{\sigma_{ii|(p+1,\dots,n)} \ \sigma_{jj|(p+1,\dots,n)}}}. \tag{3.23}$$

⊖ **Example 3.10**   (Example 3.9 cont.) To compute $\rho_{13|2}$, first write

$$\begin{bmatrix} Y_1 \\ Y_3 \\ Y_2 \end{bmatrix} \sim N\left(\begin{bmatrix} \boldsymbol{\mu}_{(1)} \\ \boldsymbol{\mu}_{(2)} \end{bmatrix}, \begin{bmatrix} \Sigma_{11} & \Sigma_{12} \\ \Sigma_{21} & \Sigma_{22} \end{bmatrix} \right),$$

where

$$\begin{bmatrix} \boldsymbol{\mu}_{(1)} \\ \boldsymbol{\mu}_{(2)} \end{bmatrix} := \mathbb{E}\begin{bmatrix} Y_1 \\ Y_3 \\ \cdots \\ Y_2 \end{bmatrix} = \begin{bmatrix} \mu_1 \\ \mu_3 \\ \cdots \\ \mu_2 \end{bmatrix} = \begin{bmatrix} 2 \\ 0 \\ \cdots \\ 1 \end{bmatrix}$$

and

$$\begin{bmatrix} \Sigma_{11} & \Sigma_{12} \\ \Sigma_{21} & \Sigma_{22} \end{bmatrix} := \mathbb{V}\left(\begin{bmatrix} Y_1 \\ Y_3 \\ \cdots \\ Y_2 \end{bmatrix}\right) = \begin{bmatrix} \sigma_{11} & \sigma_{13} & \sigma_{12} \\ \sigma_{31} & \sigma_{33} & \sigma_{32} \\ \sigma_{21} & \sigma_{23} & \sigma_{22} \end{bmatrix} = \begin{bmatrix} 2 & 1 & 1 \\ 1 & 1 & 0 \\ 1 & 0 & 3 \end{bmatrix},$$

so that

$$\mathbf{C} = \Sigma_{11} - \Sigma_{12}\Sigma_{22}^{-1}\Sigma_{21} = \begin{bmatrix} 2 & 1 \\ 1 & 1 \end{bmatrix} - \begin{bmatrix} 1 \\ 0 \end{bmatrix} [\, 3 \,]^{-1} [\, 1 \quad 0 \,] = \begin{bmatrix} 5/3 & 1 \\ 1 & 1 \end{bmatrix}$$

and

$$\rho_{13|(2)} = \frac{1}{\sqrt{5/3 \cdot 1}} = \sqrt{\frac{3}{5}}.$$

In general terms,

$$\mathbf{C} = \Sigma_{11} - \Sigma_{12}\Sigma_{22}^{-1}\Sigma_{21}$$

$$= \begin{bmatrix} \sigma_{11} & \sigma_{13} \\ \sigma_{31} & \sigma_{33} \end{bmatrix} - \begin{bmatrix} \sigma_{12} \\ \sigma_{32} \end{bmatrix} [\, \sigma_{22} \,]^{-1} [\, \sigma_{12} \quad \sigma_{32} \,]$$

$$= \begin{bmatrix} \sigma_{11} - \sigma_{12}^2/\sigma_{22} & \sigma_{13} - \sigma_{12}\sigma_{32}/\sigma_{22} \\ \sigma_{31} - \sigma_{32}\sigma_{12}/\sigma_{22} & \sigma_{33} - \sigma_{32}^2/\sigma_{22} \end{bmatrix}$$

and

$$\rho_{13|(2)} = \frac{\sigma_{13} - \sigma_{12}\sigma_{32}/\sigma_{22}}{\sqrt{(\sigma_{11} - \sigma_{12}^2/\sigma_{22})(\sigma_{33} - \sigma_{32}^2/\sigma_{22})}} = \frac{\sigma_{22}\sigma_{13} - \sigma_{12}\sigma_{32}}{\sqrt{\sigma_{22}\sigma_{11} - \sigma_{12}^2}\sqrt{\sigma_{22}\sigma_{33} - \sigma_{32}^2}}$$

$$= \frac{\sigma_{22}\sigma_{13} - \sigma_{12}\sigma_{32}}{\sqrt{\sigma_{22}\sigma_{11}\left(1 - \dfrac{\sigma_{12}^2}{\sigma_{22}\sigma_{11}}\right)}\sqrt{\sigma_{22}\sigma_{33}\left(1 - \dfrac{\sigma_{32}^2}{\sigma_{22}\sigma_{33}}\right)}},$$

or

$$\rho_{13|(2)} = \frac{\sigma_{22}\sigma_{13} - \sigma_{12}\sigma_{32}}{\sqrt{\sigma_{22}\sigma_{11}\sigma_{22}\sigma_{33}}\sqrt{\left(1 - \rho_{12}^2\right)\left(1 - \rho_{23}^2\right)}} = \frac{\dfrac{\sigma_{13}}{\sqrt{\sigma_{11}\sigma_{33}}} - \dfrac{\sigma_{12}}{\sqrt{\sigma_{22}\sigma_{11}}}\dfrac{\sigma_{32}}{\sqrt{\sigma_{22}\sigma_{33}}}}{\sqrt{\left(1 - \rho_{12}^2\right)\left(1 - \rho_{23}^2\right)}}$$

$$= \frac{\rho_{13} - \rho_{12}\rho_{23}}{\sqrt{\left(1 - \rho_{12}^2\right)\left(1 - \rho_{23}^2\right)}}. \tag{3.24}$$

Using the previous numbers, $\rho_{13} = 1/\sqrt{2}$, $\rho_{12} = 1/\sqrt{6}$ and $\rho_{23} = 0$, so that (3.24) gives

$$\rho_{13|(2)} = \frac{\rho_{13} - \rho_{12}\rho_{23}}{\sqrt{\left(1 - \rho_{12}^2\right)\left(1 - \rho_{23}^2\right)}} = \frac{1/\sqrt{2}}{\sqrt{\left(1 - \left(1/\sqrt{6}\right)^2\right)}} = \sqrt{\frac{3}{5}},$$

as before. ∎

⊚ **Example 3.11**   Let $\mathbf{Y} = (Y_1, \ldots Y_4)' \sim N(\mathbf{0}, \Sigma)$ with

$$\Sigma = \frac{1}{1-a^2} \begin{bmatrix} 1 & a & a^2 & a^3 \\ a & 1 & a & a^2 \\ a^2 & a & 1 & a \\ a^3 & a^2 & a & 1 \end{bmatrix}$$

for a value of a such that $|a| < 1$, so that

$$\begin{bmatrix} Y_1 \\ Y_3 \\ Y_4 \\ Y_2 \end{bmatrix} \sim N(\mathbf{0}, \Omega), \quad \Omega = \frac{1}{1-a^2} \begin{bmatrix} 1 & a^2 & a^3 & a \\ a^2 & 1 & a & a \\ a^3 & a & 1 & a^2 \\ a & a & a^2 & 1 \end{bmatrix}.$$

Then, with the appropriate partitions for $\boldsymbol{\mu}$ and Ω,

$$(Y_1, Y_3, Y_4 \mid Y_2)' \sim N(\boldsymbol{\nu}, \mathbf{C}),$$

where

$$\boldsymbol{\nu} = \boldsymbol{\mu}_{(1)} + \Omega_{12}\Omega_{22}^{-1}(\mathbf{y}_{(2)} - \mu_{(2)}) = \begin{bmatrix} 0 \\ 0 \\ 0 \end{bmatrix} + \begin{bmatrix} a \\ a \\ a^2 \end{bmatrix} [1]^{-1}(y_2 - 0) = \begin{bmatrix} ay_2 \\ ay_2 \\ a^2 y_2 \end{bmatrix}$$

and

$$\mathbf{C} = \Omega_{11} - \Omega_{12}\Omega_{22}^{-1}\Omega_{21}$$

$$= \frac{1}{1-a^2} \begin{bmatrix} 1 & a^2 & a^3 \\ a^2 & 1 & a \\ a^3 & a & 1 \end{bmatrix} - \frac{1}{1-a^2} \begin{bmatrix} a \\ a \\ a^2 \end{bmatrix} [1]^{-1} \begin{bmatrix} a & a & a^2 \end{bmatrix}$$

$$= \frac{1}{1-a^2} \left(\begin{bmatrix} 1 & a^2 & a^3 \\ a^2 & 1 & a \\ a^3 & a & 1 \end{bmatrix} - \begin{bmatrix} a^2 & a^2 & a^3 \\ a^2 & a^2 & a^3 \\ a^3 & a^3 & a^4 \end{bmatrix} \right)$$

$$= \frac{1}{1-a^2} \left(\begin{bmatrix} 1-a^2 & 0 & 0 \\ 0 & 1-a^2 & a-a^3 \\ 0 & a-a^3 & 1-a^4 \end{bmatrix} \right)$$

$$= \begin{bmatrix} 1 & 0 & 0 \\ 0 & 1 & a \\ 0 & a & a^2+1 \end{bmatrix}. \tag{3.25}$$

It follows that

$$\rho_{13|(2)} = \frac{\sigma_{13|(2)}}{\sqrt{\sigma_{11|(2)}\,\sigma_{33|(2)}}} = \frac{0}{1} = 0, \qquad \rho_{14|(2)} = \frac{\sigma_{14|(2)}}{\sqrt{\sigma_{11|(2)}\,\sigma_{44|(2)}}} = \frac{0}{\sqrt{1+a^2}} = 0,$$

and

$$\rho_{34|(2)} = \frac{\sigma_{34|(2)}}{\sqrt{\sigma_{33|(2)}\,\sigma_{44|(2)}}} = \frac{a}{\sqrt{1+a^2}},$$

which are results we will make use of when studying the partial autocorrelation function for time series analysis. ∎

3.7 Joint distribution of $\overline{X}$ and S^2 for i.i.d. normal samples

An interesting and very useful property of the normal distribution already alluded to at the end of Section I.8.2.5 is the independence of the sample mean and variance for an i.i.d. sample. That is, the *statistics* (functions of the data) $\overline{X}_n = \overline{X} = n^{-1}\sum_{i=1}^{n} X_i$ and $S_n^2(X) = S^2 = (n-1)^{-1}\sum_{i=1}^{n}\left(X_i - \overline{X}\right)^2$ are independent, i.e.,

$$\text{if } \quad X_i \stackrel{\text{i.i.d.}}{\sim} N\left(\mu, \sigma^2\right), \quad \text{then} \quad \overline{X} \perp S^2.$$

This is most easily demonstrated by showing the stronger result that $\overline{X} \perp \left(X_i - \overline{X}\right)$ for all i or, as $\overline{X}$ and $X_i - \overline{X}$ are both *jointly* normally distributed, $\text{Cov}\left(\overline{X}, X_i - \overline{X}\right) = 0$. In particular, with $\mathbb{V}\left(\overline{X}\right) = \sigma^2/n$ following as a special case of (I.6.4) or (3.5) and using (3.6),

$$\text{Cov}\left(\overline{X}, X_i - \overline{X}\right) = \text{Cov}\left(\overline{X}, X_i\right) - \mathbb{V}\left(\overline{X}\right) = n^{-1}\sum_{j=1}^{n} \text{Cov}\left(X_j, X_i\right) - \frac{\sigma^2}{n} = 0,$$

as $\text{Cov}\left(X_j, X_i\right) = 0$ for $i \neq j$ and σ^2 for $i = j$. Because S^2 can be expressed in terms of a function strictly of the $X_i - \overline{X}$, it follows that $\overline{X} \perp S^2$.

Interestingly enough, the converse also holds: if $X_1, \ldots, X_n$ are i.i.d., $n \geq 2$, and $\overline{X} \perp S^2$, then the X_i are normally distributed. For proof, see Bryc (1995, Section 7.4) and the references therein. It can also be shown that, for $n \geq 3$, if $\overline{X} \sim N(\mu, \sigma^2/n)$ and $(n-1)S^2/\sigma^2 \sim \chi_{n-1}^2$, then $X_i \stackrel{\text{i.i.d.}}{\sim} N\left(\mu, \sigma^2\right)$; see Abadir and Magnus (2003).

It is important to realize that the $\left\{X_i - \overline{X}\right\}$ are not independent; for $i \neq j$,

$$\text{Cov}\left(X_i - \overline{X}, X_j - \overline{X}\right) = \text{Cov}\left(X_i, X_j\right) - 2\,\text{Cov}\left(\overline{X}, X_i\right) + \mathbb{V}\left(\overline{X}\right) = -\frac{\sigma^2}{n}.$$

That this covariance is negative is intuitive: as the $X_i - \overline{X}$ are deviations from their mean, a positive $X_i - \overline{X}$ implies the existence of at least one negative $X_j - \overline{X}$. As n grows, the covariance weakens.

Remark: Notice that $\text{Cov}\left(\overline{X}, X_i - \overline{X}\right) = 0$ holds even without the normality assumption, but to conclude independence of $\overline{X}$ and S^2 does require normality. It turns out, however, that the i.i.d. assumption can be relaxed, albeit in a very restricted way: if the $\mathbf{X}_i$ are equicorrelated normal, then $\overline{X} \perp S^2$ still holds, i.e., when $\mathbf{X} \sim N_n\left(\mu\mathbf{1}, \Sigma\right)$ for

$$\Sigma = \begin{bmatrix} \sigma^2 & \rho\sigma^2 & \cdots & \rho\sigma^2 \\ \rho\sigma^2 & \sigma^2 & \cdots & \rho\sigma^2 \\ \vdots & \vdots & \ddots & \vdots \\ \rho\sigma^2 & \rho\sigma^2 & \cdots & \sigma^2 \end{bmatrix} = \rho\sigma^2\mathbf{J}_n + (1 - \rho)\sigma^2\mathbf{I}_n, \tag{3.26}$$

where $\mathbf{J}_n$ is the $n \times n$ matrix of ones, $\sigma^2 > 0$, and correlation ρ is such that $\Sigma \geq 0$. Example I.6.1 discussed when such a model might be realistic, and Problem 1.8 shows that $-1/(n - 1) < \rho < 1$ is necessary in order for $\Sigma > 0$. To show independence, first note that

$$\mathbb{V}\left(\overline{X}\right) = n^{-2}\sum_{i=1}^{n}\sum_{j=1}^{n}\text{Cov}\left(X_i, X_j\right) = n^{-2}\left(\left(n^2 - n\right)\rho\sigma^2 + n\sigma^2\right) \tag{3.27}$$

and

$$\text{Cov}\left(\overline{X}, X_i\right) = n^{-1}\sum_{j=1}^{n}\text{Cov}\left(X_j X_i\right) = n^{-1}\left((n - 1)\rho\sigma^2 + \sigma^2\right)$$

giving $\text{Cov}\left(\overline{X}, X_i - \overline{X}\right) = \text{Cov}\left(\overline{X}, X_i\right) - \mathbb{V}\left(\overline{X}\right) = 0$. This result can be extended to the context of the linear regression model; see Knautz and Trenkler (1993), Bhatti (1995), and the references therein. ∎

With independence established, the joint density of $\overline{X}$ and S^2 can be factored. Clearly,

$$\overline{X} \sim N\left(\mu, \sigma^2/n\right) \tag{3.28}$$

in the i.i.d. case and, from (3.27) in the equicorrelated case, $\overline{X} \sim N\left(\mu, \mathbb{V}\left(\overline{X}\right)\right)$. The density f_{S^2} in the i.i.d. case is derived next; f_{S^2} in the equicorrelated or more general cases is considerably more complicated and is dealt with in a future chapter dedicated to the distribution of (ratios of) quadratic forms.

With $X_i \overset{\text{i.i.d.}}{\sim} N\left(\mu, \sigma^2\right)$, recall from (2.4) that $\sum_{i=1}^{n}(X_i - \mu)^2/\sigma^2 \sim \chi_n^2$. Define

$$A := \sum_{i=1}^{n}\sigma^{-2}(X_i - \mu)^2, \quad B := \sigma^{-2}\sum_{i=1}^{n}(X_i - \overline{X})^2, \quad C := \left(\frac{\overline{X} - \mu}{\sigma\sqrt{n}}\right)^2,$$

so that

$$
A = \sum_{i=1}^{n} \sigma^{-2} (X_i - \mu)^2 = \sigma^{-2} \sum_{i=1}^{n} (X_i - \overline{X} + \overline{X} - \mu)^2
$$

$$
= \sigma^{-2} \sum_{i=1}^{n} \left[(X_i - \overline{X})^2 + (\overline{X} - \mu)^2 + 2 (X_i - \overline{X}) (\overline{X} - \mu) \right]
$$

$$
= \sigma^{-2} \sum_{i=1}^{n} (X_i - \overline{X})^2 + \left(\frac{\overline{X} - \mu}{\sigma \sqrt{n}} \right)^2
$$

$$
= B + C.
$$

We have $A \sim \chi_n^2$ and $C \sim \chi_1^2$ and note that $B \perp C$ because $(X_i - \overline{X}) \perp \overline{X}$. Therefore,

$$
\left(\frac{1}{1 - 2t} \right)^{n/2} = \mathbb{M}_A(t) = \mathbb{M}_B(t) \, \mathbb{M}_C(t) = \mathbb{M}_B(t) \left(\frac{1}{1 - 2t} \right)^{1/2},
$$

implying that $\mathbb{M}_B(t) = (1 - 2t)^{-(n-1)/2}$ and, hence, $B \sim \chi_{n-1}^2$. In terms of S^2, $B = (n-1) S^2 / \sigma^2$, i.e., $S^2 \sim \sigma^2 \chi_{n-1}^2 / (n-1)$, a scaled chi-square random variable, or

$$
\boxed{\frac{(n - 1) S^2}{\sigma^2} \sim \chi_{n-1}^2.} \tag{3.29}
$$

Let $Z = (\overline{X} - \mu) / (\sigma / \sqrt{n}) \sim N(0, 1)$. From the above results and Example 2.15, the random variable

$$
\boxed{T = \frac{Z}{\sqrt{B/(n-1)}} = \frac{(\overline{X} - \mu) / (\sigma / \sqrt{n})}{\sqrt{\sigma^{-2} \sum_{i=1}^{n} (X_i - \overline{X})^2 / (n-1)}} = \frac{\overline{X} - \mu}{S_n / \sqrt{n}} \sim t_{n-1},} \tag{3.30}
$$

i.e., T follows a Student's t distribution with $n - 1$ degrees of freedom. Observe that its distribution does not depend on σ.

The density of $S = \sqrt{S^2}$ could be derived by transformation, from which we could, among other things, investigate the extent to which S is biased for σ (meaning that $\mathbb{E}[S] \neq \sigma$). That it is not unbiased follows because the square root is not a linear function. We know from (I.4.36) that the expected value of S can be obtained directly without having to transform. This gives

$$
\mathbb{E}[S] = \mathbb{E}\left[\sqrt{S^2} \right] = \mathbb{E}\left[\frac{\sigma}{\sqrt{n-1}} \sqrt{\frac{(n-1) S^2}{\sigma^2}} \right] = \frac{\sigma}{\sqrt{n-1}} \mathbb{E}\left[U^{1/2} \right], \tag{3.31}
$$

where $U = (n-1)S^2/\sigma^2 \sim \chi^2_{n-1}$. From Example I.7.5,

$$\mathbb{E}[S] = K\sigma, \qquad K = \frac{\sqrt{2}}{\sqrt{n-1}} \frac{\Gamma\left(\frac{n}{2}\right)}{\Gamma\left(\frac{n-1}{2}\right)}, \tag{3.32}$$

so that an unbiased estimate of σ would be given by $\sqrt{S^2}/K$. Plotting K reveals that S is downward biased for σ, i.e., $K < 1$ for $n \geq 2$. In fact, the direction of the bias could have been determined without calculation using Jensen's inequality (Section I.4.4.3) and the fact that $x^{1/2}$ is a concave function.

3.8   Matrix algebra

Many who have never had the occasion to discover more about mathematics confuse it with arithmetic and consider it a dry and arid science. In reality, however, it is a science which demands the greatest imagination.

(Sonia Kovalevsky)

Some tools from matrix algebra are repeated here, although it is certainly not intended for a first exposure to the subject. We assume that the reader is familiar with vector and matrix notation, transposition, symmetric matrices, matrix addition and multiplication, diagonal matrices, computation of determinant and matrix inverses, rank, span, and solutions of systems of linear equations. See the books mentioned in Section I.A.0 for an introduction.

If $\mathbf{A}$ is an $n \times n$ symmetric real matrix, then $\mathbf{A}$ is said to be *positive definite*, denoted $\mathbf{A} > 0$, if $\mathbf{x}'\mathbf{A}\mathbf{x} > 0$ for all $\mathbf{x} \in \mathbb{R}^n \setminus \mathbf{0}$. If $\mathbf{x}'\mathbf{A}\mathbf{x} \geq 0$ for all $\mathbf{x} \in \mathbb{R}^n$, then $\mathbf{A}$ is said to be *positive semi-definite*, denoted $\mathbf{A} \geq 0$.

If $\mathbf{A}$ is a square matrix, then its trace, $\text{tr}(\mathbf{A})$, is the sum of its diagonal elements. A useful fact is that, for $\mathbf{B}$ an $n \times m$ matrix $\mathbf{C}$ an $m \times n$ matrix,

$$\text{tr}(\mathbf{BC}) = \text{tr}(\mathbf{CB}). \tag{3.33}$$

Now let $\mathbf{B}$ and $\mathbf{C}$ be $n \times n$ matrices. The determinant of the product is

$$\det(\mathbf{BC}) = \det(\mathbf{B})\det(\mathbf{C}) = \det(\mathbf{CB}). \tag{3.34}$$

Let $\mathbf{A} > 0$ be a real, symmetric $n \times n$ matrix. The *principal minors* of order k, $1 \leq k \leq n$, are determinants of all the $k \times k$ matrices obtained by deleting the same $n - k$ rows and columns of $\mathbf{A} > 0$, and the *leading principal minor* of order k, $1 \leq k \leq n$, is the principal minor obtained by deleting the first k rows and columns of $\mathbf{A} > 0$. A useful result is that, if $\mathbf{A} > 0$, i.e., $\mathbf{A}$ is positive definite, then all the leading principal minors are positive (see Graybill, 1983, p. 397). (In fact, more generally, if $\mathbf{A} > 0$, then all principal minors are positive.)

For matrix $\mathbf{A} \in \mathbb{R}^{n \times n}$, there are at most n distinct roots of the *characteristic equation* $|\lambda\mathbf{I}_n - \mathbf{A}| = 0$, and these values are referred to as the *eigenvalues* of $\mathbf{A}$, the set of which is denoted by Eig($\mathbf{A}$). If two or more roots coincide, then we say that the eigenvalues have multiplicities. Each matrix $\mathbf{A} \in \mathbb{R}^{n \times n}$ has exactly n eigenvalues, counting multiplicities. For example, the identity matrix $\mathbf{I}_n$ has one unique eigenvalue (unity), but has n eigenvalues, counting multiplicities. Denote the n eigenvalues, counting multiplicities, of matrix $\mathbf{A}$ as $\lambda_1, \ldots, \lambda_n$. Vector $\mathbf{x}$ is a (column) *eigenvector* of eigenvalue λ if it satisfies $\mathbf{Ax} = \lambda \mathbf{x}$. Eigenvectors are usually normalized to have norm one, i.e., $\mathbf{x}'\mathbf{x} = 1$. Of great importance is the fact that eigenvalues of a symmetric matrix are real. If, in addition, $\mathbf{A} > 0$, then all its eigenvalues are positive. This follows because $\mathbf{Ax} = \lambda \mathbf{x} \Rightarrow \mathbf{x}'\mathbf{Ax} = \lambda \mathbf{x}'\mathbf{x}$ and both $\mathbf{x}'\mathbf{Ax}$ and $\mathbf{x}'\mathbf{x}$ are positive.

The square matrix $\mathbf{U}$ with columns $\mathbf{u}_1, \ldots, \mathbf{u}_n$ is *orthogonal* if $\mathbf{U}'\mathbf{U} = \mathbf{I}$. It is straightforward to check[2] that $\mathbf{UU}' = \mathbf{I}$ and $\mathbf{U}' = \mathbf{U}^{-1}$.

As a special case of *Schur's decomposition theorem*, if $\mathbf{A}$ is an $n \times n$ symmetric matrix, then there exist an orthogonal matrix $\mathbf{U}$ and a diagonal matrix $\mathbf{D} = \text{diag}(\lambda_1, \ldots, \lambda_n)$ such that $\mathbf{A} = \mathbf{UDU}'$. We refer to this as the *spectral decomposition*. It implies that $\mathbf{AU} = \mathbf{UD}$ or $\mathbf{Au}_i = \lambda_i \mathbf{u}_i$, $i = 1, \ldots, n$, i.e., $\mathbf{u}_i$ is an eigenvector of $\mathbf{A}$ corresponding to eigenvalue λ_i. The theorem is easy to verify if the n eigenvalues have no multiplicities, i.e., they are mutually distinct. Let λ_i be an eigenvalue of $\mathbf{A}$ associated with normed eigenvector $\mathbf{u}_i$, so $\mathbf{Au}_i = \lambda_i \mathbf{u}_i$. Then, for $i \neq j$, as $\mathbf{u}_i'\mathbf{Au}_j$ is a scalar and $\mathbf{A}$ is symmetric, $\mathbf{u}_i'\mathbf{Au}_j = \mathbf{u}_j'\mathbf{Au}_i$, but $\mathbf{u}_i'\mathbf{Au}_j = \lambda_j \mathbf{u}_i'\mathbf{u}_j$ and $\mathbf{u}_j'\mathbf{Au}_i = \lambda_i \mathbf{u}_j'\mathbf{u}_i$, i.e., $\lambda_i \mathbf{u}_i'\mathbf{u}_j = \lambda_j \mathbf{u}_i'\mathbf{u}_j$. Thus $\mathbf{u}_i'\mathbf{u}_j = 0$ because $\lambda_i \neq \lambda_j$, i.e., the eigenvectors are orthogonal to one another and $\mathbf{U} = (\mathbf{u}_1 \; \mathbf{u}_2 \; \ldots \mathbf{u}_n)$ is an orthogonal matrix.

Let the $n \times n$ matrix $\mathbf{A}$ have spectral decomposition $\mathbf{UDU}'$ (where $\mathbf{D} = \text{diag}(\lambda_1, \ldots, \lambda_n)$ and λ_i are the eigenvalues of $\mathbf{A}$). Then, from (3.33),

$$\text{tr}\,(\mathbf{A}) = \text{tr}\left(\mathbf{UDU}'\right) = \text{tr}\left(\mathbf{U}'\mathbf{UD}\right) = \text{tr}\,(\mathbf{D}) = \sum_{i=1}^{n} \lambda_i.$$

Similarly, from (3.34),

$$\det(\mathbf{A}) = \det\left(\mathbf{UDU}'\right) = \det\left(\mathbf{U}'\mathbf{UD}\right) = \det(\mathbf{D}) = \prod_{i=1}^{n} \lambda_i,$$

recalling that the determinant of a diagonal matrix is the product of the diagonal elements. Thus, if $\mathbf{A}$ is not full rank, i.e., is *singular*, then $|\mathbf{A}| = 0$, implying that at least one eigenvalue is zero.

---

[2] If the $\mathbf{u}_i$ are orthonormal, then it is obvious that $\mathbf{U} = (\mathbf{u}_1 \; \mathbf{u}_2 \; \cdots \mathbf{u}_n)$ satisfies $\mathbf{U}'\mathbf{U} = \mathbf{I}$, but not immediately clear that $\mathbf{UU}' = \mathbf{I}$. We offer two simple proofs of this.

First proof: $\mathbf{U}'\mathbf{U} = \mathbf{I} \Rightarrow \det(\mathbf{U}')\det(\mathbf{U}) = \det(\mathbf{U}'\mathbf{U}) = \det(\mathbf{I}) = 1 \Rightarrow \det(\mathbf{U}) \neq 0 \Rightarrow \exists \mathbf{U}^{-1}$. Then

$$\mathbf{U}' = \mathbf{U}'\left(\mathbf{UU}^{-1}\right) = \left(\mathbf{U}'\mathbf{U}\right)\mathbf{U}^{-1} = \mathbf{IU}^{-1} = \mathbf{U}^{-1},$$

and in particular, $\mathbf{UU}' = \mathbf{I}$.

Second proof: This uses some basic matrix tools which will arise when studying the linear model. Begin by premultiplying $\mathbf{U}'\mathbf{U} = \mathbf{I}$ by $\mathbf{U}$ and postmultiplying by $\mathbf{U}'$ to get $\mathbf{UU}'\mathbf{UU}' = \mathbf{UU}'$, which shows that $\mathbf{UU}'$ is idempotent. Next, $\mathbf{UU}'$ is clearly symmetric (recall that $(\mathbf{AB})' = \mathbf{B}'\mathbf{A}'$). Lastly, the well-known fact that, in general, rank$(\mathbf{A}) = $ rank$(\mathbf{AA}')$ and that $\mathbf{U}$ is obviously full rank implies that rank$(\mathbf{UU}') = n$. Thus, $\mathbf{UU}'$ is a full rank projection matrix, which means it is the identity matrix. (The latter follows because $\mathbf{I}_n - \mathbf{UU}'$ must be a rank zero projection matrix yielding the orthogonal complement.)

For $n \times n$ matrix $\mathbf{A} \geq 0$ with spectral decomposition $\mathbf{A} = \mathbf{UDU}'$, the *Cholesky decomposition* of $\mathbf{A}$, denoted $\mathbf{A}^{1/2}$, is a matrix such that $\mathbf{A} = \mathbf{A}^{1/2}\mathbf{A}^{1/2}$. It can be computed as $\mathbf{A}^{1/2} = \mathbf{UD}^{1/2}\mathbf{U}'$, where $\mathbf{D}^r := \text{diag}\left(\lambda_1^r, \ldots, \lambda_n^r\right)$ for $r \in \mathbb{R}_{>0}$, and for $r = 1/2$, $\lambda_i^{1/2}$ is the nonnegative square root of λ_i. Indeed,

$$\mathbf{A}^{1/2}\mathbf{A}^{1/2} = \mathbf{UD}^{1/2}\mathbf{U}'\, \mathbf{UD}^{1/2}\mathbf{U}' = \mathbf{UD}^{1/2}\mathbf{D}^{1/2}\mathbf{U}' = \mathbf{UDU}'.$$

Notice that, if $\mathbf{A} > 0$, then $\mathbf{A}^{1/2} > 0$.

If $\mathbf{A} > 0$, then $\min(\lambda_i) > 0$ and r can be any real number. In particular, for $r = -1$, $\mathbf{A}^{-1} = \mathbf{UD}^{-1}\mathbf{U}'$. The notation $\mathbf{A}^{-1/2}$ refers to the inverse of $\mathbf{A}^{1/2}$.

The *rank* of $m \times n$ matrix $\mathbf{A}$ is the number of linearly independent columns, which is equivalent to the number of linearly independent rows. Clearly, $\text{rank}(\mathbf{A}) \leq \min(m, n)$; if $\text{rank}(\mathbf{A}) = \min(m, n)$, then $\mathbf{A}$ is said to be *full rank*. It can be shown that

$$\text{rank}\,(\mathbf{A}) = \text{rank}\,(\mathbf{A}') = \text{rank}\,(\mathbf{AA}') = \text{rank}\,(\mathbf{A}'\mathbf{A})$$

and, for conformable matrices $\mathbf{A}$, $\mathbf{B}$ and $\mathbf{C}$, if $\mathbf{B}$ and $\mathbf{C}$ are full rank, then $\text{rank}\,(\mathbf{A}) = \text{rank}\,(\mathbf{BAC})$.

Let r be the number of nonzero eigenvalues of $\mathbf{A}$, counting multiplicities. Then $r \leq \text{rank}\,(\mathbf{A})$, with equality holding when $\mathbf{A}$ is symmetric.

If $\mathbf{A}$ is an $m \times n$ matrix and $\mathbf{B}$ is an $n \times m$ matrix, $n \geq m$, then the nonzero eigenvalues of $\mathbf{AB}$ and $\mathbf{BA}$ are the same, and $\mathbf{BA}$ will have at least $n - m$ zeros (see Abadir and Magnus, 2005, p. 167, for proof). For $n \times n$ symmetric matrices $\mathbf{A}$ and $\mathbf{B}$, all $\text{Eig}(\mathbf{AB})$ are real if either $\mathbf{A}$ or $\mathbf{B}$ is positive semi-definite. To see this, let $\mathbf{A} \geq 0$ so that it admits a Cholesky decomposition $\mathbf{A}^{1/2}$. Then $\text{Eig}(\mathbf{AB}) = \text{Eig}(\mathbf{A}^{1/2}\mathbf{BA}^{1/2})$, but the latter matrix is symmetric, so its eigenvalues are real. See Graybill (1983, Thm. 12.2.11) for another proof.

3.9 Problems

Take a chance! All life is a chance. The man who goes the furthest is generally the one who is willing to do and dare. The 'sure thing' boat never gets far from shore.
(Dale Carnegie)

Courage is the first of human qualities because it is the quality that guarantees all the others.
(Sir Winston Churchill)

3.1. Let X and Y be bivariate normally distributed and define $D = X - Y$, $S = X + Y$. Compute $\text{Cov}\,(D, S)$ and derive the condition for which D and S are independent.

3.2. Let $X \sim \text{N}\left(\mu, \sigma^2\right)$, i.e., a normal r.v. with location μ and scale σ and let $Y \sim \text{N}\,(\mu, 1)$, i.e., location μ, scale one. We know that $\mathbb{E}\,[X] = \mu$. Also, $\mathbb{E}\,[\sigma Y] = \sigma \mathbb{E}\,[Y] = \sigma \mu$. But σY is just a rescaling of Y, so that the mean should be the same as that of Y, namely μ. Which is correct? Is there a conflict here?

3.3. Compute $\Pr(X_1 > X_2)$, where

$$\mathbf{X} = \left[\begin{array}{c} X_1 \\ X_2 \end{array} \right] \sim \mathrm{N}\left(\left[\begin{array}{c} 3 \\ 2 \end{array} \right], \left[\begin{array}{cc} 1 & -0.4 \\ -0.4 & 2 \end{array} \right] \right).$$

3.4. ★★ Let $\mathbf{X} \sim \mathrm{N}(\boldsymbol{\mu}, \boldsymbol{\Sigma})$, where $\boldsymbol{\mu} = (\mu_1, \ldots, \mu_n)'$ and $\boldsymbol{\Sigma} > 0$. Derive the distribution of $\mathbf{X}$ conditional on $\overline{X} = m$. Simplify for $X_i \overset{\text{i.i.d.}}{\sim} \mathrm{N}(\mu, 1)$. Hint: Use (3.22).

3.5. ★ ★ Let $U_t \overset{\text{i.i.d.}}{\sim} \mathrm{N}(0, \sigma^2)$. Let $\{X_t\}$ be a sequence of random variables such that, at time t, X_t is observed, where t is an integer referring to a constant increment of time, e.g., seconds, days, years. Sequence $\{X_t\}$ is referred to as a *time series*. Assume there exist constants $c \in \mathbb{R}$ and $a \in (-1, 1)$ such that, for all $t \in \mathbb{Z}$,

$$X_t = c + aX_{t-1} + U_t. \tag{3.35}$$

   (a) Conditional on the time $t - 1$, i.e., that $X_{t-1} = x_{t-1}$, what is the distribution of X_t?

   (b) (Independent of the previous question.) Assume that X_t has an *unconditional* expectation, i.e., that $\mathbb{E}[X_t] = \mu$ for all t with $|\mu| < \infty$. Compute μ. Similarly, assume $v = \mathbb{V}(X_t)$ for all t. Compute v.

   (c) Substitute $X_{t-1} = c + aX_{t-2} + U_{t-1}$ into (3.35) and simplify. Continue with X_{t-2}, X_{t-3}, etc. Use your result to compute the unconditional mean and variance.

   (d) Now set $c = 0$ in (3.35). Also let $\{Y_t\}$ be another time series, given by $Y_t = bY_{t-1} + U_t$. (Notice that X_t and Y_t both involve U_t.) Using your results in part (c), compute $\mathbb{E}[X_t Y_t]$ and specify when it exists.

3.6. Show that, if $\mathbf{Y} \sim \mathrm{N}(\boldsymbol{\mu}, \boldsymbol{\Sigma})$, then $\mathbf{X} := \mathbf{a} + \mathbf{BY} \sim \mathrm{N}(\boldsymbol{v}, \boldsymbol{\Omega})$, where $\boldsymbol{v} = \mathbf{a} + \mathbf{B}\boldsymbol{\mu}$ and $\boldsymbol{\Omega} = \mathbf{B}\boldsymbol{\Sigma}\mathbf{B}'$.

3.7. Let $\mathbf{Y} \sim \mathrm{N}(\boldsymbol{\mu}, \boldsymbol{\Sigma})$ with $\boldsymbol{\Sigma}$ positive semi-definite and $\det(\boldsymbol{\Sigma}) = 0$. Construct a nonzero location–scale transform of $\mathbf{Y}$ which is identically zero.

3.8. ★ Calculate the eigenvalues of the equicorrelated matrix (3.26) to show that it is positive definite when

$$-\frac{1}{n-1} < \rho < 1.$$

3.9. ★ Let X, Y have a joint bivariate normal distribution

$$\left(\begin{array}{c} X \\ Y \end{array} \right) \sim \mathrm{N}(\boldsymbol{\mu}, \boldsymbol{\Sigma}), \quad \text{where } \boldsymbol{\mu} = \left[\begin{array}{c} \mu_1 \\ \mu_2 \end{array} \right], \quad \boldsymbol{\Sigma} = \left[\begin{array}{cc} \sigma_1^2 & \rho\sigma_1\sigma_2 \\ \rho\sigma_1\sigma_2 & \sigma_2^2 \end{array} \right].$$

From the general p.d.f. expression for the multivariate normal distribution,

$$f_{\mathbf{Y}}(\mathbf{y}) = \frac{1}{|\mathbf{\Sigma}|^{1/2}(2\pi)^{n/2}} \exp\left\{-\frac{1}{2}\left((\mathbf{y}-\boldsymbol{\mu})'\mathbf{\Sigma}^{-1}(\mathbf{y}-\boldsymbol{\mu})\right)\right\},$$

with $\tilde{x} = (x - \mu_1)/\sigma_1$ and $\tilde{y} = (y - \mu_2)/\sigma_2$, $f_{X,Y}(x,y)$ is given by

$$\frac{1}{2\pi\sigma_1\sigma_2\left(1-\rho^2\right)^{1/2}} \exp\left\{-\frac{1}{2}\left(\begin{matrix} x - \mu_1 \\ y - \mu_2 \end{matrix}\right)'\right.$$

$$\left. \times \left[\begin{matrix} \sigma_1^2 & \rho\sigma_1\sigma_2 \\ \rho\sigma_1\sigma_2 & \sigma_2^2 \end{matrix}\right]^{-1}\left(\begin{matrix} x - \mu_1 \\ y - \mu_2 \end{matrix}\right)\right\}$$

$$= \frac{1}{2\pi\sigma_1\sigma_2\left(1-\rho^2\right)^{1/2}} \exp\left\{-\frac{\tilde{x}^2 - 2\rho\tilde{x}\tilde{y} + \tilde{y}^2}{2\left(1-\rho^2\right)}\right\},$$

as in (3.12).

(a) Let $U = (X - \mu_1)/\sigma_1$ and $V = (X - \mu_2)/\sigma_2$. Show that

$$f_{U,V}(u,v) = \frac{1}{2\pi\left(1-\rho^2\right)^{1/2}} \exp\left\{-\frac{u^2 - 2\rho uv + v^2}{2\left(1-\rho^2\right)}\right\}.$$

(b) ★ ★ Show that the m.g.f. of (U, V) is

$$\mathbb{M}_{U,V}(s,t) = \exp\left\{\frac{1}{2}\left(s^2 + 2\rho st + t^2\right)\right\}$$

and that

$$\mathbb{M}_{X,Y}(s,t) = \exp\left\{\frac{1}{2}\left[\sigma_1^2 s^2 + 2\rho\sigma_1\sigma_2 st + \sigma_2^2 t^2\right] + \mu_1 s + \mu_2 t\right\}. \quad (3.36)$$

(c) Show that $\rho = \mathrm{Corr}(X, Y)$. Hint: One way uses (3.36) and a result from (1.20), namely

$$\mathbb{E}[XY] = \frac{\partial^2 \mathbb{M}_{X,Y}(s,t)}{\partial s \partial t}\bigg|_{s=t=0}.$$

The calculations are trivial but tedious; the use of a symbolic mathematics software package is recommended, and gives the answer almost immediately.

(d) Show that X and $Z = Y - \rho\sigma_2 X/\sigma_1$ are independent.

PART II

ASYMPTOTICS AND OTHER APPROXIMATIONS

4

Convergence concepts

If you need to use asymptotic arguments, do not forget to let the number of observations tend to infinity.

(Lucien Le Cam, 1990)

Many readers will have taken an introductory course in statistics, and encountered some 'rules of thumb', such as when the normal distribution can be used to approximate calculations involving the random variable $X \sim \text{Bin}(n, p)$. The basic idea was that, as n grows, (and for p not too far from 0.5), the c.d.f. of X can be approximated by that of a normal distribution with the same mean and variance as X. This is a special case of one of many so-called central limit theorems, which are fundamental results of great theoretical importance in probability theory. In addition to helping mathematically explain why the variation in so many natural phenomena follows the 'bell curve', these results have great practical value in statistical inference.

The above example is just one notion of convergence which involved the sum of appropriately standardized (in this case, i.i.d. Bernoulli) r.v.s. Another type of convergence which the reader has seen is that, in the limit as the degrees of freedom parameter increases, the distribution of a Student's t random variable 'approaches' that of a normal. Yet another type of convergence we have already encountered (in Section 1.1.3) is that convergence of m.g.f.s implies convergence in distribution. These concepts will be detailed in this chapter.

Before discussing the central limit theorem and convergence of r.v.s and m.g.f.s, we examine some basic inequalities involving r.v.s, and some rudimentary notions involving convergence of sequences of sets. Our goal with the chapter as a whole is to emphasize and provide intuition for the basic concepts which are adequate for following the presentation in application-driven textbooks and research papers such as in applied statistics and econometrics. It is also to help prepare for a more advanced course in probability, in which these topics are of supreme importance, discussed in far greater detail, and with much more mathematical depth.

Intermediate Probability: A Computational Approach M. Paolella
© 2007 John Wiley & Sons, Ltd

4.1   Inequalities for random variables

Typically, in order for a particular inequality to hold, it is necessary that certain (absolute) moments of the r.v.s are finite. As such, it is convenient (and standard) to let

$$L_r = \{\text{r.v.s } X : \mathbb{E}[|X|^r] < \infty\}. \tag{4.1}$$

We have already encountered some inequalities for r.v.s in earlier chapters. One was Jensen's inequality (I.4.53), which states that, for any r.v. $X \in L_1$, i.e., with finite mean,

$$\mathbb{E}\big[g(X)\big] \geq g(\mathbb{E}[X]), \quad \text{if } g(\cdot) \text{ is convex,}$$
$$\mathbb{E}\big[g(X)\big] \leq g(\mathbb{E}[X]), \quad \text{if } g(\cdot) \text{ is concave,}$$

recalling that a function f is concave on $[a, b]$ if,

$$\forall x, y \in [a, b] \text{ and } \forall s \in [0, 1] \text{ with } t = 1 - s, \quad f(sx + ty) \geq sf(x) + tf(y). \tag{4.2}$$

In particular, a differentiable function f is concave on an interval if its derivative f' is decreasing on that interval; a twice-differential function f is concave on an interval if $f'' \leq 0$ on that interval.

Another was the Cauchy–Schwarz inequality (I.5.22), which states that, for any two r.v.s $U, V \in L_2$ (i.e., with finite variance),

$$\mathbb{E}[|UV|] \leq +\sqrt{\mathbb{E}[U^2]\,\mathbb{E}[V^2]}. \tag{4.3}$$

A basic, but useful, inequality involving expectation is the following. Let $U, V \in L_r$ for $r > 0$. Then

$$\mathbb{E}[|U + V|^r] \leq \mathbb{E}[(|U| + |V|)^r] \leq 2^r\left(\mathbb{E}[|U|^r] + \mathbb{E}[|V|^r]\right). \tag{4.4}$$

Proof: For $a, b \in \mathbb{R}$, and using the triangle inequality (I.A.6),

$$|a + b|^r \leq \big(|a| + |b|\big)^r \leq \big(2\max(|a|, |b|)\big)^r$$
$$= 2^r \max\big(|a|^r, |b|^r\big) \leq 2^r\big(|a|^r + |b|^r\big).$$

Because this holds for any $a, b \in \mathbb{R}$, it also holds for all possible realizations of r.v.s U and V, so replacing a with U, b with V, and taking expectations (which is inequality preserving; see Section I.4.4.2) yields (4.4). ■

The bound in (4.4) can be sharpened (i.e., reduced) to

$$\mathbb{E}[|U + V|^r] \leq c_r\left(\mathbb{E}[|U|^r] + \mathbb{E}[|V|^r]\right), \quad c_r = \begin{cases} 1, & \text{if } 0 < r \leq 1, \\ 2^{r-1}, & \text{if } r \geq 1. \end{cases} \tag{4.5}$$

See Gut (2005, p. 127) for proof. ■

The next three inequalities are most easily stated using the following useful bit of notation:

> the *k-norm* of r.v. X is defined to be $\|X\|_k = \left(\mathbb{E}[|X|^k]\right)^{1/k}$ for $k \geq 1$.

We will write $\|X\|_k$ as $\mathbb{E}[|X|^k]^{1/k}$ without the extra parentheses.

Hölder's inequality (after Otto Hölder, 1859–1937), generalizes the Cauchy–Schwarz inequality (4.3) to

$$\|UV\|_1 \leq \|U\|_p \|V\|_q, \qquad p, q > 1, \quad p^{-1} + q^{-1} = 1, \tag{4.6}$$

for r.v.s $U \in L_p$ and $V \in L_q$.

Proof: Taking derivatives confirms that the function $\ln x$ is concave for $x \in (0, \infty)$ so that, for $a, b \in (0, \infty)$ and $\lambda \in [0, 1]$, (4.2) and the strict monotonicity of $\exp(\cdot)$ imply

$$\lambda a + (1 - \lambda) b \geq a^\lambda b^{1-\lambda}. \tag{4.7}$$

For $p, q > 1$ and $p^{-1} + q^{-1} = 1$, let $a = |U|^p / \mathbb{E}\left[|U|^p\right]$ and $b = |V|^q / \mathbb{E}\left[|V|^q\right]$ so that (4.7) implies

$$\frac{1}{p} \frac{|U|^p}{\mathbb{E}\left[|U|^p\right]} + \frac{1}{q} \frac{|V|^q}{\mathbb{E}\left[|V|^q\right]} \geq \left(\frac{|U|^p}{\mathbb{E}\left[|U|^p\right]}\right)^{1/p} \left(\frac{|V|^q}{\mathbb{E}\left[|V|^q\right]}\right)^{1/q}. \tag{4.8}$$

The result then follows by taking expectations of (4.8), noting that the expected value of the l.h.s. is unity, and using the inequality-preserving nature of expectation. ∎

Using (4.6), it is easy to prove *Lyapunov's inequality* (after Aleksandr Lyapunov, 1857–1918) which states that

$$\|X\|_r \leq \|X\|_s, \qquad 1 \leq r \leq s, \tag{4.9}$$

if $X \in L_s$. See Problem 4.2 for proof.

For r.v.s $U, V \in L_1$, the *triangle inequality* states that $\|U + V\| \leq \|U\| + \|V\|$, i.e.,

$$\mathbb{E}[|U + V|] \leq \mathbb{E}[|U|] + \mathbb{E}[|V|]. \tag{4.10}$$

Proof: This follows from (I.A.6) and the properties of the expectation operator. ∎

Minkowski's inequality (after Hermann Minkowski, 1864–1909) generalizes the triangle inequality to[1]

$$\|U + V\|_p \leq \|U\|_p + \|V\|_p, \qquad p \geq 1, \tag{4.11}$$

for r.v.s $U, V \in L_p$.

---

[1] Lyapunov's inequality (4.9) can be applied to give $\|U + V\| \leq \|U + V\|_p$ for $p \geq 1$, so that Minkowski's inequality is sometimes stated as $\|U + V\| \leq \|U\|_p + \|V\|_p$, $p \geq 1$, though this is a weaker statement.

Proof: As $\mathbb{E}\left[|U|^p\right]$ and $\mathbb{E}\left[|V|^p\right]$ are finite, (4.4) implies that $\mathbb{E}\left[|U+V|^p\right] < \infty$ and, thus, that $\|U+V\|_p < \infty$.

Next, from the triangle inequality (4.10),

$$\mathbb{E}\left[|U+V|^p\right] = \mathbb{E}\left[|U+V|^{p-1}|U+V|\right] \le \mathbb{E}\left[|U+V|^{p-1}|U|\right] \\ + \mathbb{E}\left[|U+V|^{p-1}|V|\right].$$

Then, with $q = p/(p-1)$ so that $p^{-1} + q^{-1} = 1$, Hölder's inequality (4.6) implies, substituting $(p-1)q = p$ and $1/q = 1 - 1/p$,

$$\mathbb{E}\left[|U+V|^{p-1}|U|\right] \le \left\||U+V|^{p-1}\right\|_q \|U\|_p = \mathbb{E}\left[\left(|U+V|^{p-1}\right)^q\right]^{1/q}\|U\|_p \\ = \mathbb{E}\left[|U+V|^p\right]^{1-1/p}\|U\|_p,$$

and, similarly, $\mathbb{E}\left[|U+V|^{p-1}|V|\right] \le \mathbb{E}\left[|U+V|^p\right]^{1-1/p}\|V\|_p$. Thus

$$\mathbb{E}\left[|U+V|^p\right] \le \mathbb{E}\left[|U+V|^p\right]^{1-1/p}\left(\|U\|_p + \|V\|_p\right),$$

and (4.11) follows by dividing by $\mathbb{E}\left[|U+V|^p\right]^{1-1/p}$. ∎

The next two inequalities, named after two highly influential Russian mathematicians (Markov and Chebyshev), are quite simple, yet very useful.

Markov's inequality (after Andrey Markov, 1856–1922) states that if $X \in L_r$ for some $r > 0$, then, for all $a > 0$,

$$\Pr(|X| \ge a) \le \frac{\mathbb{E}\left[|X|^r\right]}{a^r}. \tag{4.12}$$

The most common special case, also referred to as Markov's inequality, is if $X \in L_1$ is nonnegative, then for all $a > 0$,

$$\Pr(X \ge a) \le \frac{\mathbb{E}[X]}{a}. \tag{4.13}$$

Proof: For (4.13), using the notation from (I.4.31) for discrete and continuous r.v.s,

$$\mathbb{E}[X] = \int_0^\infty x\,dF_X = \int_0^a x\,dF_X + \int_a^\infty x\,dF_X$$

$$\ge \int_a^\infty x\,dF_X \ge \int_a^\infty a\,dF_X = a\int_a^\infty dF_X = a\Pr(X \ge a).$$

The general case is no harder: As in Gut (2005, p. 119), let $g : \mathbb{R}_{\ge 0} \mapsto \mathbb{R}_{\ge 0}$ be a nondecreasing function such that $\mathbb{E}\left[g(|X|)\right] < \infty$. Then

$$\mathbb{E}\left[g(|X|)\right] \ge \mathbb{E}\left[g(|X|)\mathbb{I}(|X| > a)\right] \ge g(a)\mathbb{E}\left[\mathbb{I}(|X| > a)\right] = g(a)\Pr(|X| > a).$$

The result follows by taking $g(x) = x^r$, which, for $x \geq 0$ and $r > 0$, satisfies the constraints on g. ∎

⊖ **Example 4.1** By using (4.13), Problem 4.3 verifies *Chernoff's inequality* (after Herman Chernoff, b. 1923) for random variable X and $c > 0$,

$$\Pr(X \geq c) \leq \inf_{t>0} \mathbb{E}\left[e^{t(X-c)}\right]. \tag{4.14}$$

From this, it is easy to show the *Chernoff bound*

$$\Pr\left(\overline{X}_n \geq c\right) \leq \inf_{t>0} \exp\left(n \log \mathbb{M}\left(\frac{t}{n}\right) - tc\right), \tag{4.15}$$

where $\overline{X}_n := n^{-1} \sum_{i=1}^{n} X_i$, $\mathbb{M}$ is the moment generating function of each of the X_i, and the X_i are i.i.d. r.v.s. This bound is particularly useful in the tail of the distribution, i.e., $c \gg \mathbb{E}\left[\overline{X}_n\right]$. See Problems 4.3–4.5 and 4.17. ∎

Chebyshev's inequality (after Pafnuty Chebyshev, 1821–1894) states that, for $X \in L_2$ with mean μ and variance σ^2, for any $b > 0$,

$$\boxed{\Pr\left(|X - \mu| \geq b\right) \leq \frac{\sigma^2}{b^2}.} \tag{4.16}$$

Proof 1: Apply (4.13) to the nonnegative r.v. $(X - \mu)^2$ and $a = b^2$ to get

$$\Pr\left(|X - \mu| \geq b\right) = \Pr\left((X - \mu)^2 \geq b^2\right) \leq \frac{\mathbb{E}\left[(X - \mu)^2\right]}{b^2} = \frac{\sigma^2}{b^2}.$$

Proof 2: Let $E = \{|X - \mu| \geq b\}$ and $I = \mathbb{I}(E)$, so that $\mathbb{E}[I] = \Pr(E)$. Observe that $I \leq |X - \mu|^2/b^2 = \sigma^2/b^2$ (check both cases, that E occurs, and E does not occur), so taking expectations of both sides (and recalling that expectation preserves inequalities; see Section I.4.4.2) yields (4.16). ∎

⊖ **Example 4.2** Let $X \in L_2$ with $\mathbb{V}(X) = 0$. As $\mathbb{V}(X)$ is finite, $\mu = \mathbb{E}[X]$ exists, as was shown in Section I.4.4.2. It seems intuitive that $\mathbb{V}(X) = 0$ implies $\Pr(X = \mu) = 1$. To prove this, for $n \geq 1$, Chebyshev's inequality (4.16) implies $\Pr\left(|X - \mu| > n^{-1}\right) = 0$, and taking limits of both sides yields

$$0 = \lim_{n \to \infty} \Pr\left(|X - \mu| > n^{-1}\right) = \Pr\left(\lim_{n \to \infty} \{|X - \mu| > n^{-1}\}\right) = \Pr(X \neq \mu),$$

where the exchange of lim and Pr is justified in (I.2.23). ∎

⊖ **Example 4.3** Let $X \in L_2$ with $\mathbb{E}[X] = \mu$ and $\mathbb{V}(X) = \sigma^2$. For some $a > 0$, as $\{X - \mu > a\} \Rightarrow \{|X - \mu| > a\}$, i.e., $\Pr(X - \mu > a) \leq \Pr(|X - \mu| > a)$, Chebyshev's inequality (4.16) implies $\Pr(X - \mu > a) \le \Pr(X - \mu \ge a) \le \sigma^2/a^2$. However, this bound can be sharpened to

$$\Pr(X > \mu + a) \le \frac{\sigma^2}{\sigma^2 + a^2}, \quad \Pr(X < \mu - a) \le \frac{\sigma^2}{\sigma^2 + a^2}, \qquad (4.17)$$

which is known as the *one-sided Chebyshev inequality*, or *Cantelli's inequality* (Gut, 2005, p. 154). To see this, first let $\mu = 0$, so that $\mathbb{E}[X] = 0$, or

$$-a = \int_{-\infty}^{\infty} (x - a)\, dF_X \ge \int_{-\infty}^{\infty} (x - a)\mathbb{I}_{(-\infty, a)}(x)\, dF_X = \mathbb{E}\big[(X - a)\mathbb{I}_{(-\infty, a)}(X)\big]$$

or, multiplying by -1, squaring, and applying (the squares of both sides of) the Cauchy–Schwarz inequality (4.3),

$$a^2 \le \big(\mathbb{E}[(a - X)\mathbb{I}_{(-\infty, a)}(X)]\big)^2 \le \mathbb{E}\big[(a - X)^2\big]\mathbb{E}\big[\mathbb{I}_{(-\infty, a)}^2(X)\big].$$

As $\mathbb{E}\big[\mathbb{I}_{(-\infty, a)}^2(X)\big] = \mathbb{E}\big[\mathbb{I}_{(-\infty, a)}(X)\big] = F_X(a)$ and, expanding and evaluating each term, $\mathbb{E}\big[(a - X)^2\big] = a^2 + \sigma^2$, we get $a^2 \le (a^2 + \sigma^2)F_X(a)$ or

$$\Pr(X > a) \le \frac{\sigma^2}{a^2 + \sigma^2}. \qquad (4.18)$$

Now assume $\mu \ne 0$. Observe that $X - \mu$ and $\mu - X$ have mean zero, so that both statements in (4.17) follow from (4.18). ∎

Chebyshev's order inequality states that, for discrete r.v. X and nondecreasing real functions f and g,

$$\mathbb{E}\big[f(X)\big]\mathbb{E}\big[g(X)\big] \le \mathbb{E}\big[f(X)g(X)\big]. \qquad (4.19)$$

Proof: Following Steele (2004, p. 76), let the support of X be $x_1, \ldots, x_n$ with $x_1 \le x_2 \le \cdots \le x_n$ and $\Pr(X = x_i) = p_i$, $i = 1, \ldots, n$, so that we need to show

$$\left(\sum_{k=1}^{n} f(x_k)\, p_k\right)\left(\sum_{k=1}^{n} g(x_k)\, p_k\right) \le \sum_{k=1}^{n} f(x_k)\, g(x_k)\, p_k.$$

Note that, as f and g are nondecreasing, $0 \le \{f(x_k) - f(x_j)\}\{g(x_k) - g(x_j)\}$ for any $1 \le j \le n$ and $1 \le k \le n$, or

$$f(x_k)g(x_j) + f(x_j)g(x_k) \le f(x_j)g(x_j) + f(x_k)g(x_k). \qquad (4.20)$$

Being probabilities, $0 \leq p_j$, so multiplying by $p_j p_k$ and summing over both j and k yields, for the l.h.s. of inequality (4.20),

$$\sum_{j=1}^{n} \sum_{k=1}^{n} \left[f(x_k) g(x_j) + f(x_j) g(x_k) \right] p_j p_k = 2 \sum_{j=1}^{n} \sum_{k=1}^{n} f(x_k) g(x_j) p_j p_k$$

$$= 2 \left\{ \sum_{j=1}^{n} f(x_j) p_j \right\} \left\{ \sum_{k=1}^{n} f(x_k) p_k \right\}.$$

$$(4.21)$$

Doing the same for the r.h.s. gives $2 \sum_{k=1}^{n} f(x_k) g(x_k) p_k$, from which (4.19) follows. ∎

In the following, we use the abbreviated notation $X_i \overset{\text{ind}}{\sim} (0, \sigma_i^2)$ to indicate that the X_i are independent r.v.s in L_2, each with mean zero and variance $\sigma_i^2 < \infty$.

For $X_i \overset{\text{ind}}{\sim} (0, \sigma_i^2)$ and $n \in \mathbb{N}$, let $S_n = \sum_{i=1}^{n} X_i$, so that $\mathbb{E}[S_n] = 0$, $\mathbb{V}(S_n) = \sum_{i=1}^{n} \sigma_i^2$. Then Chebyshev's inequality (4.16) implies $\Pr(|S_n| \geq a) \leq \mathbb{V}(S_n)/a^2$, i.e.,

$$\Pr(|X_1 + \cdots + X_n| \geq a) \leq \frac{1}{a^2} \sum_{i=1}^{n} \sigma_i^2.$$

However, it turns out that this bound applies to the larger set

$$A_{a,n} := \bigcup_{j=1}^{n} \{ |S_j| \geq a \} = \left\{ \max_{1 \leq j \leq n} |S_j| \geq a \right\}, \qquad (4.22)$$

instead of just $\{ |S_n| \geq a \}$, which is the statement of *Kolmogorov's inequality* (after Andrey Kolmogorov, 1903–1987). Let $X_i \overset{\text{ind}}{\sim} (0, \sigma_i^2)$, $S_j := \sum_{i=1}^{j} X_i$ and $A_{a,n}$ as in (4.22). For any $a > 0$ and $n \in \mathbb{N}$,

$$\boxed{\Pr(A_{a,n}) \leq \frac{1}{a^2} \sum_{i=1}^{n} \sigma_i^2} \qquad (4.23)$$

and, if there exists a c such that $\Pr(|X_k| \leq c) = 1$ for each k, then

$$\boxed{\Pr(A_{a,n}) \geq 1 - \frac{(c+a)^2}{\sum_{i=1}^{n} \sigma_i^2}}, \qquad (4.24)$$

sometimes referred to as the 'other' Kolmogorov inequality (Gut, 2005, p. 123).

Proof: Let a and n be fixed. To prove (4.23), let $S_{j,k} := \sum_{i=j}^{k} X_i$ (so that $S_j = S_{1,j}$), and let N be the smallest value of i such that $S_i^2 > a^2$, and if $S_j^2 \leq a^2$, $j = 1, \dots, n$, then let $N = n$. With this definition of N, it is easy to verify that the two events $\{\max_j S_j^2 > a^2\}$ and $\{S_N^2 > a^2\}$ are equivalent, so that, from Markov's inequality (4.13),

$$\Pr\left(\max_j S_j^2 > a^2\right) = \Pr\left(S_N^2 > a^2\right) \leq \frac{\mathbb{E}[S_N^2]}{a^2}.$$

This is not (4.23), but we will see that $\mathbb{E}[S_N^2] \leq \mathbb{E}[S_n^2] = \mathbb{V}(S_n) = \sum_{i=1}^{n} \sigma_i^2$, thus proving (4.23). To confirm $\mathbb{E}[S_N^2] \leq \mathbb{E}[S_n^2]$, observe that $\mathbb{E}[S_n^2 \mid N = n] = \mathbb{E}[S_N^2 \mid N = n]$ and, for $i = 1, 2, \dots, n-1$, write

$$\mathbb{E}[S_n^2 \mid N = i] = \mathbb{E}[(S_i + S_{i+1,n})^2 \mid N = i]$$
$$= \mathbb{E}[S_i^2 \mid N = i] + 2\mathbb{E}[S_i S_{i+1,n} \mid N = i] + \mathbb{E}[S_{i+1,n}^2 \mid N = i].$$

We have

$$\mathbb{E}[S_i S_{i+1,n} \mid N = i] = \mathbb{E}[S_i \mid N = i]\mathbb{E}[S_{i+1,n} \mid N = i] = \mathbb{E}[S_i \mid N = i]\mathbb{E}[S_{i+1,n}] = 0,$$

because the occurrence or nonoccurrence of event $\{N = i\}$ contains no information about $X_{i+1}, \dots, X_n$, so that, conditional on $N = i$, S_i and $S_{i+1,n}$ are independent. Thus,

$$\mathbb{E}[S_n^2 \mid N = i] \geq \mathbb{E}[S_i^2 \mid N = i] = \mathbb{E}[S_N^2 \mid N = i],$$

i.e., for all values of N, $\mathbb{E}[S_n^2 \mid N] \geq \mathbb{E}[S_N^2 \mid N]$, and taking expectations and using the law of the iterated expectation (I.8.31), $\mathbb{E}[S_n^2] \geq \mathbb{E}[S_N^2]$.

The proof of (4.24) is similar, and can be found in Karr (1993, p. 184) and Gut (2005, p. 123). ∎

4.2   Convergence of sequences of sets

Let Ω denote the sample space of a random experiment, i.e., the set of all possible outcomes, and let $\{A_n \in \Omega, n \in \mathbb{N}\}$ be an infinite sequence $A_1, A_2, \dots$ of subsets of Ω, which we abbreviate to $\{A_n\}$. Recall that the union and intersection of $\{A_n\}$ are given by

$$\bigcup_{n=1}^{\infty} A_n = \{\omega : \omega \in A_n \text{ for some } n \in \mathbb{N}\}, \quad \bigcap_{n=1}^{\infty} A_n = \{\omega : \omega \in A_n \text{ for all } n \in \mathbb{N}\},$$

respectively. The sequence $\{A_n\}$ is monotone increasing if $A_1 \subset A_2 \subset \cdots$, monotone decreasing if $A_1 \supset A_2 \supset \cdots$, and monotone if it is either monotone increasing or monotone decreasing.

Section I.2.3.1 presented the basic properties of probability spaces, two of which we repeat here. First, for sets A, $B \subset \Omega$,

$$A \subset B \Rightarrow \Pr(A) \leq \Pr(B). \tag{4.25}$$

Second, for the sequence of sets $\{A_n\}$,

$$\Pr\left(\bigcup_{n=1}^{\infty} A_n\right) \leq \sum_{n=1}^{\infty} \Pr(A_n), \tag{4.26}$$

which is Boole's inequality, or the property of countable subadditivity. Another useful and easily verified fact is that

$$A \subset B \Leftrightarrow B^c \subset A^c \tag{4.27}$$

or, when combined with (4.25),

$$A \subset B \quad \Rightarrow \quad \Pr(B^c) \leq \Pr(A^c). \tag{4.28}$$

We will also make use of the the continuity property of $\Pr(\cdot)$ for a sequence of monotone events, as given in (I.2.23), i.e., if $A_1, A_2, \ldots$ is a monotone sequence of events, then

$$\lim_{n \to \infty} \Pr(A_n) = \Pr\left(\lim_{n \to \infty} A_n\right). \tag{4.29}$$

Recall from (I.A.1) that, if $\{A_n\}$ is a monotone increasing sequence, then $\lim_{n \to \infty} A_n = A := \bigcup_{n=1}^{\infty} A_n$. This is commonly written as $A_n \uparrow A$. Similarly, from (I.A.2), if the A_i are monotone decreasing, then $\lim_{n \to \infty} A_n = A := \bigcap_{n=1}^{\infty} A_n$, written as $A_n \downarrow A$.

The question arises as to the limits of sets which are not monotone. Let $\{A_n\}$ be an arbitrary (not necessarily monotone) sequence of sets. Analogous to the limit of a deterministic sequence of real numbers (see the beginning of Section I.A.2.4), the *limit supremum* (or *limit superior*) of $\{A_n\}$, and the *limit infimum* (or *limit inferior*) of A_n are denoted and defined as

$$A^* = \limsup_{i \to \infty} A_i = \bigcap_{k=1}^{\infty} \bigcup_{n=k}^{\infty} A_n, \qquad A_* = \liminf_{i \to \infty} A_i = \bigcup_{k=1}^{\infty} \bigcap_{n=k}^{\infty} A_n. \tag{4.30}$$

To better interpret what A^* contains, observe that, for an $\omega \in \Omega$, if $\omega \in A^*$, then $\omega \in \bigcup_{n=k}^{\infty} A_n$ for *every* k. In other words, for any k, no matter how large, there exists an $n \geq k$ with $\omega \in A_n$. This means that $\omega \in A_n$ for infinitely many values of n. Likewise, if $\omega \in \Omega$ belongs to A_*, then it belongs to $\bigcap_{n=k}^{\infty} A_n$ for *some* k, i.e., there exists a k such that $\omega \in A_n$ for all $n \geq k$. Thus, definitions (4.30) are equivalent to

$$A^* = \{\omega : \omega \in A_n \text{ for infinitely many } n \in \mathbb{N}\}, \tag{4.31}$$

$$A_* = \{\omega : \omega \in A_n \text{ for all but finitely many } n \in \mathbb{N}\},$$

and are thus sometimes abbreviated as $A^* = \{A_n \text{ i.o.}\}$ and $A_* = \{A_n \text{ ult.}\}$, where i.o. stands for 'infinitely often' and ult. stands for 'ultimately'.

As a definition, the sequence $\{A_n\}$ converges to A, written $A_n \to A$, iff $A = A^* = A_*$, i.e.,

$$\boxed{A_n \to A \quad \text{iff} \quad A = \limsup A_n = \liminf A_n.}$$
(4.32)

⊖ **Example 4.4** For events $\{A_n\}$, (I.A.3) gives De Morgan's laws as

$$\left(\bigcup_{n=1}^{\infty} A_n \right)^c = \bigcap_{n=1}^{\infty} A_n^c \quad \text{and} \quad \left(\bigcap_{n=1}^{\infty} A_n \right)^c = \bigcup_{n=1}^{\infty} A_n^c.$$

With $B_k = \bigcup_{n=k}^{\infty} A_n$, these imply $B_k^c = \bigcap_{n=k}^{\infty} A_n^c$ and, thus,

$$(A^*)^c = \left(\bigcap_{k=1}^{\infty} \bigcup_{n=k}^{\infty} A_n \right)^c = \left(\bigcap_{k=1}^{\infty} B_k \right)^c = \bigcup_{k=1}^{\infty} B_k^c = \bigcup_{k=1}^{\infty} \bigcap_{n=k}^{\infty} A_n^c,$$
(4.33)

and, similarly, $(A_*)^c = \bigcap_{k=1}^{\infty} \bigcup_{n=k}^{\infty} A_n^c$.   ∎

⊖ **Example 4.5** For $\{A_n\}$ an arbitrary sequence of sets, and with $B_k := \bigcup_{n=k}^{\infty} A_n$, $k = 1, 2, \ldots$, $\{B_k\}$ is a monotone decreasing sequence of events, so that

$$\bigcup_{n=k}^{\infty} A_n = B_k \downarrow \bigcap_{k=1}^{\infty} B_k = \bigcap_{k=1}^{\infty} \bigcup_{n=k}^{\infty} A_n = A^*.$$

That is, as $k \to \infty$, $\bigcup_{n=k}^{\infty} A_n \downarrow A^*$, so that, from (4.29),

$$\Pr(A^*) = \lim_{k \to \infty} \Pr\left(\bigcup_{n=k}^{\infty} A_n \right).$$
(4.34)

Similarly, with $B_k := \bigcap_{n=k}^{\infty} A_n$ a monotone increasing sequence of events,

$$\bigcap_{n=k}^{\infty} A_n = B_k \uparrow \bigcup_{k=1}^{\infty} B_k = \bigcup_{k=1}^{\infty} \bigcap_{n=k}^{\infty} A_n = A_*,$$

i.e., as $k \to \infty$, $\bigcap_{n=k}^{\infty} A_n \uparrow A_*$, and

$$\Pr(A_*) = \lim_{k \to \infty} \Pr\left(\bigcap_{n=k}^{\infty} A_n \right),$$
(4.35)

which are results we use below.   ∎

⊖ ***Example 4.6*** Let $\{A_n\}$ be monotone increasing and let $U = \bigcup_{n=1}^{\infty} A_n$. Because the A_n are monotone increasing, $U = \bigcup_{n=k}^{\infty} A_n$ for any $k \in \mathbb{N}$, so that

$$A^* = \bigcap_{k=1}^{\infty} \bigcup_{n=k}^{\infty} A_n = \bigcap_{k=1}^{\infty} U = U.$$

Likewise, for each $k \in \mathbb{N}$, $A_k = \bigcap_{n=k}^{\infty} A_n$, so that

$$A_* = \bigcup_{k=1}^{\infty} \bigcap_{n=k}^{\infty} A_n = \bigcup_{k=1}^{\infty} A_k = U.$$

Thus, $A^* = A_* = U$, and, from definition (4.32), $A_n \to U$, which is (I.A.1). Similarly, let $\{A_n\}$ be monotone decreasing and let $C = \bigcap_{n=1}^{\infty} A_n$. Because the A_n are monotone decreasing, $C = \bigcap_{n=k}^{\infty} A_n$ for any $k \in \mathbb{N}$, so that

$$A_* = \bigcup_{k=1}^{\infty} \bigcap_{n=k}^{\infty} A_n = \bigcup_{k=1}^{\infty} C = C.$$

Likewise, for each $k \in N$, $A_k = \bigcup_{n=k}^{\infty} A_n$, so that

$$A^* = \bigcap_{k=1}^{\infty} \bigcup_{n=k}^{\infty} A_n = \bigcap_{k=1}^{\infty} A_k = C,$$

and $A_* = A^* = C$, which is (I.A.2). ∎

⊙ ***Example 4.7*** Let $\{A_n\}$ be a sequence of events which is not necessarily monotone. We wish to show that

$$\Pr(A_*) \leq \liminf_n \Pr(A_n) \quad \text{and} \quad \limsup_n \Pr(A_n) \leq \Pr(A^*). \tag{4.36}$$

For the former, let $B_k = \bigcap_{n=k}^{\infty} A_n$. As $B_k \subset A_k$ for each k, (4.25) implies that $\Pr(B_k) \leq \Pr(A_k)$ for each k, and, as B_k is a monotone increasing sequence, $B_k \uparrow \bigcup_{k=1}^{\infty} B_k$, and $\bigcup_{k=1}^{\infty} B_k = \bigcup_{k=1}^{\infty} \bigcap_{n=k}^{\infty} A_n = \liminf_n A_n$. Then, from (4.35),

$$\Pr(\liminf A_k) = \lim_{k \to \infty} \Pr(B_k) \leq \liminf \Pr(A_k),$$

where the last inequality follows from the following facts. Recall from analysis that (i) if sequences b_k and a_k are such that $b_k \leq a_k$ for all k, then $\lim_{k \to \infty} b_k \leq \lim_{k \to \infty} a_k$, and (ii) while $\lim_{k \to \infty} a_k$ may not exist, $\liminf_k a_k$ always does, so that $\lim_{k \to \infty} b_k \leq \liminf_{k \to \infty} a_k$. The second inequality in (4.36) is similar: let $B_k = \bigcup_{n=k}^{\infty} A_k$, so $A_k \subset B_k$ and $B_k \downarrow \limsup_n A_n$. Then, from (4.34) and the aforementioned facts on real sequences, $\Pr(\limsup A_n) = \lim_{k \to \infty} \Pr(B_k) \geq \limsup \Pr(A_k)$. ∎

We can now show the fundamental result which extends the convergence result for monotone sequences. Let $\{A_n\}$ be a sequence of events which is not necessarily monotone. We wish to show that

$$\boxed{\text{if } A_n \to A, \text{ then } \lim_{n\to\infty} \Pr(A_n) \text{ exists, and } \lim_{n\to\infty} \Pr(A_n) = \Pr(A).}$$

First recall some facts from Section I.A.2.4 on analysis of real sequences. If s_n is a deterministic sequence of real numbers, then $U = \limsup s_n$ and $L = \liminf s_n$ exist, and $\lim s_n$ exists iff $U = L$, in which case $\lim s_n = U = L$. From (I.A.84), for any $\epsilon > 0$, $\exists\, N_U \in \mathbb{N}$ such that, $\forall\, n \geq N_U$, $s_n < U + \epsilon$. Likewise, $\exists\, N_L \in \mathbb{N}$ such that, for all $n \geq N_L$, $s_n > L - \epsilon$. Thus, for all $n \geq \max(N_U, N_L)$, $L - \epsilon < s_n < U + \epsilon$, and as $\epsilon > 0$ is arbitrary, it must be the case that $L \leq U$. In particular, if A_n is a sequence of events, and $s_n = \Pr(A_n)$, then $\liminf_n \Pr(A_n) \leq \limsup_n \Pr(A_n)$. Now, from this and (4.36),

$$\Pr\left(\liminf_n A_n\right) \leq \liminf_n \Pr(A_n) \leq \limsup_n \Pr(A_n) \leq \Pr\left(\limsup_n A_n\right). \qquad (4.37)$$

From the assumption that $A_n \to A$ and definition (4.32), we know that $A = \lim_n A_n = \liminf_n A_n = \limsup_n A_n$, so that (4.37) implies

$$\Pr(A) \leq \liminf_n \Pr(A_n) \leq \limsup_n \Pr(A_n) \leq \Pr(A),$$

i.e., $p := \liminf_n \Pr(A_n) = \limsup_n \Pr(A_n)$. Thus, $\lim_n \Pr(A_n)$ exists and $\lim_n \Pr(A_n) = p$. Again from (4.37), we have $\Pr(A) \leq p \leq \Pr(A)$, or $\lim_n \Pr(A_n) = \Pr(A)$, as was to be shown.

The two standard *Borel–Cantelli lemmas*, named after work of Émile Borel and Francesco Cantelli around 1909, are also fundamental results. They are as follows. First, for a sequence $\{A_n\}$ of arbitrary events,

$$\boxed{\sum_{n=1}^{\infty} \Pr(A_n) < \infty \quad \Rightarrow \quad \Pr(A_n \text{ i.o.}) = 0.} \qquad (4.38)$$

Second, for a sequence $\{A_n\}$ of *independent* events,

$$\boxed{\sum_{n=1}^{\infty} \Pr(A_n) = \infty \quad \Rightarrow \quad \Pr(A_n \text{ i.o.}) = 1.} \qquad (4.39)$$

To prove (4.38), use (4.34), (4.26) and the Cauchy criterion for convergent sums (I.A.86) to get

$$\Pr(A_n \text{ i.o.}) = \lim_{k\to\infty} \Pr\left(\bigcup_{n=k}^{\infty} A_n\right) \leq \lim_{k\to\infty} \sum_{n=k}^{\infty} \Pr(A_n) = 0.$$

To prove (4.39), use (4.33) and (4.35) to get

$$\Pr(A_n \text{ i.o.}) = 1 - \Pr\left(\bigcup_{k=1}^{\infty}\bigcap_{n=k}^{\infty} A_n^c\right) = 1 - \lim_{k\to\infty}\Pr\left(\bigcap_{n=k}^{\infty} A_n^c\right).$$

As the A_n are independent, so are the events A_n^c (see Section I.3.2), so, continuing,

$$\Pr(A_n \text{ i.o.}) = 1 - \lim_{k\to\infty}\prod_{n=k}^{\infty}\Pr\left(A_n^c\right) = 1 - \lim_{k\to\infty}\prod_{n=k}^{\infty}\left[1 - \Pr\left(A_n\right)\right].$$

As $1 - x \le e^{-x}$ for $x \ge 0$,[2]

$$\Pr(A_n \text{ i.o.}) \ge 1 - \lim_{k\to\infty}\exp\left\{-\sum_{n=k}^{\infty}\Pr(A_n)\right\} = 1 - 0 = 1,$$

because $\sum_{n=1}^{\infty}\Pr(A_n) = \infty$ implies that, for any $k \in \mathbb{N}$, $\sum_{n=k}^{\infty}\Pr(A_n) = \infty$.

By imposing independence, the two lemmas can be combined to give a so-called *zero–one law*: for a sequence $\{A_n\}$ of independent events, $\Pr(A_n \text{ i.o.}) = 0$ when $\sum_{n=1}^{\infty}\Pr(A_n)$ is finite, and equals one otherwise. This, implies, for example, that if one shows $\Pr(A_n \text{ i.o.}) < 1$, then $\Pr(A_n \text{ i.o.}) = 0$.

As an example of the first lemma, let X_n be a sequence of r.v.s with $\Pr(X_n = 0) = n^{-2}$, $n \ge 1$. Then, from (I.A.88) (see also Example 1.26), $\sum_{n=1}^{\infty}\Pr(X_n = 0) = \pi^2/6 < \infty$, so that, from (4.31) and the first lemma, the probability of event $\{X_n = 0\}$ occurring for infinitely many n is zero.

A famous illustration of the second lemma is the *infinite monkey theorem*, which states that a monkey typing at random (each keystroke is independent of the others) on a typewriter keyboard for an infinite time will, with probability one, type the collected works of William Shakespeare. Limiting the goal to *Hamlet*, the intuition is that, if there are k keys on the keyboard and a sequence of s letters and punctuation in the desired text, then with n monkeys working independently of one another on their own typewriters, the probability of not a single monkey typing *Hamlet* is $(1 - k^{-s})^n$, which, in the limit as $n \to \infty$, is zero. For the lemma, let $p = \Pr(A_n)$, where A_n is the event that the nth monkey succeeds. As $\sum_{n=1}^{\infty}\Pr(A_n) = \sum_{n=1}^{\infty} p = \infty$, (4.39) states that infinity many monkeys will accomplish *Hamlet*! Further discussion of this, and an interesting digression of the relevance of this lemma to the so-called Bible Code (finding hidden messages in ancient religious books), as well as much more detail on the lemmas, can be found in Gut (2005, Section 2.18).

---

[2] To see this, with $f(x) = e^{-x}$ and $g(x) = 1 - x$, $f(0) = g(0) = 1$, and $g'(x) \le f'(x)$ because

$$x \ge 0 \Leftrightarrow 0 \ge -x \Leftrightarrow 1 \ge e^{-x} \Leftrightarrow -1 \le -e^{-x} \Leftrightarrow g'(x) \le f'(x).$$

This is similar to the result shown in Section I.A.2.2.3 (just below Example I.A.12) that, for $x > 0$, $\ln(1 + x) < x$.

4.3 Convergence of sequences of random variables

Recall from Section I.4.1.1 that, for the general probability space $\{\Omega, \mathcal{A}, \Pr(\cdot)\}$, where the σ-field $\mathcal{A}$ is the class of subsets of $\mathbb{R}$ which can be assigned a probability, the function $X : \Omega \to \mathbb{R}$ is a random variable (relative to the collection of measurable events $\mathcal{A}$) iff, for every $x \in \mathbb{R}$, $\{\omega \in \Omega \mid X(\omega) \le x\} \in \mathcal{A}$. The random variable X induces the probability space $\{\mathbb{R}, \mathcal{B}, \Pr(\cdot)\}$, where $\mathcal{B}$ is the Borel σ-field, generated by the collection of intervals $(a, b]$, $a, b \in \mathbb{R}$.

There are several different notions of convergence when it comes to random variables, and some imply one or more of the others. Of critical importance before commencing is to understand the difference between two *random variables*, say X and Y, 'being close' (meaning that, when both are observed, their values coincide with probability one) and their *distributions* being close. Let X and Y be r.v.s defined on the same probability space $\{\mathbb{R}, \mathcal{B}, \Pr(\cdot)\}$. As in (1.17), if $\Pr(X \in A) = \Pr(Y \in A)$ for all $A \in \mathcal{B}$, then X and Y are said to be *equal in distribution*, written $X \overset{d}{=} Y$. If the set $\{\omega : X(\omega) \ne Y(\omega)\}$ is an event in $\mathcal{B}$ having probability zero (termed a *null event*), then X and Y are said to be *equal almost surely* or *almost surely equal*, written $X \overset{a.s.}{=} Y$. To emphasize the difference, let X and Y be i.i.d. standard normal. They then have *exactly* the same distribution, but, as they are independent, they are equal with probability zero.

4.3.1 Convergence in probability

The sequence of (univariate) random variables $\{X_n\}$ is said to *converge in probability* to the r.v. X iff, for all $\epsilon > 0$,

$$\lim_{n \to \infty} \Pr(\{\omega \in \Omega : |X_n(\omega) - X(\omega)| > \epsilon\}) = 0, \qquad (4.40)$$

and we write $X_n \overset{p}{\to} X$. More commonly, this is written without reference to ω as

$$\lim_{n \to \infty} \Pr(|X_n - X| > \epsilon) = 0 \quad \text{or, equivalently,} \quad \lim_{n \to \infty} \Pr(|X_n - X| < \epsilon) = 1, \qquad (4.41)$$

for all $\epsilon > 0$. This can also be expressed by saying $X_n \overset{p}{\to} X$ iff, for all $\epsilon > 0$ and $\delta > 0$, there exists $N \in \mathbb{N}$ such that $\Pr(|X_n - X| > \epsilon) < \delta$ for all $n \ge N$. In the following, we will write 'Assume $X_n \overset{p}{\to} X$' to mean 'Let $\{X_n\}$ be a sequence of r.v.s which converges in probability to X'.

Remark: Another common notation for convergence in probability besides $X_n \overset{p}{\to} X$, particularly in time series analysis and econometrics, is plim $X_n = X$ (read: the *probability limit of X_n is X*), which was introduced into the literature by the influential paper of Mann and Wald (1943). ∎

It is often the case in applications that X is degenerate, i.e., a constant, as in the next example.

⊛ **Example 4.8** (*Weak law of large numbers* (WLLN) for uncorrelated r.v.s with same, finite, first and second moments) Let $\{X_n\}$ be a sequence of uncorrelated r.v.s in L_2, each with mean μ and variance σ^2, and let $\overline{X}_n = n^{-1} \sum_{i=1}^n X_i$, the average of the first n elements of the sequence. The WLLN states that $\overline{X}_n \overset{p}{\to} \mu$. To prove this, as $\overline{X}_n$ has mean μ and variance σ^2/n, it follows immediately from Chebyshev's inequality (4.16) that, for any $\epsilon > 0$, $\Pr\left(|\overline{X}_n - \mu| \geq \epsilon\right) \leq \sigma^2/\left(n\epsilon^2\right)$, so that, in the limit, from definition (4.41), $\overline{X}_n \overset{p}{\to} \mu$. ∎

Remark: In the previous example, it was required that the variance of the X_n was finite. This assumption can be relaxed, but then it is necessary to impose that $\{X_n\}$ be i.i.d. r.v.s. In particular, if $\{X_n\}$ is a sequence of i.i.d. r.v.s in L_1 with expectation μ, then $\overline{X}_n \overset{p}{\to} \mu$. For proof, see Bierens (2004, p. 140) or Rohatgi and Saleh (2001, p. 278). Finally, the existence of the first moment can be relaxed as well; see Resnick (1999, Section 7.2). ∎

⊖ **Example 4.9** Let $c, k \in \mathbb{R}$ and assume $X_n \overset{p}{\to} X$. If $c = 0$, then it is immediate from (4.41) that $(cX_n + k) \overset{p}{\to} cX + k$, while for $c \neq 0$, observe that, for any $\epsilon > 0$,

$$\lim_{n \to \infty} \Pr\left(|(cX_n + k) - (cX + k)| \geq \epsilon\right) = \lim_{n \to \infty} \Pr\left(|X_n - X| \geq \epsilon/|c|\right) = 0,$$

so that, for any $c, k \in \mathbb{R}$, $(cX_n + k) \overset{p}{\to} cX + k$. ∎

⊚ **Example 4.10** Assume $X_n \overset{p}{\to} a$, let $A \subset \mathbb{R}$, and let $g : A \to \mathbb{R}$ be a function continuous at point a with $a \in A$. We wish to confirm that $g(X_n) \overset{p}{\to} g(a)$. Recall the definition of continuity from Section I.A.2.1: the function g is continuous at a if, for a given $\epsilon > 0$, there exists $\delta > 0$ (with δ being a function of a and ϵ) such that, if $|x - a| < \delta$ and $x \in A$, then $|g(x) - g(a)| < \epsilon$. The contrapositive of this is: if g is continuous at a, then, for a given $\epsilon > 0$, there exists $\delta > 0$ such that, if $|g(x) - g(a)| \geq \epsilon$ then $\{|x - a| \geq \delta\}$. This implies that (recalling that r.v. X_n is a function of $\omega \in \Omega$),

$$\{\omega : |g(X_n(\omega)) - g(a)| \geq \epsilon\} \subset \{\omega : |X_n(\omega) - a| \geq \delta\}. \tag{4.42}$$

From (4.25), this implies that, for a given $\epsilon > 0$, there exists $\delta > 0$ such that

$$\Pr\{|g(X_n) - g(a)| \geq \epsilon\} \leq \Pr\{|X_n - a| \geq \delta\}.$$

The r.h.s. probability tends to zero for all δ, including the one corresponding to the choice of ϵ, so that $\lim_{n \to \infty} \Pr\{|g(X_n) - g(a)| \geq \epsilon\} = 0$, i.e., $g(X_n) \overset{p}{\to} g(a)$. ∎

⊚ **Example 4.11** In Example 4.10, we would like to be able to replace a with X, i.e., to be able to assume that $X_n \overset{p}{\to} X$, where X is a (possibly nondegenerate) random variable, thus generalizing the result in Example 4.9 to the nonlinear case:

$$\boxed{X_n \overset{p}{\to} X, \ g \in \mathcal{C}^0 \quad \Rightarrow \quad g(X_n) \overset{p}{\to} g(X).} \tag{4.43}$$

The problem is that, unlike a, $X(\omega)$ varies according to ω. This is not a problem if g is uniformly continuous: the function $g : A \to \mathbb{R}$ is uniformly continuous on $A \subset \mathbb{R}$ if, given $\epsilon > 0$, $\exists\, \delta > 0$ such that, for all $x, y \in A$ such that $|x - y| < \delta$, $|g(x) - g(y)| < \epsilon$. Also, a continuous, real-valued function on a closed and bounded interval is uniformly continuous (see Stoll, 2001, p. 146).

To prove (4.43), as in Gut (2005, p. 245), for $g \in C^0$ (that is, g is continuous but not necessarily uniform continuous), and given $\epsilon > 0$ and $\eta > 0$, choose $L \in \mathbb{R}$ such that $\Pr(|X| > L) < \eta/2$ (note that, because $F_X(-\infty) = 0$ and $F_X(\infty) = 1$, L exists). Then g is uniformly continuous on $[-L, L]$, and, for $S := \{\omega : |X(\omega)| \le L\}$, (4.42) now reads

$$\{\omega \in S : |g(X_n(\omega)) - g(X(\omega))| \ge \epsilon\} \subset \{\omega : |X_n(\omega) - X(\omega)| \ge \delta\}. \qquad (4.44)$$

Then, from (4.44) and (4.25),

$$\begin{aligned}
\Pr(|g(X_n) - g(X)| \ge \epsilon) &= \Pr(\{|g(X_n) - g(X)| \ge \epsilon\} \cap \{|X| \le L\}) \\
&\quad + \Pr(\{|g(X_n) - g(X)| \ge \epsilon\} \cap \{|X| > L\}) \\
&\le \Pr(|X_n - X| \ge \delta) + \Pr(|X| > L). \qquad (4.45)
\end{aligned}$$

As $X_n \overset{p}{\to} X$, $\exists N \in \mathbb{N}$ such that, for any δ (in particular, the one corresponding to the chosen ϵ), for all $n > N$, $\Pr(|X_n - X| \ge \delta) < \eta/2$. Thus, for the chosen ϵ and η, continuing (4.45),

$$\Pr(|g(X_n) - g(X)| \ge \epsilon) \le \Pr(|X_n - X| \ge \delta) + \Pr(|X| > L) \le \frac{\eta}{2} + \frac{\eta}{2} = \eta,$$

thus verifying (4.43). ∎

⊙ **Example 4.12** Assume $X_n \overset{p}{\to} X$, $Y_n \overset{p}{\to} Y$, and $\epsilon > 0$. Let $d_X = X_n - X$, $d_Y = Y_n - Y$, $S_n = X_n + Y_n$, and $S = X + Y$. From the triangle inequality,

$$\{|S_n - S| > \epsilon\} = \{|d_X + d_Y| > \epsilon\} \subset \{|d_X| + |d_Y| > \epsilon\} =: C. \qquad (4.46)$$

With $A = \{|d_X| > \epsilon/2\}$, $B = \{|d_Y| > \epsilon/2\}$, Figure 4.1 confirms that $C \subset \{A \cup B\}$ (Problem 4.11 shows this algebraically) in which case (4.25) implies

$$\Pr(|S_n - S| > \epsilon) \le \Pr(C) \le \Pr(A \cup B) \le \Pr(A) + \Pr(B) \to 0,$$

so that

$$\boxed{X_n \overset{p}{\to} X, \; Y_n \overset{p}{\to} Y \;\; \Rightarrow \;\; X_n + Y_n \overset{p}{\to} X + Y.} \qquad (4.47)$$

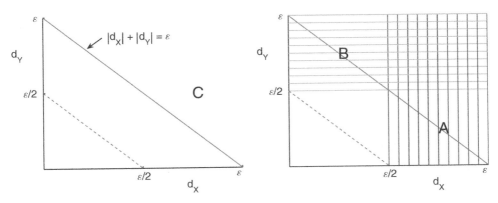

Figure 4.1 Graphically verifies that $C \subset \{A \cup B\}$, where C is the region above the line $|d_X| + |d_Y| = \epsilon$ (left), B is the region indicated by horizontal lines, and A is the region indicated by vertical lines

Combining (4.47) and the result of Example 4.9, we see that convergence in probability is closed under linear transformations, i.e., if $X_n \overset{p}{\to} X$ and $Y_n \overset{p}{\to} Y$, then, for constants $a, b \in \mathbb{R}$, $aX_n + bY_n \overset{p}{\to} aX + bY$. More generally, from (4.43), if $X_n \overset{p}{\to} X$, $Y_n \overset{p}{\to} Y$, and $g, h \in C^0$, then $g(X_n) + h(Y_n) \overset{p}{\to} g(X) + g(Y)$. For example,

$$X_n Y_n = \frac{1}{2}X_n^2 + \frac{1}{2}Y_n^2 - \frac{1}{2}(X_n + Y_n)^2 \overset{p}{\to} \frac{1}{2}X^2 + \frac{1}{2}Y^2 - \frac{1}{2}(X + Y)^2 = XY,$$

i.e., if $X_n \overset{p}{\to} X$ and $Y_n \overset{p}{\to} Y$, then $X_n Y_n \overset{p}{\to} XY$. ∎

The concept of convergence in probability is easily extended to sequences of multivariate r.v.s. In particular, the sequence $\{\mathbf{X}_n\}$ of k-dimensional r.v.s converges in probability to k-dimensional r.v. $\mathbf{X}$ iff

$$\lim_{n \to \infty} \Pr\left(\|\mathbf{X}_n - \mathbf{X}\| > \epsilon\right) = 0, \tag{4.48}$$

and we write $\mathbf{X}_n \overset{p}{\to} \mathbf{X}$. Letting X_{nj} denote the jth marginal random variable in the vector $\mathbf{X}_n$ and X_j denote the jth element of vector $\mathbf{X}$, Problem 4.8 verifies that

$$\mathbf{X}_n \overset{p}{\to} \mathbf{X} \quad \Leftrightarrow \quad X_{nj} \overset{p}{\to} X_j, \quad j = 1, 2, \ldots, k. \tag{4.49}$$

4.3.2 Almost sure convergence

Sequence $\{X_n\}$ is said to converge *almost surely* to r.v. X iff

$$\Pr\left(\omega : \lim_{n \to \infty} X_n(\omega) = X(\omega)\right) = 1, \tag{4.50}$$

and we write $\lim_{n\to\infty} X_n = X$ a.s. or $X_n \overset{a.s.}{\to} X$. Observe how this definition differs from (4.40) for convergence in probability (more on this below).

Remark: Other expressions for almost sure convergence are convergence *with probability one* and convergence *almost everywhere*. ∎

Almost sure convergence is similar to pointwise convergence of functions (recall Section I.A.2.4.4); however, it does not impose the condition that $\lim_{n\to\infty} X_n(\omega) = X(\omega)$ for *all* $\omega \in \Omega$, but rather only on a set of ω with probability one. In particular, $X_n \overset{a.s.}{\to} X$ iff there exists a null event $E \in \mathcal{A}$ (often termed the *exception set*) with $\Pr(E) = 0$ and for all $\omega \in E^c$, $\lim_{n\to\infty} X_n(\omega) = X(\omega)$. Observe that the definition allows E to be empty.

⊖ **Example 4.13**   Let ω be a random number drawn from the probability space $\{\Omega, \mathcal{A}, \Pr(\cdot)\}$ with $\Omega = [0, 1]$, $\mathcal{A}$ the Borel σ-field given by the collection of intervals $[a, b]$, $a \le b$, $a, b \in [0, 1]$, and $\Pr(\cdot)$ to be uniform, i.e., for event $A = [a, b] \in \mathcal{A}$, $\Pr(A) = b - a$. (In short, let ω be a uniformly distributed r.v. on the interval $[0, 1]$.) Let $X_n(\omega) = n\mathbb{I}_{[0,1/n]}(\omega)$ and let $E = \{0\}$ be the exception set, with $\Pr(E) = 0$. For $\omega \in E^c = (0, 1]$, $\lim_{n\to\infty} X_n(\omega) \to 0$, but as $X_n(0) = n$, $\lim_{n\to\infty} X_n(\omega) \nrightarrow 0$ for all $\omega \in \Omega$. Thus, the sequence $\{X_n\}$ converges almost surely to zero. See Resnick (1999, p. 168) for further discussion and a less trivial example. ∎

⊙ **Example 4.14**   We wish to confirm that (i) X in (4.50) is unique up to a set of measure zero, (ii) almost sure convergence is preserved under addition, and (iii) almost sure convergence is preserved under continuous transformation. We use the notation $N(\{Z_n\}, Z)$ to denote the exception set with respect to sequence $\{Z_n\}$ and random variable Z, i.e., $N(\{Z_n\}, Z) = \{\omega : \lim_{n\to\infty} Z_n(\omega) \ne Z(\omega)\}$.

(i) Assume $X_n \overset{a.s.}{\to} X$ and $X_n \overset{a.s.}{\to} Y$. We wish to show that $\Pr(X = Y) = 1$. Let $N_X = N(\{X_n\}, X)$ and $N_Y = N(\{X_n\}, Y)$, with $\Pr(N_X) = \Pr(N_Y) = 0$. Choose an $\omega \in (N_X \cup N_Y)^c$, so that, using the triangle inequality and taking the limit as $n \to \infty$, definition (4.50) implies

$$|X(\omega) - Y(\omega)| \le |X(\omega) - X_n(\omega)| + |X_n(\omega) - Y(\omega)| \to 0.$$

Thus, $(N_X \cup N_Y)^c \subset \{\omega : X(\omega) = Y(\omega)\}$, and using (4.28) then gives

$$\Pr(X \ne Y) \le \Pr(N_X \cup N_Y) \le \Pr(N_X) + \Pr(N_Y) = 0,$$

or $\Pr(X = Y) = 1$.

(ii) Now assume $X_n \overset{a.s.}{\to} X$ and $Y_n \overset{a.s.}{\to} Y$. To show that $X_n + Y_n \overset{a.s.}{\to} X + Y$, let $N_X = N(\{X_n\}, X)$ and $N_Y = N(\{Y_n\}, Y)$. Then, for all $\omega \in (N_X \cup N_Y)^c$,

$$X_n(\omega) + Y_n(\omega) = (X_n + Y_n)(\omega) \to (X + Y)(\omega) = X(\omega) + Y(\omega).$$

As $\Pr(N_X \cup N_Y) \le \Pr(N_X) + \Pr(N_Y) = 0$, $X_n + Y_n \overset{a.s.}{\to} X + Y$.

(iii) We wish to show that

$$\text{if } X_n \overset{a.s.}{\to} X \text{ and } g \in C^0, \text{ then } g(X_n) \overset{a.s.}{\to} g(X). \tag{4.51}$$

For $g \in C^0$, similar to (4.44),

$$\{\omega : g(X_n(\omega)) \nrightarrow g(X(\omega))\} \subset \{\omega : X_n(\omega) \nrightarrow X(\omega)\},$$

which, from (4.27), is equivalent to

$$\{\omega : X_n(\omega) \to X(\omega)\} \subset \{\omega : g(X_n(\omega)) \to g(X(\omega))\},$$

and as the first set has probability one, so must the second, thus proving (4.51). Alternatively, let $N_X = N(\{X_n\}, X)$ and $\omega \notin N_X$, so that

$$\lim_{n \to \infty} g(X_n(\omega)) = g\left(\lim_{n \to \infty} X_n(\omega)\right) = g(X(\omega)),$$

from the continuity of g; see also (I.A.12). This is true for all $\omega \notin N_X$ and $\Pr(N) = 0$, which establishes almost sure convergence. ∎

To help exemplify the difference between almost sure convergence and convergence in probability, we use a statement equivalent to (4.50). Sequence $X_n \overset{a.s.}{\to} X$ iff, for every $\epsilon > 0$,

$$\lim_{m \to \infty} \Pr\left(\bigcup_{n=m}^{\infty} \{|X_n - X| > \epsilon\}\right) = \lim_{m \to \infty} \Pr\left(\sup_{n \geq m} |X_n - X| > \epsilon\right) = 0, \tag{4.52}$$

or, equivalently,

$$\lim_{m \to \infty} \Pr\left(\bigcap_{n \geq m} \{|X_n - X| \leq \epsilon\}\right) = 1. \tag{4.53}$$

Using (4.34) with $A_n = |X_n - X| > \epsilon$, (4.52) states that

$$X_n \overset{a.s.}{\to} X \quad \Leftrightarrow \quad \Pr(A^*) = \lim_{m \to \infty} \Pr\left(\bigcup_{n=m}^{\infty} A_n\right) = 0, \tag{4.54}$$

while (4.35) implies that

$$X_n \overset{a.s.}{\to} X \quad \Leftrightarrow \quad \Pr(A_*) = \lim_{m \to \infty} \Pr\left(\bigcap_{n=m}^{\infty} A_n^c\right) = 1.$$

Informally speaking, the latter statement, for example, says that, with probability one, an ω occurs such that, for any $\epsilon > 0$, $|X_n - X| \leq \epsilon$ for all n sufficiently large. A detailed proof of (4.52) can be found in Gut (2005, Section 5.1.2).

Based on (4.52), almost sure convergence can also be expressed by saying $X_n \overset{a.s.}{\to} X$ iff $\forall \epsilon > 0$ and $\forall \delta \in (0, 1)$, $\exists N \in \mathbb{N}$ such that, $\forall m \geq N$, $\Pr\left(\bigcap_{n \geq m}\{|X_n - X| < \epsilon\}\right) < 1 - \delta$. This is easily used to show that $X_n \overset{a.s.}{\to} X$ implies $X_n \overset{p}{\to} X$ as follows. Let $X_n \overset{a.s.}{\to} X$ and define $A_n := \{|X_n - X| > \epsilon\}$. Clearly, for all $n \in \mathbb{N}$, $A_n \subset \bigcup_{k=n}^{\infty} A_k$. Using (4.25) on, and taking limits of, the latter expression, and using (4.52), we have

$$\lim_{n \to \infty} \Pr\left(|X_n - X| > \epsilon\right) = \lim_{n \to \infty} \Pr\left(A_n\right) \leq \lim_{n \to \infty} \Pr\left(\bigcup_{k=n}^{\infty} A_k\right) = 0,$$

i.e.,

$$\boxed{X_n \overset{a.s.}{\to} X \quad \Rightarrow \quad X_n \overset{p}{\to} X.} \tag{4.55}$$

Remark:

(a) A converse of (4.55) exists: if $\{X_n\}$ is a *monotone* sequence of r.v.s, then $X_n \overset{p}{\to} X \Rightarrow X_n \overset{a.s.}{\to} X$. See Gut (2005, p. 213) for proof.

(b) There is an even stronger form of convergence called *complete convergence*, $X_n \overset{c.c.}{\to} X$. Problem 4.10 shows that it implies almost sure convergence. ∎

⊖ ***Example 4.15*** To see that the converse of (4.55) does not hold in general, as in Gut (1995, p. 156), let $\{X_n, n \in \mathbb{N}\}$ be a sequence of *independent* r.v.s such that $\Pr(X_n = 1) = 1 - n^{-1}$ and $\Pr(X_n = n) = n^{-1}$. Then, for every $\epsilon > 0$,

$$\lim_{n \to \infty} \Pr(|X_n - 1| > \epsilon) = \lim_{n \to \infty} \Pr(X_n = n) = \lim_{n \to \infty} \frac{1}{n} = 0,$$

so that $X_n \overset{p}{\to} 1$. Now choose an $\epsilon > 0$, $\delta \in (0, 1)$, and $n \in \mathbb{N}$. In light of (4.53) and using the continuity property (4.29) and the independence of the X_n, with $A_m := \{|X_m - 1| < \epsilon\}$,

$$\Pr\left(\bigcap_{m=n+1}^{\infty} A_m\right) = \Pr\left(\lim_{N \to \infty} \bigcap_{m=n+1}^{N} A_m\right) = \lim_{N \to \infty} \Pr\left(\bigcap_{m=n+1}^{N} A_m\right)$$

$$= \lim_{N \to \infty} \prod_{m=n+1}^{N} \Pr(A_m) = \lim_{N \to \infty} \prod_{m=n+1}^{N} \Pr(X_m = 1)$$

$$= \lim_{N \to \infty} \frac{n}{n+1} \frac{n+1}{n+2} \cdots \frac{N-1}{N} = \lim_{N \to \infty} \frac{n}{N} = 0 \neq 1,$$

so that $X_n \overset{a.s.}{\nrightarrow} 1$. ∎

⊖ ***Example 4.16*** (Example 4.15 cont.) Now let $\{X_n, n \geq 1\}$ be a sequence of r.v.s, not necessarily independent, such that $\Pr(X_n = 1) = 1 - n^{-\alpha}$ and $\Pr(X_n = n) = n^{-\alpha}$ for

$\alpha > 1$. Combining the first Borel–Cantelli lemma (4.38) (using $A_n = |X_n - 1| > \epsilon$) with (4.54) and the convergence result in Example I.A.34,

$$\sum_{n=1}^{\infty} \Pr(A_n) = \sum_{n=1}^{\infty} \frac{1}{n^{\alpha}} < \infty \quad \Rightarrow \quad \Pr(A_n \text{ i.o.}) = 0 \quad \Leftrightarrow \quad X_n \overset{a.s.}{\to} 1.$$

Observe that this holds for any $\alpha > 1$, and also the independence of the X_n is not required. If we assume independence, and take $\alpha \in (0, 1]$, then the second Borel–Cantelli lemma (4.39) can similarly be used to prove that $X_n \overset{a.s.}{\nrightarrow} 1$, thus generalizing the result in Example 4.15. ∎

⊛ **Example 4.17**   (*Strong law of large numbers* (SLLN)) Let $\{X_n\}$ be a sequence of i.i.d. r.v.s in L_4 with expected value μ and $K := \mathbb{E}[X_1^4]$, and let $S_r = \sum_{i=1}^{r} X_i$ and $\overline{X}_r = r^{-1} S_r$. The SLLN states that $\overline{X}_n \overset{a.s.}{\to} \mu$. To prove this, first take $\mu = 0$. As in Karr (1993, Section 5.5.1) and Ross (2006, Section 8.4), from the multinomial theorem (I.1.34), expanding S_r^4 yields terms

$$S_r^4 = \sum_{\substack{(n_1,\dots,n_r):n_i \geq 0, \\ n_1+\cdots+n_r=4}} \binom{4}{n_1, \dots, n_r} \prod_{i=1}^{r} X_i^{n_i} = \sum_{i=1}^{r} X_i^4 + \binom{4}{2} \sum_{i=1}^{r} \sum_{j=i+1}^{r} X_i^2 X_j^2 + \cdots,$$

where the second sum is over all $\binom{r}{2}$ pairs of (i, j) subscripts such that $i \neq j$, and the additional terms are of the form $X_i^3 X_j$, $X_i^2 X_j X_k$ and $X_i X_j X_k X_\ell$, where all subscripts are different. The independence assumption and zero mean of the $\{X_n\}$ imply that the expected values of these terms are all zero. Thus, from the i.i.d. assumption,

$$\mathbb{E}\left[S_r^4\right] = r \mathbb{E}\left[X_1^4\right] + \binom{4}{2}\binom{r}{2}\mathbb{E}\left[X_1^2 X_2^2\right] = rK + 3r(r-1)\left(\mathbb{E}\left[X_1^2\right]\right)^2.$$

From (I.4.46), $\left(\mathbb{E}\left[X_1^2\right]\right)^2 \leq \mathbb{E}\left[X_1^4\right] = K$, so $\mathbb{E}\left[S_r^4\right] \leq rK + 3r(r-1)K \leq rK + 3r^2 K$ and

$$\mathbb{E}\left[\frac{S_r^4}{r^4}\right] \leq K\left[\frac{1}{r^3} + \frac{3}{r^2}\right].$$

Since $\sum_{r=1}^{\infty} r^{-\alpha}$ is a convergent series for $\alpha > 1$ (see Section I.A.2.4) and since, for the sequence $\{Y_n\}$ of nonnegative r.v.s in L_1, $\mathbb{E}\left[\sum_{n=1}^{\infty} Y_n\right] = \sum_{n=1}^{\infty} \mathbb{E}[Y_n]$ (see Fristedt and Gray, 1997, p. 50, or Resnick, 1999, p. 131), we have, with S so defined,

$$\mathbb{E}[S] = \mathbb{E}\left[\sum_{r=1}^{\infty} \frac{S_r^4}{r^4}\right] = \sum_{r=1}^{\infty} \mathbb{E}\left[\frac{S_r^4}{r^4}\right] < \infty.$$

Then

$$\mathbb{E}[S] < \infty \Rightarrow \Pr(S < \infty) = 1 \Rightarrow \Pr\left(\lim_{r \to \infty}\left(\frac{S_r}{r}\right)^4 = 0\right) = 1$$

$$\Leftrightarrow \Pr\left(\lim_{r \to \infty}\overline{X}_r = 0\right) = 1 \Leftrightarrow \overline{X}_r \overset{a.s.}{\to} 0,$$

where the first implication follows from the fact that, for nonnegative r.v. S, $\mathbb{E}[S] < \infty \Rightarrow \Pr(S < \infty) = 1$ (which is intuitively seen by considering its contrapositive, and proven in, for example, Gut, 2005, p. 52), and the second implication follows from properties of convergence series (see Section I.A.2.4). For general $\mu \in \mathbb{R}$, apply the proof to $\{X_n - \mu\}$ to see that $\overline{X}_n \overset{a.s.}{\to} \mu$. ∎

Remark: As with the WLLN, the SLLN holds if just the first moment exists: if $\{X_n\}$ is a sequence of i.i.d. r.v.s in L_1 with expectation μ, then $\overline{X}_n \overset{a.s.}{\to} \mu$. The proof when assuming $\{X_n\}$ are i.i.d. and in L_2 is instructive and still straightforward; see Jacod and Protter (2000, p. 169). The general proof is more advanced and can be found in virtually all advanced probability textbooks. ∎

4.3.3 Convergence in r-mean

So far, convergence in probability and almost sure convergence have involved statements of probability. Convergence in r-mean involves expectations and is common in the literature on time series analysis.

The sequence $\{X_n\}$ in L_r is said to *converge in r-mean* to $X \in L_r$ iff

$$\lim_{n \to \infty} \mathbb{E}\left[|X_n - X|^r\right] = 0. \tag{4.56}$$

In this case, we write $X_n \overset{L_r}{\to} X$, alternative popular notation being $X_n \overset{L^r}{\to} X$ or just $X_n \overset{r}{\to} X$.

Remark: Convergence in r-mean is also referred to as *convergence in L_r*. A common case is when $r = 2$, in which case one speaks of *mean square convergence* or *convergence in quadratic mean* and sometimes writes $X_n \overset{q.m.}{\to} X$. ∎

Problem 4.9 shows that: (i) r.v. X in (4.56) is unique up to a set of measure zero, i.e., if $X_n \overset{r}{\to} X$ and $X_n \overset{r}{\to} Y$, then $\Pr(X = Y) = 1$; (ii) if $X_n \overset{r}{\to} X$ and $Y_n \overset{r}{\to} Y$, then $X_n + Y_n \overset{r}{\to} X + Y$.

⊙ **Example 4.18** With $\{X_n\}$ an i.i.d. sequence in L_2 with $\mathbb{E}[X_n] = \mu$ and $\mathbb{V}(X_n) = \sigma^2$, and $\overline{X} = n^{-1}S_n$, $S_n = \sum_{i=1}^{n} X_i$, recall that the WLLN and SLLN in Examples 4.8 and

4.17 state that $\overline{X}_n \overset{p}{\to} \mu$ and $\overline{X}_n \overset{a.s.}{\to} \mu$, respectively, with the latter implying the former from (4.55). Now observe that

$$\lim_{n \to \infty} \mathbb{E}\left[\left|\overline{X}_n - \mu\right|^2\right] = \lim_{n \to \infty} \mathbb{V}(\overline{X}_n) = \lim_{n \to \infty} \frac{\sigma^2}{n} = 0,$$

so that $\overline{X} \overset{L_2}{\to} \mu$.   ■

⊙ **Example 4.19**   Directly from Markov's inequality (4.13), for any $\epsilon > 0$,

$$\Pr\left(|X_n - X| \geq \epsilon\right) \leq \frac{\mathbb{E}[|X_n - X|]}{\epsilon},$$

from which it follows that $X_n \overset{L_1}{\to} X \Rightarrow X_n \overset{p}{\to} X$.

If $X_n \overset{L_2}{\to} X$, the Cauchy–Schwarz inequality (4.3) with $U = |X_n - X|$ and $V = 1$ gives $\mathbb{E}[|X_n - X|] \leq \sqrt{\mathbb{E}\left[|X_n - X|^2\right]}$, so that $X_n \overset{L_2}{\to} X \Rightarrow X_n \overset{L_1}{\to} X$. Recall that, if function f is Riemann integrable over $[a, b]$, then $\left|\int_a^b f\right| \leq \int_a^b |f|$. Thus, $\left|\mathbb{E}[X_n - X]\right| \leq \mathbb{E}[|X_n - X|]$, implying $X_n \overset{L_1}{\to} X \Rightarrow \mathbb{E}[X_n] \to \mathbb{E}[X]$. Similarly, writing $\mathbb{E}[X_n^2] = \mathbb{E}[(X_n - X)^2] + \mathbb{E}[X^2] + 2\mathbb{E}[X(X_n - X)]$, (4.3) implies

$$\mathbb{E}[X(X_n - X)] \leq \left|\mathbb{E}[X(X_n - X)]\right| \leq \mathbb{E}[|X(X_n - X)|] \leq \sqrt{\mathbb{E}[X^2]\mathbb{E}[(X_n - X)^2]}$$

and taking limits, we see that $X_n \overset{L_2}{\to} X \Rightarrow \mathbb{E}[X_n^2] \to \mathbb{E}[X^2]$.   ■

The results from the previous example might suggest that

$$X_n \overset{L_r}{\to} X \quad \Rightarrow \quad X_n \overset{p}{\to} X, \quad r > 0, \tag{4.57}$$

$$X_n \overset{L_s}{\to} X \quad \Rightarrow \quad X_n \overset{L_r}{\to} X, \quad s \geq r \geq 1, \tag{4.58}$$

and

$$X_n \overset{L_r}{\to} X \quad \Rightarrow \quad \mathbb{E}[|X_n|^r] \to \mathbb{E}[|X|^r], \quad r > 0. \tag{4.59}$$

These are true, and the first result (4.57) follows directly from Markov's inequality (4.12). The second result (4.58) follows easily from Lyapunov's inequality (4.9), i.e., $\|X\|_r \leq \|X\|_s$, recalling that the r-norm of r.v. X is $\|X\|_r = \mathbb{E}[|X|^r]^{1/r}$, $r \geq 1$. To prove the third result (4.59), first take $r \geq 1$. Recall that Minkowski's inequality (4.11) is $\|U + V\|_r \leq \|U\|_r + \|V\|_r$. Then with $U = X_n - X$ and $V = X$, this is

$\|X_n\|_r \leq \|X_n - X\|_r + \|X\|_r$, and with $U = X_n$ and $V = X - X_n$, we get $\|X\|_r \leq \|X_n - X\|_r + \|X_n\|_r$, and combining,

$$\big| \|X_n\|_r - \|X\|_r \big| \leq \|X_n - X\|_r.$$

As $X_n \overset{L_r}{\to} X$, the r.h.s. converges to zero in the limit, showing that $\|X_n\|_r \to \|X\|_r$, or $\mathbb{E}[|X_n|^r] \to \mathbb{E}[|X|^r]$.

For $r \in (0, 1]$, inequality (4.5) with $U = X_n - X$ and $V = X$ gives $\mathbb{E}[|X_n|^r] - \mathbb{E}[|X|^r] \leq \mathbb{E}[|X_n - X|^r]$. Using $U = X_n$ and $V = X - X_n$ then leads to

$$\big| \mathbb{E}[|X|^r] - \mathbb{E}[|X_n|^r] \big| \leq \mathbb{E}[|X - X_n|^r],$$

showing the result.

The next two examples are standard for showing that $X_n \overset{a.s.}{\to} X \not\Rightarrow X_n \overset{L_r}{\to} X$.

⊖ **Example 4.20**   Recall Example 4.13, in which ω is a uniform r.v. on $[0, 1]$ and $X_n(\omega) = n\mathbb{I}_{[0,1/n]}(\omega)$ converges almost surely to zero. Then, for all $n \in \mathbb{N}$,

$$\mathbb{E}[|X_n - 0|] = \mathbb{E}[X_n] = n\Pr(0 \leq \omega \leq 1/n) = 1,$$

so that $\{X_n\}$ does not converge in L_1. Note that, if instead $X_n(\omega) = \sqrt{n}\mathbb{I}_{[0,1/n]}(\omega)$, then $\{X_n\}$ still converges almost surely to zero, and $\mathbb{E}[X_n] = n^{-1/2}$, which in the limit is zero, so that $X_n \overset{L_1}{\to} 0$. However, $\mathbb{E}[|X_n - 0|^2] = \mathbb{E}[X_n^2] = 1$, so that $\{X_n\}$ does not converge in L_2.   ∎

⊖ **Example 4.21**   As in Example 4.16, let $\{X_n\}$ be a sequence of r.v.s with $\Pr(X_n = 1) = 1 - n^{-\alpha}$ and $\Pr(X_n = n) = n^{-\alpha}$ for $\alpha > 0$. Then

$$\mathbb{E}[|X_n - 1|^r] = 0^r \Pr(X_n = 1) + |n - 1|^r \Pr(X_n = n) = \frac{(n-1)^r}{n^\alpha},$$

and, with $\ell_r = \lim_{n \to \infty} \mathbb{E}[|X_n - 1|^r]$, $\ell_r = 0$ for $r < \alpha$, $\ell_r = 1$ for $r = \alpha$, and $\ell_r = \infty$ for $r > \alpha$, showing that $X_n \overset{L_r}{\to} 1$ for $r < \alpha$. If $\alpha > 1$, then $X_n \overset{a.s.}{\to} 1$, but does not converge in r-mean for any $r \geq \alpha$.   ∎

Examples demonstrating that $X_n \overset{p}{\to} X \not\Rightarrow X_n \overset{L_r}{\to} X$ and $X_n \overset{L_r}{\to} X \not\Rightarrow X_n \overset{a.s.}{\to} X$ can be found in Resnick (1999, p. 182) and Gut (2005, p. 211). Observe that almost sure convergence and r-mean convergence cannot be ranked in terms of which is stronger (i.e., one implies the other), unless further conditions are imposed (see Gut, 2005, p. 221).

⊙ **Example 4.22**   To show that

$$\boxed{X_n \overset{L_2}{\to} X, \quad Y_n \overset{L_2}{\to} Y \quad \Rightarrow \quad X_n Y_n \overset{L_1}{\to} XY,} \tag{4.60}$$

use the triangle inequality (4.10) and the Cauchy–Schwarz inequality (4.3) twice to write

$$
\begin{aligned}
\mathbb{E}\big[|X_n Y_n - XY|\big] &= \mathbb{E}\big[|X_n Y_n - X_n Y + X_n Y - XY|\big] \\
&\le \mathbb{E}\big[|X_n Y_n - X_n Y|\big] + \mathbb{E}\big[|X_n Y - XY|\big] \\
&\le \sqrt{\mathbb{E}[X_n^2]\mathbb{E}[(Y_n - Y)^2]} + \sqrt{\mathbb{E}[Y^2]\mathbb{E}[(X_n - X)^2]}.
\end{aligned}
$$

The first term goes to zero in the limit because $Y_n \overset{L_2}{\to} Y$ and because, from Example 4.19, $X_n \overset{L_2}{\to} X \Rightarrow \mathbb{E}[X_n^2] \to \mathbb{E}[X^2]$. The second term converges to zero in the limit because $X_n \overset{L_2}{\to} X$. ∎

4.3.4 Convergence in distribution

For a given c.d.f. F, let $C(F) = \{x : F(x) \text{ is continuous at } x\}$. The sequence $\{X_n\}$ is said to *converge in distribution* to X iff

$$
\lim_{n\to\infty} F_{X_n}(x) = F_X(x) \quad \forall x \in C(F_X), \tag{4.61}
$$

and we write $X_n \overset{d}{\to} X$. Convergence in distribution is the weakest form of convergence.

Remark: In somewhat older texts, convergence in distribution may be referred to as *convergence in law* or *weak convergence*. ∎

Similar to the other types of convergence, if $X_n \overset{d}{\to} X$, then X is unique. For suppose that $X_n \overset{d}{\to} X$ and $X_n \overset{d}{\to} Y$. Then, for an $x \in C(F_X) \cap C(F_Y)$, the triangle inequality (4.10) implies that

$$
|F_X(x) - F_Y(x)| \le |F_X(x) - F_{X_n}(x)| + |F_{X_n}(x) - F_Y(x)|,
$$

and in the limit, the r.h.s. goes to zero. It can also be shown that $F_X(x) = F_Y(x)$ for all $x \in \mathbb{R}$; see Gut (2005, p. 208).

The next two examples verify, respectively, that $X_n \overset{d}{\to} X \not\Rightarrow X_n \overset{L_r}{\to} X$ and $X_n \overset{p}{\to} X \Rightarrow X_n \overset{d}{\to} X$.

Example 4.23 Let $\{X_n\}$ be a sequence of discrete r.v.s such that $\Pr(X_n = 0) = 1 - n^{-1}$ and $\Pr(X_n = n) = n^{-1}$, so that $F_{X_n}(x) = \left(1 - n^{-1}\right)\mathbb{I}_{[0,n)}(x) + \mathbb{I}_{[n,\infty)}(x)$. Then $\lim_{n\to\infty} F_{X_n}(x) = F_X(x)$ at all points of continuity of F_n, where $F_X(x) = \mathbb{I}_{[0,\infty)}(x)$, and F_X is the c.d.f. of the r.v. X which is degenerate at zero, i.e., $\Pr(X = 0) = 1$. Then, for all $k \in \mathbb{N}$, $\mathbb{E}[X^k] = 0$ but $\mathbb{E}[X_n^k] = n^{k-1}$, so that, for $r > 0$, $\mathbb{E}\big[|X_n|^r\big] \not\to \mathbb{E}\big[|X|^r\big]$, and thus, from (4.59), X_n does not converge in r-mean to X. ∎

⊙ **Example 4.24** To show that $X_n \overset{p}{\to} X \Rightarrow X_n \overset{d}{\to} X$, let F_n and F denote the c.d.f.s of X_n and X respectively, and let x be a point of continuity of F. As in Fristedt and Gray (1997, p. 249), let $\delta > 0$ and event $B = \{|X_n - X| \le \delta\}$. We can split $F_{X_n}(x)$ into the two disjoint events

$$\Pr(X_n \le x) = \Pr(\{X_n \le x\} \cap B) + \Pr(\{X_n \le x\} \cap B^c),$$

and observe (draw a line) that

$$\{X_n \le x\} \cap B \subset \{X \le x + \delta\} \cap B \subset \{X \le x + \delta\}$$

and, trivially, $\{X_n \le x\} \cap B^c \subset B^c$. Thus,

$$\Pr(X_n \le x) \le \Pr(X \le x + \delta) + \Pr(B^c).$$

For any $\epsilon > 0$, choose a $\delta > 0$ such that $F(x + \delta) - F(x - \delta) < \epsilon/2$, and choose $N \in \mathbb{N}$ such that, for $n \ge N$, $\Pr(B^c) = \Pr(|X_n - X| > \delta) < \epsilon/2$. Then

$$\Pr(X_n \le x) \le \Pr(X \le x + \delta) + \Pr(B^c)$$
$$< \left[F(x) + \epsilon/2\right] + \epsilon/2.$$

Similarly,

$$F(x) = \Pr(X \le x) \le \Pr(X + \delta \le x) \le \Pr(X \le x - \delta) + \epsilon/2$$
$$\le \Pr(X_n \le x) + \Pr(|X_n - X| > \delta) + \epsilon/2$$
$$< \left[F_n(x) + \epsilon/2\right] + \epsilon/2.$$

As $\epsilon > 0$ is arbitrary, we have shown that

$$\boxed{X_n \overset{p}{\to} X \quad \Rightarrow \quad X_n \overset{d}{\to} X.} \tag{4.62}$$

A converse of (4.62) exists if the limiting random variable is degenerate: if $\{X_n\}$ is a sequence of r.v.s and X is a degenerate r.v. with $\Pr(X = c) = 1$ for some $c \in \mathbb{R}$, then $X_n \overset{d}{\to} X \Rightarrow X_n \overset{p}{\to} X$. To prove this, for any $\epsilon > 0$, as $c \pm \epsilon \in C(F_X) = \{x : x \ne c\}$, we have

$$\Pr(|X_n - c| > \epsilon) = 1 - \Pr(c - \epsilon \le X_n \le c + \epsilon)$$
$$= 1 - F_{X_n}(c + \epsilon) + F_{X_n}(c - \epsilon) - \Pr(X_n = c - \epsilon)$$
$$\le 1 - F_{X_n}(c + \epsilon) + F_{X_n}(c - \epsilon)$$
$$\to 1 - 1 + 0 = 0,$$

so that $\lim_{n \to \infty} \Pr(|X_n - c| > \epsilon) = 0$.

Another converse of significant theoretical importance exists; see Fristedt and Gray (1997, p. 250) or Gut (2005, p. 258).   ∎

Similar to (4.48) for convergence in probability, the concept of convergence in distribution is easily extended in a natural way to sequences of multivariate r.v.s: the sequence $\{\mathbf{X}_n\}$ of k-dimensional r.v.s, with distribution functions $F_{\mathbf{X}_n}$, converges in distribution to the k-dimensional r.v. $\mathbf{X}$ with distribution $F_{\mathbf{X}}$ if

$$\lim_{n \to \infty} F_{\mathbf{X}_n}(\mathbf{x}) = F_{\mathbf{X}}(\mathbf{x}) \quad \forall \mathbf{x} \in C(F_{\mathbf{X}}), \tag{4.63}$$

and we write $\mathbf{X}_n \xrightarrow{d} \mathbf{X}$.

Some further important results regarding convergence in distribution are as follows.

1. If $X_n \xrightarrow{d} X$, $Y_n \xrightarrow{d} Y$, $X \perp Y$ (X is independent of Y), and $X_n \perp Y_n$ for all n, then $X_n + Y_n \xrightarrow{d} X + Y$.

2. Suppose $X_n \xrightarrow{d} X$. If $h \in C^0[a, b]$ with $a, b \in C(F_X)$, then $\mathbb{E}[h(X_n)] \to \mathbb{E}[h(X)]$.

3. Suppose $X_n \xrightarrow{d} X$. If $h \in C^0$ and bounded, then $\mathbb{E}[h(X_n)] \to \mathbb{E}[h(X)]$.

4. (Continuous mapping theorem) Suppose $X_n \xrightarrow{d} X$. If $g \in C^0$, then $g(X_n) \xrightarrow{d} g(X)$.

5. (Scheffé's lemma) Suppose that $X, X_1, X_2, \ldots$ are continuous r.v.s with respective p.d.f.s $f_X, f_{X_1}, f_{X_2}, \ldots$, and such that $f_{X_n} \to f_X$ for 'almost all' x (for all $x \in \mathbb{R}$ except possibly on a set of measure zero). Then $X_n \xrightarrow{d} X$.

6. (Slutsky's theorem) Suppose $X_n \xrightarrow{d} X$ and $Y_n \xrightarrow{p} a$ for some $a \in \mathbb{R}$. Then

$$X_n \pm Y_n \xrightarrow{d} X \pm a, \tag{4.64}$$

$$X_n \cdot Y_n \xrightarrow{d} X \cdot a, \quad \text{and} \quad X_n / Y_n \xrightarrow{d} X/a, \quad a \neq 0. \tag{4.65}$$

7. (Continuity theorem for c.f.) For r.v.s $X, X_1, X_2, \ldots$ with respective characteristic functions $\varphi_X, \varphi_{X_1}, \varphi_{X_2}, \ldots$,

$$\lim_{n \to \infty} \varphi_{X_n}(t) = \varphi_X(t) \, \forall t \quad \Leftrightarrow \quad X_n \xrightarrow{d} X. \tag{4.66}$$

Moreover, if the c.f.s of the X_n converge, for all $t \in \mathbb{R}$, to some function φ which is continuous at zero, then there exists an r.v. X with c.f. φ such that $X_n \xrightarrow{d} X$.

8. (Continuity theorem for m.g.f.) Let X_n be a sequence of r.v.s such that the corresponding m.g.f.s $\mathbb{M}_{X_n}(t)$ exist for $|t| < h$, for some $h > 0$, and all $n \in \mathbb{N}$. If X is an r.v. whose m.g.f. $\mathbb{M}_X(t)$ exists for $|t| \leq h_1 < h$ for some $h_1 > 0$, then

$$\lim_{n \to \infty} \mathbb{M}_{X_n}(t) = \mathbb{M}_X(t) \text{ for } |t| < h_1 \quad \Rightarrow \quad X_n \xrightarrow{d} X. \tag{4.67}$$

9. (Cramér–Wold device) Let $\mathbf{X}$ and $\{\mathbf{X}_n\}$ be k-dimensional r.v.s. Then

$$\mathbf{X}_n \xrightarrow{d} \mathbf{X} \quad \Leftrightarrow \quad \mathbf{t}'\mathbf{X}_n \xrightarrow{d} \mathbf{t}'\mathbf{X}, \quad \forall \mathbf{t} \in \mathbb{R}^k. \tag{4.68}$$

Proofs of these results can be found in Gut (2005) on pp. 247, 222, 223, 246, 227, 249, 238, 242, and 246, respectively. Some examples of the use of (4.67) were given in Section 1.1.3.

⊖ ***Example 4.25*** Let $X_i \overset{\text{i.i.d.}}{\sim} N(2,2)$, $i = 1, 2, \ldots$, with $S_n = \sum_{i=1}^n X_i$ and $Q_n = \sum_{i=1}^n X_i^2$. Then $\mathbb{E}[S_n] = n\mathbb{E}[X_1] = 2n$ and $\mathbb{E}[Q_n] = n\mathbb{E}[X_1^2] = n(2^2 + 2) = 6n$. From the SLLN applied to both the numerator and denominator of S_n/Q_n and using Slutsky's theorem (4.65),

$$\lim_{n\to\infty} \frac{S_n}{Q_n} = \lim_{n\to\infty} \frac{S_n/n}{Q_n/n} = \frac{\lim_{n\to\infty} S_n/n}{\lim_{n\to\infty} Q_n/n} = \frac{1}{3}.$$

Now let $X_i \overset{\text{i.i.d.}}{\sim} N(0,1)$. With $K_n = \sum_{i=1}^n X_i^4$, $\mathbb{E}[Q_n] = n\mathbb{E}[X_1^2] = n$ and $\mathbb{E}[K_n] = n\mathbb{E}[X_1^4] = 3n$, a similar application of the SLLN and Slutsky's theorem gives

$$\lim_{n\to\infty} \frac{Q_n}{K_n} = \frac{\lim_{n\to\infty} Q_n/n}{\lim_{n\to\infty} K_n/n} = \frac{1}{3}.$$

Finally, again with $X_i \overset{\text{i.i.d.}}{\sim} N(0,1)$, $\lim_{n\to\infty} Q_n/n = 1$ and

$$\lim_{n\to\infty} \sqrt{n}\frac{S_n}{Q_n} = \frac{\lim_{n\to\infty} S_n/\sqrt{n}}{\lim_{n\to\infty} Q_n/n} \overset{d}{=} N(0,1),$$

as $S_n/\sqrt{n} \sim N(0,1)$, which implies, trivially, that $S_n/\sqrt{n} \xrightarrow{d} Z \sim N(0,1)$. ∎

⊖ ***Example 4.26*** Recall the construction of a Student's t random variable in Example I.9.7: if $G \sim N(0,1)$ independent of $C_n \sim \chi_n^2$, then $T_n := G/\sqrt{C_n/n} \sim t_n$. From result (ii) of Example 2.3, $C_n \overset{d}{=} \sum_{i=1}^n X_i$, where $X_i \overset{\text{i.i.d.}}{\sim} \chi_1^2$, so that the WLLN implies that $C_n/n \xrightarrow{p} \mathbb{E}[X_1] = 1$. Next, the result of Example 4.10 implies that $\sqrt{C_n/n} \xrightarrow{p} \sqrt{1} = 1$. Thus, applying Slutsky's theorem (4.65) to T and using the fact that, trivially, $G \xrightarrow{d} N(0,1)$, we see that that $T_n \xrightarrow{d} N(0,1)$. ∎

All the relationships discussed above (or in the problems below) between the various methods of convergence are embodied in the following diagram, for a sequence $\{X_n\}$ and constant $c \in \mathbb{R}$.

4.4 The central limit theorem

> Although this may seem a paradox, all exact science is dominated by the idea of approximation.
>
> (Bertrand Russell)

Recalling Example 1.9 in Section 1.1.3 and the fact that, for $B_i \overset{\text{i.i.d.}}{\sim} \text{Ber}(p)$, $\sum_{i=1}^{n} B_i \sim \text{Bin}(n, p)$, we can conclude that, as $n \to \infty$, $\sum_{i=1}^{n} B_i \overset{d}{\to} \text{N}(np, np(1-p))$. Similarly, Example 1.10 and result (i) of Example 2.3 imply that, if $X_i \overset{\text{i.i.d.}}{\sim} \text{Exp}(\lambda)$, then, as $n \to \infty$, $\sum_{i=1}^{n} X_i \overset{d}{\to} \text{N}(n/\lambda, n/\lambda^2)$. This is interesting, for it says that, properly standardized, the sums of i.i.d. Bernoulli and the sums of i.i.d. exponential r.v.s converge to the same distribution. This fact behoves us to search for other r.v.s, or, better yet, necessary conditions in general, for which this holds. It turns out that it works for *any* i.i.d. r.v.s X_i with mean μ and finite variance σ^2. This fascinating limiting result is so theoretically important and central to much applied work that it is referred to as the *central limit theorem* (CLT), a term coined by George Pólya in 1920.

A common application of the CLT, usually adequately emphasized in introductory statistics courses, is that, for certain r.v.s, such as binomial, Poisson, and negative binomial, the c.d.f. at x can be approximated by $\Phi((x - \mu)/\sigma)$, where μ and σ^2 are the mean and variance, respectively. This is put to good use when constructing confidence intervals and tests of hypotheses. With modern computing power and more recent theoretical developments, the distribution of, say, a test statistic can often be accurately approximated without appeal to asymptotic arguments. Nevertheless, the use of asymptotics is ubiquitous in virtually all branches of statistical data analysis (not to mention probability theory), and so a basic understanding of its major concepts is crucial.

Arguably more useful is that the i.i.d. assumption in the CLT can be relaxed to a certain extent. The ability to allow for certain forms of covariance between the X_i is valuable for, among other things, the analysis of time series and other stochastic processes which evolve in space or time. Being able to relax the identical assumption in the CLT implies, for example, that the distribution of $\sum_{i=1}^{n} X_i$ for $X_i \overset{\text{ind}}{\sim} \text{Bin}(n_i, p_i)$ or $\sum_{i=1}^{n} Y_i$ for $Y_i \overset{\text{ind}}{\sim} \text{NBin}(r_i, p_i)$ can be approximated, or even $\sum_{i=1}^{n} X_i + \sum_{i=1}^{m} Y_i$, etc. All that is required is the mean and variance of the components in the sum. This justifies the normal approximation entertained in Problem I.6.7.

One of the most fundamental applications of the CLT makes use of the fact that the distribution of the components need not be known, but only that they have finite variance. In many statistical models, the error term is assumed to be the accumulation of all other factors of possible influence that are not explicitly taken into account. As an example, the change in the closing price P_t on day t of a heavily traded stock consists of the buying and selling actions of thousands of market participants, many of them acting independently, but certainly not all with identical 'trading inclinations'. Ironically in this example, empirical evidence shows that, even with the very large number of components that contribute to the change in the price of the financial asset, the daily returns, $(P_{t+1} - P_t)/P_t$, do not follow a normal distribution! There are, however, other limit theorems which can better accommodate this phenomenon – but we are getting ahead of ourselves (see Chapter 8 for more information).

We now outline the usual proof assuming that the m.g.f. of the X_i exist. Let $X_i \overset{\text{i.i.d.}}{\sim}$ (μ, σ^2), i.e., $\{X_i\}$ is a sequence of i.i.d. r.v.s with common mean μ and variance σ^2, and let $Z_i = (X_i - \mu)/\sigma$ with common m.g.f. $\mathbb{M}_Z(t)$, which exists for t in a neighbourhood of zero. Let $S_n = n^{-1/2} \sum_{i=1}^{n} Z_i = n^{1/2} \overline{Z}_n$, where $\overline{Z}_n = n^{-1} \sum_{i=1}^{n} Z_i$. As the Z_i are also i.i.d. $\mathbb{M}_{S_n}(t) = \left(\mathbb{M}_Z(tn^{-1/2})\right)^n$, and, from (1.3),

$$\mathbb{M}_Z(tn^{-1/2}) = \sum_{k=0}^{\infty} \frac{\mu'_k}{k!} \left(tn^{-1/2}\right)^k = 1 + \mu'_1 \frac{t}{n^{1/2}} + \frac{\mu'_2}{2} \frac{t^2}{n} + \frac{\mu'_3}{3!} \frac{t^3}{n^{3/2}} + \cdots,$$

where $\mu'_k = \mathbb{M}_Z^{(k)}(t)\Big|_{t=0}$ with $\mu'_0 = 1$, $\mu'_1 = \mathbb{E}[Z_i] = 0$, and $\mu'_2 = \mathbb{V}(Z_i) + (\mathbb{E}[Z_i])^2 = 1$. As

$$\lim_{n \to \infty} \left(1 + \frac{k_1}{n} + \frac{k_2}{n^p}\right)^n = e^{k_1}$$

for $p > 1$,

$$\lim_{n \to \infty} \mathbb{M}_{S_n}(t) = \lim_{n \to \infty} \left(\mathbb{M}_Z(tn^{-1/2})\right)^n = \lim_{n \to \infty} \left(1 + \frac{t^2}{2n} + \cdots\right)^n = \exp\left(\frac{t^2}{2}\right).$$

Then, assuming the characterizing property of m.g.f.s, as discussed in Section 1.1.3, the result follows. (Problem 4.15 shows another method of proof.)

Summarizing,

$$\boxed{\text{if } X_i \overset{\text{i.i.d.}}{\sim} (\mu, \sigma^2), \quad \text{then} \quad \frac{\overline{X}_n - \mu}{\sigma/\sqrt{n}} \overset{d}{\to} N(0, 1).} \qquad (4.69)$$

This can also be expressed informally as $\sum_{i=1}^{n} X_i \overset{d}{\to} N(n\mu, n\sigma^2)$ as $n \to \infty$, but with the understanding that (4.69) is the correct statement.

Although for many applications of interest, the CLT 'kicks in' quickly, i.e., it approximates quite accurately the random variable of interest, there are situations for which this will not be the case, particularly in the tails of the distribution. This is illustrated in the following example.

⊖ **Example 4.27**   Consider the sum of i.i.d. Student's t r.v.s with 3 degrees of freedom. As the m.g.f. of each component does not exist, the previous proof does not apply, but a similar analysis can be applied to the c.f. Let $X_i \overset{\text{i.i.d.}}{\sim} t_3$, each with $\mathbb{E}[X] = 0$, $\mathbb{V}(X) = 3/(3-2) = 3$ and c.f.

$$\varphi_X(t) = \left(1 + |t\sqrt{3}|\right) \exp\left(-|t\sqrt{3}|\right)$$

(see Johnson, Kotz and Balakrishnan, 1994, p. 367), which is real because X is symmetric. The c.f. of the standardized sum $S = S(n) = (3n)^{-1/2} \sum_{i=1}^{n} X_i$ is then

$$\varphi_S(t) = \left(1 + t\sqrt{1/n}\right)^n \exp\left(-nt\sqrt{1/n}\right), \quad t > 0, \tag{4.70}$$

or, with $m = n^{-1/2}$,

$$\log \varphi_S(t) = m^{-2} \log(1 + tm) - \frac{t}{m} = -\frac{1}{2}t^2 + \frac{1}{3}t^3 m - \frac{1}{4}t^4 m^2 + O\left(m^3\right), \quad t > 0,$$

which, for constant t, approaches $-t^2/2$ as $n \to \infty$, which is the log of the c.f. of a standard normal random variable. A similar result holds for $t \leq 0$.

The density of S can be computed by applying (1.59) to (4.70); Figure 4.2 shows the density in the right tail region for $n = 1, 3, 5, 7, 9, 11, 40$ and 100. As n increases, the density indeed decreases in the tails, albeit extremely slowly. Even for $n = 100$ the density in the tails is far closer to a t_3 than a normal density, which is 2.1×10^{-32} at $x = 12$. See also Problem 4.19.

Nevertheless, the density where most of the mass lies is quite close to the normal for much smaller values of n. Figure 4.3 shows, for $n = 5, 10, 20$ and 100, the true density (dashed) and the standard normal (solid) along with a histogram (truncated at -6 and 6) based on 30 000 simulated replications. ∎

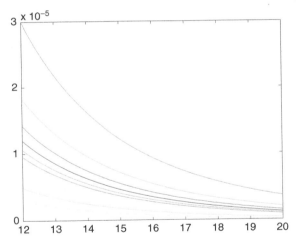

Figure 4.2   Density for scaled sums of $t(3)$ r.v.s; the solid lines from top to bottom correspond to $n = 1, 3, 5, 7, 9, 11, 40$ and 100

⊖ **Example 4.28**   To illustrate a random variable which does not converge to a normal, consider Banach's matchbox problem from Example I.4.13. For given N, the p.m.f. of r.v. K is

$$f_K(k; N) = \binom{2N - k}{N}\left(\frac{1}{2}\right)^{2N-k}, \tag{4.71}$$

and interest centres on what happens as $N \to \infty$. As K is not a sum of N r.v.s, the CLT is not applicable and the limiting form of the density is not obvious. Plots of the

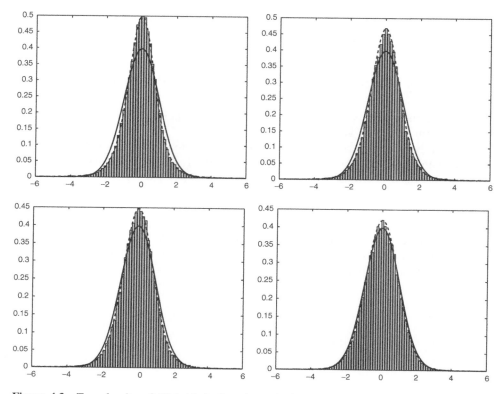

Figure 4.3   True density of $S(n)$ (dashed) and standard normal (solid) for $n = 5$ (top left), 10 (top right), 20 (bottom left) and 100 (bottom right). Histograms based on 30 000 simulated values of $S(n)$

mass function reveal that its shape resembles that of a half bell curve, particularly as N grows. To illustrate, Figure 4.4 shows part of the exact mass function (solid lines) for $N = 50$ and $N = 500$ overlaid with the density of the scaled folded normal distribution

$$g\,(x;c) = c^{-1}\sqrt{\frac{2}{\pi}}\,\exp\left\{-\frac{x^2}{2c^2}\right\}\,\mathbb{I}\,(x \geq 0)\,, \quad c > 0. \tag{4.72}$$

The choice of c used in the plots is that obtained by equating $\mathbb{E}\,[X] = c\sqrt{2/\pi}$ with $\mathbb{E}\,[K]$, where the latter was given in Problem I.4.13. For large N, $c \approx \sqrt{2N} - \sqrt{\pi/2}$ or $c \approx \sqrt{2N}$. In addition, $\mathbb{E}\left[X^2\right] = c^2$.

The absolute relative error of the density approximation, $ARE = (E - A)\,/E$, where E denotes the exact and A the approximate density (4.72), appears to take the same shape for all N. Figure 4.5 shows ARE for N-values 50, 500, 5000 and 50 000. It takes on a 'W' shape for lower values of k and then continues to increase 'without bound' (for $N = 50$, $\max(ARE) = 10^7$, for $N = 500$, $\max(ARE) = 10^{90}$). (An explanation for the poor accuracy as k grows will be given below.)

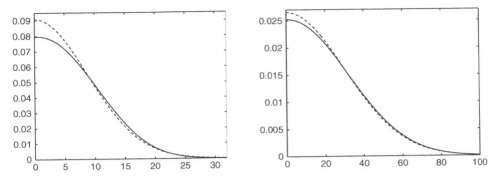

Figure 4.4 The mass function f_k (solid) and approximate density $g(x; c)$ (dashed) for $N = 50$ (left) and $N = 500$ (right)

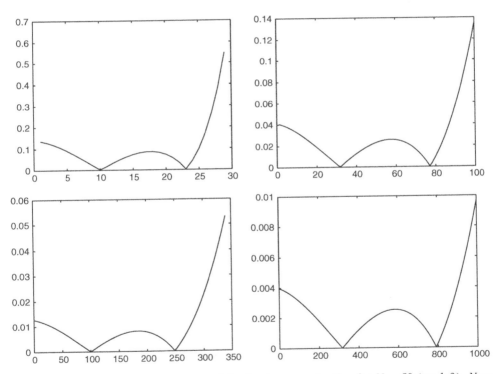

Figure 4.5 The absolute relative error of the density approximation for $N = 50$ (top left), $N = 500$ (top right), $N = 5000$ (bottom left) and $N = 50\,000$ (bottom right)

To see that the density $g(x; c)$ is indeed the limiting form of (4.71) as $N \to \infty$, begin with Stirling's formula, which yields

$$\binom{2N}{N} \simeq \frac{2^{2N}}{\sqrt{\pi N}},$$

so that

$$\Pr(K = 0) \simeq \frac{1}{\sqrt{N\pi}}.$$

For $k = 2$, using (I.4.22),

$$\Pr(K = 2) = \frac{N-1}{N-\frac{1}{2}} \Pr(K = 1) \simeq \frac{N-1}{N-\frac{1}{2}} \frac{1}{\sqrt{N\pi}}.$$

Then, similar to a method to show the CLT for binomial random variables (see Feller, 1968), expand the log of $\frac{N-1}{N-1/2}$ in a Taylor series and drop the higher-order terms to obtain

$$\log\left(\frac{N-1}{N-1/2}\right) = -\log\left(\frac{N-1/2}{N-1}\right) = -\log\left(1 + \frac{1/2}{N-1}\right)$$

$$= -\left\{\frac{1}{2}\frac{1}{N-1} + \frac{1}{2}\left(\frac{1}{2}\frac{1}{N-1}\right)^2 + \cdots\right\} \simeq -\frac{1}{2}\frac{1}{N-1}$$

or

$$\Pr(K = 2) \simeq \frac{1}{\sqrt{N\pi}} \exp\left(-\frac{1}{2}\frac{1}{N-1}\right) \simeq \frac{1}{\sqrt{N\pi}} \exp\left(-\frac{1}{2}\frac{1}{N}\right).$$

Similarly for $k = 3$, from (I.4.22),

$$\Pr(K = 3) = \frac{N-2}{N-1} \Pr(K = 2) \simeq \frac{N-2}{N-1}\frac{N-1}{N-1/2}\frac{1}{\sqrt{N\pi}}$$

and, with

$$-\log\left(\frac{N-1}{N-2}\right) = -\log\left(1 + \frac{1}{2}\frac{2}{N-2}\right) \simeq -\frac{1}{2}\frac{2}{N-2} \simeq -\frac{1}{2}\frac{2}{N},$$

$$\Pr(K = 3) \simeq \frac{1}{\sqrt{N\pi}} \exp\left(-\frac{1}{2}\frac{1}{N}\right) \exp\left(-\frac{1}{2}\frac{2}{N}\right)$$

$$= \frac{1}{\sqrt{N\pi}} \exp\left(-\frac{1}{2N}\sum_{i=1}^{2} i\right) = \frac{1}{\sqrt{N\pi}} \exp\left(-\frac{3}{2N}\right).$$

The form for general k is now straightforward: with the first factor in (I.4.22) expressible as

$$\frac{N - \frac{k-1}{2}}{N-k+1} = 1 + \frac{1}{2}\frac{k-1}{N-k+1},$$

it follows that

$$- \log \left(1 + \frac{1}{2} \frac{k-1}{N-k+1} \right) \simeq -\frac{1}{2} \frac{k}{N-k} \simeq -\frac{1}{2} \frac{k}{N} \tag{4.73}$$

and

$$\Pr(K = k) \simeq \frac{1}{\sqrt{N\pi}} \exp \left(-\frac{1}{2N} \sum_{i=1}^{k-1} i \right) = \frac{1}{\sqrt{N\pi}} \exp \left(-\frac{1}{2N} \frac{k(k-1)}{2} \right)$$

$$\simeq \frac{1}{\sqrt{N\pi}} \exp \left(-\frac{k^2}{4N} \right) =: g(k)$$

is established. As a check,

$$\int_0^\infty \frac{1}{\sqrt{N\pi}} e^{-\frac{k^2}{4N}} \, dk = \int_0^\infty \frac{1}{2\sqrt{N\pi}} \frac{e^{-\frac{1}{4}\frac{x}{N}}}{\sqrt{x}} \, dx = \frac{1}{\sqrt{\pi}} \int_0^\infty \frac{e^{-y}}{\sqrt{y}} \, dy = \pi^{-1/2} \Gamma\left(\frac{1}{2}\right) = 1.$$

Also, using this approximation,

$$\mathbb{E}[K] \approx \int_0^\infty kg(k) \, dk = 2\sqrt{\frac{N}{\pi}} \quad \text{and} \quad \mathbb{E}[K^2] \approx 2N,$$

so that, as $N \to \infty$, $\mathbb{V}(K) \approx 2N - 4N/\pi = (2 - 4/\pi) N$, as in (I.4.60).

Finally, note that the second approximation in (4.73) breaks down for k relatively large, so that the density approximation will fail in the right tail. This explains the results in Figure 4.5. ∎

4.5 Problems

Most of the important things in the world have been accomplished by people who have kept on trying when there seemed to be no help at all.

(Dale Carnegie)

Experience is a hard teacher because she gives the test first, the lesson afterwards.

(Vernon Law)

4.1. Show that

$$S = \lim_{\theta \to \infty} \left(\frac{\theta - 1}{\theta + 1} \right)^\theta = e^{-2}.$$

4.2. Using Hölder's inequality, prove Lyapunov's inequality.

4.3. ★ Verify (4.14) and (4.15).

4.4. ★ With $X_i \sim \text{Ber}(p)$, show $\Pr(\overline{X}_n \geq c) \leq \exp(nK(c))$, $p \leq c < 1$, where

$$K(c) = d \ln q + c \ln p - c \ln c - d \ln d$$

with $q := 1 - p$ and $d := 1 - c$.

4.5. Find the Chernoff bound when $X_i \overset{\text{i.i.d.}}{\sim} \text{Poi}(\lambda)$, and specify for which values of c the bound is valid.

4.6. ★ Assume $X_n \overset{p}{\to} X$ and $X_n \overset{p}{\to} Y$. Show that (i) $\Pr(X = Y) = 1$ and (ii) as $n, m \to \infty$, $X_n - X_m \overset{p}{\to} 0$. Hint: Use a result in Example 4.12.

4.7. Show that, for a set of nonnegative r.v.s $Y_1, \ldots, Y_k$, for any $\epsilon > 0$,

$$\Pr\left(\sum_{j=1}^{k} Y_j \geq \epsilon\right) \leq \sum_{j=1}^{k} \Pr\left(Y_j \geq \frac{\epsilon}{k}\right). \tag{4.74}$$

4.8. ★ Show (4.49), i.e., with $\mathbf{X}_n = (X_{n1}, X_{n2}, \ldots, X_{nk})$, $n = 1, 2, \ldots$, and $\mathbf{X} = (X_1, X_2, \ldots, X_k)$, $\mathbf{X}_n \overset{p}{\to} \mathbf{X} \Leftrightarrow X_{nj} \overset{p}{\to} X_j$, $j = 1, 2, \ldots, k$.
Hint: Recall from (I.A.9) that the norm of vector $\mathbf{z} = (z_1, \ldots, z_m) \in \mathbb{R}^m$ is $\|\mathbf{z}\| = \sqrt{z_1^2 + \cdots + z_m^2}$. It is easy to verify (as was done in the proof of (I.A.140)), that, for vector $\mathbf{z} = (z_1, \ldots, z_m) \in \mathbb{R}^m$,

$$|z_i| \leq \|\mathbf{z}\| \leq |z_1| + \cdots + |z_m|, \quad i = 1, \ldots, m.$$

4.9. ★ First, show that, if $\{X_n\}$ converges in r-mean, then the limiting random variable is unique, i.e., assume $X_n \overset{r}{\to} X$ and $X_n \overset{r}{\to} Y$ and show that $\Pr(X = Y) = 1$. Hint: Use (4.4). Second, show that, if $X_n \overset{r}{\to} X$ and $Y_n \overset{r}{\to} Y$, then $X_n + Y_n \overset{r}{\to} X + Y$.

4.10. ★ As in Gut (2005, p. 203), the sequence $\{X_n\}$ is said to *converge completely* to r.v. X (i.e., $X_n \overset{c.c.}{\to} X$) iff, for all $\epsilon > 0$,

$$\sum_{n=1}^{\infty} \Pr(|X_n - X| > \epsilon) < \infty. \tag{4.75}$$

Prove that, for $\{X_n\}$ with the X_n not necessarily independent,

$$X_n \overset{c.c.}{\to} X \quad \Rightarrow \quad X_n \overset{a.s.}{\to} X. \tag{4.76}$$

Prove also, for constant c and X_n independent r.v.s, that

$$X_n \overset{c.c.}{\to} c \quad \Leftrightarrow \quad X_n \overset{a.s.}{\to} c \quad \text{if the } X_n \text{ are independent.} \tag{4.77}$$

Why is it not possible to extend (4.77) to $X_n \overset{c.c.}{\to} X$ instead of using a degenerate random variable?

4.11. In Example 4.12, with $A = \{|d_X| > \epsilon/2\}$, $B = \{|d_Y| > \epsilon/2\}$, and $C = \{|d_X| + |d_Y| > \epsilon\}$, show algebraically that C implies $A \cup B$.

4.12. Recall the construction of an $F(n_1, n_2)$ random variable in Example I.9.8: if $X_i \overset{\text{ind}}{\sim} \chi^2_{n_i}$, then $F_{n_1, n_2} = (X_1/n_1) / (X_2/n_2) \sim F(n_1, n_2)$. Show that, as $n_2 \to \infty$,

$$n_1 F_{n_1, n_2} \overset{d}{\to} \chi^2_{n_1}.$$

4.13. ★ ★ Let $X_1, X_2, \ldots,$ be i.i.d. Bernoulli random variables with

$$p = \Pr(X_i = 0) = 1 - \Pr(X_i = 1),$$

$i \geq 1$ and let $L_1, L_2, \ldots$ denote the respective lengths of the first run, second run, etc. As an example, for the sequence $1110110010\cdots$, $L_1 = 3$, $L_2 = 1$, $L_3 = 2$, $L_4 = 2$, $L_5 = 1$, etc. It follows that the L_i are i.i.d. random variables.

(a) Assume $p = 1/2$. What is the distribution of L_1?

(b) Calculate the distribution of L_1 for any p, $0 < p < 1$.

(c) It is of interest to find the distribution of the number of runs, K, in the first n observed values of X. First argue why the relation $\Pr(K > m) = \Pr\left(\sum_{i=1}^{m} L_i < n\right)$ is valid. Then approximate $\Pr(K > 55)$ using the central limit theorem with $n = 100$ and $p = 1/2$.

(d) Let M_n denote the maximum length of the runs in the first n observed values of X. An expression for the c.d.f. of M_n is straightforward if we condition on the number of runs. For example, assume that there are exactly 50 runs and compute $\Pr(M_n \leq 6)$ for $p = 1/2$.[3]

4.14. ★ A coin with probability p of 'heads' will be tossed n times. We want the smallest value n such that $\Pr(|\hat{p} - p| < \epsilon_1) \geq 1 - \epsilon_2$, for some $\epsilon_1 > 0$ and $\epsilon_2 > 0$. In terms of the c.d.f. of the binomial, $F(x; n, p)$, and the standard normal, $\Phi(x)$, give (possibly implicit) expressions for n:

(a) using the CLT without continuity correction;

(b) using the CLT with continuity correction;

(c) exactly.

4.15. Recall the proof of the CLT. An alternative proof uses $\mathbb{K}_{S_n}(t) = \ln \mathbb{M}_{S_n}(t)$ and shows that $\lim_{n \to \infty} \mathbb{K}_{S_n}(t) = t^2/2$. Use this method to derive the CLT. Hint: Apply l'Hôpital's rule twice. (Tardiff, 1981)

---

[3] Notice that this calculation does not use n. With more work, one could compute the c.d.f. of M_n via (I.8.16), i.e., as a sum of conditional (on the number of runs) probabilities, weighted by the probability of a total number of runs, as given in Section I.6.3. The latter requires, however, knowing (i.e., conditioning on) the total number of successes and failures. This, in turn, could also be used to compute the c.d.f. of M_n via a nested use of (I.8.16), so that only specification of n and p is required.

4.16. ★ From the CLT, a binomial random variable X with n trials and success probability p can be asymptotically expressed as $Y \sim N(np, npq)$, $q = 1 - p$. For finite samples, the 'continuity correction'

$$\Pr(X \le x) \approx \Pr(Y \le x + 0.5), \tag{4.78}$$

$$\Pr(X \ge x) \approx \Pr(Y \ge x - 0.5) \tag{4.79}$$

can be applied.

Let $X \sim \text{Bin}(20, 0.05)$.

(a) Calculate the percentage error of approximating $\Pr(X > 3)$ using the Poisson and the normal approximation, the latter both with and without continuity correction.

(b) Using a computer, calculate the optimal continuity correction value $\nu = \nu(n, p)$ such that

$$1 - \Phi\left(\frac{3 + \nu - 1}{\sqrt{0.95}}\right)$$

is as accurate as possible.

4.17. ★ Augment Problem 4.16 using the usual Chebyshev inequality, the one-sided Chebyshev inequality, and Chernoff's bound. Optionally apply the saddlepoint approximation developed later in Example 5.4.

4.18. ★ A bank employee is responsible for making sacks of coins, n coins per sack. However, in each sack, she puts m fake coins, $m < n$, and gives the rest to charity. Her boss suspects something and checks a single random coin from each of n bags. Let X_n be the number of fake coins found.

(a) What is $\Pr(X_n = 0)$? What is it as $n \to \infty$?

(b) What is the p.d.f. and expected value of X_n?

(c) Evaluate $\Pr(X_n = \mathbb{E}[X_n])$ for $m = 1$ and $m = 2$.

(d) Calculate the limiting values as $n \to \infty$ in part **(c)**.

(e) In part **(c)**, using Stirling's approximation, find n such that $\Pr(X_n = m)$ is close to $\alpha = 0.05$, for $m = \left[\frac{n}{10}\right]$, where $[a]$ denotes the integer portion of a.

(f) Repeat part **(e)** using both the Poisson and Stirling approximations *together*.

4.19. In Example 4.27, $X_i \overset{\text{i.i.d.}}{\sim} t_3$ and $S = S(n) = (3n)^{-1/2} \sum_{i=1}^{n} X_i$. We would like to use simulation to confirm the appearance of the p.d.f. of S. For $n = 40$ and a grid of x-values between 2.5 and 4.5, simulate 20 000 realizations of S and plot their kernel density, overlaid with the true density computed by inverting the c.f.,

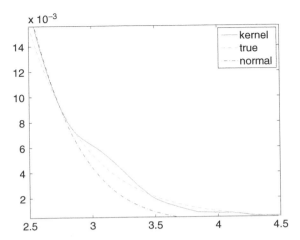

Figure 4.6 Kernel density of 20 000 simulations of S (solid curve), true density of S (dashed) and standard normal density (dash-dotted)

and the standard normal distribution. The resulting plot should look something that in Figure 4.6, showing that the calculation of the true density of S is indeed accurate, and that S, even with $n = 40$, still has much fatter tails than the normal distribution.

5

Saddlepoint approximations

Any sufficiently advanced technology is indistinguishable from magic.
(Arthur C. Clarke)

For many situations occurring in statistics, the distribution of a random variable X arising from a function of other random variables, or as a test statistic or parameter estimate, is either unknown or is algebraically intractable. Recalling the central limit theorem, if X can be represented as a sum of n (usually independent) r.v.s with finite variance, then, as n increases, the normal distribution is a valid candidate for approximating f_X and F_X.

The normal approximation tend to be adequate near the mean of X, but typically breaks down as one moves into the tails. To see this, consider the following two examples. Let $X \sim t(\nu)$, the Student's t distribution with $\nu > 2$ degrees of freedom. Recalling its variance from (I.7.46), its c.d.f. can be approximated by that of $Y \sim N(0, \nu/(\nu - 2))$. Figure 5.1(a) shows the RPE of the approximation for $\nu = 6$; it is zero at the mean, while it increases without bound as $|x|$ increases.

Now let $X \sim \chi^2(\nu)$, a chi-square r.v. with $\nu > 0$ degrees of freedom. Its distribution is approximated by that of $Y \sim N(\nu, 2\nu)$. Figure 5.1(b) shows the associated c.d.f. percentage error for $\nu = 6$. As Y is not symmetric, the error is not zero at the mean. Also, because of the truncated true left tail, the percentage error will blow up as x moves towards zero. It also increases without bound as x increases.

Another illustration of the inaccuracy of the normal approximation was given in Problem I.6.7, where interest centred on the sum of three independent negative binomial random variables with different success probabilities. The exact p.m.f. was calculated by straightforward application of the convolution formula, but, in general, as the number of components in the sum increases, the computation becomes slower and slower. The method of approximation developed in this chapter is applicable to this case (see Example 5.3 and Problem 5.4), and, like the normal approximation, does not become any more complicated as the number of negative binomial components in the sum increases, but – as will be seen – is vastly more accurate.

Several results in this chapter will be stated without proof. In addition to the original articles we mention, several surveys, textbook chapters, and monographs on the subject

Intermediate Probability: A Computational Approach   M. Paolella
© 2007 John Wiley & Sons, Ltd

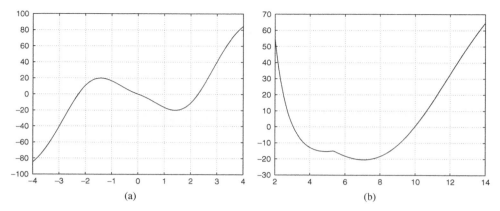

Figure 5.1   Relative percentage error, defined as $100 \, (\text{Approx} - \text{True}) \, / \min (\text{True}, 1 - \text{True})$, (a) for the normal approximation to the $t \, (6)$ c.d.f., (b) for $\chi^2 \, (6)$

are available, including Barndorff-Nielsen and Cox (1989), Field and Ronchetti (1990), Jensen (1995), Reid (1988, 1996, 1997), Kolassa (1997), Pace and Salvan (1997), and Butler (2007).

We begin with a detailed exposition of the univariate case in Section 5.1, cover some aspects of the multivariate case in Section 5.2, and provide an introduction to the hypergeometric functions $_1F_1$ and $_2F_1$ in Section 5.3, which are needed in several exercises and later chapters.

5.1   Univariate

5.1.1   Density saddlepoint approximation

A modern mathematical proof is not very different from a modern machine, or a modern test setup: the simple fundamental principles are hidden and almost invisible under a mass of technical details.   (Hermann Weyl)

With the above quote in mind, we provide only a simple, informal derivation of the density saddlepoint approximation, and refer the reader to the aforementioned texts, particularly Butler (2007), for more formal derivations and much deeper insight into the nature and quality of the approximation.

The basis of this method is to overcome the inadequacy of the normal approximation in the tails by 'tilting' the random variable of interest in such a way that the normal approximation is evaluated at a point near the mean. Let X be a r.v. with p.d.f. or p.m.f. f_X and m.g.f. $\mathbb{M}_X$ existing in a neighbourhood U around zero. From (1.9), the mean and variance of X can be expressed as $\mathbb{K}_X'(0)$ and $\mathbb{K}_X''(0)$, respectively, where $\mathbb{K}_X(s) = \ln \mathbb{M}_X(s)$ is the c.g.f. of X at $s \in U$. If T_s is an r.v. having density

$$f_{T_s} (x; s) = \frac{e^{xs} f_X (x)}{\mathbb{M}_X(s)}, \tag{5.1}$$

for some $s \in U$, then T_s is referred to as an *exponentially tilted* random variable. Notice that its density integrates to one. Its m.g.f. and c.g.f. are easily seen to be

$$\mathbb{M}_{T_s}(t) = \frac{\mathbb{M}_X(t+s)}{\mathbb{M}_X(s)}, \quad \mathbb{K}_{T_s}(t) = \mathbb{K}_X(t+s) - \mathbb{K}_X(s),$$

so that $\mathbb{E}[T_s] = \mathbb{K}'_T(0) = \mathbb{K}'_X(s)$ and $\mathbb{V}(T_s) = \mathbb{K}''_X(s)$. Let $s_0 \in U$. Now consider using the normal distribution to approximate the true distribution of T_{s_0}; it must have mean $x_0 := \mathbb{K}'_X(s_0)$ and variance $v_0 := \mathbb{K}''_X(s_0)$, and is thus given by $x \mapsto \phi(x; x_0, v_0)$, where ϕ is the normal p.d.f. Use of (5.1) then yields an approximation for f_X as

$$x \mapsto \phi(x; x_0, v_0) \, \mathbb{M}_X(s_0) \, e^{-s_0 x} = \frac{1}{\sqrt{2\pi \, v_0}} \exp\left\{-\frac{1}{2v_0}(x - x_0)^2\right\} \mathbb{M}_X(s_0) \, e^{-s_0 x}.$$

The accuracy of this approximation to f_X, for a fixed x, depends crucially on the choice of s_0. We know that, in general, the normal approximation to the distribution of an r.v. X is accurate near the mean of X, but degrades in the tails. As such, we are motivated to choose an s_0 such that x is close to the mean of the tilted distribution. In particular, we would like to find a value $\hat{s}$ such that

$$\mathbb{K}'_X(\hat{s}) = x, \tag{5.2}$$

for which it can be shown that a unique solution exists when $\hat{s}$ is restricted to U, the convergence strip of the m.g.f. of X. The normal density approximation to the tilted r.v. with mean x at the point x is then $\phi(x; \mathbb{K}'_X(\hat{s}), \mathbb{K}''_X(\hat{s}))$, and the approximation for f_X simplifies to

$$\boxed{\hat{f}_X(x) = \frac{1}{\sqrt{2\pi \, \mathbb{K}''_X(\hat{s})}} \exp\{\mathbb{K}_X(\hat{s}) - x\hat{s}\}, \quad x = \mathbb{K}'_X(\hat{s}).} \tag{5.3}$$

Approximation $\hat{f}_X$ is referred to as the *(first-order) saddlepoint approximation* (s.p.a.) to f_X, where $\hat{s} = \hat{s}(x)$ is the solution to the *saddlepoint equation* and is referred to as the *saddlepoint* at x. It is valid for all values of x in the interior of the support of X; for example, if X follows a gamma distribution, then the s.p.a. is valid only for $x > 0$, and if $X \sim \text{Bin}(n, p)$, then the s.p.a. is valid for $x = 1, 2, \ldots, n - 1$.

It is important to emphasize the aforementioned fact that there always exists a unique root to the saddlepoint equation when $\hat{s}$ is restricted to the convergence strip of the m.g.f. of X. Numerically speaking, for a variety of practical problems in which the s.p.a. has been implemented, only very close to the borders of the support of X do numerical problems arise in solving the saddlepoint equation. Of course, no problems arise if a closed-form expression for $\hat{s}$ is obtainable, as occurs in some simple examples below and in the important context of the noncentral distributions detailed in Chapter 10.

The saddlepoint method of approximation is attributed to Daniels (1954), though, via its similarity to the *Laplace method* of approximation, there is evidence that the idea can be traced back to Georg Bernhard Riemann (see the discussion and references in Kass, 1988, p. 235; and also Tierney, 1988).

We now illustrate the application of (5.3) using some simple examples for which exact results are available.

⊙ ***Example 5.1***   Let $X \sim \text{Gam}(a, b)$ with $f_X = b^a x^{a-1} e^{-xb} \mathbb{I}_{(0,\infty)}(x) / \Gamma(a)$ and m.g.f. $\mathbb{M}_X(s) = \left(\frac{b}{b-s}\right)^a$ (see Example 1.5). It is easy to verify that the tilted r.v. T is also gamma distributed, though we do not need to calculate it explicitly. We require the solution to the saddlepoint equation $x = \mathbb{K}_X'(\hat{s}) = a/(b - \hat{s})$, which is available as the closed-form expression $\hat{s} = b - a/x$, valid for $x > 0$ (which corresponds to the interior of the support of X). Substituting this into (5.3) and simplifying yields

$$\hat{f}_X(x) = b^a \frac{a^{-a+1/2} e^a}{\sqrt{2\pi}} x^{a-1} e^{-xb} \mathbb{I}_{(0,\infty)}(x). \tag{5.4}$$

Approximation $\hat{f}_X$ has exactly the same kernel as f_X; both constant terms contain term b^a; and, recalling that Stirling's approximation to $n!$ is $(2\pi)^{1/2} n^{n+1/2} e^{-n}$, we see that

$$\frac{a^{-a+1/2} e^a}{\sqrt{2\pi}} \approx \frac{1}{\Gamma(a)},$$

where $\approx$ means that the ratio of the l.h.s. and the r.h.s. converges to one as $a \to \infty$.

Thus, the saddlepoint approximation to the gamma density will be extremely accurate for all x and values of a not too small. As a increases, the relative accuracy increases. This is quite plausible, recalling that $X = \sum_{i=1}^{a} E_i \sim \text{Gam}(a, b)$, where $E_i \overset{\text{i.i.d.}}{\sim} \text{Exp}(b)$; as a increases, the accuracy of the normal approximation to X via the CLT increases. ∎

The s.p.a. (5.3) will not, in general, integrate to one, although it will usually not be far off. It will often be possible to renormalize it, i.e.,

$$\boxed{\overline{\hat{f}}_X(x) = \frac{\hat{f}_X(x)}{\int \hat{f}_X(x)\, dx},} \tag{5.5}$$

which is a proper density. In Example 5.1, the renormalized density $\overline{\hat{f}}_X$ is exactly equal to f_X.

⊖ ***Example 5.2***   Let $X_i \overset{\text{ind}}{\sim} \text{Gam}(a_i, b_i)$, $i = 1, 2$, and $Q = X_1 + X_2$. If the a_i are integer values, then (2.15) can be used. Otherwise, the convolution formula (2.7), or

$$f_Q(q) = \int_0^q f_{X_1}(x) f_{X_2}(q - x)\, dx,$$

can be used to exactly evaluate the p.d.f., although this will be relatively slow, as it involves numeric integration for each point of the density. For the s.p.a., the m.g.f. is

$$\mathbb{M}_Q(t) = \mathbb{M}_{X_1}(t)\, \mathbb{M}_{X_2}(t) = \left(\frac{b_1}{b_1 - t}\right)^{a_1} \left(\frac{b_2}{b_2 - t}\right)^{a_2}, \quad t \in U := (-\infty, \min(b_1, b_2)),$$

and straightforward computation yields

$$\mathbb{K}_Q'(t) = \frac{a_1}{b_1 - t} + \frac{a_2}{b_2 - t} \quad \text{and} \quad \mathbb{K}_Q''(t) = \frac{a_1}{(b_1 - t)^2} + \frac{a_2}{(b_2 - t)^2}.$$

The saddlepoint equation is, for a fixed q, $\mathbb{K}_Q'(t) = q$, or $t^2 A + tB + C = 0$, where

$$A = q,$$
$$B = (a_1 + a_2) - q(b_1 + b_2),$$
$$C = b_1 b_2 q - (a_2 b_1 + a_1 b_2).$$

The two solutions are $\left(-B \pm \sqrt{B^2 - 4AC}\right)/(2A)$, denoted t_- and t_+, both of which are real.[1] For a given set of parameter values q, a_1, b_1, a_2, b_2, only one of the two roots t_- and t_+ can be in U, the convergence strip of Q, because we know that, in general, there exists a unique solution to the saddlepoint equation which lies in U. If $b_1 = b_2 =: b$, then it is easy to verify that $t_+ = b \notin U$, so that t_- must be the root. Computation with several sets of parameter values suggests that t_- is always the correct root (the reader is invited to prove this algebraically), so that

$$\hat{t} = \frac{1}{2A}\left(-B - \sqrt{B^2 - 4AC}\right),$$

and the s.p.a. for f_Q is expressible in closed form. ∎

○ **Example 5.3** (Example 2.8 cont.) We return to the occupancy distribution, whereby Y_k is the number of cereal boxes to purchase in order to get at least one of k different prizes, $2 \le k \le r$, and r is the total number of different prizes available. From the decomposition $Y_k = \sum_{i=0}^{k-1} G_i$, with $G_i \overset{ind}{\sim} \text{Geo}(p_i)$, the c.f. is simple, and was inverted in Example 2.8 to get the p.m.f. of Y_k.

We now consider the saddlepoint approximation. From the m.g.f. (2.6), the c.g.f. is easily seen to be

$$\mathbb{K}_{Y_k}(t) = kt + \sum_{i=0}^{k-1} \ln p_i - \sum_{i=0}^{k-1} \ln\left(1 - q_i e^t\right),$$

---

[1] That both roots are real follows because if both were complex, the saddlepoint would not exist. Another way of seeing this is as follows. Let

$$D := B^2 - 4AC = q^2 (b_2 - b_1)^2 + q(-2)(b_2 - b_1)(a_2 - a_1) + (a_1 + a_2)^2$$
$$=: \alpha q^2 + \beta q + \gamma$$

with $\alpha > 0$, so that D could be negative when q lies between the two roots of D, or, as $q > 0$, when

$$0 < q < \left(-\beta + \sqrt{\beta^2 - 4\alpha\gamma}\right)/(2\alpha).$$

But, as $\beta^2 - 4\alpha\gamma = -16a_1 a_2 (b_2 - b_1)^2 < 0$, we see that both roots of D are complex, i.e., that $D > 0$.

from which the saddlepoint equation is

$$\mathbb{K}'_{Y_k}(t) = k + \sum_{i=0}^{k-1} \frac{q_i e^t}{1 - q_i e^t} = y,$$

which needs to be solved numerically in general. Convergence of the m.g.f. requires $1 - q_i e^t > 0$ for each i or $1 - \max\left(q_i e^t\right) > 0$ or

$$t < -\ln\left(\max\left(q_i\right)\right) = -\ln\left(1 - \min\left(p_i\right)\right).$$

Next, a simple computation reveals that

$$\mathbb{K}''_{Y_k}(t) = \sum_{i=0}^{k-1} \frac{q_i e^t}{\left(1 - q_i e^t\right)^2},$$

so that (5.3) is straightforward to calculate for ordinates in the interior of the support, i.e., $y = k+1, k+2, \ldots$. The program in Listing 5.1 computes the s.p.a., and Figure 5.2 shows the exact and s.p.a. p.m.f. for $k = r = 20$, and the RPE, given by $100 \times$ (s.p.a. $-$ exact) /exact. Figure 5.3 is similar, but with $k = 15$. In both cases, the s.p.a. is highly accurate over the whole (interior of the) support.

```
function pdf = occupancyspa(yvec,r,k)
ivec=0:(k-1); pvec=(r-ivec)/r; yl=length(yvec); pdf=zeros(yl,1);
for loop=1:yl
  y=yvec(loop); shat = occspe(y,pvec);   qet=(1-pvec)*exp(shat);
  Ks=k*shat+sum(log(pvec)) - sum(log(1-qet));
  Kpp = sum(qet./(1-qet).^2);
  pdf(loop) = exp(Ks-y*shat) / sqrt(2*pi*Kpp);
end

function spe=occspe(y,pvec)
q=1-pvec; opt=optimset('Display','none','TolX',1e-6);
uplim = -log(max(q)); lolim = -1e1;
spe=fzero(@speeq,[lolim,0.9999*uplim],opt,q,y);

function dd=speeq(t,q,y)
k=length(q); et=exp(t); kp = k+et*sum(q./(1-q*et)); dd = y-kp;
```

Program Listing 5.1   Computes the p.m.f. s.p.a. of Y_k for a vector of values of y (yvec). The lower limit for the saddlepoint, lolim, was arbitrarily chosen (and is really $-\infty$), and might need to be changed for other k and r

Remarks:

(a) The degenerate case $k = 1$ corresponds to $p_0 = 1$, or $Y_1 \equiv 1$, and the saddlepoint solution

$$t = \ln \frac{y-1}{y(1-p)}$$

is not valid. But $Y_1 = 1$ is not in the interior of the support, where the saddlepoint is not defined.

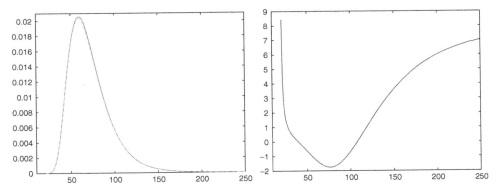

Figure 5.2 Exact (solid) and s.p.a. (dashed) of the p.m.f. of Y_k with $k = r = 20$ (left) and the s.p.a. relative percentage error (right)

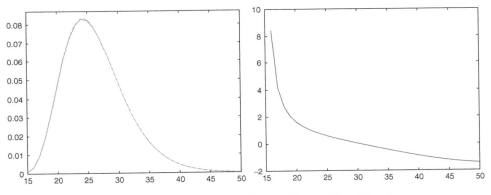

Figure 5.3 Same as Figure 5.2, but with $r = 20$ and $k = 15$

(b) The accuracy of the s.p.a. in this example can be increased without much more work: (5.10) below gives a higher-order s.p.a. expansion, which the reader is encouraged to use and compare.
(c) Kuonen (2000) has also looked at the use of the s.p.a. in this context, where further detail can be found. ∎

5.1.2 Saddlepoint approximation to the c.d.f.

The approximate c.d.f. of the r.v. X could be obtained by numerically integrating $\hat{f}$ or $\overline{\hat{f}}$. However, in a celebrated paper, Lugannani and Rice (1980) derived a simple expression for the s.p.a. to the c.d.f. of X, given by

$$\hat{F}_X\left(x^-\right) = \Phi\left(\hat{w}\right) + \phi\left(\hat{w}\right)\left\{\frac{1}{\hat{w}} - \frac{1}{\hat{u}}\right\}, \qquad x \neq \mathbb{E}\left[X\right], \tag{5.6}$$

where $F_X(x^-) = \Pr(X < x)$ (strict inequality), Φ and ϕ are the c.d.f. and p.d.f. of the standard normal distribution respectively,

$$\hat{w} = \text{sgn}(\hat{s})\sqrt{2\hat{s}x - 2\mathbb{K}_X(\hat{s})}, \tag{5.7a}$$

and

$$\hat{u} = \begin{cases} \hat{s}\sqrt{\mathbb{K}_X''(\hat{s})}, & \text{if } x \text{ is continuous,} \\ \left(1 - e^{-\hat{s}}\right)\sqrt{\mathbb{K}_X''(\hat{s})}, & \text{if } x \text{ is discrete.} \end{cases} \tag{5.7b}$$

Remarks:
(a) It is important to keep in mind that (5.6) is an approximation to $\Pr(X < x)$ and not $F_X(x) = \Pr(X \leq x)$. This is, of course, only relevant when X is discrete. There are other expressions for the c.d.f. approximation in the discrete case, and these exhibit different accuracies depending on the distribution of X and choice of x; see Butler (2007) for a very detailed and accessible discussion.
(b) If $x = \mathbb{E}[X]$, then $\mathbb{K}_X'(0) = \mathbb{E}[X]$ and $\hat{s} = 0$ is the saddlepoint for $\mathbb{E}[X]$. Thus, at the mean, $\mathbb{K}_X(0) = 0$, so that $\hat{w} = 0$, rendering $\hat{F}$ in (5.6) useless. This singularity is removable, however, and can be shown to be

$$\hat{F}_X(\mathbb{E}[X]) = \frac{1}{2} + \frac{\mathbb{K}_X'''(0)}{6\sqrt{2\pi}\mathbb{K}_X''(0)^{3/2}}. \tag{5.8}$$

For practical use, it is numerically wiser to use linear interpolation based on the s.p.a. to $\mathbb{E}[X] \pm \epsilon$, where ϵ is chosen small enough to ensure high accuracy, but large enough to ensure numerical stability of (5.7) and (5.6).
(c) Approximation (5.6) could be used in conjunction with Example 5.1 to deliver an accurate approximation to the incomplete gamma function $\gamma(x, a)$, discussed in Section I.1.5. ∎

⊖ ***Example 5.4***   (Binomial s.p.a.) Let $X \sim \text{Bin}(n, p)$, so that, with $q = 1 - p$, $\mathbb{M}_X(s) = (q + pe^s)^n$, as was shown in Problem 1.4. With

$$\mathbb{K}_X'(s) = \frac{d}{ds} \ln \mathbb{M}_X(s) = np\frac{e^s}{1 - p + pe^s},$$

we get the closed-form solution for $\hat{s}$ as

$$\hat{s} = \ln\left(\frac{x - xp}{np - xp}\right) = \ln\left(\frac{xq}{mp}\right),$$

where $m = n - x$. Also,

$$\mathbb{K}_X(\hat{s}) = n(\ln n + \ln q - \ln m) \quad \text{and} \quad \mathbb{K}_X''(\hat{s}) = npe^s q(q + pe^s)^{-2},$$

so that $\mathbb{K}''_X(\hat{s}) = xm/n$. The s.p.a. density (5.3) simplifies to

$$\hat{f}_X(x) = \frac{1}{\sqrt{2\pi \, (xm/n)}} \exp \left\{ n \, (\ln n + \ln q - \ln m) - \ln \left(\frac{xq}{mp} \right) x \right\}$$

$$= \widehat{\binom{n}{x}} p^x \, (1-p)^{n-x} \, , \tag{5.9}$$

where the hatted quantity is Stirling's formula applied to each factorial, i.e.,

$$\frac{\widehat{n!}}{(n-x)! \widehat{x!}} = \frac{\sqrt{2\pi} e^{-n} n^{n+1/2}}{\sqrt{2\pi} e^{-(n-x)} (n-x)^{n-x+1/2} \sqrt{2\pi} e^{-x} x^{x+1/2}}$$

$$= \frac{n^{n+1/2}}{\sqrt{2\pi} \, (n-x)^{n-x+1/2} \, x^{x+1/2}} \, .$$

It is quite remarkable that the form of (5.9) is so close to the true mass function.
For the c.d.f., with

$$\hat{s} = \ln \left(\frac{x}{n-x} \right) - \ln \left(\frac{p}{1-p} \right)$$

and using the fact that $\ln (\cdot)$ is monotonic,

$$\mathrm{sgn} \, (\hat{s}) = \mathrm{sgn} \left(\frac{x}{n-x} - \frac{p}{1-p} \right) = \mathrm{sgn} \left(\frac{x}{n} - p \right),$$

because

$$\frac{x}{n-x} < \frac{p}{1-p} \Leftrightarrow \frac{n-x}{x} > \frac{1-p}{p} \Leftrightarrow \frac{n}{x} > \frac{1}{p} \Leftrightarrow \frac{x}{n} < p.$$

Then $\hat{u} = (x-np) \sqrt{nxm}/(xqn)$ and

$$\hat{w} = \mathrm{sgn} \left(\frac{x}{n} - p \right) \sqrt{2} \sqrt{x \, (\ln q - \ln p + \ln x - \ln m) - n \, (\ln q + \ln n - \ln m)}$$

so that (5.6) is readily computable.

Remark: Of course, the value of this example is not in offering an approximation to the p.m.f. or c.d.f. of a binomially distributed random variable, for which readily computed exact expressions exist (although the c.d.f. approximation will indeed be much faster than an exact calculation for large n), but rather to further demonstrate the mechanics of the saddlepoint procedure and the high accuracy of the approximation, as is evident from (5.9). Moreover, this hints at the possible high accuracy of the s.p.a. when working with *sums* of independent binomial r.v.s, which involve only a simple generalization of the above results (though the saddlepoint is no longer available as a closed-form solution). The reader is encouraged to investigate this case, and compare the accuracy to the exact results obtained via c.f. inversion developed in Example 2.7.∎

Expression (5.3) is the leading term in an asymptotic expansion; the second-order approximation is given by (see Daniels, 1987)

$$\tilde{f}(x) = \hat{f}(x) \left(1 + \frac{\hat{\kappa}_4}{8} - \frac{5}{24}\hat{\kappa}_3^2 \right),$$ (5.10)

where

$$\hat{\kappa}_i = \frac{\mathbb{K}_X^{(i)}(\hat{s})}{\left[\mathbb{K}_X''(\hat{s}) \right]^{i/2}}.$$

Similarly,

$$\tilde{F}(x) = \hat{F}(x) - \phi(\hat{w}) \left\{ \hat{u}^{-1} \left(\frac{\hat{\kappa}_4}{8} - \frac{5}{24}\hat{\kappa}_3^2 \right) - \hat{u}^{-3} - \frac{\hat{\kappa}_3}{2\hat{u}^2} + \hat{w}^{-3} \right\}$$ (5.11)

for $x \neq \mathbb{E}[X]$. As with the first-order approximation (5.6), linear interpolation around $x = \mathbb{E}[X]$ is most effective for obtaining the continuity point of the approximation at the mean of X.

We next present several examples in which an exact solution would be more challenging and/or computationally more intensive than the saddlepoint approximation.

⊖ **Example 5.5**   (Differences of Poisson means) Let $X_i \overset{\text{i.i.d.}}{\sim} \text{Poi}(\lambda_1)$, $i = 1, \ldots, n_1$ independent of $Y_i \overset{\text{i.i.d.}}{\sim} \text{Poi}(\lambda_2)$, $i = 1, \ldots, n_2$. From (1.11), $\mathbb{M}_{X_i}(s) = \exp(\lambda_1 (e^s - 1))$, from which it follows that the m.g.f. of $\overline{X} - \overline{Y}$ is

$$\mathbb{M}_{\overline{X}-\overline{Y}}(s) = \exp\left(n_1 \lambda_1 \left(e^{s/n_1} - 1 \right) \right) \exp\left(n_2 \lambda_2 \left(e^{-s/n_2} - 1 \right) \right),$$

implying $\mathbb{K}_{\overline{X}-\overline{Y}}'(s) = \lambda_1 e^{s/n_1} - \lambda_2 e^{-s/n_2}$, with higher-order derivatives easily derived. For $n_1 = n_2 = n$, a closed-form solution to $x = \mathbb{K}_{\overline{X}-\overline{Y}}'(s)$ is

$$s = n \ln \left(\frac{x + \sqrt{x^2 + 4\lambda_1\lambda_2}}{2\lambda_1} \right).$$

For $n_1 = n_2 = 1$, the exact density of $X - Y$ was given in (I.6.35).   ∎

⊖ **Example 5.6**   (Differences of Bernoulli means) Let $X_i \overset{\text{i.i.d.}}{\sim} \text{Ber}(p_1)$, $i = 1, \ldots, n_1$, independent of $Y_i \overset{\text{i.i.d.}}{\sim} \text{Ber}(p_2)$, $i = 1, \ldots, n_2$. Then

$$\mathbb{M}_{\overline{X}-\overline{Y}}(s) = \left(q_1 + p_1 e^{s/n_1} \right)^{n_1} \left(q_2 + p_2 e^{-s/n_2} \right)^{n_2},$$

with

$$\mathbb{K}'_{X-Y}(s) = \frac{p_1 e^{s/n_1}}{1 - p_1 + p_1 e^{s/n_1}} - \frac{p_2 e^{-s/n_2}}{1 - p_2 + p_2 e^{-s/n_2}}$$

and

$$\mathbb{K}''_{X-Y}(s) = \frac{p_1 e^{s/n_1}(1-p_1)}{n_1 (1 - p_1 + p_1 e^{s/n_1})^2} + \frac{p_2 e^{-s/n_2}(1-p_2)}{n_2 (1 - p_2 + p_2 e^{-s/n_2})^2}.$$

In this case, the saddlepoint equation does not have a closed-form solution, but is readily computed numerically. ∎

Example 5.7 (Saddlepoint approximation to sum of normal and gamma) As in Example 2.19, let $Z \sim N(0, \sigma^2)$ independent of $G \sim \text{Gam}(a, c)$, for $\sigma, a, c \in \mathbb{R}_{>0}$, and let $X = Z + G$. We wish to derive the s.p.a. to $f_X(x)$ for $x > 0$. The m.g.f. of X is given by

$$\mathbb{M}_X(s; \sigma, a, c) = \left(\frac{c}{c - s}\right)^a e^{(\sigma^2 s^2)/2}, \quad -\infty < s < c,$$

which easily leads to

$$\mathbb{K}'_X(s; \sigma, a, c) = \frac{a}{c - s} + \sigma^2 s,$$

implying that the saddlepoint $\hat{s}$ is given by the single real root of the quadratic

$$\sigma^2 s^2 - (\sigma^2 c + x) s + (xc - a) = 0, \quad \text{such that } s < c.$$

Both roots are given by

$$s_\pm = \frac{\sigma^2 c + x \pm \sqrt{C}}{2\sigma^2}, \quad C = (\sigma^2 c - x)^2 + 4\sigma^2 a.$$

Clearly, $C > 0$, and $\sqrt{C} \geq \sigma^2 c - x$. From the constraint $s < c$, this rules out use of s_+, i.e., $\hat{s} = s_- = (\sigma^2 c + x - \sqrt{C})/2\sigma^2$, and the saddlepoint equation gives rise to a closed-form solution. Higher-order derivatives of $\mathbb{K}_X(s)$ are easily obtained; in particular, $\mathbb{K}''_X(s) = a(s - c)^{-2} + \sigma^2$. ∎

5.1.3 Detailed illustration: the normal–Laplace sum

Recall Example 2.16 in which $Z \sim \text{NormLap}(c)$, i.e., $Z = X + Y$, whereby $X \sim N(0, c^2)$ independent of $Y \sim \text{Lap}(0, k)$, $k = 1 - c$, $0 \leq c \leq 1$. To compute the s.p.a. of f_Z, note that the m.g.f. of X is $\exp(c^2 s^2/2)$, while the m.g.f. of Y is

$$\mathbb{M}_Y(s) = \mathbb{E}\left[e^{sY}\right] = \int_{-\infty}^{\infty} e^{sy} \frac{1}{2k} e^{-|y|/k} dy$$

$$= \frac{1}{2k} \int_{-\infty}^{0} e^{y(s+1/k)} dy + \frac{1}{2k} \int_{0}^{\infty} e^{-y(-s+1/k)} dy$$

$$= \frac{1}{2} \frac{1}{sk+1} - \frac{1}{2} \frac{1}{sk-1}$$

$$= \frac{1}{1 - s^2 k^2}, \tag{5.12}$$

provided $sk > -1$ and $sk < 1$, i.e., $|s| < 1/k$. It follows that

$$\mathbb{K}_{X+Y}(s) = \frac{1}{2} c^2 s^2 - \ln\left(1 - s^2 k^2\right), \quad |s| < 1/k, \tag{5.13}$$

so that, for $0 < c < 1$, the solution to the saddlepoint equation

$$\mathbb{K}'_Z(s) = c^2 s + \frac{2sk^2}{1 - s^2 k^2} = z$$

involves finding the single real root of the cubic

$$s^3 c^2 k^2 - s^2 k^2 z - s\left(2k^2 + c^2\right) + z = 0 \tag{5.14}$$

which satisfies $|s| < 1/k$. There are three cases of interest:

- If $c = 1$, then $Z \sim N(0, 1)$ and $\hat{s} = z$.

- If $c = 0$, then $Z \sim \text{Lap}(0, 1)$ and $\mathbb{K}'_Z(s) = 2s/(1 - s^2)$. For $z = 0$, $\hat{s} = 0$ is the only solution to the saddlepoint equation, while for $z \neq 0$, the two roots are $s_\pm = -\left(1 \pm \sqrt{1 + z^2}\right)/z$. As $s_+ < -1$, the valid solution is $\hat{s} = s_-$ when $z \neq 0$ and $c = 0$.

- For $0 < c < 1$, (5.14) needs to be solved. One way is, for each value of z, to use Matlab's `roots` command to compute the three roots and then decide which one is correct. The following piece of code shows how this might be done:

```
shat = roots([(c*k)^2,-x*k^2,-(2*k^2+c^2),x])';
s1=shat(1); s2=shat(2); s3=shat(3);
bool1 = (  abs(imag(s1)) < 1e-8  ) & (  abs(real(s1))<1/k );
bool2 = (  abs(imag(s2)) < 1e-8  ) & (  abs(real(s2))<1/k );
bool3 = (  abs(imag(s3)) < 1e-8  ) & (  abs(real(s3))<1/k );
bs=bool1+bool2+bool3;
if (bs>1),  disp(['Multiple roots for x=',int2str(x)]), end
if (bs==0), disp(['No roots for x=',int2str(x)]), end
s=bool1.*real(s1) + bool2.*real(s2) + bool3.*real(s3);
```

This is inefficient for two reasons. Firstly, the fact that the solution of a cubic can be expressed algebraically means that we could avoid the general polynomial root calculation, thus saving some time, and, more importantly, also compute the *vector* of $\hat{s}$-values corresponding to the vector of z-values 'in one shot' using the vector-calculation capabilities of Matlab (thus securing an enormous time saving).

Secondly, trial and error demonstrates that, of the three roots, the desired saddlepoint is always the same one, so that the boolean checking in the above code segment can also be avoided, thus saving even more time and programming complexity.[2] This is

---

[2] This can, in fact, be proven. The method will be discussed in the context of the noncentral distributions in Chapter 10.

another example where the saddlepoint takes on a closed-form solution which can be 'vectorized', so that the p.d.f. s.p.a. calculation is 'practically instantaneous'.[3]

This is implemented in Listing 5.2. The code for the saddlepoint is based on the notation of Abramowitz and Stegun (1972, p. 17), as the solution to the cubic $s^3 + a_2 s^2 + a_1 s + a_0$.

```
function [pdf,cdf]=normlapvec(x,c,acclevel)
% uses s.p.a. to compute pdf (and cdf if nargout>1) of mixture of
% c*normal plus (1-c)*laplace, 0<=c<=1 at x, which can be a vector.
% Set acclevel to 2 to use the higher order s.p.a.

if nargin<3, acclevel=2; end

if c==0 % pure Laplace
  pdf=exp(-abs(xvec))/2; return
elseif c==1 % pure normal
  pdf=normpdf(xvec); return
end

b1 = [0.98288003175962; -0.00098815145737; 0.04020574131157; ...
      0.01053174029526; 0.00580768773849;  0.67360537495647; ...
     -0.96740007771465];

b2=[19.321676; -127.31933; 351.36036; -481.0206;  326.6766;   -88.0728];
b3=[15.020794; -73.994190; 156.20436; -164.87816; 87.016516; -18.36932];

k=1-c; X2 = [1 c c^2 c^3 c^4 c^5]; X1=[X2 c^6];

if      c<=0.65,  kon=X1*b1;
elseif c<=0.85,  kon=X2*b2;
elseif c<=0.946, kon=X2*b3;
else kon=1;
end % 3rd regression gives 1 at 0.946

a2  = (-x*k^2)./(c*k)^2; a1 = -(2*k^2+c^2)/(c*k)^2; a0 = x./(c*k)^2;
q   = a1/3 -a2.^2/9; r = (a1.*a2-3*a0)/6 - a2.^3/27; m = q.^3 + r.^2;
s1  = (r+sqrt(m)).^(1/3); s2 = (r-sqrt(m)).^(1/3);
sps = s1+s2; sms = s1-s2; z3 = -sps/2 - a2/3 - sqrt(-3).*sms/2;
% z1 = sps - a2/3; z2 = -sps/2 - a2/3 + sqrt(-3).*sms/2;
s = z3;

K=0.5*s.^2.*c^2 - log(1-s.^2.*k.^2);
Kpp=c.^2+2*k.^2 * (s.^2*k.^2+1) ./ (1-s.^2*k.^2).^2;
pdf = exp(K - s.*x)./sqrt(2*pi*Kpp);
```

Program Listing 5.2 Computes the first-order and renormalized second-order s.p.a. to the p.d.f. and c.d.f. of the normal–Laplace weighted sum; continued in Listing 5.3

The first-order s.p.a. density is then given by (5.3), with

$$\mathbb{K}''_Z(s) = c^2 + 2k^2 \frac{s^2 k^2 + 1}{\left(1 - s^2 k^2\right)^2}.$$

---

[3] With 10 000 data points, about 0.1 second is required on a 1 GHz Pentium machine.

```
if acclevel==2
  K3 = 4*s.*k.^4 .* (3+s.^2.*k.^2) ./ (1-s.^2.*k.^2).^3;
  K4 = 12*k.^4 .* (1+6*s.^2.*k.^2+s.^4.*k.^4) ./ (1-s.^2.*k.^2).^4;
  kap3= K3./(Kpp).^(3/2); kap4=K4./(Kpp).^(4/2);
  pdf = pdf .* (1 + kap4/8 - 5*(kap3.^2)/24);
  pdf = pdf / kon;
else
  pdf = pdf;
  % No correction done.  The constant of integration
  %  drops to about 0.85 as c -> 0.
end

if nargout>1
  bad=find(abs(s)<1e-8); cdf(bad)=0.5;
  good=find(abs(s)>=1e-8);
  x=x(good); s=s(good);  K=K(good); Kpp=Kpp(good);
  w=sign(s).*sqrt(2*(s.*x-K)); u=s.*sqrt(Kpp);
  npdf=normpdf(w);
  if acclevel==2
    kap3=kap3(good); kap4=kap4(good);
    O1 = -npdf.*((kap4/8-5*kap3.^2/24)./u ...
         - 1./u.^3 - kap3/2./u.^2 + 1./w.^3);
  else
    O1=0;
  end
  cdf(good)=normcdf(w) + npdf.*(1./w - 1./u) + O1;

  bad=find(abs(x)<0.02); % some polishing near the mean.
  if any(bad)
    xx=[-0.02 0 0.02];
    [garb,cdf02]=normlapvec(0.02,c,acclevel);
    yy=[1-cdf02 0.5 cdf02];
    cdf(bad) = interp1(xx',yy',x(bad));
  end
end
```

Program Listing 5.3   Continuation of the program in Listing 5.2

The second-order approximation entails use of (5.10), where

$$\mathbb{K}_Z^{(3)}(s) = 4sk^4 \frac{3+s^2k^2}{\left(1-s^2k^2\right)^3}, \qquad \mathbb{K}_Z^{(4)}(s) = 12k^4 \frac{1+6s^2k^2+s^4k^4}{\left(1-s^2k^2\right)^4}.$$

Near the mean of Z (which is obviously zero), the c.d.f. approximations (5.6) and (5.11) cannot be used. Of course, (5.8) could be used to approximate the c.d.f. at $z = 0$, but there is no need, as the true value is clearly 1/2. However, numeric problems with (5.6) and (5.11) will occur for values of z *close to* zero. One straightforward way to circumvent this is simply to set the c.d.f. to 1/2 if $Z \approx 0$, which we take to be when $|\hat{s}| < 10^{-8}$. Alternatively (and better), for $|z| < 0.02$, the c.d.f. is approximated by linear interpolation, based on the s.p.a. c.d.f. values at -0.02, 0, and 0.02. This is implemented at the end of the program.

Because the s.p.a. is exact for the normal, we would expect the accuracy of $\hat{f}_Z$ to increase as $c \to 1$. One way of seeing this is to examine the values of $I_1(c) =$

$\int \hat{f}_Z(z; c) \, dz$. These were computed for several c and are shown as the circles in Figure 5.4. Indeed, as c moves away from one, $I_1(c)$ deviates considerably from unity. A simple function in c could be fitted to the curve and used to obtain the correct integrating constant. However, given that $\mathbb{K}_Z^{(3)}$ and $\mathbb{K}_Z^{(4)}$ are easily calculated, the second-order approximation could be used instead. The crosses in Figure 5.4 show $I_2(c)$ based on (5.10); they are indeed much closer to one. Using a finer grid of values, piecewise polynomials in c were fitted to $I_2(c)$ via least squares. The results are implemented in the program given in Listing 5.2. The difference between unity and the integrated density using this method is quite small and is shown in Figure 5.5.[4]

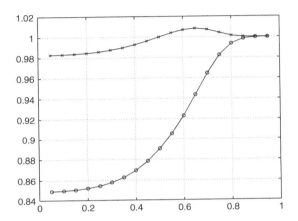

Figure 5.4 Numerically computed integral of $\hat{f}_Z$ based on the first-order (circles) and second-order (crosses) s.p.a. as a function of c

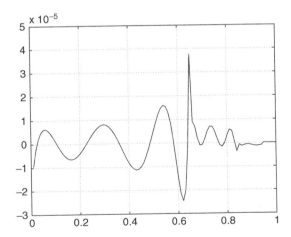

Figure 5.5 The constant of integration minus one of the second-order s.p.a. normalized with a piecewise polynomial approximation, as a function of c

---

[4] This discrepancy implies an integration accuracy which is higher than the accuracy of the s.p.a. itself and so would seem unnecessary. It turns out to be helpful when numerically maximizing the likelihood function associated with the model $Z_i \overset{\text{i.i.d.}}{\sim} \text{NormLap}(c)$.

It is illustrative to compare the accuracy of the first- and second-order s.p.a.s after both have been appropriately normalized. Figure 5.6(a) shows the percentage error of the normalized first- and second-order saddlepoint approximations of $\hat{f}_Z$ for $c = 0.3$. It is clear that the second-order density is vastly more accurate. Figure 5.6(b) is similar, but uses $c = 0.05$; as c moves towards zero, even the renormalized second-order s.p.a. suffers near the mean.

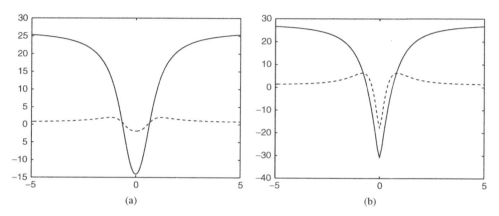

(a) (b)

Figure 5.6 Percentage error of normalized first-order s.p.a. of $f_Z(z)$ (solid) and normalized second-order s.p.a. of $f_Z(z)$ (dashed) as a function of z, for (a) $c = 0.3$ and (b) $c = 0.05$

5.2 Multivariate

The p.m.f. or p.d.f. saddlepoint approximation (5.3) generalizes naturally to the multivariate case. For a d-dimensional random vector $\mathbf{X}$ having joint c.g.f. $\mathbb{K}$ with gradient $\mathbb{K}'$ and Hessian $\mathbb{K}''$, the approximation is given by

$$\hat{f}_{\mathbf{X}}(\mathbf{x}) = (2\pi)^{-d/2} |\mathbb{K}''(\hat{\mathbf{s}})|^{-1/2} \exp\left(\mathbb{K}(\hat{\mathbf{s}}) - \hat{\mathbf{s}}'\mathbf{x}\right), \tag{5.15}$$

where the multivariate saddlepoint satisfies $\mathbb{K}'(\hat{\mathbf{s}}) = \mathbf{x}$ for $\hat{\mathbf{s}}$ in the convergence region of the m.g.f. of $\mathbf{X}$. Saddlepoint c.d.f. approximations for the general multivariate setting are less straightforward, and we restrict ourselves to the bivariate case. Section 5.2.1 illustrates the s.p.a. for the conditional distribution $X \mid Y$, Section 5.2.2 details the c.d.f. approximation in the continuous bivariate case, and Section 5.2.3 shows a method for approximating the marginal distribution.

5.2.1 Conditional distributions

Let $\mathbb{K}(s, t)$ denote the joint c.g.f. for continuous random variables X and Y, assumed convergent over S, an open neighbourhood of $(0, 0)$. The gradient of $\mathbb{K}$ is $\mathbb{K}'(s, t) = (\mathbb{K}'_s(s, t), \mathbb{K}'_t(s, t))'$, where

$$\mathbb{K}'_s(s, t) := \frac{\partial}{\partial s} \mathbb{K}(s, t), \quad \mathbb{K}'_t(s, t) := \frac{\partial}{\partial t} \mathbb{K}(s, t),$$

and

$$\mathbb{K}''(s,t) := \left[\begin{array}{cc} \mathbb{K}''_{ss}(s,t) & \mathbb{K}''_{st}(s,t) \\ \mathbb{K}''_{ts}(s,t) & \mathbb{K}''_{tt}(s,t) \end{array} \right], \quad \mathbb{K}''_{ss}(s,t) = \frac{\partial^2}{\partial s^2} \mathbb{K}(s,t), \quad \text{etc.,} \quad (5.16)$$

is the Hessian.

Let $\mathcal{X}$ be the interior of the convex hull[5] of the joint support of (X, Y). Skovgaard (1987) derived a *double-saddlepoint* approximation for the conditional c.d.f. of X at x given $Y = y$, for $(x, y) \in \mathcal{X}$. In this case, the gradient is a one-to-one mapping from the convergence strip $\mathcal{S}$ onto $\mathcal{X}$. As the name implies, there are two saddlepoints to compute when using this approximation. The first is the unique pre-image of (x, y) in $\mathcal{S}$, denoted $(\tilde{s}, \tilde{t})$, computed as the solutions to

$$\mathbb{K}'_s(\tilde{s}, \tilde{t}) = x, \qquad \mathbb{K}'_t(\tilde{s}, \tilde{t}) = y. \qquad (5.17)$$

This is the numerator saddlepoint in the approximation. The second saddlepoint is found by fixing $s = 0$ and solving $\mathbb{K}'_t(0, \tilde{t}_0) = y$ for the unique value of $\tilde{t}_0$ in $\{t : (0, t) \in \mathcal{S}\}$. This is the denominator saddlepoint. The c.d.f. approximation is then given by

$$\Pr(X \leq x \mid Y = y) \approx \Phi(\tilde{w}) + \phi(\tilde{w}) \left\{ \tilde{w}^{-1} - \tilde{u}^{-1} \right\}, \qquad \tilde{s} \neq 0, \qquad (5.18)$$

where

$$\tilde{w} = \operatorname{sgn}(\tilde{s}) \sqrt{2} \sqrt{\tilde{s}x + \tilde{t}y - \mathbb{K}(\tilde{s}, \tilde{t}) - \tilde{t}_0 y + \mathbb{K}(0, \tilde{t}_0)}, \qquad (5.19)$$

$$\tilde{u} = \tilde{s} \sqrt{\left| \mathbb{K}''(\tilde{s}, \tilde{t}) \right| / \mathbb{K}''_{tt}(0, \tilde{t}_0)}. \qquad (5.20)$$

Remarks:

(a) Because of the name 'double-saddlepoint approximation', the Lugannani–Rice c.d.f. approximation (5.6) is sometimes referred to as the 'single-saddlepoint approximation'.

(b) To see that $\tilde{w}$ makes sense, note that

$$\sup_{(s,t)\in\mathcal{S}} \left[sx + ty - \mathbb{K}(s,t) \right] \qquad (5.21)$$

occurs at $(\tilde{s}, \tilde{t})$, as (5.17) is its critical value. Also, the supremum (5.21) must be greater than the constrained supremum over $\{(0, t) \in \mathcal{S}\}$, which occurs at $\tilde{t}_0$. Thus, the term inside the square root is positive. The term $\tilde{u}$ is also well defined because $\mathbb{K}''(s, t)$ is positive definite for all $(s, t) \in \mathcal{S}$.

(c) Expression (5.18) is not meaningful when $\tilde{s} = 0$. In this case, Skovgaard (1987) has shown that the singularity of (5.18) is removable and its value is defined by continuity. ∎

○ *Example 5.8* As in Example 1.11, let $f_{X,Y}(x, y) = e^{-y} \mathbb{I}_{(0,\infty)}(x) \mathbb{I}_{(x,\infty)}(y)$. The conditional distribution is easily seen to be $f_{X|Y}(x; y) = y^{-1} \mathbb{I}_{(0,y)}(x)$, i.e., uniform on

---

[5] Recall that a set S is *convex* if, for every pair of points $P, Q \in S$, the entire line segment PQ is also in S. The *convex hull* of a set S is the smallest convex set that contains S.

$(0, y)$, with c.d.f. $F_{X|Y}(x; y) = (x/y) \mathbb{I}_{(0,y)}(x) + \mathbb{I}_{[y,\infty)}(x)$. From (1.22) and using s and t instead of t_1 and t_2,

$$\mathbb{K}(s, t) = -\ln(1 - s - t) - \ln(1 - t), \quad s + t < 1, \quad t < 1,$$

so that

$$\mathbb{K}'_s(s, t) = \frac{1}{1 - s - t}, \quad \mathbb{K}'_t(s, t) = \frac{1}{1 - s - t} + \frac{1}{1 - t},$$

and

$$\mathbb{K}''(s, t) = \begin{bmatrix} (s + t - 1)^{-2} & (s + t - 1)^{-2} \\ (s + t - 1)^{-2} & (t - 1)^{-2} + (s + t - 1)^{-2} \end{bmatrix},$$

with $\det \mathbb{K}'' = (t - 1)^{-2}(s + t - 1)^{-2} > 0$. From (5.17), the numerator saddlepoint solves $\mathbb{K}'_s(\tilde{s}, \tilde{t}) = x$ and $\mathbb{K}'_t(\tilde{s}, \tilde{t}) = y$, which gives

$$\tilde{s} = \frac{2x - y}{x(y - x)}, \quad \tilde{t} = \frac{x - y + 1}{x - y},$$

and the denominator saddlepoint solves $\mathbb{K}'_t(0, \tilde{t}_0) = y$, or $\tilde{t}_0 = 1 - 2/y$. The reader is encouraged to verify that $\mathbb{K}''(\tilde{s}, \tilde{t})$ simplifies to

$$\mathbb{K}''(\tilde{s}, \tilde{t}) = \begin{bmatrix} x^2 & x^2 \\ x^2 & (y - x)^2 + x^2 \end{bmatrix}, \quad \det \mathbb{K}'' = x^2(y - x)^2 > 0,$$

while $\tilde{w}$ and $\tilde{u}$ simplify to

$$\tilde{w} = \text{sgn}(2x - y) \sqrt{2}\sqrt{-\ln(x) - \ln(y - x) - 2\ln(2/y)}, \quad \tilde{u} = \sqrt{2}\frac{2x - y}{y}.$$

With $x = 0.8$ and $y = 1$, (5.18) returns 0.7970, which compares well with the true c.d.f. value of 0.8. Figure 5.7 shows the relative absolute percentage error of (5.18) for $y = 1$. ∎

An example of an important and far less trivial application of the conditional s.p.a. will be given in a chapter on ratios of quadratic forms in normal variables in a forthcoming book.

5.2.2 Bivariate c.d.f. approximation

Wang (1990) shows that a saddlepoint approximation to the c.d.f. of $\mathbf{X} = (X, Y)$ at $\mathbf{x} = (x, y)$ is given by

$$\hat{F}_{\mathbf{X}}(\mathbf{x}) = \Phi_2(\tilde{x}_1, \tilde{y}_1, \tilde{\rho}) + \Phi(\tilde{w}_0)\tilde{n} + \Phi(\tilde{w})\tilde{n}_0 + \tilde{n}\tilde{n}_0, \tag{5.22}$$

where

$$\tilde{x}_1 = \text{sgn}(\tilde{t}_0)\sqrt{2(\tilde{t}_0 y - \mathbb{K}(0, \tilde{t}_0))}, \quad \tilde{w}_0 = \text{sgn}(\tilde{t})\sqrt{2(\mathbb{K}(\tilde{s}, 0) - \mathbb{K}(\tilde{s}, \tilde{t}) + \tilde{t}y)},$$

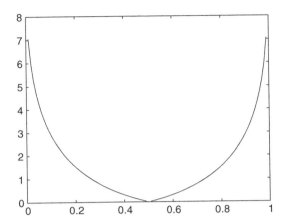

Figure 5.7 The relative absolute percentage error of the conditional c.d.f. approximation (5.18) of $X \mid (Y = 1)$

$$\tilde{y}_1 = \frac{\tilde{w} - b\tilde{x}_1}{\sqrt{1 + b^2}}, \quad \tilde{\rho} = -\frac{b}{\sqrt{1 + b^2}}, \quad b = \frac{\tilde{w}_0 - \tilde{x}_1}{\tilde{w}}, \quad \tilde{n} = \phi(\tilde{w}) \left[\frac{1}{\tilde{w}} - \frac{1}{\tilde{u}} \right],$$

$$\tilde{n}_0 = \phi(\tilde{x}_1) \left[\frac{1}{w_0} - \frac{1}{\tilde{u}_0} \right], \quad \tilde{u} = \tilde{s} \sqrt{\frac{|\mathbb{K}''(\tilde{s}, \tilde{t})|}{\mathbb{K}''_{tt}(\tilde{s}, \tilde{t})}}, \quad \tilde{u}_0 = \tilde{t} \sqrt{\mathbb{K}''_{tt}(\tilde{s}, \tilde{t})},$$

$\tilde{w}$ and the saddlepoints $(\tilde{s}, \tilde{t})$ and $\tilde{t}_0$ are as in the Skovgaard approximation from Section 5.2.1, and $\Phi_2(x, y, \rho)$ denotes the bivariate standard normal c.d.f. with correlation ρ.

Remarks:
(a) If $\tilde{t} = 0$, then $\tilde{n}_0$ should be replaced by its limiting value, given by

$$\frac{\phi(\tilde{x}_1)}{6} \frac{\mathbb{K}_{ttt}(\tilde{s}, 0)}{\mathbb{K}_{tt}(\tilde{s}, 0)^{3/2}}.$$

The same applies for quantities $\tilde{n}$ and b if $\tilde{s} = 0$; while their limiting value can, in theory, be derived by applying l'Hôpital's rule twice to $(\tilde{w} - \tilde{u})/(\tilde{w}\tilde{u})$ and b, respectively, this is an extremely tedious task, so that it appears more practical to instead apply approximation (5.22) to the vector $(X^*, Y^*) = (Y, X)$ with c.g.f. $\mathbb{K}^*(s, t) = \mathbb{K}(t, s)$, which will result in $\tilde{t}^* = 0$. Wang also recommends use of (5.22) with the order X and Y interchanged when $\tilde{w}_0$ is 'large' and positive. One possible strategy is to do so whenever $\tilde{w}_0$ is bigger than the respective quantity of the reversed-variable approximation,

$$\tilde{w}_0^* = \text{sgn}(\tilde{s}) \sqrt{2 \left(\mathbb{K}(0, \tilde{t}) - \mathbb{K}(\tilde{s}, \tilde{t}) + \tilde{s}x \right)}.$$

(b) Wang derives approximation (5.22) only for the case where $\mathbb{E}[X] = \mathbb{E}[Y] = 0$. To circumvent this, note that the c.d.f. of X can be approximated by

$$\Pr(X < x) = \hat{F}_{X_0}(x_0, y_0),$$

i.e., approximation (5.22) applied to the demeaned random variable

$$\mathbf{X}_0 \equiv (X - \mathbb{E}[X], Y - \mathbb{E}[Y]) = (X - \mathbb{K}'_s(0, 0), Y - \mathbb{K}'_s(0, 0)),$$

and evaluated at $(x_0, y_0) = (x - \mathbb{E}[X], y - \mathbb{E}[Y])$. As $\mathbf{X}_0$ has c.g.f.

$$\mathbb{K}_{\mathbf{X}_0} = \mathbb{K}_{\mathbf{X}} - s\mathbb{K}'_s(0, 0) - t\mathbb{K}'_t(0, 0),$$

the saddlepoint equation becomes

$$\mathbb{K}'_s(\tilde{s}, \tilde{t}) - \mathbb{K}'_s(0, 0) = x_0 \quad \text{and} \quad \mathbb{K}'_t(\tilde{s}, \tilde{t}) - \mathbb{K}'_t(0, 0) = y_0.$$

Plugging in x_0 and y_0 from their definition shows that the saddlepoint remains unchanged. The same is true for $\tilde{t}_0$. Finally, evaluating $\tilde{x}_1$, $\tilde{w}$, and $\tilde{w}_0$ in (5.22) using $\mathbb{K}_{\mathbf{X}_0}$, x_0, and y_0, the terms containing $\tilde{s}\mathbb{K}'_t(0, 0)$, $\tilde{t}\mathbb{K}'_t(0, 0)$, and $\tilde{t}_0\mathbb{K}'_t(0, 0)$ cancel, so that $\hat{F}_{X_0}(x_0, y_0) = \hat{F}_X(x, y)$.
(c) Kolassa (2003) gives an approximation valid for $d > 2$. ∎

⊖ **Example 5.9**   As in Wang (1990), let $X = U_1 + U_2$ and $Y = U_2 + U_3$, where $U_i \overset{\text{i.i.d.}}{\sim}$ Exp(1). Note that, being sums of i.i.d. exponentials, the marginal distributions of X and Y are gamma. Their joint m.g.f. is

$$\begin{aligned}
\mathbb{M}_{X,Y}(s, t) &= \mathbb{E}\big[\exp(sX + tY)\big] = \mathbb{E}\big[\exp(sU_1 + (s + t)U_2 + tU_3)\big] \\
&= \int_0^\infty \exp(-u_1 - u_2 - u_3)\exp(su_1 + (s + t)u_2 + tu_3)\,du_1\,du_2\,du_3 \\
&= \frac{1}{1 - s}\frac{1}{1 - t}\frac{1}{1 - s - t}, \qquad s < 1, t < 1, s + t < 1,
\end{aligned}$$

and the c.g.f. is

$$\mathbb{K}(s, t) = -\log(1 - s) - \log(1 - t) - \log(1 - s - t),$$

with gradient

$$\mathbb{K}'_s(s, t) = (1 - s)^{-1} + (1 - s - t)^{-1}, \qquad \mathbb{K}'_t(s, t) = (1 - t)^{-1} + (1 - s - t)^{-1},$$

and Hessian

$$\mathbb{K}''(s, t) = \begin{bmatrix} (1 - s)^{-2} + (1 - s - t)^{-2} & (1 - s - t)^{-2} \\ (1 - s - t)^{-2} & (1 - t)^{-2} + (1 - s - t)^{-2} \end{bmatrix}.$$

The saddlepoint $(\tilde{s}, \tilde{t})$ solves

$$\frac{1}{(1 - \tilde{s})} + \frac{1}{(1 - \tilde{s} - \tilde{t})} = x, \qquad \frac{1}{(1 - \tilde{t})} + \frac{1}{(1 - \tilde{s} - \tilde{t})} = y.$$

With the convenient help of Maple, it is found that $\tilde{t}$ is a root of the cubic

$$a_3 z^3 + a_2 z^2 + a_1 z + a_0 = 0,$$

where

$$a_3 = -y^2 + xy, \quad a_2 = -4y + 2y^2 - 2xy + 2x, \quad a_1 = -3 - 3x + 6y + xy - y^2,$$

and $a_0 = 2 + x - 2y$. Define

$$c_2 = \frac{a_2}{a_3}, \quad c_1 = \frac{a_1}{a_3}, \quad c_0 = \frac{a_0}{a_3}, \quad q = \frac{1}{3}c_1 - \frac{1}{9}c_2^2,$$

and

$$r = \frac{1}{6}(c_1 c_2 - 3c_0) - \frac{1}{27}c_2^3, \quad m = q^3 + r^2, \quad s_{1,2} = (r \pm \sqrt{m})^{1/3}.$$

Then, if $m < 0$, as is the case here, all three roots z_j, $j \in \{1, 2, 3\}$, are real and given by

$$z_i = \sqrt{-4q} \cos\left\{\left[\cos^{-1}\left(r/\sqrt{-q^3}\right) + \pi j\right]/3\right\} - \frac{c_2}{3},$$

from which the one satisfying $\tilde{s} < 1$, $\tilde{t} < 1$, and $\tilde{s} + \tilde{t} < 1$ has to be selected. The second saddlepoint is given by $\tilde{t}_0 = y/(y - 2)$, so that all quantities entering (5.22) are readily computed; Listing 5.4 contains the requisite Matlab code, and Figure 5.8 illustrates the accuracy of the approximation. Evaluation of (5.22) requires a means to compute the bivariate standard normal p.d.f.; this is achieved by the program in Listing 3.3. ∎

5.2.3 Marginal distributions

Let $\mathbf{X} = (X_1, X_2)$ be a continuous bivariate random variable with joint cumulant generating function $\mathbb{K}(\mathbf{t}) \equiv \mathbb{K}(s, t)$. Consider a bijection

$$\mathbf{Y} = (Y_1, Y_2) = g^{-1}(\mathbf{X}) = (g_1^{-1}(\mathbf{X}), g_2^{-1}(\mathbf{X}))',$$

so that $\mathbf{X} = g(\mathbf{Y}) = (g_1(\mathbf{Y}), g_2(\mathbf{Y}))'$, and denote by

$$\nabla_{y_i} g(\mathbf{Y}) = \left(\frac{\partial g_1}{\partial Y_i}, \frac{\partial g_2}{\partial Y_i}\right)', \quad \nabla_{y_i}^2 g(\mathbf{Y}) = \left(\frac{\partial^2 g_1}{\partial Y_i^2}, \frac{\partial^2 g_2}{\partial Y_i^2}\right)' \quad i \in \{1, 2\},$$

the vectors of first and second derivatives of g with respect to Y_i. Interest centres on the marginal distribution of Y_1. Daniels and Young (1991) show that saddlepoint approximations to the marginal p.d.f. and c.d.f. of y_1 are given by

$$\tilde{f}_{Y_1}(y_1) = \frac{\phi(\tilde{w})}{\tilde{u}} \tag{5.23}$$

and

$$\tilde{F}_{Y_1}(y_1) = \Phi(\tilde{w}) + \phi(\tilde{w})\left(\frac{1}{\tilde{w}} - \frac{\tilde{d}}{\tilde{u}}\right), \tag{5.24}$$

```
function cdf= bivgammaspa(xvec,yvec)
cdf=zeros(length(xvec),length(yvec));
for xloop=1:length(xvec), for yloop=1:length(yvec)
  x=xvec(xloop); y=yvec(yloop);
  if x==0 | y==0, cdf(xloop,yloop)=0;
  else
    a3=(-y.^2  + x.*y); h=(x-2*y);a2=(2*h -2*a3);
    a1=(-3 - 3*h + a3); a0=(2 + h);
    if a3==0 % we have a quadratic
      c1=a1/a2;c0=a0/a2; ts=-c1/2+(-1).^[1:2]*sqrt(c1^2/4-c0);
    else
      c2=a2/a3;c1=a1/a3;c0=a0/a3; q=c1/3-(c2.^2)/9;
      r=1/6*(c1.*c2-3.*c0)-1/27.*c2.^3;
      ts=sqrt(-4*q).*cos((acos(r./sqrt(-q.^3))+2*[1:3]*pi)/3)-c2/3;
    end
    ss=1-1./(x-y+1./(1-ts)); which=(ts<1 & ss<1 & ts+ss<1);
    t=ts(which); s=ss(which); %choose right sp
    t0=(y-2)/y; Kt0=-2*log(1-t0); Kst=-log(1-s)-log(1-t)-log(1-s-t);
    K0t=-2*log(1-t); K0s=-2*log(1-s);
    x1=real(sign(t0)*sqrt(2*(t0*y-Kt0)));
    w0=real(sign(t)*sqrt(2*(t*y-Kst+K0s)));
    w0star=real(sign(s)*sqrt(2*(s*x-Kst+K0t)));
    if (abs(s)<1e-6 & abs(t)>1e-6) | (abs(t)>1e-6 & w0>w0star+1e-6),
      cdf(xloop,yloop)=bivgammaspa(y,x);
    else
      w=real(sign(s)*sqrt(2*((t-t0)*y+s*x-Kst+Kt0)));
      b=(w0-x1)/w; rho=-b/sqrt(1+b^2); y1=(w-b*x1)/sqrt(1+b^2);
      K_ss=1/(1-s)^2+1/(1-s-t)^2;K_tt=1/(1-t)^2+1/(1-s-t)^2;
      K_st=1/(1-s-t)^2;K_ttt=2*(1-s)^-3+2*(1-s)^-3;
      u=s*sqrt((K_ss-(K_st^2)/K_tt));
      if abs(t)<1e-6
        w1=0; n0=normpdf(x1)*K_ttt/K_tt^(3/2)/6;
      else, n0=normpdf(x1)*(1/w0-1/(t*sqrt(K_tt))); end
      n=normpdf(w)*(1/w-1/u);
      cdf(xloop,yloop)=bvncdf(x1,y1,rho) + ...
                  normcdf(w0)*n+normcdf(w)*n0+n*n0;
    end
  end
 end
end, end
```

Program Listing 5.4   Matlab code for the bivariate gamma s.p.a.

respectively, where

$$\tilde{w} = \sqrt{2\big(\tilde{\mathbf{t}}'g(\tilde{\mathbf{y}}) - \mathbb{K}(\tilde{\mathbf{t}})\big)}\,\mathrm{sgn}(y_1 - \alpha), \quad \tilde{\mathbf{y}} = (y_1, \tilde{y}_2),$$

$$\alpha = g_1^{-1}(\mathbb{K}'(\mathbf{0})), \quad \tilde{d} = \big(\tilde{\mathbf{t}}'\nabla_{y_1}g(\tilde{\mathbf{y}})\big)^{-1},$$

$$\tilde{u} = \frac{\sqrt{\det\big(\mathbb{K}''(\tilde{\mathbf{t}})\big)\Big[\nabla_{y_2}g(\tilde{\mathbf{y}})'\big(\mathbb{K}''(\tilde{\mathbf{t}})\big)^{-1}\nabla_{y_2}g(\tilde{\mathbf{y}}) + \tilde{\mathbf{t}}'\nabla_{y_2}^2 g(\tilde{\mathbf{y}})\Big]}}{\det\big(\partial g/\partial\mathbf{y}(\tilde{\mathbf{y}})\big)},$$

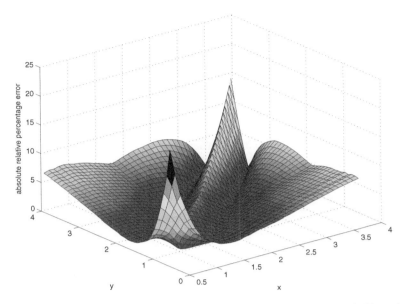

Figure 5.8   Absolute relative percentage error, defined as (s.p.a. − True)/ min(True, 1 − True), of the bivariate gamma s.p.a. The true values have been obtained from simulation with 500 000 replications

and, for each value of y_1, $\tilde{\mathbf{t}}$ and $\tilde{y}_2$ solve the system

$$\mathbb{K}'(\tilde{\mathbf{t}}) = g(\tilde{\mathbf{y}})$$

$$\tilde{\mathbf{t}}' \nabla_{y_2} g(\tilde{\mathbf{y}}) = 0.$$

As usual in saddlepoint applications, if additional accuracy is desired, the p.d.f. approximation can be renormalized by numerically integrating it over its support. A formula valid for the general multivariate case can be found in DiCiccio and Martin (1991).

⊖   ***Example 5.10***   Let $X_1 \sim \mathrm{N}(0, 1)$, independent of $X_2 \sim \chi_n^2$, and let $\mathbf{X} = g(\mathbf{Y}) = (Y_1 Y_2, nY_2^2)$, so that $(Y_1, Y_2) = g^{-1}(X_1, X_2) = \left(X_1 / \sqrt{X_2/n}, \sqrt{X_2/n}\right)'$, and Y_1 has a Student's t distribution with n degrees of freedom. The joint cumulant generating function of (X_1, X_2) is, from independence,

$$\mathbb{K}(\mathbf{t}) = \mathbb{K}_{X_1}(s) + \mathbb{K}_{X_2}(t) = \frac{1}{2}s^2 - \frac{n}{2}\log(1 - 2t),$$

where $\mathbb{K}_{X_1}$ and $\mathbb{K}_{X_2}$ are the c.g.f.s of X_1 and X_2, respectively. The saddlepoint $(\tilde{\mathbf{t}}, \tilde{y}_2) = (\tilde{s}, \tilde{t}, \tilde{y}_2)$, $\tilde{t} < \frac{1}{2}$, $\tilde{y}_2 > 0$, solves the system of equations

$$s = y_1 y_2$$

$$\frac{1}{1 - 2t} = y_2^2$$

$$sy_1 + 2nty_2 = 0,$$

which can be solved to give

$$\tilde{y}_2 = \sqrt{\frac{n}{(y_1^2 + n)}}, \quad \tilde{s} = y_1 \tilde{y}_2, \quad \tilde{t} = -\frac{y_1 \tilde{s}}{2n\tilde{y}_2}.$$

The other required quantities are given by

$$\tilde{d} = (y_1 \tilde{y}_2^2)^{-1}, \quad \tilde{u} = \tilde{y}_2^{-1}, \quad w = \sqrt{-2n \log(\tilde{y}_2)} \operatorname{sgn}(y_1),$$

and, plugging in, the p.d.f. approximation becomes

$$\hat{f}_t(y_1; n) = \frac{1}{\sqrt{2\pi}} \left(\frac{n}{y_1^2 + n}\right)^{\frac{1}{2}(n+1)},$$

which is exact after renormalization. Figure 5.9 shows that the c.d.f. approximation, which was computed using the program in Listing 5.5, is remarkably accurate even for $n = 2$.

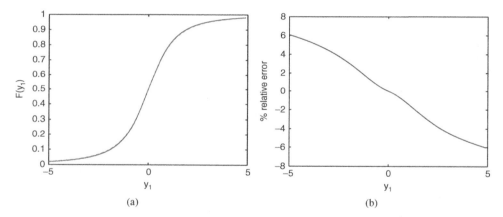

Figure 5.9 (Left) Exact (solid) and approximate (dashed) t_2 c.d.f. (Right) Relative percentage error of the approximation

```
function [pdf,cdf]=tspa(y,n)
y2=sqrt(n./(y.^2+n)); w=sqrt(-2*n*log(y2)).*sign(y);
u=1./y2;d=1./(y.*y2.^2); pdf=normpdf(w)./u;
cdf=normcdf(w)+normpdf(w).*(1./w-d./u);
```

Program Listing 5.5 Matlab code to compute the s.p.a. to the Student's t distribution

Of course, via the c.d.f. expression (I.7.47) involving the incomplete beta ratio, fast and reliable algorithms are available for computing the Student's t c.d.f. to machine precision, so this approximation, while extraordinary accurate, is not needed. The real benefit of this exercise will be seen when working with the singly and doubly noncentral t distributions, which are important extension of the t, but whose exact c.d.f. calculation entails (particularly in the doubly noncentral case) a very large amount of computation. For these, the s.p.a. is not only also applicable, but also results in a closed-form solution for the saddlepoint equation, thus yielding an easily computed and highly accurate method of calculation which is considerably faster than exact calculations. ∎

5.3 The hypergeometric functions $_1F_1$ and $_2F_1$

A number of exercises below (Problems 5.7, 5.8, 6.6, and 7.2) will require knowledge of the $_1F_1$ and $_2F_1$ functions, and these also arise prominently in the study of noncentral distributions, discussed in detail in Chapter 10. The following brief outline discusses the basics of these functions and how to compute and approximate them.

The *generalized hypergeometric function* is denoted $_jF_k$, with the low-order cases $_1F_1$ and $_2F_1$ being the most popular. For $a, b, c, z \in \mathbb{R}$, they are given by

$$_1F_1(a, b; z) = \sum_{n=0}^{\infty} \frac{a^{[n]}}{b^{[n]}} \frac{z^n}{n!} \quad \text{and} \quad _2F_1(a, b; c; z) = \sum_{n=0}^{\infty} \frac{a^{[n]}b^{[n]}}{c^{[n]}} \frac{z^n}{n!}, \tag{5.25}$$

respectively, where

$$a^{[j]} = \begin{cases} a(a+1)\cdots(a+j-1), & \text{if } j \geq 1, \\ 1, & \text{if } j = 0, \end{cases} \tag{5.26}$$

is the ascending factorial. The function $_1F_1$ is also referred to as the *confluent hypergeometric function*. Lebedev (1972, Ch. 9) provides a detailed, accessible discussion of these functions, as does Abadir (1999), who also discusses some applications in economics and finance. For a more detailed, high-level textbook treatment of special functions, including the hypergeometric functions, see Andrews, Askey and Roy (1999).

Recall from Section I.1.5.2 that the beta function is

$$B(a, b) := \int_0^1 x^{a-1}(1-x)^{b-1}\, dx, \quad a, b \in \mathbb{R}_{>0},$$

with the relation to the gamma function given by $B(a, b) = \Gamma(a)\Gamma(b) / \Gamma(a+b)$. Valuable both analytically and for numerical evaluation are the integral equations

$$_1F_1(a, b; z) = \frac{1}{B(a, b-a)} \int_0^1 y^{a-1}(1-y)^{b-a-1} e^{zy}\, dy, \tag{5.27}$$

$$\text{for} \quad a > 0, \ b - a > 0,$$

and

$$_2F_1(a, b; c; z) = \frac{1}{B(a, c-a)} \int_0^1 y^{a-1}(1-y)^{c-a-1}(1-zy)^{-b}\, dy, \tag{5.28}$$

$$\text{for} \quad a > 0, \ c - a > 0, \ z < 1,$$

but note their parameter restrictions. These were first given by Euler in 1769, and can be derived as follows. For $_1F_1 (a, b; z)$, expand e^{zy} to get

$$\int_0^1 y^{a-1}(1 - y)^{b-a-1} \left(1 + zy + \left(\frac{z^2}{2!} \right) y^2 + \left(\frac{z^3}{3!} \right) y^3 + \cdots \right) dy$$

$$= \int_0^1 y^{a-1}(1 - y)^{b-a-1} \, dy + \int_0^1 zy y^{a-1}(1 - y)^{b-a-1} \, dy$$

$$+ \frac{1}{2!} \int_0^1 z^2 y^2 y^{a-1}(1 - y)^{b-a-1} \, dy + \cdots$$

$$= B (a, b - a) + \frac{a}{b} z B (a, b - a) + \frac{(a + 1) a}{(b + 1) b} \frac{z^2}{2} B (a, b - a) + \cdots$$

$$= \sum_{n=0}^\infty \frac{a^{[n]}}{b^{[n]}} \frac{z^n}{n!},$$

where, for example,

$$\int_0^1 y^a (1 - y)^{b-a-1} \, dy = \int_0^1 y^{(a+1)-1} (1 - y)^{(b+1)-(a+1)-1} \, dy$$

$$= B (a + 1, b - a)$$

$$= \frac{\Gamma (a + 1) \Gamma (b - a)}{\Gamma (b + 1)} = \frac{a \Gamma (a) \Gamma (b - a)}{b \Gamma (b)}$$

$$= \frac{a}{b} B (a, b - a),$$

with higher-order terms given similarly. (See Lebedev, 1972, p. 239, for justification for reversing the summation and integral.) Similarly, for $_2F_1 (a, b; c; z)$, (I.1.12) shows that, for $|z| < 1$,

$$(1 - zy)^{-b} = 1 + bzy + \frac{b (b + 1) (zy)^2}{2!} + \cdots = \sum_{n=0}^\infty \frac{b^{[n]} z^n y^n}{n!}.$$

Then the r.h.s. of (5.28) is given by

$$\frac{\Gamma (c)}{\Gamma (a) \Gamma (c - a)} \sum_{n=0}^\infty \int_0^1 y^{n+a-1} (1 - y)^{c-a-1} \frac{b^{[n]} z^n}{n!} \, dy$$

$$= \frac{\Gamma (c)}{\Gamma (a) \Gamma (c - a)} \sum_{n=0}^\infty \frac{\Gamma (n + a) \Gamma (c - a)}{\Gamma (n + c)} \frac{b^{[n]} z^n}{n!} \, dy$$

$$= \sum_{n=0}^\infty \frac{\Gamma (c)}{\Gamma (n + c)} \frac{\Gamma (n + a) \Gamma (n + b)}{\Gamma (a)} \frac{z^n}{\Gamma (b)} \frac{z^n}{n!} = \sum_{n=0}^\infty \frac{a^{[n]} b^{[n]}}{c^{[n]}} \frac{z^n}{n!} = {}_2F_1 (a, b; c; z) .$$

⊖ **Example 5.11**   Substituting $w = 1 - y$ in (5.28) and noting that

$$(1 - z(1 - w)) = (1 - z)\left(1 - w\frac{z}{z - 1}\right),$$

we can write

$$_2F_1(a, b; c; z) = -\frac{1}{B(a, c - a)} \int_1^0 (1 - w)^{a-1} w^{c-a-1}(1 - z(1 - w))^{-b} \, dw$$

$$= (1 - z)^{-b} \frac{\Gamma(c)}{\Gamma(a)\Gamma(c - a)} \int_0^1 w^{c-a-1}(1 - w)^{a-1}\left(1 - w\frac{z}{z - 1}\right)^{-b} dw$$

$$= (1 - z)^{-b} {}_2F_1\left(c - a, b; c; \frac{z}{z - 1}\right). \qquad ■$$

Of value is *Kummer's (first) transformation*, given by

$$_1F_1(a, b, x) = e^x {}_1F_1(b - a, b, -x), \qquad (5,29)$$

as in Abramowitz and Stegun (1972, eq. 13.1.27), and confirmed by expanding both sides and comparing coefficients of powers of x; see Andrews, Askey and Roy (1999, p. 191) for derivation.[6]

For the first derivative of $_1F_1(a, b; z)$,

$$\frac{d}{dz}\sum_{n=0}^{\infty} \frac{a^{[n]}}{b^{[n]}}\frac{z^n}{n!} = \frac{d}{dz}\left(1 + \frac{a}{b}z + \frac{a(a + 1)}{b(b + 1)}\frac{z^2}{2!} + \frac{a^{[3]}}{b^{[3]}}\frac{z^3}{3!} + \cdots\right)$$

$$= \frac{a}{b} + \frac{a(a + 1)}{b(b + 1)}z + \frac{a^{[3]}}{b^{[3]}}\frac{z^2}{2!} + \cdots$$

$$= \frac{a}{b}\left(1 + \frac{a + 1}{b + 1}z + \frac{(a + 1)(a + 2)}{(b + 1)(b + 2)}\frac{z^2}{2!} + \cdots\right)$$

$$= \frac{a}{b}\sum_{n=0}^{\infty} \frac{(a + 1)^{[n]}}{(b + 1)^{[n]}}\frac{z^n}{n!}$$

$$= \frac{a}{b}{}_1F_1(a + 1, b + 1; z).$$

Similar calculations verify that

$$\frac{d^j}{dz^j}{}_1F_1(a, b; z) = \frac{a^{[j]}}{b^{[j]}}{}_1F_1(a + j, b + j; z) \qquad (5.30)$$

_____

[6] But note the typographical error in their eq. 4.1.11.

and

$$\frac{d^j}{dz^j} \, {}_2F_1\,(a, b; c; z) = \frac{a^{[j]}b^{[j]}}{c^{[j]}} \, {}_2F_1\,(a + j, b + j; c + j; z). \tag{5.31}$$

The functions ${}_1F_1$ and ${}_2F_1$ can be approximated by truncating the infinite sum or applying numerical integration to (5.27) and (5.28). The Maple toolbox in Matlab can be used to compute them by calling `mfun('Hypergeom',a,b,x)` and `mfun('Hypergeom',[a,b],c,x)` for ${}_1F_1(a, b; x)$ and ${}_2F_1(a, b; c; x)$, respectively. This is much faster than 'doing it yourself' and the results are near machine precision.

Of some use is the approximation

$$_1F_1\left(r, \frac{n_2}{2}, \frac{-\theta_2}{2}\right) \approx \left(1 + \frac{\theta_2}{n_2}\right)^{-r}, \tag{5.32}$$

which increases in accuracy as $n_2 \to \infty$; see Problem 5.8.

Closed-form approximations to ${}_1F_1$ and ${}_2F_1$ based on a Laplace approximation have been derived by Butler and Wood (2002) and are given by

$$_1\hat{F}_1\,(a, b; z) = \frac{{}_1\tilde{F}_1\,(a, b; z)}{{}_1\tilde{F}_1\,(a, b; 0)}, \qquad {}_2\hat{F}_1\,(a, b, c; z) = \frac{{}_2\tilde{F}_1\,(a, b, c; z)}{{}_2\tilde{F}_1\,(a, b, c; 0)}, \tag{5.33}$$

where

$$_1\tilde{F}_1(a; b; x) = w^{-1/2}\hat{y}^a\,(1 - \hat{y})^{b-a}\,e^{x\hat{y}},$$

$$w = a\,(1 - \hat{y})^2 + (b - a)\,\hat{y}^2,$$

$$\hat{y} = \begin{cases} \left[(x - b) + \sqrt{(x - b)^2 + 4ax}\right]/2x, & \text{if } x \neq 0, \\ a/b, & \text{if } x = 0, \end{cases}$$

and

$$_2\tilde{F}_1(a, b, c; x) = w^{-1/2}\hat{y}^a\,(1 - \hat{y})^{c-a}\,(1 - x\hat{y})^{-b},$$

$$w = a\,(1 - \hat{y})^2 + (c - a)\,\hat{y}^2 - bx^2\hat{y}^2\,(1 - \hat{y})^2 / (1 - x\hat{y})^2,$$

$$\hat{y} = \begin{cases} \left[\tau + \sqrt{\tau^2 - 4ax\,(c - b)}\right] / [2x\,(b - c)], & \text{if } x \neq 0, \\ a/c, & \text{if } x = 0, \end{cases}$$

$$\tau = x\,(b - a) - c.$$

Programs for their computation are given in Listings 5.6 and 5.7.

⊚ **Example 5.12**   Using the Taylor series expansion $e^{-t} = 1 - t + \frac{1}{2}t^2 - \frac{1}{6}t^3 + \cdots$ and integrating termwise,

$$\Gamma_x\,(a) = \int_0^x t^{a-1}e^{-t}\,dt$$

$$= \int_0^x t^{a-1}\,dt - \int_0^x t^a\,dt + \frac{1}{2}\int_0^x t^{a+1}\,dt - \frac{1}{6}\int_0^x t^{a+2}\,dt + \cdots$$

$$= \sum_{k=-1}^{\infty} \frac{(-1)^{k+1}}{(k+1)!}\frac{x^{a+k+1}}{a+k+1} \overset{i=k+1}{=} \frac{x^a}{a}\sum_{i=0}^{\infty}\frac{(-x)^i}{i!}\frac{a}{a+i}.$$

From (5.26), for $i \in \mathbb{N}_0 = \{0, 1, 2, \ldots\}$,

$$\frac{a^{[i]}}{(a+1)^{[i]}} = \frac{a\,(a+1)\,(a+2)\cdots(a+i-1)}{(a+1)\,(a+2)\cdots(a+i)} = \frac{a}{a+i},$$

and we can write

$$\Gamma_x\,(a) = \frac{x^a}{a}\sum_{i=0}^{\infty}\frac{a^{[i]}}{(a+1)^{[i]}}\frac{(-x)^i}{i!} = {}_1F_1\,(a, a+1; -x),$$

which can be approximated with (5.33).   ■

```
function out=f11(a,b,xvec)

raw=[]; for loop=1:length(xvec)
  raw(loop)= doit(a,b,xvec(loop));
end out= raw / doit(a,b,0);

function loc=doit(a,b,x)
  if abs(x)<1e-8
    y=a/b;
  else
    y=( (x-b) + sqrt((x-b)^2 + 4*a*x) ) / (2*x);
  end
  w=a*(1-y)^2 + (b-a)*y^2;
  loc= (1 / sqrt(w)) * y^a * (1-y)^(b-a) * exp(x*y);
```

Program Listing 5.6   The Laplace approximation to the $_1F_1$ function. The last argument in the function call can be a vector

```
function out=f21(a,b,c,xvec)

raw=[]; for loop=1:length(xvec)
  raw(loop)= doit21(a,b,c,xvec(loop));
end out = raw / doit21(a,b,c,0);

function loc=doit21(a,b,c,x)
  tau=x*(b-a)-c;
  if abs(x)<1e-8
    y=a/c;
  else
    y=( tau+sqrt(tau^2-4*a*x*(c-b)) ) / ( 2*x*(b-c) );
  end
  w=a*(1-y)^2 + (c-a)*y^2 - b*x^2*y^2*(1-y)^2 / (1-x*y)^2;
  loc= (1 / sqrt(w)) * y^a * (1-y)^(c-a) / (1-x*y)^b;
```

Program Listing 5.7   The Laplace approximation to the $_2F_1$ function

⊖ **Example 5.13**   From (I.7.34), the standard normal c.d.f. can be expressed as

$$\Phi(z) = \frac{1}{2} + \frac{z}{\sqrt{2\pi}} \sum_{i=0}^{\infty} \frac{(-z^2/2)^i}{(2i+1)\, i!}.$$

Writing out a couple of terms confirms that choosing $a = 1/2$ and $b = 3/2$ gives

$$\Phi(z) = \frac{1}{2} + \frac{z}{\sqrt{2\pi}}\, {}_1F_1\left(\frac{1}{2}, \frac{3}{2}; -\frac{z^2}{2}\right),$$

or, applying (5.29),

$$\Phi(z) = \frac{1}{2} + z\,\phi(z)\, {}_1F_1\left(1, \frac{3}{2}; \frac{z^2}{2}\right),$$

where $\phi(z)$ is the standard normal p.d.f. The approximation (5.33) could be used with this, though for z around 2.8 and larger, it returns a value larger than one.   ∎

5.4 Problems

Success usually comes to those who are too busy to be looking for it.
(Henry David Thoreau)

I'm a great believer in luck, and I find the harder I work the more I have of it.
(Thomas Jefferson)

5.1. Calculate the p.d.f. and c.d.f. s.p.a. for the standard normal distribution evaluated at x.

5.2. ★ Let Y be a continuous random variable whose p.d.f. f_Y is symmetric about 0, so that $f_Y(y) = f_Y(-y)$ for all $y \in \mathbb{R}$, and whose m.g.f. exists. Show that

$$\boxed{\hat{f}_Y(-y) = \hat{f}_Y(y) \quad \text{when} \quad f_Y(y) = f_Y(-y),}$$

where $\hat{f}$ is the p.d.f. s.p.a. (5.3). Also show that the c.d.f. s.p.a. (5.6) also obeys the symmetry property.

5.3. ★ Let X be a continuous r.v. with support $\mathcal{S}_X$ and whose m.g.f. exists. Define $Y = aX + b$, for $a > 0$. Show that the first-order s.p.a. to the p.d.f. is *equivariant*, i.e., for each x in the interior of $\mathcal{S}_X$ and $y = ax + b$,

$$\boxed{\hat{f}_Y(y) = \hat{f}_X(x)\left|\frac{\mathrm{d}x}{\mathrm{d}y}\right|.}$$

Also show that

$$\hat{F}_Y(y) = \hat{F}_X\left(\frac{y-b}{a}\right),$$

where $\hat{F}$ refers to the first-order saddlepoint approximation to the c.d.f.

5.4. ★ Let $S = X_1 + X_2$, where $X_i \overset{\text{ind}}{\sim} \text{NBin}(r_i, p_i)$, with $r_i = 2i$, $p_i = i/3$, $i = 1, 2$. Compare (via plots) the performance of the normal approximation and the s.p.a. to the exact mass function of S.

5.5. ★ ★ (Saddlepoint approximation to the beta distribution) Let

$$X := \frac{Y_\alpha}{Y_\alpha + Y_\beta}, \quad Y_\alpha \sim \chi^2(2\alpha), \quad Y_\beta \sim \chi^2(2\beta), \tag{5.34}$$

for $\alpha, \beta \in \mathbb{R}_{>0}$, so that, from Example I.9.11 and its subsequent remark, $X \sim \text{Beta}(\alpha, \beta)$.

(a) Write an expression for r.v. Z_x in terms of x, Y_α and Y_β, where Z_x satisfies

$$\Pr(X < x) = \Pr(Z_x < 0). \tag{5.35}$$

(b) Derive the m.g.f. $\mathbb{M}_{Z_x}(s)$ and the c.g.f. $\mathbb{K}_{Z_x}(s) = \ln \mathbb{M}_{Z_x}(s)$.

(c) Verify that the solution to the equation $\mathbb{K}'_{Z_x}(s) = 0$ is given by

$$\hat{s} = \frac{1}{2} \frac{\alpha - \alpha x - \beta x}{\alpha x^2 - \alpha x - \beta x + \beta x^2}. \tag{5.36}$$

(d) Using the s.p.a., approximate $\Pr(X < 0.3)$, where $X \sim \text{Beta}(10, 5)$ and calculate the RPE with the exact answer 0.00166566.

(e) Express the c.d.f. of $R \sim F(n_1, n_2)$ at r, where R follows an F distribution with $n_1 = 2\alpha$ and $n_2 = 2\beta$ degrees of freedom, in terms of the c.d.f. of X, a beta r.v., at x.

(f) Use the previous result and (5.35) to express the c.d.f. of R in terms of that of Z and provide simplified expressions for $\hat{s}$, $\mathbb{K}_{Z_x}(s)$, $\mathbb{K}'_{Z_x}(\hat{s})$, $\mathbb{K}''_{Z_x}(\hat{s})$ and thus verify that

$$\hat{w} = \text{sgn}(\hat{s}) \sqrt{n_1 \ln\left(\frac{n_1 r + n_2}{r(n_1 + n_2)}\right) + n_2 \ln\left(\frac{n_1 r + n_2}{n_1 + n_2}\right)}, \tag{5.37}$$

$$\hat{u} = \frac{1}{\sqrt{2}} \frac{(r-1)\sqrt{n_1 n_2 (n_1 + n_2)}}{n_1 r + n_2}. \tag{5.38}$$

(g) Using these results, calculate the s.p.a. of $F(1, 12)$ at $r = 4.75$ and compute its RPE with the true answer 0.950057.

5.6. ★ (Saddlepoint approximation to the central F distribution) Instead of using the approach in Problem 5.5 to get the c.d.f. of the F distribution indirectly (via the beta), one could proceed directly. Let $C_i \overset{\text{ind}}{\sim} \chi^2(n_i)$, $i = 1, 2$,

$$R = \frac{C_1/n_1}{C_2/n_2},$$

and $r > 0$.

(a) Write an expression for r.v. Z such that

$$\Pr(R < r) = \Pr(Z < 0). \tag{5.39}$$

Then derive $\mathbb{M}_Z(s)$, $\mathbb{K}_Z(s)$, $\mathbb{K}'_Z(s)$ and $\mathbb{K}''_Z(s)$.

(b) Derive a closed-form expression for the saddlepoint $\hat{s}$ which solves $\mathbb{K}'_Z(s) = 0$.

(c) Recompute the example in Problem 5.5() using this direct method.

(d) As the manipulation in (5.35) and (5.39) are the same, one would expect that the two s.p.a. approaches are identical. Show this algebraically, i.e., that $\hat{w}$ and $\hat{u}$ are the same.

Remark: When working with products of (independent) r.v.s, or distributions which do not possess an m.g.f., it is convenient to work with the *Mellin transform*, given by

$$m(s) = \mathbb{E}\left[X^s\right] = \int_0^\infty x^s f(x) \, dx,$$

which is the m.g.f. of $\ln(X)$ when it converges over a neighbourhood of s around zero. (Actually, the Mellin transform is defined as $\int_0^\infty x^{s-1} f(x) \, dx$, but the above form will be more convenient for our purposes.) ■

5.7. ★ ★ Let $X_i \overset{\text{ind}}{\sim} B(p_i, q_i)$, i.e., independently distributed beta random variables, $i = 1, \ldots, n$, and consider the distribution of $P = \prod_{i=1}^n X_i$.

This product is connected to the likelihood ratio test for hypothesis testing in MANOVA models. Dennis (1994) has derived infinite-sum recursive expressions for both the p.d.f. and c.d.f. of P by inverting the Mellin transform, while, for p_i and q_i integer-valued, Springer (1978) gives a method for expressing the density as a finite-order polynomial in x and $\log x$. Pham-Gia and Turkkan (2002) derive the density of the product and quotient of two location- and scale-independent beta r.v.s and discuss applications in operations reseach and management science.

(a) Derive the distribution of $P = X_1 X_2$ for the special case with $p_1 = p_2 = q_1 = q_2 = 1$. For general p_1, p_2, q_1 and q_2, Steece (1976) showed that $f_P(p)$ is given by

$$\frac{\Gamma(p_1 + q_1)\Gamma(p_2 + q_2)}{\Gamma(p_1)\Gamma(p_2)\Gamma(q_1 + q_2)} p^{p_1 - 1} (1 - p)^{q_1 + q_2 - 1} \, {}_2F_1$$

$$\times (q_2, p_1 - p_2 + q_1, q_1 + q_2; 1 - p), \tag{5.40}$$

where $_2F_1$ is the hypergeometric function; see Section 5.3 for details.

(b) Derive the s.p.a. of P via the Mellin transform and simplify the result for the special case with $n = 2$ and $p_1 = p_2 = q_1 = q_2 = 1$. (For further details, see Srivastava and Yao, 1989; and Butler, Huzurbazar and Booth, 1992.)

The s.p.a. could also be used to provide much easier expressions than those for the product considered in Pham-Gia and Turkkan (2002), and also generalize their results to the product of more than two r.v.s.

(c) By equating the first two moments, Fan (1991) derived a simple approximation to the density of P. He showed that

$$P \stackrel{\text{app}}{\sim} \text{Beta}(p, q), \tag{5.41}$$

i.e., P approximately follows a beta distribution, where

$$p = S(S - T)(T - S^2)^{-1}, \quad q = (1 - S)(S - T)(T - S^2)^{-1},$$

and

$$S = \prod_{i=1}^{n} \frac{p_i}{p_i + q_i}, \quad T = \prod_{i=1}^{n} \frac{p_i(p_i + 1)}{(p_i + q_i)(p_i + q_i + 1)}$$

(see also Johnson, Kotz and Balakrishnan, 1995, p. 262, though note that the expression for p contains a typo). Using the method in Springer (1978), it can be shown (see Johnson, Kotz and Balakrishnan, 1995, p. 261) that the true density of P for $n = 3$, $p_1 = 9$, $p_2 = 8$, $p_3 = 4$, $q_1 = 3$, $q_2 = 3$ and $q_3 = 2$ is given by

$$f_P(x) = \frac{3960}{7} x^3 - 1980x^4 + 99\,000x^7 + (374\,220 + 356\,400 \log x) x^8$$

$$- (443\,520 - 237\,600 \log x) x^9 - \frac{198\,000}{7} x^{10}. \tag{5.42}$$

Compare the accuracy of the s.p.a. and (5.41) for this case by plotting the RPE, where

$$\text{RPE} = 100 \left(\hat{f}(x) - f(x) \right) \begin{cases} 1/f(x), & \text{if } x < 0.5, \\ 1/(1 - f(x)), & \text{if } x \geq 0.5. \end{cases}$$

(d) Discuss how you would develop a saddlepoint approximation to the distribution of $S = \sum_{i=1}^{n} X_i$.

5.8. ★ Show that $_1F_1(a, b/2; -z/2) \approx (1 + z/b)^{-a}$ (in the sense that the ratio of the two terms approaches one) as $b \to \infty$.

6

Order statistics

Congratulations. I knew the record would stand until it was broken.

(Yogi Berra)

Recall the statistics $\overline{X}_n$ and S_n^2 discussed in Section 3.7. While these are of great importance for ascertaining basic properties about the underlying distribution from which an i.i.d. sample has been observed, there are many occasions on which the *extremes* of the sample, or the smallest and largest observations, are more relevant. One might imagine an electrical device which is triggered when the current exceeds a certain threshold. How often this is expected to occur may not have much to do with the average current, but rather the behaviour (or probability distribution) of the largest values that are anticipated to occur. The choice of height of a water dam, the size of a military force, or an optimal financial strategy to avoid bankruptcy will be at least influenced, if not dictated, not by average or typical behaviour, but by consideration of maximal occurrences. Similar arguments can be made for the behaviour of the minimum; the saying about a chain's strength being only that of its weakest link comes to mind. There are other ordered values besides the extremes which are useful for inference. For example, another *measure of central tendency* besides the sample mean is the sample median, or the middle observation of the ordered set of observations. It is useful when working with heavy-tailed data, such as Cauchy, and plays a prominent role in *robust* statistical inference, which is designed to minimize the influence of contaminated or suspect data.

The *order statistics* of a random i.i.d. sample X_i, $i = 1, \ldots, n$, are the n values arranged in ascending order. Several common notations exist, including $X_{1:n} \leq X_{2:n} \leq \ldots \leq X_{n:n}$ or $X_{(1)} \leq X_{(2)} \leq \ldots \leq X_{(n)}$ or $Y_1 \leq Y_2 \leq \ldots \leq Y_n$, the first of these being preferred when the sample size n is to be emphasized in the analysis. The first and second notational options tend to make complicated expressions even less readable, while the third can become cumbersome when the discussion involves two or more sets of random variables. In what follows, we concentrate on univariate samples and so use mostly use the Y notation.

Intermediate Probability: A Computational Approach   M. Paolella
© 2007 John Wiley & Sons, Ltd

6.1 Distribution theory for i.i.d. samples

6.1.1 Univariate

The ith order statistic of the i.i.d. sample X_i, $i = 1, \ldots, n$, from distribution $F = F_X$ (and density $f = f_X$) is distributed with c.d.f.

$$F_{Y_i}(y) = \Pr(Y_i \le y) = \sum_{j=i}^{n} \binom{n}{j} [F(y)]^j [1 - F(y)]^{n-j}. \tag{6.1}$$

To understand this, let $Z_i = \mathbb{I}_{(-\infty, y]}(X_i)$ so that the Z_i are i.i.d. Bernoulli and $S_y = \sum_{i=1}^{n} Z_i \sim \text{Bin}(n, F(y))$ is the number of X_i which are less than or equal to y. Then

$$\Pr(Y_i \le y) = \Pr(\text{at least } i \text{ of the } Xs \text{ are } \le y) = \Pr(S_y \ge i). \tag{6.2}$$

Special cases of interest are the sample minimum

$$F_{Y_1}(y) = 1 - [1 - F(y)]^n \tag{6.3}$$

and sample maximum

$$F_{Y_n}(y) = [F(y)]^n. \tag{6.4}$$

Expressions (6.3) and (6.4) for the case with $n = 2$ were derived using basic principles of conditioning in Example I.8.19.

⊖ **Example 6.1**   Let $X_i \overset{\text{i.i.d.}}{\sim} \text{Par II}(1)$, $i = 1, \ldots, n$, each with c.d.f. $F(x) = [1 - (1 + x)^{-1}]\mathbb{I}_{(0,\infty)}(y)$. Then, with $S_n = nX_{1:n}$ and using (6.3),

$$F_{S_n}(y) = \Pr(S_n \le y) = \Pr\left(X_{1:n} \le \frac{y}{n}\right) = F_{X_{1:n}}\left(\frac{y}{n}\right)$$

$$= 1 - \left[1 - F\left(\frac{y}{n}\right)\right]^n = 1 - \left(\frac{1}{1 + y/n}\right)^n = \left[1 - \left(1 + \frac{y}{n}\right)^{-n}\right]\mathbb{I}_{(0,\infty)}(y)$$

and, from (I.A.46), $\lim_{n\to\infty} F_{S_n}(y) = (1 - e^{-y})\mathbb{I}_{(0,\infty)}(y)$, so that $S_n \overset{d}{\to} \text{Exp}(1)$.
Now let $L_n = X_{n:n}/n$, so that, from (6.4),

$$F_{L_n}(y) = F_{X_{n:n}}(ny) = [F(ny)]^n = \left(1 + \frac{1}{ny}\right)^{-n}\mathbb{I}_{(0,\infty)}(y)$$

and $\lim_{n\to\infty} F_{L_n}(y) = \exp(-1/y)\mathbb{I}_{(0,\infty)}(y)$.   ∎

Example 6.2   Let $X_i \overset{\text{i.i.d.}}{\sim} \text{Exp}\,(\lambda), i = 1, \ldots, n$, with typical c.d.f. $F\,(x) = \left(1 - e^{-\lambda x}\right)$ $\mathbb{I}_{(0,\infty)}\,(x)$. Let $L_n = \lambda X_{n:n} - \ln n$ so that, from (6.4),

$$F_{L_n}\,(y) = \Pr\left(X_{n:n} \leq \frac{y + \ln n}{\lambda}\right) = \left[F\left(\frac{y + \ln n}{\lambda}\right)\right]^n = \left(1 - \frac{e^{-y}}{n}\right)^n \mathbb{I}_{(-\ln n, \infty)}\,(y)$$

and $\lim_{n\to\infty} F_{L_n}\,(y) = \exp\left(-e^{-y}\right)$, or $L_n \overset{d}{\to} \text{Gum}\,(0, 1)$, i.e., the Gumbel, or extreme value, distribution. The construction of L_n suggests a way of simulating a $\text{Gum}\,(0, 1)$ random variable: take the maximum of a set of (say, $n = 20\,000$) i.i.d. $\text{Exp}\,(1)$ r.v.s and subtract $\ln n$. Alternatively, the use of the probability integral transform for simulation, as discussed in Section I.7.4, is applicable and vastly simpler; we solve $x = \exp\left(-e^{-y}\right)$ to get $y = -\ln\left(-\ln x\right)$. Both methods were used and produced the histograms shown in Figure 6.1. They do indeed appear indistinguishable.   ∎

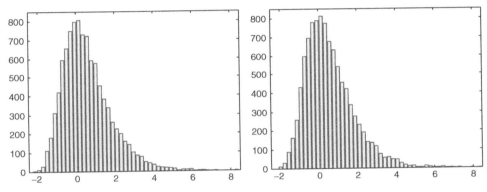

Figure 6.1   Histogram of simulated Gum(0, 1) r.v.s using the probability integral transform (left) and using the sample maximum of $n = 20\,000$ i.i.d. Exp(1) r.v.s (right)

Remark: The previous example is a special case of the general asymptotic distributional result for maxima of i.i.d. samples, known as the Fisher–Tippett theorem. It is the cornerstone of what is referred to as *extreme value theory*, incorporating both distributional limit theory, and statistical methods for working with real data sets. To do the subject the proper justice it deserves, a book-length treatment is required, and a handful of good ones exist. Books which emphasize applications include Beirlant, Teugels and Vynckier (1996), Coles (2001), and the edited volume by Finkenstädt and Rootzén (2004), while Embrechts, Klüppelberg and Mikosch (2000) and Beirlant *et al.* (2004) cover both applications as well as the theoretical underpinnings.   ∎

For computing f_{Y_i}, we consider the discrete and continuous cases separately. If X is discrete with k distinct support points $\{x_1, \ldots, x_k\}$ such that $x_i < x_j$ for $i < j$, then the mass function f_{Y_i} can be evaluated as

$$f_{Y_i}\,(y) = \begin{cases} F_{Y_i}\,(x_1), & \text{if } y = x_1, \\ F_{Y_i}\,(x_i) - F_{Y_i}\,(x_{i-1}), & \text{if } y = x_i \in \{x_2, \ldots, x_k\}, \\ 0, & \text{otherwise,} \end{cases} \qquad (6.5)$$

which we justify using the following example as a special case.

⊖ **_Example 6.3_**   Recall the discrete uniform distribution, with c.d.f. given in (I.4.4). Let $X_i \overset{\text{i.i.d.}}{\sim} \text{DUnif}(6)$, $i = 1, \ldots, 20$, as would arise, for example, when a fair, six-sided die is tossed 20 times, and denote their order statistics as Y_i, $i = 1, \ldots, 20$. To calculate $f_{Y_3}(y) = \Pr(Y_3 = y)$ for $y = 1$, note that

$$\{Y_3 = 1\} = \{\text{at least 3 of the } X_i \text{ are equal to } 1\} = \{\text{at least 3 of the } X_i \text{ are } \leq 1\},$$

and using (6.2), it follows that $f_{Y_3}(1) = F_{Y_3}(1)$. For $y \in \{2, \ldots, 6\}$, the p.m.f. is defined in the usual way as the difference between the c.d.f. at y and the c.d.f. at $y - 1$, though this can also be motivated as follows. To compute $f_{Y_{10}}(4) = \Pr(Y_{10} = 4)$, let

$$C_1 = \{\text{at least 10 of the } X_i \text{ are } \leq 4\} \quad \text{and} \quad C_2 = \{\text{at most 9 of the } X_i \text{ are } \leq 3\}.$$

Observe that $C_1 \Leftrightarrow \{Y_{10} \leq 4\}$ and $C_2 \Leftrightarrow \{Y_{10} \geq 4\}$ so that $\{Y_{10} = 4\} \Leftrightarrow C_1 \cap C_2$. But

$$C_2 = C_3^c, \quad \text{where } C_3 = \{\text{at least 10 of the } X_i \text{ are } \leq 3\},$$

so that $\Pr(Y_{10} = 4) = \Pr(C_1 C_3^c)$, and as (the occurrence of the event) C_3 implies C_1, it follows from (I.2.8)[1] that $\Pr(C_1 C_3^c) = \Pr(C_1) - \Pr(C_3) = F_{Y_{10}}(4) - F_{Y_{10}}(3)$. ■

When X is a continuous r.v., the easiest way to derive the p.d.f. of Y_i is by using a limiting argument based on the multinomial distribution, i.e.,

$$f_{Y_i}(y)$$

$$= \lim_{\Delta y \to 0} \frac{\Pr\left(i-1 \; X_i \in (-\infty, y]; \text{ one } X_i \in (y, y + \Delta y); \; n-i \; X_i \in (y + \Delta y, \infty)\right)}{\Delta y}$$

shown graphically as the partition

where $z = y + \Delta y$. This implies

$$\boxed{f_{Y_i}(y) = i \binom{n}{i} F(y)^{i-1} \left[1 - F(y)\right]^{n-i} f(y).} \tag{6.6}$$

In addition to (6.1), the c.d.f. of Y_i can be expressed as

$$F_{Y_i}(y) = \frac{n!}{(n-i)!(i-1)!} \int_0^{F(y)} t^{i-1} (1-t)^{n-i} \, dt \tag{6.7}$$

---

[1] Equation (I.2.8) states that, for _sets_ A and B such that $B \subset A$, $\Pr(AB^c) = \Pr(A) - \Pr(B)$. Let A be the set of all outcomes of $X_1, \ldots, X_{20}$ such that event C_1 is true, and let B be the set of all outcomes such that C_3 is true. Then $B \subset A$ because $C_3 \Rightarrow C_1$.

(see Problem 6.3) so that differentiating (6.7) with respect to y immediately gives (6.6). The p.d.f. of Y_i could also be derived directly by differentiating (6.1) (see Casella and Berger, 1990, p. 232) or by use of a multivariate transformation (see Roussas, 1997, p. 248).

Two special cases are of great interest:

$$f_{Y_1}(y) = n\left[1 - F(y)\right]^{n-1} f(y) \quad \text{and} \quad f_{Y_n}(y) = n\left[F(y)\right]^{n-1} f(y). \qquad (6.8)$$

⊖ **Example 6.4** Let $X_i \overset{\text{i.i.d.}}{\sim} \mathrm{DUnif}(\theta)$. From (6.5) and (6.4), the p.m.f. of the maximum, $X_{n:n}$, is easily seen to be

$$\Pr(X_{n:n} = t) = \Pr(X_{n:n} \le t) - \Pr(X_{n:n} \le t - 1)$$
$$= \theta^{-n}\left[t^n - (t-1)^n\right]\mathbb{I}_{\{1,2,\dots,\theta\}}(t).$$

Note that

$$\sum_{t=1}^{\theta} \Pr(X_{n:n} = t) = \frac{1}{\theta^n}\left[\sum_{t=1}^{\theta} t^n - \sum_{t=1}^{\theta}(t-1)^n\right]$$
$$= \frac{1}{\theta^n}\left[\sum_{j=1}^{\theta} j^n - \sum_{j=0}^{\theta-1} j^n\right] = 1.$$

Also, for fixed θ,

$$\lim_{n\to\infty} \left.\frac{t^n - (t-1)^n}{\theta^n}\right|_{t=\theta} = 1, \qquad \lim_{n\to\infty} \left.\frac{t^n - (t-1)^n}{\theta^n}\right|_{1\le t<\theta} = 0,$$

i.e., the p.m.f. of $X_{n:n}$ 'piles up' at $t = \theta$ as $n \to \infty$, as would be expected. ■

⊙ **Example 6.5** For $X_i \overset{\text{i.i.d.}}{\sim} \mathrm{Unif}(0,1)$, it is straightforward to verify that $Y_i \sim$ Beta $(i, n - i + 1)$, where Y_i denotes the ith order statistic. Hence, the c.d.f. of Y_i can be expressed using (6.1) or integrating (6.6), i.e., for $0 \le y \le 1$,

$$F_{Y_i}(y) = \sum_{j=i}^{n} \binom{n}{j} y^j (1-y)^{n-j} = \frac{n!}{(i-1)!\,(n-i)!}\int_0^y x^{i-1}(1-x)^{n-i}\,dx,$$

(6.9)

which gives rise to an interesting identity as well as a computation method for evaluating the incomplete beta function. ■

⊖ **Example 6.6** Let $Y = X_{r:n}$ denote the median of an i.i.d. $\mathrm{N}(\mu, \sigma^2)$ sample for n odd, i.e., $r = (n+1)/2$. Then

$$f_Y(y) = \frac{n!}{(r-1)!\,(n-r)!}\left[\Phi(y)\right]^{r-1}\left[1 - \Phi(y)\right]^{n-r}\phi(y),$$

which can only be numerically evaluated but is (intuitively) symmetric about μ. Figure 6.2(a) plots the density for $\mu = 0$, $\sigma^2 = 1$ and $n = 3$; it looks just like a normal density! The difference between f_Y and $\phi(x; 0, 0.448)$, a normal density with mean zero and variance 0.448, is shown in Figure 6.2(b) and verifies that Y is indeed reasonably close to being normally distributed.

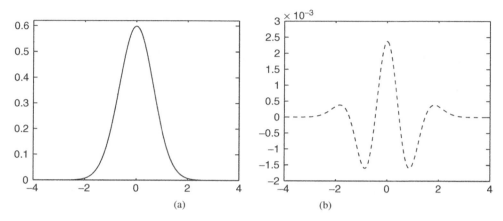

(a)   (b)

Figure 6.2 (a) Exact density of $X_{2:3}$ and (b) discrepancy from a normal distribution with mean zero and variance 0.448

However, examination of the relative error $(f - \phi)/f$ verifies that the approximation worsens as we go farther into the tails. It can be shown that Y is asymptotically normal; but apparently, even for minuscule sample sizes, the normal approximation to the median is quite good. The value 0.448 can be obtained by numerically integrating $y^2 f_Y(y)$ or by simulation, the latter being faster, both to program and to run, for small n.

The following table reports the ratio $\rho = \mathbb{V}(\overline{X})/\mathbb{V}(Y)$ for several values of n, where $\mathbb{V}(\overline{X}) = n^{-1}$ and $\mathbb{V}(Y)$ is approximated by the sample variance of 50,000 sample medians, each 50 000 sample medians, each computed from a simulated n-length vector of random draws.

n	1	3	7	11	21	31	51	101	201	501
ρ	1	0.744	0.681	0.664	0.653	0.648	0.643	0.643	0.640	0.638

The asymptotic variance of Y is $\pi/(2n)$ so that $\rho \to 0.6366$. ∎

Example 6.7 Recall that the quantile ξ_p of the continuous r.v. X is that value such that $F_X(\xi_p) = p$ for given probability $0 < p < 1$. To calculate $\Pr(Y_i \leq \xi_p \leq Y_j)$ for $1 \leq i < j \leq n$, let $U_i = F_X(Y_i)$ and use a similar derivation to that for the probability integral transform (I.7.66), i.e.,

$$F_{U_i}(u) = \Pr(U_i \leq u) = \Pr(F_X(Y_i) \leq u) = \Pr\left(Y_i \leq F_X^{-1}(u)\right) = F_{Y_i}\left(F_X^{-1}(u)\right).$$

Continuing with (6.1), $F_{U_i}(u)$ is given by

$$F_{Y_i}\left(F_X^{-1}(u)\right) = \sum_{j=i}^{n} \binom{n}{j} \left[F_X\left(F_X^{-1}(u)\right)\right]^j \left[1 - F_X\left(F_X^{-1}(u)\right)\right]^{n-j}$$

$$= \sum_{j=i}^{n} \binom{n}{j} u^j (1-u)^{n-j} .$$

Now we see from the l.h.s. of (6.9) that the $F_X(Y_1), \ldots, F_X(Y_n)$ have the same distribution as order statistics from an i.i.d. uniform sample. Thus, with $\xi_p = F_X^{-1}(p)$,

$$\Pr\left(Y_i \leq \xi_p \leq Y_j\right) = \Pr\left(F_X(Y_i) \leq p \leq F_X(Y_j)\right)$$

$$= \Pr\left(U_i \leq p \leq U_j\right)$$

$$= \Pr\left(U_i \leq p \cap p \leq U_j\right)$$

$$= \Pr\left(U_i \leq p\right) + \left(1 - \Pr\left(U_j < p\right)\right) - \Pr\left(U_i \leq p \cup p \leq U_j\right)$$

$$= \sum_{k=i}^{n} \binom{n}{k} p^k (1-p)^{n-k} + \left(1 - \sum_{k=j}^{n} \binom{n}{k} p^k (1-p)^{n-k}\right) - 1$$

because, with $i < j$, $\Pr\left(U_i \leq p \cup p \leq U_j\right) = 1$, i.e.,

$$\boxed{\Pr\left(Y_i \leq \xi_p \leq Y_j\right) = \sum_{k=i}^{j-1} \binom{n}{k} p^k (1-p)^{n-k} = F_B\left(j-1, n, p\right) - F_B\left(i, n, p\right),}$$

(6.10)

where $B \sim \text{Bin}(n, p)$ and F_B is the c.d.f. of B.

Remark: A natural (and popular) use of (6.10) is to construct a *nonparametric con-fidence interval* for ξ_p. While confidence intervals will not be formally defined and used in this book, it is worth mentioning at this point that this method is attractive because it is nonparametric, meaning that f_X and F_X are not needed (and in reality, F_X is not known!) For example, to construct a 95 % confidence interval for the popu-lation median based on $n = 100$ i.i.d. samples of data, we require values i and j such that

$$F_B\left(j-1, 100, 0.5\right) - F_B\left(i, 100, 0.5\right) \approx 0.95,$$

which need to be found numerically. With

$$F_B^{-1}\left(0.025, 100, 0.5\right) = 40 \quad \text{and} \quad F_B^{-1}\left(1 - 0.025, 100, 0.5\right) = 60$$

(computed using the `binoinv` function in Matlab), we calculate

$$F_B (59, 100, 0.5) - F_B (40, 100, 0.5) = 0.943.$$

Thus, $\Pr (Y_{40} < \xi_{0.5} < Y_{59}) = 0.94$ based on $n = 100$. ■

⊖ **Example 6.8**   Let X and Y be independent, continuous r.v.s, but not necessarily identically distributed. Then the survivor function of $M = \min(X, Y)$ is

$$\Pr(M > m) = \Pr(\min(X, Y) > m) = \Pr(X > m \text{ and } Y > m)$$
$$= \Pr(X > m) \Pr(Y > m) = \left(1 - F_X(m)\right)\left(1 - F_Y(m)\right),$$

and the p.d.f. of M is

$$f_M (m) = f_X (m)\left(1 - F_Y (m)\right) + f_Y (m)\left(1 - F_X (m)\right),$$

as $f_M (m) = -\mathrm{d} \Pr(M > m)/\mathrm{d}m.$ ■

6.1.2   Multivariate

The joint c.d.f. and p.d.f. of several order statistics are also of interest. For the bivariate case, consider the order statistics Y_i and Y_j with $i < j$. For $x \geq y$, $Y_i \leq Y_j$ and $\Pr\left(Y_i \leq x, Y_j \leq y\right) = \Pr\left(Y_j \leq y\right)$, so that the c.d.f. is $F_{Y_i, Y_j} (x, y) = F_{Y_j} (y)$, also easily seen by placing Y_i, Y_j, x and y along a line. For $x < y$ (again, draw a line marked with x and y),

$$F_{Y_i, Y_j} (x, y) = \Pr \text{(at least } i \text{ of the } Xs \leq x \ \cap \ \text{at least } j \text{ of the } Xs \leq y)$$

$$= \sum_{a=j}^{n} \sum_{b=i}^{a} \Pr \text{(exactly } b \text{ of the } Xs \leq x \ \cap \ \text{exactly } a \text{ of the } Xs \leq y)$$

or, using a multinomial argument as in the univariate case (6.1),

$$F_{Y_i, Y_j} (x, y) = \sum_{a=j}^{n} \sum_{b=i}^{a} \frac{n! \, [F (x)]^b \left[F (y) - F (x)\right]^{a-b} \left[1 - F (y)\right]^{n-a}}{b! \, (a - b)! \, (n - a)!}. \qquad (6.11)$$

⊖ **Example 6.9**   Let $n = 2$, $i = 1$, $j = 2$, $V = \min(X_1, X_2)$ and $W = \max(X_1, X_2)$. For $v < w$, (6.11) gives

$$F_{V,W} (v, w) = \sum_{a=j}^{n} \sum_{b=i}^{a} \frac{n!}{b! \, (a - b)! \, (n - a)!} [F (v)]^b [F (w) - F (v)]^{a-b} [1 - F (w)]^{n-a}$$

$$= \sum_{b=1}^{2} \frac{2}{b! \, (2 - b)! \, (2 - 2)!} [F (v)]^b [F (w) - F (v)]^{2-b} [1 - F (w)]^{2-2}$$

$$= 2\,[F\,(v)]\,[F\,(w) - F\,(v)] + [F\,(v)]^2$$
$$= 2F\,(v)\,F\,(w) - [F\,(v)]^2$$

which can also be written as

$$2F\,(v)\,F\,(w) - [F\,(v)]^2 = [F\,(w)]^2 - \big([F\,(w)]^2 + [F\,(v)]^2 - 2F\,(v)\,F\,(w)\big)$$
$$= [F\,(w)]^2 - [F\,(w) - F\,(v)]^2. \tag{6.12}$$

Problem 6.9 extends this to the case where the two r.v.s are independent but not identically distributed. ∎

The extension of (6.11) to the general kth-order case, i.e.,

$$F_{Y_{i_1}, Y_{i_2}, \ldots, Y_{i_k}}, \quad 1 \leq i_1 < i_2 < \cdots < i_k \leq n,$$

follows along similar lines. For the joint density when F_X is discrete (continuous), appropriate differencing (differentiation) of the c.d.f. will yield the density. For the continuous case, use of the multinomial method considered above proves to be easiest. In particular, for two order statistics Y_i and Y_j with $i < j$ and $x < y$, we obtain

$$f_{Y_i, Y_j}\,(x, y) = \frac{n!}{(i-1)!\,(j-i-1)!\,(n-j)!}\,[F\,(x)]^{i-1}$$
$$\times \big[F\,(y) - F\,(x)\big]^{j-i-1}\,\big[1 - F\,(y)\big]^{n-j}\,f\,(x)\,f\,(y)\,\mathbb{I}_{(x,\infty)}\,(y). \tag{6.13}$$

For three order statistics Y_i, Y_j and Y_k, with $i < j < k$ and $x < y < z$, an expression corresponding to (6.13) can be found in a similar way. The exponents in the expression are most easily obtained with the help of the line graph, i.e.,

where $A = i - 1$ is the number of values strictly less than x, and B is the number of values strictly between x and y, this being given by the total of $j - 1$ which are strictly less than y, minus the i values less than or equal to x, or $B = j - i - 1$. Similarly, $C = k - 1 - (j - i - 1) - (i - 1) - 1 - 1 = k - j - 1$, and D is the number of values strictly greater than z, or $D = n - k$. As a check, $A + B + C + D = n - 3$.

Example 6.10   Let $X_i \overset{\text{i.i.d.}}{\sim} \text{Exp}\,(1)$, $i = 1, \ldots, 6$, with order statistics $Y_1, \ldots, Y_6$. Then, as $F_X\,(x) = 1 - e^{-x}$, (6.13) is easily computed. Figure 6.3 shows a contour plot of the joint p.d.f. of Y_3 and Y_4 computed by calling

```
biv=bivorder(0:0.01:2,0:0.01:2,6,3,4);
contour(0:0.01:2,0:0.01:2,biv',20)
```

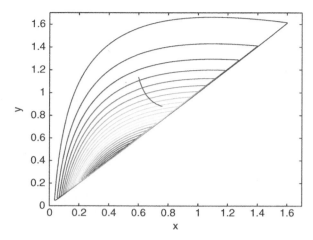

Figure 6.3   Joint p.d.f. of Y_3 and Y_4 for $n = 6$ i.i.d. Exp(1) r.v.s; see Example 6.10. The 'extra' line segment starting at $(x, y) = (0.6, 1.14)$ plots x against y, where y solves $F_{Y_3,Y_4}(x, y) = 0.5$

which uses the program in Listing 6.1. From (6.11), the joint c.d.f. of Y_3 and Y_4 is given by

$$F_{Y_3,Y_4}(x, y) = \sum_{a=4}^{n} \sum_{b=3}^{a} \frac{6!}{b!\,(a-b)!\,(6-a)!} \left[1 - e^{-x}\right]^b \left[e^{-x} - e^{-y}\right]^{a-b} \left[e^{-y}\right]^{6-a}.$$

Just for fun, if $x = 0.6$, then numerically solving $F_{Y_3,Y_4}(0.6, y) = 0.5$ yields $y = 1.13564$. This was computed for several values of x (using Maple's numeric solver) and is shown as a line in Figure 6.3.   ■

```
function biv=bivorder(xvec,yvec,n,i,j)
lx=length(xvec); ly=length(yvec); biv=zeros(lx,ly);
logkon= gammaln(n+1) - gammaln(i) - gammaln(j-i) - gammaln(n-j+1);
for xl=1:lx
  x=xvec(xl);
  for yl=1:lx
    y=yvec(yl);
    if (x<y) & (x>0) & (y>0) % for Exp r.v.s, we need x>0 and y>0
       main=(i-1)*log(F(x)) + (j-i-1)*log(F(y)-F(x)) + (n-j)*log(1-F(y));
       main=main+log(f(x))+log(f(y)); biv(xl,yl)=exp(logkon+main);
    else
       biv(xl,yl)=0;
    end
  end
end
end

function den=f(x) % the pdf of the underlying i.i.d.\ r.v.s
lambda=1; den=lambda*exp(-lambda*x);

function cdf=F(x) % the cdf of the underlying i.i.d.\ r.v.s
lambda=1; cdf=1-exp(-lambda*x);
```

Program Listing 6.1   Computes (6.13)

Density (6.13) will also be algebraically tractable when working with uniformly distributed r.v.s, as shown in the next example.

○ **Example 6.11**   Let $X_i \overset{\text{i.i.d.}}{\sim} \text{Unif}(0, 1)$, $i = 1, \ldots, n$, with order statistics $Y_1, \ldots, Y_n$. From (6.13),

$$f_{Y_i, Y_j}(x, y) = K\, x^{i-1}\, (y - x)^{j-i-1}\, (1 - y)^{n-j}\, \mathbb{I}_{(0,1)}(x)\, \mathbb{I}_{(x,1)}(y),$$

where

$$K = \frac{n!}{(i - 1)!\, (j - i - 1)!\, (n - j)!}.$$

For $i = 1$ and $j = n$, $f_{Y_1, Y_n}(x, y) = n\,(n - 1)\, (y - x)^{n-2}\, \mathbb{I}_{(0,1)}(x)\, \mathbb{I}_{(x,1)}(y)$. The density of $P = Y_1 Y_n$ follows from (2.9), i.e.,

$$f_P(p) = n\,(n - 1) \int_p^{\sqrt{p}} x^{-1} \left(\frac{p}{x} - x \right)^{n-2} dx, \tag{6.14}$$

where the lower bound on Y_1 is p because

$$0 < Y_n < 1 \Leftrightarrow 0 < P/Y_1 < 1 \Leftrightarrow 0 < P < Y_1,$$

while the upper bound is $\sqrt{p}$ because $Y_1^2 < Y_1 Y_n = P$. Now let $v = x^2/p$, for which $0 < v < 1$, so that $x = p^{1/2} v^{1/2}$ and

$$f_P(p) = \frac{n\,(n - 1)}{2}\, p^{(n-2)/2} \int_p^1 v^{-n/2}\,(1 - v)^{n-2}\, dv. \tag{6.15}$$

Either of these formulae could be used in conjunction with numerical integration to evaluate $f_P(p)$. Figure 6.4 illustrates f_P for $n = 4$ and a (scaled) histogram of 10 000 simulated values of P. Problem 6.6 considers the more general case of $P = Y_i Y_j$ (and also gives the Matlab code to produce the figure).   ■

Taking $i = 1$ and $j = 2$ in (6.13) gives the joint density of the first two order statistics as

$$f_{Y_1, Y_2}(x, y) = \frac{n!}{(n - 2)!}\, [1 - F(y)]^{n-2}\, f(x)\, f(y)\, \mathbb{I}_{(x,\infty)}(y),$$

and generalizing this to the first k order statistics gives

$$f_{Y_1, \ldots, Y_k}(y_1, \ldots, y_k) = \frac{n!}{(n - k)!}\, [1 - F(y_k)]^{n-k} \prod_{i=1}^{k} f(y_i), \quad y_1 < y_2 < \cdots < y_k. \tag{6.16}$$

Taking $k = n$ in (6.16) gives the p.d.f. of the whole sample of order statistics,

$$f_{Y_1, \ldots, Y_n}(y_1, y_2, \ldots, y_n) = n! \prod_{i=1}^{n} f(y_i), \quad y_1 < y_2 < \cdots < y_n. \tag{6.17}$$

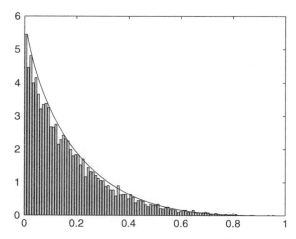

Figure 6.4 Density (6.14) for $n = 4$ overlaid with a scaled histogram of simulated values of P

This seems reasonable, as there are $n!$ different ways of arranging the observed **x** vector to give rise to the observed **y** vector. It is straightforward to formally prove this fact using a multivariate transformation; see e.g., Roussas (1997, pp. 245–247).

Remark: Inspection of (6.17) sometimes leads students to erroneously believe that $\int \cdots \int f_{\mathbf{Y}}(\mathbf{y}) \, d\mathbf{y} = n!$. It is worth emphasizing that, while

$$\int_{\mathbb{R}^n} \cdots \int f_{\mathbf{X}}(\mathbf{x}) \, d\mathbf{x} = \int_{-\infty}^{\infty} \cdots \int_{-\infty}^{\infty} \prod_{i=1}^{n} f_X(x_i) \, dx_n \cdots dx_1 = \prod_{i=1}^{n} \int_{-\infty}^{\infty} f_X(x_i) \, dx_i = 1,$$

it is not the case that $\int \cdots \int f_{\mathbf{Y}}(\mathbf{y}) \, d\mathbf{y} = n!$ because **Y** has support

$$S := \left\{ \mathbf{y} \in \mathbb{R}^n : y_1 < \cdots < y_n \right\} \in \mathbb{R}^n,$$

i.e.,

$$1 = \int_S \cdots \int f_{\mathbf{Y}}(\mathbf{y}) \, d\mathbf{y} = \int_{-\infty}^{\infty} \int_{y_1}^{\infty} \int_{y_2}^{\infty} \cdots \int_{y_{n-1}}^{\infty} f_{\mathbf{Y}}(\mathbf{y}) \, dy_n \cdots dy_1.$$

As an illustration, let $X_i \overset{\text{i.i.d.}}{\sim} \text{Exp}(\lambda)$, $i = 1, \ldots, n$. Then the joint density of the corresponding order statistics $\mathbf{Y} = (Y_1, \ldots, Y_n)$ is, with $s = \sum_{i=1}^{n} x_i = \sum_{i=1}^{n} y_i$, given by

$$f_{\mathbf{Y}}(\mathbf{y}) = n! \prod_{i=1}^{n} \lambda e^{-\lambda y_i} = n! \lambda^n e^{-\lambda s}, \qquad 0 < y_1 < y_2 < \cdots < y_n.$$

To verify that $f_{\mathbf{Y}}$ indeed integrates to one, note that, for $n = 3$,

$$\int \cdots \int f_{\mathbf{Y}}(\mathbf{y}) \, d\mathbf{y} = 3! \int_0^{\infty} \lambda e^{-\lambda y_1} \int_{y_1}^{\infty} \lambda e^{-\lambda y_2} \int_{y_2}^{\infty} \lambda e^{-\lambda y_3} \, dy_3 \, dy_2 \, dy_1$$

$$= 3! \int_0^\infty \lambda e^{-\lambda y_1} \int_{y_1}^\infty \lambda e^{-\lambda y_2} \left(e^{-\lambda y_2} \right) dy_2\, dy_1$$

$$= 3! \int_0^\infty \lambda e^{-\lambda y_1} \left(\frac{1}{2} e^{-2\lambda y_1} \right) dy_1$$

$$= 3! \times \frac{1}{2} \times \frac{1}{3} = 1,$$

with the pattern for higher n being clear. ∎

6.1.3 Sample range and midrange

Let X_i, $i = 1, \ldots, n$, be an i.i.d. sample from a population with density f and c.d.f. F, and denote the order statistics as $Y_1, \ldots, Y_n$. The *sample range* is defined to be $R = Y_n - Y_1$, and provides a measure of the length of the support, or range, of the underlying distribution. The *sample midrange* is defined to be $T = (Y_1 + Y_n)/2$, and is a measure of central tendency. Depending on the presence and extent of asymmetry and tail thickness in f, the central tendency measures $\overline{X}$, med(X), and T could be quite different from one another.

The joint distribution of the sample range and midrange is given by

$$f_{R,T}(r, t) = n(n-1)\left[F\left(t + \frac{r}{2}\right) - F\left(t - \frac{r}{2}\right) \right]^{n-2} f\left(t - \frac{r}{2}\right) f\left(t + \frac{r}{2}\right) \mathbb{I}_{(0,\infty)}(r),$$

(6.18)

which will be derived in Problem 6.4. The marginals are computed in the usual way as

$$f_R(r) = \int_{-\infty}^\infty f_{R,T}(r, t)\, dt \quad \text{and} \quad f_T(t) = \int_0^\infty f_{R,T}(r, t)\, dr.$$

(6.19)

The mth raw moment of R can be expressed as

$$\mathbb{E}[R^m] = \int_0^\infty r^m f_R(r)\, dr = \int_0^\infty \int_{-\infty}^\infty r^m f_{R,T}(r, t)\, dt\, dr.$$

(6.20)

Of course, from (I.6.1), the expected values of R and T can be computed without recourse to double integration, e.g., using (6.8),

$$\mathbb{E}[R] = \mathbb{E}[Y_n] - \mathbb{E}[Y_1] = \int_0^\infty y f_{Y_n}(y)\, dy - \int_0^\infty y f_{Y_1}(y)\, dy.$$

(6.21)

Moreover, recalling (I.7.71), which states that, for continuous r.v. X,

$$\mathbb{E}[X] = \int_0^\infty (1 - F_X(x))\, dx - \int_{-\infty}^0 F_X(x)\, dx = \int_0^\infty \overline{F}_X(x)\, dx - \int_{-\infty}^0 F_X(x)\, dx,$$

we have, from (6.3) and (6.4), i.e.,

$$F_{Y_1}(y) = 1 - \left[1 - F(y)\right]^n =: 1 - \overline{F}^n(y) \quad \text{and} \quad F_{Y_n}(y) =: F^n(y),$$

that

$$\mathbb{E}[R] = \mathbb{E}[Y_n] - \mathbb{E}[Y_1]$$

$$= \int_0^\infty \overline{F}_{Y_n}(y)\, dy - \int_{-\infty}^0 F_{Y_n}(y)\, dy - \int_0^\infty \overline{F}_{Y_1}(y)\, dy + \int_{-\infty}^0 F_{Y_1}(y)\, dy$$

$$= \int_0^\infty \left[1 - F^n(y)\right] dy - \int_{-\infty}^0 F^n(y)\, dy$$

$$\quad - \int_0^\infty \left[1 - \left\{1 - \overline{F}^n(y)\right\}\right] dy + \int_{-\infty}^0 \left[1 - \overline{F}^n(y)\right] dy$$

$$= \int_{-\infty}^0 \left[-F^n(y) + \left(1 - \overline{F}^n(y)\right)\right] dy + \int_0^\infty \left[\left(1 - F^n(y)\right) - \overline{F}^n(y)\right] dy,$$

or

$$\mathbb{E}[R] = \int_{-\infty}^\infty \left[1 - F^n(y) - \overline{F}^n(y)\right] dy, \tag{6.22}$$

first reported in Tippett (1925). Figure 6.5 plots $\mathbb{E}[R]$ as a function of sample size n for normal and Student's t i.i.d. data.

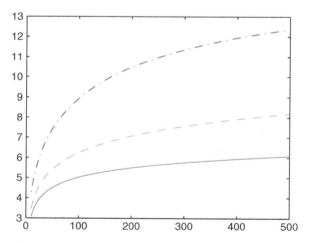

Figure 6.5   Expected range $\mathbb{E}[R]$ as a function of sample size n for i.i.d. standard normal (bottom curve) and standard Student's t i.i.d. data with 10 (middle) and 4 (top) degrees of freedom

Remarks:
(a) Tippett (1925) proved (6.22) in two ways, both differing from the previous derivation. The first starts with the result given by Pearson in 1902 for the expected spacing

$S_i := Y_{i+1} - Y_i,$

$$\mathbb{E}[S_i] = \mathbb{E}[Y_{i+1} - Y_i] = \binom{n}{i} \int_{-\infty}^{\infty} F^i(y) \overline{F}^{n-i}(y) \, dy,$$

which is interesting in its own right (see also David, 1981; and Jones and Balakrishnan, 2002). Based on this, the result is practically trivial: we have

$$\mathbb{E}[R] = \mathbb{E}[Y_n - Y_1] = \mathbb{E}[S_{n-1} + S_{n-2} + \cdots + S_1]$$

$$= \sum_{i=1}^{n-1} \binom{n}{i} \int_{-\infty}^{\infty} F^i(y) \overline{F}^{n-i}(y) \, dy$$

or, with $p = p(y) = F(y)$,

$$\mathbb{E}[R] = \int_{-\infty}^{\infty} \sum_{i=1}^{n-1} \binom{n}{i} p^i (1-p)^{n-i} \, dy = \int_{-\infty}^{\infty} \left[1 - (1-p)^n - p^n \right] dy$$

from the binomial theorem. The second proof (which was suggested to Tippett by Pearson) is considerably more involved and will not be shown here.

(b) Expression (6.22) can be generalized to the mth moment, $m = 1, 2, \ldots$, as

$$\mathbb{E}[R^m] = m! \int \cdots \int_{-\infty < y_1 < \cdots < y_m < \infty} G(y_1, y_m) \, dy_1 \cdots dy_m, \tag{6.23}$$

where

$$G(y_1, y_m) = \left[1 - F^n(y_m) - \overline{F}^n(y_1) + \{ F(y_m) - F(y_1) \}^n \right].$$

This can further be generalized to the independent but not (necessarily) identical case, by replacing $G(y_1, y_m)$ in (6.23) by

$$1 - \prod_{i=1}^{n} F_i(y_m) - \prod_{i=1}^{n} \overline{F}_i(y_1) + \prod_{i=1}^{n} \{ F_i(y_m) - F_i(y_1) \},$$

as shown by Jones and Balakrishnan (2002).

Example 6.12 Let $X_i \overset{\text{i.i.d.}}{\sim} N(0, 1)$, $i = 1, \ldots, n$. Then the joint distribution of R and T is given by (6.18), but replacing f with ϕ and F with Φ. From this, the marginals (6.19) can be computed by using numerical integration, which the reader is encouraged to try. Also, a standard numeric computation using (6.21) and (6.22) with $n = 10$ yields $\mathbb{E}[R] = 3.0775$ in each case.

The mth raw moment of R can be computed from (6.20) and calculated by bivariate numerical integration. A program to carry this out using Matlab's built-in function dblquad is shown in Listing 6.2. With $n = 10$, and assuming the existence of the first

```
function m=srm(n,mom)
xmin=0.2; xmax=8;      % range of R, the sample range.
ymin=-5; ymax=-ymin; % range of the sample midrange
tol=1e-6; m=dblquad(@srmf,xmin,xmax,ymin,ymax,tol,@quadl,n,mom);

function z=srmf(r,t,n,mom)
part1=( normcdf(t+r/2)-normcdf(t-r/2) ).^(n-2);
z = n*(n-1)* (r.^mom) .* part1.*normpdf(t-r/2).*normpdf(t+r/2);
```

Program Listing 6.2   Computes (6.20)

three moments, this yields $\mathbb{E}[R] = 3.0775$, $\mathbb{E}[R^2] = 10.1063$ (so that $\mathbb{V}(R) = 0.6353$ and the standard deviation is 0.7971) and $\mathbb{E}[R^3] = 35.2134$.[2]

As n grows, one would expect the dependency of Y_1 and Y_n to fade. Thus, for example,

$$\mathbb{E}\left[R^2\right] = \mathbb{E}\left[Y_n^2 - 2Y_nY_1 + Y_1^2\right] \approx \mathbb{E}\left[Y_n^2\right] + \mathbb{E}\left[Y_1^2\right] - 2\mathbb{E}[Y_n]\mathbb{E}[Y_1], \qquad (6.24)$$

which only requires univariate numeric integration. For $n = 10$, (6.24) yields 10.1597, which is reasonably close to the true value.

Simulation offers an alternative method of calculation which is very fast in this case. Using code

```
n=10; r=zeros(50000,1);
for i=1:50000, s=randn(n,1); r(i)=max(s)-min(s); end
mean(r), mean(r.^2), mean(r.^3)
```

the moments are approximated by $\mathbb{E}[R] \approx 3.0797$, $\mathbb{E}[R^2] \approx 10.1219$ and $\mathbb{E}[R^3] \approx 35.2982$, which are close to the true values. Of course, the simulated values of R can also be used to construct a histogram, as shown in the top left-hand panel of Figure 6.6. The other left-hand panels use $n = 100$ (with true $\mathbb{E}[R] = 5.015$ and $\mathbb{E}[R] \approx 5.018$ empirically) and $n = 1000$ ($\mathbb{E}[R] = 6.482$ and $\mathbb{E}[R] \approx 6.485$).

We now repeat the same exercise but using Student's t data with 5 degrees of freedom. To avoid confusion with the normal case, we denote the range statistic as R_5. The resulting histograms are shown in the right-hand panels of Figure 6.6, and were truncated at $r = 40$. As would be expected, we see that the tails of R_5 are much thicker than those for the normal case, so that the existence of higher-order moments could be questioned. For $n = 10$, calculation with (6.20) (using an integration range for R_5 of 0.1 to 40, and that for T of -12 to 12) gives $\mathbb{E}[R_5] = 4.0052$, $\mathbb{E}[R_5^2] = 18.4316$ and $\mathbb{E}[R_5^3] = 99.3011$, while the empirical values are $\mathbb{E}[R_5] \approx 4.0029$, $\mathbb{E}[R_5^2] \approx 18.4094$ and $\mathbb{E}[R_5^3] \approx 99.6153$.   ∎

---

[2] These calculations take a matter of seconds, as does writing the little program to do them. It is interesting, if not painful, to look at how such calculations were performed in 1925. As Tippett wrote regarding the calculation of $\mathbb{E}[R^2]$ in the normal case, 'The work is very laborious, as it involves cubature, and even so, the result can only be given to a few figures. It is believed that the values given in Table IV are correct to the last figure.' Indeed, for $n = 10$, Tippett reports a standard deviation for R of 0.797, which we see to be correct to three decimal places. Presumably, he was lucky that a referee did not ask him to repeat the calculations based on Student's t data with various degrees of freedom!

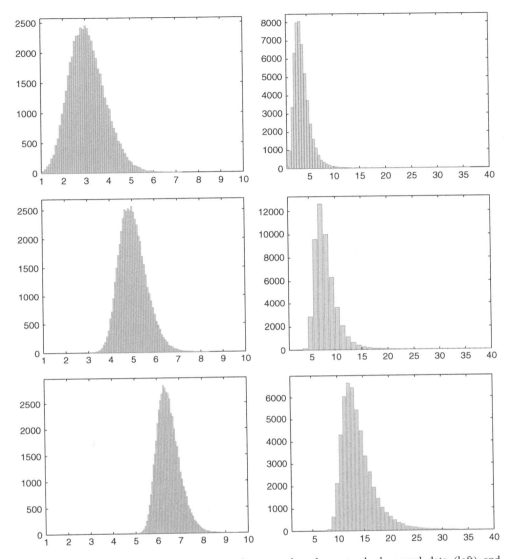

Figure 6.6 Empirical density of the sample range, based on standard normal data (left) and Student's t data (right) for sample sizes $n = 10$ (top), $n = 100$ (middle) and $n = 1000$ (bottom)

6.2 Further examples

The examples in this section combine several techniques just introduced and also further emphasize the importance of simulation for situations in which tractable algebraic expressions are not available.

⊖ *Example 6.13*   The results in this example are based on Kamara and Siegel (1987), having arisen in the context of optimal hedging strategies in futures markets.[3] Let

$$\begin{bmatrix} X \\ Y \end{bmatrix} \sim N\left(\begin{bmatrix} \mu_X \\ \mu_Y \end{bmatrix}, \begin{bmatrix} \sigma_X^2 & \sigma_{XY} \\ \sigma_{XY} & \sigma_Y^2 \end{bmatrix}\right),$$

and define $T = Y - X$, which is normally distributed with mean $\mu_T = \mu_Y - \mu_X$ and variance $\sigma_T^2 = \sigma_X^2 + \sigma_Y^2 - 2\sigma_{XY}$. With $t := \mu_T/\sigma_T$ and $M := \min(X, Y)$, we wish to show that

$$\mathbb{E}[M] = \mu_X - \sigma_T \phi(t) + \mu_T \Phi(-t), \tag{6.25}$$

$$\mathbb{V}(M) = \sigma_X^2 + (\sigma_Y^2 - \sigma_X^2)\Phi(-t) + \frac{\mu_T^2}{4} - \left[\frac{\mu_T}{2} - \mu_T \Phi(-t) + \sigma_T \phi(t)\right]^2, \tag{6.26}$$

and

$$\text{Cov}(X, M) = \sigma_X^2 - (\sigma_X^2 - \sigma_{XY})\Phi(-t), \tag{6.27}$$

where Φ and ϕ are the standard normal c.d.f. and p.d.f., respectively.

We begin by deriving some preliminary results. Let $R \sim N(0, 1)$ and $\alpha \in \mathbb{R}$. Then

$$\mathbb{E}[R\,\mathbb{I}(R < \alpha)] = -\phi(\alpha) \quad \text{and} \quad \mathbb{E}[R(R - \alpha)\,\mathbb{I}(R < \alpha)] = \Phi(\alpha). \tag{6.28}$$

For the first expression in (6.28), similar to the numerator in the definition of expected shortfall (I.8.38),

$$\mathbb{E}[R\,\mathbb{I}(R < \alpha)] = \int_{-\infty}^{\infty} r\,\mathbb{I}(r < \alpha)\phi(r)\,dr = \int_{-\infty}^{\alpha} r\phi(r)\,dr$$

$$= \int_{-\infty}^{\alpha} \frac{r}{\sqrt{2\pi}} \exp(-r^2/2)\,dr = \left[-\frac{1}{\sqrt{2\pi}} \exp(-r^2/2)\right]_{-\infty}^{\alpha} = -\phi(\alpha).$$

Similarly, for the second expression in (6.28), $\mathbb{E}[R(R - \alpha)\,\mathbb{I}(R < \alpha)]$ is

$$\int_{-\infty}^{\infty} r(r - \alpha)\mathbb{I}(r < \alpha)\phi(r)\,dr = \int_{-\infty}^{\alpha} \frac{r(r - \alpha)}{\sqrt{2\pi}} \exp(-r^2/2)\,dr,$$

and using integration by parts with $u = (r - \alpha)/\sqrt{2\pi}$ and $dv = r \exp(-r^2/2)\,dr$,

$$\mathbb{E}[R(R - \alpha)\,\mathbb{I}(R < \alpha)] = -\left[\frac{(r - \alpha)}{\sqrt{2\pi}} \exp(-r^2/2)\right]_{-\infty}^{\alpha} + \int_{-\infty}^{\alpha} \frac{1}{\sqrt{2\pi}} \exp(-r^2/2)\,dr$$

$$= \Phi(\alpha),$$

having used l'Hôpital's rule for the value at $-\infty$.

---

[3] The results were elegantly extended to the general multivariate normal distribution by Siegel (1993), and further extended by Liu (1994) to other order statistics and other distributions. The moment generating function of the minimum of bivariate normal r.v.s is derived in Cain (1994).

The last preliminary result we need is

$$\mathbb{E}\left[UV\,\mathbb{I}(V < 0)\right] = \sigma_{UV}\,\Phi\left(-\frac{\mu_V}{\sigma_V}\right), \tag{6.29}$$

where

$$\begin{bmatrix} U \\ V \end{bmatrix} \sim N\left(\begin{bmatrix} 0 \\ \mu_V \end{bmatrix}, \begin{bmatrix} \sigma_U^2 & \sigma_{UV} \\ \sigma_{UV} & \sigma_V^2 \end{bmatrix}\right).$$

To prove (6.29), write U as a linear combination of $(V - \mu_V)$ and W, where W is a random variable independent of V. In particular, let $c_1, c_2 \in \mathbb{R}$ such that $U = c_1(V - \mu_V) + c_2 W$. Then, as $\mathbb{E}[U] = 0$, we have that

$$c_1\left(\mathbb{E}[V] - \mu_V\right) + c_2\mathbb{E}[W] = 0 \Rightarrow \mathbb{E}[W] = 0.$$

From the definition of covariance, it is simple to confirm that $\mathbb{E}[UV] = \sigma_{UV}$, and, as $W \perp V$ and $\mathbb{E}[W] = 0$, that $\mathbb{E}[WV] = 0$. Thus, with $\mathbb{V}(V) = \mathbb{E}[V^2] - \mu_V^2$,

$$\mathbb{E}[UV] = \sigma_{UV} \Rightarrow c_1\mathbb{E}[V^2 - \mu_V V] + c_2\mathbb{E}[WV] = \sigma_{UV} \Rightarrow c_1\sigma_V^2$$
$$= \sigma_{UV} \Rightarrow c_1 = \frac{\sigma_{UV}}{\sigma_V^2}.$$

Let $(V - \mu_V)/\sigma_V =: R \sim N(0, 1)$. Then

$$\mathbb{E}\left[UV\mathbb{I}(V < 0)\right] = \mathbb{E}\left[\frac{\sigma_{UV}}{\sigma_V^2}(V - \mu_V)V\,\mathbb{I}(V < 0)\right] + \mathbb{E}\left[c_2 WV\,\mathbb{I}(V < 0)\right]$$

$$= \sigma_{UV}\mathbb{E}\left[\frac{V - \mu_V}{\sigma_V}\frac{V}{\sigma_V}\,\mathbb{I}(V/\sigma_V < 0)\right] + c_2\mathbb{E}[W]\mathbb{E}\left[V\,\mathbb{I}(V < 0)\right]$$

$$= \sigma_{UV}\mathbb{E}\left[R\left(R + \frac{\mu_V}{\sigma_V}\right)\mathbb{I}\left(R < -\frac{\mu_V}{\sigma_V}\right)\right],$$

and (6.29) follows from (6.28).

It is now straightforward to show (6.25)–(6.27). First, observe that $M = \min(X, Y) = X + T\,\mathbb{I}(T < 0)$, $T = Y - X$. Then, with $R := (T - \mu_T)/\sigma_T \sim N(0, 1)$,

$$\mathbb{E}[M] = \mathbb{E}[X] + \mathbb{E}[T\,\mathbb{I}(T < 0)] \tag{6.30}$$

$$= \mu_X + \mathbb{E}\left[\sigma_T\left(\frac{T - \mu_T}{\sigma_T}\right)\mathbb{I}(T < 0)\right] + \mathbb{E}[\mu_T\,\mathbb{I}(T < 0)]$$

$$= \mu_X + \sigma_T\mathbb{E}\left[R\,\mathbb{I}\left(R < -\frac{\mu_T}{\sigma_T}\right)\right] + \mu_T\mathbb{E}\left[\mathbb{I}\left(R < -\frac{\mu_T}{\sigma_T}\right)\right]$$

$$\stackrel{(6.28)}{=} \mu_X - \sigma_T\phi\left(\frac{\mu_T}{\sigma_T}\right) + \mu_T\,\mathrm{Pr}\left(R < -\frac{\mu_T}{\sigma_T}\right),$$

which is (6.25), using the fact that $\phi(\alpha) = \phi(-\alpha)$, $\alpha \in \mathbb{R}$.

Next, from (I.6.3),

$$\mathbb{V}(M) = \mathbb{V}(X + T\,\mathbb{I}(T < 0))$$
$$= \sigma_X^2 + 2\,\text{Cov}(X, T\,\mathbb{I}(T < 0)) + \mathbb{V}(T\,\mathbb{I}(T < 0)). \tag{6.31}$$

For the covariance term,

$$\text{Cov}(X, T\,\mathbb{I}(T < 0)) = \mathbb{E}[XT\,\mathbb{I}(T < 0)] - \mathbb{E}[X]\,\mathbb{E}[T\,\mathbb{I}(T < 0)]$$
$$= \mathbb{E}[(X - \mu_X)T\,\mathbb{I}(T < 0)].$$

Let $U := X - \mu_X$, a mean-zero normal variate, and apply (6.29) to get, with $t :=$ μ_T/σ_T,

$$\text{Cov}(X, T\,\mathbb{I}(T < 0)) = \text{Cov}(U, T)\,\Phi(-t) = (\mathbb{E}[UT] - \mathbb{E}[U]\,\mathbb{E}[T])\,\Phi(-t)$$
$$= \mathbb{E}[UT]\,\Phi(-t) = \mathbb{E}[(X - \mu_X)(Y - X)]\,\Phi(-t)$$
$$= \left(\sigma_{XY} + \mu_X\mu_Y - \sigma_X^2 - \mu_X^2 - \mu_X\mu_Y + \mu_X\mu_X\right)\Phi(-t)$$
$$= \left(\sigma_{XY} - \sigma_X^2\right)\Phi(-t). \tag{6.32}$$

The variance term in (6.31) can be expressed as

$$\mathbb{V}(T\,\mathbb{I}(T < 0)) = \mathbb{E}[(T - \mu_T)\,T\,\mathbb{I}(T < 0)] + \mu_T\mathbb{E}[T\,\mathbb{I}(T < 0)] - \{\mathbb{E}[T\,\mathbb{I}(T < 0)]\}^2$$
$$= \mathbb{E}[(T - \mu_T)\,T\,\mathbb{I}(T < 0)] - \left\{\frac{\mu_T}{2} - \mathbb{E}[T\,\mathbb{I}(T < 0)]\right\}^2 + \frac{\mu_T^2}{4},$$

where, from (6.30),

$$\mathbb{E}[T\,\mathbb{I}(T < 0)] = -\sigma_T\phi(t) + \mu_T\Phi(-t).$$

Then applying (6.29) to the first expectation gives

$$\mathbb{E}[(T - \mu_T)\,T\,\mathbb{I}(T < 0)] = \text{cov}(T - \mu_T, T)\,\Phi(-t)$$
$$= \{\mathbb{E}[T^2 - \mu_T T] - \mathbb{E}[T - \mu_T]\,\mathbb{E}[T]\}\,\Phi(-t)$$
$$= \sigma_T^2\Phi(-t).$$

Substituting these results into (6.31) and using the fact that

$$2(\sigma_{XY} - \sigma_X^2) + \sigma_T^2 = 2\sigma_{XY} - 2\sigma_X^2 + \sigma_Y^2 - 2\sigma_{XY} + \sigma_X^2 = \sigma_Y^2 - \sigma_X^2$$

yields (6.26).

Finally, from (I.6.5) and (6.32),

$$\text{Cov}(X, M) = \text{Cov}(X, X + T\,\mathbb{I}(T < 0)) = \sigma_X^2 + \text{Cov}(X, T\,\mathbb{I}(T < 0))$$
$$= \sigma_X^2 + \left(\sigma_{XY} - \sigma_X^2\right)\Phi(-t),$$

which is (6.27). ∎

⊖ **Example 6.14**   Let $X_1, X_2 \overset{\text{i.i.d.}}{\sim}$ Gam $(2, 1)$, i.e., $f_X(x) = xe^{-x}\mathbb{I}_{(0,\infty)}(x)$ and define

$$S = \min(X_1, X_2), \quad L = \max(X_1, X_2) \quad \text{and} \quad R = S/L,$$

(S standing for 'smallest', L for 'largest', and R for 'ratio'). The c.d.f. of X is

$$F_X(x) = \int_0^x te^{-t}\, dt = 1 - (x + 1)e^{-x},$$

so that, from (6.8),

$$f_S(s) = 2s(s+1)e^{-2s} \quad \text{and} \quad f_L(l) = 2\left(1 - (l+1)e^{-l}\right)le^{-l}.$$

It follows that

$$\mu_S = \mathbb{E}[S] = \int_0^\infty 2s^2(s+1)e^{-2s}\, ds = \frac{5}{4},$$

$$\sigma_S^2 = \mathbb{V}(S) = \mathbb{E}[S^2] - \mathbb{E}[S]^2 = \frac{11}{16}$$

and, similarly, $\mu_L = 11/4$ and $\sigma_L^2 = 35/16$.

From (6.13), the joint distribution of S and L is given by

$$f_{S,L}(s, l) = 2! f_X(s) f_X(l)\, \mathbb{I}_{(0,\infty)}(s)\, \mathbb{I}_{(s,\infty)}(l) = 2sle^{-(s+l)}\mathbb{I}_{(0,\infty)}(s)\, \mathbb{I}_{(s,\infty)}(l)$$

and, using the first equation in (2.10), the density of ratio R is

$$f_R(r) = \int_{-\infty}^\infty \frac{|s|}{r^2} f_{S,L}\left(s, \frac{s}{r}\right) ds = \int_0^\infty \frac{s}{r^2} 2s\left(\frac{s}{r}\right) e^{-s-s/r}\mathbb{I}_{(s,\infty)}\left(\frac{s}{r}\right) ds \qquad (6.33)$$

$$= \frac{2}{r^3}\int_0^\infty s^3 e^{-s\left(1+r^{-1}\right)}\mathbb{I}_{(0,1)}(r)\, ds = \frac{12r}{(r+1)^4}\mathbb{I}_{(0,1)}(r),$$

with straightforward integration yielding

$$\mu_R = \frac{1}{2} \quad \text{and} \quad \sigma_R^2 = -\frac{33}{4} + 12\ln 2 \approx 0.067766.$$

Figure 6.7 compares the kernel density estimate (see Section I.7.4.2) of 2000 independent draws of S/L with the theoretical density (6.33); they are indeed quite close, but notice that the estimated density is less accurate in the tails. The sample means and variances of the drawn sample were $\hat{\mu}_S = 1.28$, $\hat{\sigma}_S^2 = 0.688$, $\hat{\mu}_L = 2.78$, $\hat{\sigma}_L^2 = 2.23$, $\hat{\mu}_R = 0.506$, and $\hat{\sigma}_R^2 = 0.0664$, which are very close to their theoretical values.

The sample covariance between S and L was 0.582, which compares well with the exact value

$$\text{Cov}(S, L) = \mathbb{E}[SL] - \mathbb{E}[S]\mathbb{E}[L]$$

$$= \int_0^\infty \int_s^\infty 2s^2 l^2 e^{-(s+l)}\, dl\, ds - \frac{5}{4}\frac{11}{4} = \frac{9}{16} = 0.5625.$$

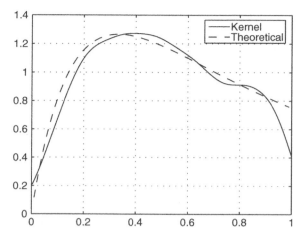

Figure 6.7 Theoretical density (6.33) and kernel density estimate from 2000 simulated Gam(2,1) draws

An approximation to $c = \text{Cov}(S, L)$ can be obtained by equating the two-term Taylor series approximation (2.32) of $\mathbb{E}[R]$ to the true value:

$$\mu_R = \frac{1}{2} \approx \frac{\mu_S}{\mu_L} - \mu_L^{-2} \text{Cov}(S, L) + \mu_S \mu_L^{-3} \sigma_L^2 = \frac{5}{11} - \left(\frac{11}{4}\right)^{-2} c + \frac{5}{4}\left(\frac{11}{4}\right)^{-3}\frac{35}{16},$$

which yields $c \approx 0.651$. Similarly, using (2.33), the Taylor series approximation to $\mathbb{V}(R)$ gives

$$\sigma_R^2 = 12\ln 2 - \frac{33}{4} \approx \left(\frac{\mu_S}{\mu_L}\right)^2 \left(\frac{\sigma_S^2}{\mu_S^2} + \frac{\sigma_L^2}{\mu_L^2} - \frac{2\,\text{Cov}(S, L)}{\mu_S \mu_L}\right)$$

or $c \approx 0.690$, with the approximation using $\mathbb{E}[R]$ being somewhat better. Usually of more interest is the correlation:

$$\text{Corr}(S, L) = \frac{\text{Cov}(S, L)}{\sqrt{\mathbb{V}(S)\,\mathbb{V}(L)}} = \frac{\frac{9}{16}}{\sqrt{\frac{11}{16}\frac{35}{16}}} \approx 0.45868,$$

as compared to the empirical value of 0.4694. ∎

⊙ **Example 6.15** Let $X_i \overset{\text{i.i.d.}}{\sim} \exp(\alpha)$, $i = 1, \ldots, n$, with Y_i the ith order statistic. From (6.6) the density of Y_1 is

$$f_{Y_1}(y) = n\left(e^{-\alpha y}\right)^{n-1}\alpha e^{-\alpha y} = (n\alpha)\,e^{-(n\alpha)y}\mathbb{I}_{(0,\infty)}(y)$$

or $Y_1 \sim \exp(n\alpha)$ while, from (6.13), the joint density of Y_1 and Y_2 is

$$f_{Y_1,Y_2}(y_1, y_2) = n(n-1)\alpha^2 e^{-\alpha(y_1+y_2)}\left(e^{-\alpha y_2}\right)^{n-2}\mathbb{I}_{(0,\infty)}(y_2)\,\mathbb{I}_{(0,y_2)}(y_1). \qquad (6.34)$$

The density of the difference $D_1 = Y_2 - Y_1$ can be obtained via transformation. In particular, with $Z = Y_2$, we have $y_2 = z$, $y_1 = z - d_1$, and

$$\text{abs } |J| = \text{abs det} \begin{bmatrix} \dfrac{\partial y_1}{\partial d_1} & \dfrac{\partial y_1}{\partial z} \\[2mm] \dfrac{\partial y_2}{\partial d_1} & \dfrac{\partial y_2}{\partial z} \end{bmatrix} = \text{abs det} \begin{bmatrix} -1 & 1 \\ 0 & 1 \end{bmatrix} = 1,$$

so that

$$f_{D_1, Z}(d_1, z) = f_{Y_1, Y_2}(z - d_1, z) = n(n-1)\alpha^2 e^{-\alpha(z - d_1 + z)}\left(e^{-\alpha z}\right)^{n-2}$$
$$= n(n-1)\alpha^2 e^{\alpha d_1} e^{-\alpha n z} \mathbb{I}_{(0,z)}(d_1),$$

where the indicator function is $\mathbb{I}_{(0,z)}(d_1)$ because D_1 is positive but bounded above by $Y_2 = Z$. Then, as $Z > D_1$,

$$f_{D_1}(d_1) = \int_{d_1}^{\infty} f_{D_1, Z}(d_1, z)\, dz = n(n-1)\alpha^2 e^{\alpha d_1} \int_{d_1}^{\infty} e^{-\alpha n z}\, dz$$
$$= n(n-1)\alpha^2 e^{\alpha d_1} \times \frac{1}{\alpha n} e^{-\alpha n d_1} = \alpha(n-1) e^{-\alpha(n-1)d_1},$$

i.e., $D_1 \sim \exp(\alpha(n-1))$.

From (6.18), the joint density of $R = Y_n - Y_1$ and $T = (Y_1 + Y_n)/2$ is given by

$$f_{R,T}(r, t) = n(n-1)\left[\left(1 - e^{-\alpha(t + \frac{r}{2})}\right) - \left(1 - e^{-\alpha(t - \frac{r}{2})}\right)\right]^{n-2}\alpha e^{-\alpha(t - \frac{r}{2})}\alpha e^{-\alpha(t + \frac{r}{2})}$$
$$= n(n-1)\alpha^2 \left[e^{-\alpha(t - \frac{r}{2})} - e^{-\alpha(t + \frac{r}{2})}\right]^{n-2} e^{-2\alpha t}$$
$$= n(n-1)\alpha^2 e^{-n\alpha t}\left[e^{r\alpha/2} - e^{-r\alpha/2}\right]^{n-2}\mathbb{I}_{(0,\infty)}(r)\,\mathbb{I}_{(r/2,\infty)}(t),$$

where the requirement $T > R/2$ follows because $Y_n = Y_1 + R$ and $Y_n = 2T - Y_1$ imply that $Y_1 + R = 2T - Y_1$ or $Y_1 = T - R/2$, but $Y_1 > 0$ so that $T > R/2$. Thus,

$$f_R(r) = \int_{r/2}^{\infty} f_{R,T}(r, t)\, dt = n(n-1)\alpha^2 \left[e^{r\alpha/2} - e^{-r\alpha/2}\right]^{n-2} \int_{r/2}^{\infty} e^{-n\alpha t}\, dt$$
$$= n(n-1)\alpha^2 \left[e^{r\alpha/2} - e^{-r\alpha/2}\right]^{n-2}\left(\frac{1}{n\alpha} e^{-n\alpha r/2}\right)\mathbb{I}_{(0,\infty)}(r)$$
$$= \alpha(n-1)\left[e^{r\alpha/2} - e^{-r\alpha/2}\right]^{n-2} e^{-n\alpha r/2}\,\mathbb{I}_{(0,\infty)}(r)$$
$$= \alpha(n-1)\left[\frac{e^{r\alpha/2} - e^{-r\alpha/2}}{e^{r\alpha/2}}\right]^{n-2} e^{(r\alpha/2)(n-2)} e^{-n\alpha r/2}\,\mathbb{I}_{(0,\infty)}(r)$$
$$= \alpha(n-1)\left[1 - e^{-r\alpha}\right]^{n-2} e^{-r\alpha}\,\mathbb{I}_{(0,\infty)}(r). \tag{6.35}$$

To check that this is a valid density, note that, with $y = e^{-r\alpha}$, $r = -(\ln y)/\alpha$ and $dr = -(y\alpha)^{-1}\, dy$,

$$\int_0^\infty \left(1 - e^{-r\alpha}\right)^{n-2} e^{-r\alpha}\, dr = -\frac{1}{\alpha} \int_1^0 (1 - y)^{n-2}\, y\frac{1}{y}\, dy = \frac{B(1, n-1)}{\alpha} = \frac{1}{\alpha(n-1)}.$$

Figure 6.8 shows the density of R for several parameter combinations. From (6.35) the m.g.f. of the range R is

$$\mathbb{M}_R(s) = \alpha(n-1) \int_0^\infty \left(1 - e^{-r\alpha}\right)^{n-2} e^{-r(\alpha-s)}\, dr,$$

which, for $n = 2$, simplifies to $\mathbb{M}_R(s) = \alpha \int_0^\infty e^{-r(\alpha-s)}\, dr = \alpha/(\alpha - s)$ when $s < \alpha$, while, for $n = 3$,

$$\mathbb{M}_R(s) = 2\alpha \int_0^\infty e^{-r(\alpha-s)}\, dr - 2\alpha \int_0^\infty e^{-r(2\alpha-s)}\, dr$$

$$= \frac{2\alpha}{\alpha - s} - \frac{2\alpha}{2\alpha - s} = \frac{\alpha}{\alpha - s}\frac{2\alpha}{2\alpha - s}$$

when $s < \alpha$ and $s < 2\alpha$, which is satisfied when $s < \alpha$. In general, using $u = e^{-r\alpha}$, $r = -\alpha^{-1}\ln u$ and $dr = -(u\alpha)^{-1}\, du$,

$$\int_0^\infty \left(1 - e^{-r\alpha}\right)^{n-2} e^{-r(\alpha-s)}\, dr = -\frac{1}{\alpha} \int_1^0 (1 - u)^{n-2}\, u e^{\left(-\alpha^{-1}\ln u\right)s} u^{-1}\, du$$

$$= \frac{1}{\alpha} \int_0^1 (1 - u)^{n-2}\, u^{-s/\alpha}\, du = \frac{1}{\alpha} B(n-1, 1 - s/\alpha),$$

so that

$$\mathbb{M}_R(s) = (n-1) \frac{\Gamma(n-1)\Gamma(1 - s/\alpha)}{\Gamma(n - s/\alpha)} = \frac{(n-1)!}{(n - s/\alpha - 1)(n - s/\alpha - 2)\cdots(1 - s/\alpha)}$$

$$= \frac{(n-1)!\alpha^{n-1}}{(\alpha(n-1) - s)(\alpha(n-2) - s)\cdots(\alpha(n - (n-1)) - s)}$$

$$= \prod_{j=1}^{n-1} \frac{\alpha(n-j)}{\alpha(n-j) - s}, \qquad \text{for } s < \alpha, \tag{6.36}$$

having used the recursion $\Gamma(x) = (x-1)\Gamma(x-1)$. ∎

⊙ **Example 6.16**   (Example 6.15 cont.) The joint density of the n order statistics $\mathbf{Y} = (Y_1, \ldots, Y_n)$ is

$$f_{\mathbf{Y}}(\mathbf{y}) = n!\alpha^n \exp\left(-\alpha \sum_{i=1}^n y_i\right).$$

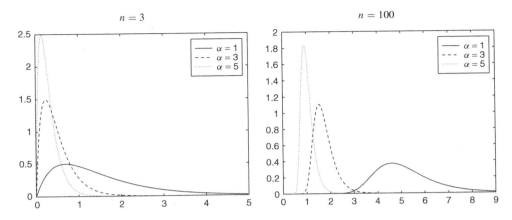

Figure 6.8 Density $f_R(r)$ of range R for $X_i \overset{\text{i.i.d.}}{\sim} \exp(\alpha)$, $i = 1, \ldots, n$

Defining

$$D_0 = Y_1,$$
$$D_1 = Y_2 - Y_1,$$
$$D_2 = Y_3 - Y_2,$$
$$\vdots$$
$$D_{n-1} = Y_n - Y_{n-1}$$

implies that $Y_j = \sum_{i=0}^{j-1} D_i$, so that the Jacobian is a lower triangular matrix with ones on the diagonal, i.e.,

$$
\mathbf{J} = \begin{vmatrix}
\dfrac{\partial y_1}{\partial d_0} & \dfrac{\partial y_1}{\partial d_1} & \cdots & \dfrac{\partial y_1}{\partial d_{n-1}} \\
\dfrac{\partial y_2}{\partial d_0} & \dfrac{\partial y_2}{\partial d_1} & \cdots & \dfrac{\partial y_2}{\partial d_{n-1}} \\
\vdots & \vdots & \ddots & \\
\dfrac{\partial y_n}{\partial d_0} & \dfrac{\partial y_n}{\partial d_1} & \cdots & \dfrac{\partial y_n}{\partial d_{n-1}}
\end{vmatrix}
=
\begin{bmatrix}
1 & 0 & 0 & \cdots & 0 \\
1 & 1 & 0 & \cdots & 0 \\
1 & 1 & 1 & \cdots & 0 \\
\vdots & \vdots & \vdots & \ddots & 0 \\
1 & 1 & 1 & \cdots & 1
\end{bmatrix}
$$

and $|\det \mathbf{J}| = 1$. From (6.17),

$$f_{\mathbf{D}}(\mathbf{d}) = n! \alpha^n \exp\left\{ -\alpha \left[n d_0 + (n-1) d_1 + \cdots + (n-j) d_j + \cdots + d_{n-1} \right] \right\},$$

which factors into n separate distributions, showing that $D_0, \ldots, D_{n-1}$ are independent (not just pairwise). From this, one also sees that $D_j \sim \text{Exp}(\alpha(n-j))$, $j = 0, \ldots, n-1$. (See Problem 6.11 as well.) These facts can be used as follows. Recalling the c.d.f. and expectation of an exponential r.v., (6.6) implies that

$$\mathbb{E}\left[Y_k^m \right] = k \binom{n}{k} \int_0^\infty y^m \left(1 - e^{-\alpha y} \right)^{k-1} e^{-\alpha y (n-k)} \alpha e^{-\alpha y} \, dy,$$

but as

$$\mathbb{E}\left[Y_k\right] = \mathbb{E}\left[\sum_{i=0}^{k-1} D_i\right] = \frac{1}{\alpha n} + \frac{1}{\alpha\,(n-1)} + \cdots + \frac{1}{\alpha\,(n-k+1)},$$

we arrive at the pleasant identity

$$\sum_{i=0}^{k-1} \frac{1}{n-i} = \alpha^2 k \binom{n}{k} \int_0^\infty y \left(1 - e^{-\alpha y}\right)^{k-1} e^{-\alpha y(n-k+1)} dy, \qquad (6.37)$$

for $1 \le k \le n$ and *any* $\alpha > 0$ (via a simple transformation, α can be removed from the integral). Likewise, for the second moment (and setting $\alpha = 1$ without loss of generality),

$$\mathbb{E}\left[Y_k^2\right] = k \binom{n}{k} \int_0^\infty y^2 \left(1 - e^{-y}\right)^{k-1} e^{-y(n-k)} e^{-y} dy$$

and, as $\mathbb{E}\left[D_i^2\right] = 2/\left(\alpha^2\,(n-i)^2\right)$, this integral is the same as

$$\mathbb{E}\left[Y_k^2\right] = \mathbb{E}\left[\left(\sum_{i=0}^{k-1} D_i\right)^2\right]$$

$$= \sum_{i=0}^{k-1} \frac{2}{(n-i)^2} + 2 \sum_{i=0}^{k-2} \sum_{j=i+1}^{k-1} \frac{1}{(n-i)\,(n-j)}.$$

The range can be expressed in terms of the D_i as

$$R = Y_n - Y_1 = \sum_{i=0}^{n-1} D_i - \sum_{i=0}^{1-1} D_i = \sum_{i=1}^{n-1} D_i$$

and, as the D_i are independent with $D_j \sim \text{Exp}\,(\alpha\,(n-j))$,

$$\mathbb{M}_R(s) = \mathbb{E}\left[e^{s \sum_{i=1}^{n-1} D_i}\right] = \mathbb{E}\left[e^{sD_1}\right] \mathbb{E}\left[e^{sD_2}\right] \cdots \mathbb{E}\left[e^{sD_{n-1}}\right]$$

$$= \prod_{j=1}^{n-1} \frac{\alpha\,(n-j)}{\alpha\,(n-j) - s}, \qquad \text{for } s < \min_{1 \le j \le n-1} \alpha\,(n-j) = \alpha,$$

which agrees with (6.36). ∎

⊙ **Example 6.17** (Example 6.16 cont.) For the continuous random variable Z, recall that its pth *quantile* ξ_p is the value such that $F_Z\left(\xi_p\right) = p$ for a given $0 < p < 1$. If $Z \sim \text{Exp}(\alpha)$, then solving $p = F_Z\left(\xi_p\right) = 1 - e^{-\alpha \xi_p}$ gives the closed-form solution $\xi_p = -\alpha^{-1} \ln\,(1-p)$. Also, $\mathbb{E}\,[Z] = \alpha^{-1}$ and $\mathbb{V}\,(Z) = \alpha^{-2}$. Thus, from the results in

Example 6.16, for $j = 1, \ldots, n$,

$$\mathbb{E}\left[Y_j\right] = \sum_{i=0}^{j-1} \mathbb{E}\left[D_i\right] = \frac{1}{\alpha}\left[\frac{1}{n} + \frac{1}{(n-1)} + \cdots + \frac{1}{(n-j+1)}\right] \tag{6.38}$$

and, because of independence,

$$\mathbb{V}\left(Y_j\right) = \sum_{i=0}^{j-1} \mathbb{V}\left(D_i\right) = \frac{1}{\alpha^2}\left[\frac{1}{n^2} + \frac{1}{(n-1)^2} + \cdots + \frac{1}{(n-j+1)^2}\right]. \tag{6.39}$$

For $n \in \mathbb{N}$, we know from Example I.A.38 that $\gamma_n := 1 + 2^{-1} + 3^{-1} + \cdots + n^{-1} - \ln n$ converges to Euler's constant $\gamma \approx 0.5772$ as $n \to \infty$. Thus, for large n and j small relative to n,

$$\mathbb{E}\left[Y_j\right] = \alpha^{-1}\left[\sum_{i=1}^{n} i^{-1} - \sum_{i=1}^{n-j} i^{-1}\right] \approx \alpha^{-1}\left[\ln(n) - \ln(n-j)\right] = -\alpha^{-1}\ln\left(1 - \frac{j}{n}\right),$$

so that $\xi_{j/n} \approx \mathbb{E}\left[Y_j\right]$. In words, as n grows, for j such that j/n is small, the Y_j serve as unbiased estimators for the quantiles $\xi_{j/n}$. Figure 6.9 confirms this by plotting $\xi_{j/n}$ versus Y_j for simulated samples of n i.i.d. r.v.s from an Exp(1) distribution. For $n = 40$ (Figure 6.9(a)) the accuracy is not great even for small j, while for $n = 1000$ (Figure 6.9(b)) $\xi_{j/n}$ and Y_j are nearly equal for $1 \le j < n/3$.

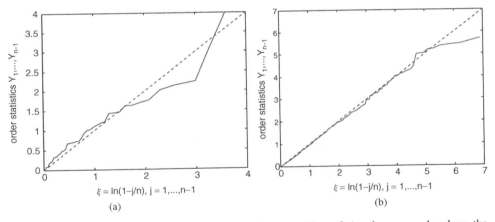

Figure 6.9 Plot of $\xi_p = -\ln(1-p)$ versus Y_j, for $p = j/n$ and $j = 1, \ldots, n-1$, where the Y_j are n order statistics from an Exp(1) distribution: (a) $n = 40$, (b) $n = 1000$

From (3.6) and the independence of the D_i, with $m = \min(i, j)$,

$$\mathrm{Cov}\left(Y_i, Y_j\right) = \sum_{p=0}^{i-1}\sum_{q=0}^{j-1}\mathrm{Cov}\left(D_p, D_q\right) = \sum_{h=0}^{m-1}\mathbb{V}\left(D_h\right) = \frac{1}{\alpha^2}\sum_{h=0}^{m-1}\frac{1}{(n-h)^2}. \tag{6.40}$$

In particular, $\mathbb{V}\left(Y_j\right) = \alpha^{-2}\sum_{h=0}^{j-1}(n-h)^{-2}$, which is small for n large and j small, also explains the behaviour of the Y_j in Figure 6.9. As $\sum_{k=1}^{\infty} k^{-2} = \pi^2/6$ (see (I.A.88) and Example 1.26), $\mathbb{V}(Y_n) \to \pi^2/(6\alpha^2)$ as $n \to \infty$. ∎

6.3 Distribution theory for dependent samples

Not surprisingly, matters become more complicated when abandoning the i.i.d. assumption. Let $\mathbf{X} = (X_1, \ldots, X_n)$ be a multivariate r.v. with c.d.f. $F_{\mathbf{X}}$, and denote the order statistics of $\mathbf{X}$ by $\mathbf{Y} = (Y_1, \ldots, Y_n)$. As before, $\Pr(Y_i \leq y) = \Pr$ (at least i of the Xs are $\leq y$). From the De Moivre–Jordan theorem (I.2.13),

$$\Pr(Y_i \leq y) = \sum_{j=i}^{n}(-1)^{j-i}\binom{j-1}{i-1}S_j(y), \tag{6.41}$$

where

$$S_j(y) = \sum_{k_1 < k_2 < \cdots < k_j} \Pr\left(X_{k_1} \leq y, \ X_{k_2} \leq y, \ \ldots, X_{k_j} \leq y\right)$$

sums over the $\binom{n}{j}$ unique combinations of j of the n elements in $\mathbf{X}$. If the X_i are i.i.d. with c.d.f. F_X, then $S_j(y) = \binom{n}{j}p_y^j$ for $p_y = F_X(y)$ and (6.41) reduces to (6.1), i.e.,

$$\Pr(Y_i \leq y) = \sum_{j=i}^{n}(-1)^{j-i}\binom{j-1}{i-1}\binom{n}{j}p_y^j = \sum_{j=i}^{n}\binom{n}{j}p_y^j(1-p_y)^{n-j},$$

thus giving rise to (yet another) nontrivial combinatoric identity for $0 < p_y < 1$. This was algebraically proven in Problem I.1.7.

⊖ ***Example 6.18*** (Zielinski, 1999) For vector $\mathbf{X} = (X_1, \ldots, X_n)$ with continuous joint p.d.f. and order statistics $\mathbf{Y}$, if c is the median of each X_i, i.e., $\Pr(X_i \leq c) = 1/2$, $i = 1, \ldots, n$, and, for every $m = 1, \ldots, n$, every set of indices $i_1, \ldots, i_m$ with $1 \leq i_1 < i_2 < \cdots < i_m \leq n$, and every $(x_1, \ldots, x_{m-1}) \in \mathbb{R}^{m-1}$,

$$\Pr\left(X_{i_m} \leq c \mid X_{i_1} = x_1, \ldots, X_{i_{m-1}} = x_{m-1}\right) = 1/2, \tag{6.42}$$

then $\Pr(\text{med}(\mathbf{X}) \leq c) = 1/2$, where $\text{med}(\mathbf{X})$ is the sample median, i.e., $\text{med}(\mathbf{X}) = Y_k$ for $n = 2k - 1$ and $\text{med}(\mathbf{X}) = VY_k + (1-V)Y_{k+1}$, where $V \sim \text{Ber}(1/2)$ independent of $\mathbf{X}$. The result clearly holds in the important special case when the X_i are i.i.d. with median c, because, from (6.7) and (I.1.49),

$$F_{Y_k}(c) = \frac{(2k-1)!}{(k-1)!\,(k-1)!}\int_0^{1/2} t^{k-1}(1-t)^{k-1}\,dt = \frac{1}{2}.$$

To prove the general case, first use (I.8.15) and (6.42) recursively to obtain

$$\Pr\left(X_{i_1} \le c, \, X_{i_2} \le c, \ldots, X_{i_m} \le c\right)$$

$$= \int_{-\infty}^{r} \cdots \int_{-\infty}^{r} \Pr\left(X_{i_m} \le c \mid X_{i_1} = x_1, \ldots, X_{i_{m-1}} = x_{m-1}\right)$$

$$\times f_{X_{i_1}, \ldots, X_{i_{m-1}}}(x_1, \ldots, x_{m-1}) \, dx_1 \cdots dx_{m-1}$$

$$= \frac{1}{2} \int_{-\infty}^{r} \cdots \int_{-\infty}^{r} f_{X_{i_1}, \ldots, X_{i_{m-1}}}(x_1, \ldots, x_{m-1}) \, dx_1 \cdots dx_{m-1}$$

$$= \frac{1}{2} \Pr\left(X_{i_1} \le c, \, \ldots, X_{i_{m-1}} \le c\right)$$

$$= \vdots$$

$$= 2^{-m}.$$

Now assume $n = 2k - 1$, so that, from (6.41) with $S_m = \binom{n}{m} 2^{-m}$ and $j = m - k$,

$$\Pr\left(\text{med}\,(\mathbf{X}) \le c\right) = \Pr\left(Y_k \le c\right) = \sum_{m=k}^{n} (-1)^{m-k} \binom{m-1}{k-1} \binom{n}{m} 2^{-m} \qquad (6.43)$$

$$= \frac{n!}{[(k-1)!]^2} \sum_{j=0}^{k-1} (-1)^j \binom{k-1}{j} \frac{1}{j+k} 2^{-(j+k)}$$

$$= \frac{n!}{[(k-1)!]^2} \sum_{j=0}^{k-1} (-1)^j \binom{k-1}{j} \int_0^{1/2} t^{j+k-1} \, dt$$

$$= \frac{n!}{[(k-1)!]^2} \int_0^{1/2} t^{k-1} \sum_{j=0}^{k-1} \binom{k-1}{j} (-t)^j \, dt$$

$$= \frac{n!}{[(k-1)!]^2} \int_0^{1/2} t^{k-1} (1-t)^{k-1} \, dt = \frac{1}{2}.$$

For $n = 2k$, let $Z = \text{med}\,(\mathbf{X}) = V Y_k + (1 - V) Y_{k+1}$, so that

$$\Pr\left(Z \le c\right) = \Pr\left(Z \le c \mid V = 0\right) \Pr\left(V = 0\right) + \Pr\left(Z \le c \mid V = 1\right) \Pr\left(V = 1\right)$$

$$= \frac{1}{2} \Pr\left(Y_{k+1} \le c\right) + \frac{1}{2} \Pr\left(Y_k \le c\right).$$

Similar calculations to (6.43) reveal that $\Pr\left(\text{med}\,(\mathbf{X}) \le c\right) = 1/2$. ∎

6.4 Problems

Human beings, who are almost unique in having the ability to learn from the experience of others, are also remarkable for their apparent disinclination to do so.   (Douglas Adams, *Last Chance to See*)

The greatest obstacle to discovery is not ignorance – it is the illusion of knowledge.
(Daniel J. Boorstin)

6.1. Let X_1, X_2, X_3 be i.i.d. r.v.s, each with continuous density f_X. Compute

(a) $p_0 = \Pr(X_1 > X_2)$,

(b) $p_1 = \Pr(X_1 > X_2 \mid X_1 > X_3)$,

(c) $p_2 = \Pr(X_1 > X_2 \mid X_1 < X_3)$.

6.2. Let $X_i \overset{\text{ind}}{\sim} \mathrm{Exp}(\lambda_i)$, $i = 1, \ldots, n$, and let $S = \min(X_i)$. Derive a simple expression for $\Pr(X_i = S)$. Hint: Use the conditional probability formula (I.8.40). The case with $n = 2$ was illustrated in Example I.8.17 (taking $r = 1$).

6.3. ★ Use repeated integration by parts to show that (6.1) is equivalent to (6.7).

6.4. Derive (6.18).

6.5. Let X be an r.v. with p.d.f. $f_X(x) = 2\theta^{-2} x \mathbb{I}_{(0,\theta)}(x)$, for $\theta > 0$.

(a) Derive $\mathbb{E}[X]$ and the c.d.f. of X.

(b) Let $X_1, \ldots, X_n$ be i.i.d. r.v.s each with p.d.f. f_X. Calculate f_M and $\mathbb{E}[M]$, where $M = \min X_i$.

(c) Repeat for $M = \max X_i$.

6.6. ★ Recall Example 6.11. The integral in (6.15) can be alternatively expressed as follows. Let $u = (1 - v) / (1 - p)$ (so that $v = 1 - u(1 - p)$ and $dv = -(1 - p)\, du$) and use the hypergeometric function (5.28) (see Section 5.3) to obtain

$$\int_p^1 v^{-n/2} (1 - v)^{n-2}\, dv = (1 - p)^{n-1} \int_0^1 u^{n-2} (1 - (1 - p)u)^{-n/2}\, du$$

$$= (1 - p)^{n-1} B(n - 1, 1)\, {}_2F_1(n - 1, n/2; n; 1 - p)$$

or[4]

$$f_P(p) = \frac{n}{2} p^{(n-2)/2} (1 - p)^{n-1}\, {}_2F_1(n - 1, n/2; n; 1 - p).$$

Following similar steps, derive the p.d.f. of P for the more general case in which $P = Y_i Y_j$, $1 \le i < j \le n$.

Also construct a program to calculate the exact density (using the built-in function quadl to do numerical integration) and one to simulate P. For $n = 8$, $i = 3$, and $j = 6$, the density and a (scaled) histogram of simulated values of P should look like those in Figure 6.10.

---

[4] Note the minor misprint in Johnson, Kotz and Balakrishnan (1994, p. 281, eq. 26.13b).

Figure 6.10 Density for $n = 8$, $i = 3$, and $j = 6$, along with a (scaled) histogram of simulated values of P

6.7. Assume X has density $f_X(x) = (1 + x)^{-2} \mathbb{I}_{(0,\infty)}(x)$. Define $Y = \min(X, 1)$ and $Z = \max(X, 1)$.

   (a) What is $F_Y(y)$, the c.d.f. of Y?

   (b) What is $\mathbb{E}[Y]$?

   (c) What is $\mathbb{E}[Z]$?

   (d) What is $f_G(g)$, where $G = \ln(1 + X)$?

6.8. Let $X_i \overset{\text{i.i.d.}}{\sim} \text{Unif}(0, 1)$ with corresponding order statistics $X_{(i)}$, $i = 1, 2, 3$. Calculate

   (a) $\Pr\left(X_{(1)} + X_{(2)} > 1\right)$,

   (b) $\Pr\left(X_{(2)} + X_{(3)} > 1\right)$,

   (c) $\mathbb{E}\left[X_{(1)}\right]$ and $\mathbb{V}\left(X_{(1)}\right)$.

6.9. ★ ★ Let X and Y be independent random variables and set $V = \min(X, Y)$ and $W = \max(X, Y)$. (Contributed by Walther Paravicini)

   (a) Derive the joint c.d.f. of (V, W).

   (b) If X and Y have a p.d.f., calculate the joint p.d.f. of (V, W).

   (c) Give explicit expressions for the joint c.d.f. and p.d.f. of (V, W) when $X \sim \exp(\alpha)$ and $Y \sim \exp(\beta)$.

6.10. ★ ★ Assume $X_1, \ldots, X_n \overset{\text{i.i.d.}}{\sim} \text{Unif}(\alpha - \beta, \alpha + \beta)$ with order statistics $X_{1:n} \leq \cdots \leq X_{n:n}$ and n is odd, say $n = 2k + 1$, $k \in \mathbb{N}^+$.

(a) From the X_i, construct random variables U_i and $U_{i:n}$, $i = 1, \ldots, n$ such that $U_i \overset{\text{i.i.d.}}{\sim} \text{Unif}(0, 1)$.

(b) Compute the mean and variance of $\hat{\alpha}_1 := \overline{X}_n$.

(c) Find the distribution of $U_{(k+1):(2k+1)}$.

(d) Compute the mean and variance of $\hat{\alpha}_2 := X_{(k+1):(2k+1)}$.

(e) Show that $\text{Cov}(U_{1:n}, U_{n:n}) = (n+1)^{-2}(n+2)^{-1}$.

(f) Compute the mean and variance of $T = \hat{\alpha}_3 := \frac{1}{2}(X_{1:n} + X_{n:n})$.

(g) The three sample statistics $\hat{\alpha}_i$ provide an estimate of the population value of α. Compare their variances for sample sizes $n = 1, 2, \ldots$.

6.11. ★ As a generalization of Example 6.15, derive the density of $D_j = Y_{j+1} - Y_j$ for any j, $j = 1, \ldots, n-1$.

6.12. ★ Again recalling Example 6.15, derive both the density of D_1 and that of R using the difference formula (2.8).

6.13. ★ ★ Let $X_1, X_2 \overset{\text{i.i.d.}}{\sim} \text{Beta}(2, 2)$, i.e., $f_X(x) = 6x(1-x)\mathbb{I}_{(0,1)}(x)$, and define $S = \min(X_1, X_2)$, $L = \max(X_1, X_2)$ and $R = S/L$.

(a) Derive the p.d.f.s of S and L and their means and variances.

(b) Explain (informally) why $\mathbb{V}(S) = \mathbb{V}(L)$.

(c) Show that $f_R(r) = \frac{6}{5}r(3 - 2r)\mathbb{I}_{(0,1)}(r)$ and compute $\mathbb{E}[R]$ and $\mathbb{V}(R)$.

(d) By simulating draws from f_R, compare the resulting kernel density to the algebraic one and examine the empirical mean and variance.

(e) Calculate $c = \text{Cov}(S, L)$ and $\text{Corr}(S, L)$.

(f) Obtain estimates of c by equating the Taylor series approximations of μ_R and σ_R^2 to their respective true values.

6.14. ★ ★ Let $X_1, \ldots, X_n \overset{\text{i.i.d.}}{\sim} f_X(x) = a\theta^{-a}x^{a-1}\mathbb{I}_{(0,\theta)}(x)$ and let Y_i be the ith order statistic.

(a) Derive the joint density $f_{\mathbf{Z}}(\mathbf{z})$ of the Z_j, $j = 1, \ldots, n$, where

$$Z_j = \begin{cases} Y_j/Y_{j+1}, & \text{if} \quad 1 \le j < n, \\ Y_j, & \text{if} \quad j = n. \end{cases}$$

(b) State the density of each Z_j and verify that they are proper.

(c) Compute the expected value of Z_j.

(d) Verify that $(Y_1/Y_n, \ldots, Y_{n-1}/Y_n)$ is independent of Y_n.

6.15. ★ Derive a closed-form expression for $\mathbb{E}[\ln X]$, where $X \sim \text{Beta}(a, b)$ and $b \geq 1$ is an integer. Confirm your result for the special case $a = b = 1$.

6.16. ★ Consider a circle with perimeter one, i.e., radius $1/2\pi$. Define the distance between two points on the circle as being the length of the shorter arc connecting the points.[5]

    (a) Two points are randomly and independently chosen on the circle. Calculate the probability that the two points are *not* within a distance of d from each other, $0 < d < 1/2$. What if, instead of a circle, the two points are randomly and independently placed on a straight line of unit length?

    (b) Similarly, n points, $n \geq 2$, are randomly and independently chosen on the circle. Calculate the probability that no two points are within a distance of d from each other, $0 < d < 1/n$. (Take $n = 3$ to begin with.) As before, what if a straight line is used?

---

[5] This is similar to Example 6a of Ross (1988, p. 225), which considers points on a straight line instead of a circle.

PART III

MORE FLEXIBLE AND ADVANCED RANDOM VARIABLES

7

Generalizing and mixing

Normality is a myth; there never was, and never will be, a normal distribution.

(Geary, 1947)

In this and the subsequent chapters, we examine a selection of univariate distributions which extend, generalize, and/or nest some of the ones previously introduced. Those 'more basic' distributions encountered earlier are still of fundamental importance and arise ubiquitously in applications, but there is also a large number of settings in which more general structures are required.

This chapter concentrates on some basic ideas concerning how distributions can be generalized. Section 7.1 illustrates some rudimentary methods, Section 7.2 considers r.v.s formed by taking a weighted sum of (usually simple and independent) r.v.s, and Section 7.3 is similar, but lets the *density* of an r.v. be a weighted sum of (usually simple) densities.

7.1 Basic methods of extension

Any Riemann integrable function f on the real line which is positive over a specific support $\mathcal{S}$ and such that $\int_{\mathcal{S}} f(x)\,\mathrm{d}x = 1$ can be interpreted as a p.d.f., and similarly for a mass function. It is, however, more appealing (and more practical) if the mass or density function f possesses a tractable algebraic representation over the whole support. This is the case for most distributions encountered, though notable exceptions include the so-called class of Tukey lambda distributions, discussed in Section 7.1.4, and the class of stable Paretian distributions, discussed in Chapter 8.[1] In order to understand

---

[1] For an overview of the hundreds of known distributions, see, for example, Johnson, Kotz and Kemp (1993), Johnson, Kotz and Balakrishnan (1994, 1995) and Stuart and Ord (1994). Kleiber and Kotz (2003) provide a very nice survey of selected distributions appropriate for certain applications in economics and actuarial science. Some of the techniques discussed in this chapter are also applicable in the multivariate setting, and the interested reader is encouraged to begin with Kotz and Nadarajah (2004) and Genton (2004) and the references therein.

Intermediate Probability: A Computational Approach   M. Paolella
© 2007 John Wiley & Sons, Ltd

how some of these distributions were derived, we present some basic methods for generalizing common distributions, such as those presented in Chapter I.7.

7.1.1   Nesting and generalizing constants

Two or more families of distributions can often be *nested* by constructing a probability density or mass function with one or more parameters such that, when these assume specific values, the p.d.f. or p.m.f. reduces to one of the special nested cases.

⊖   ***Example 7.1***   Consider the binomial and hypergeometric distributions. Their corresponding sampling schemes are quite disparate, and their p.m.f.s accordingly so. However, they can both be nested under the sampling scheme whereby, from an urn with w white and b black balls, a ball is drawn, and then replaced, *along with s balls of the same colour*. This is repeated n times. Let X be the number of black balls drawn. Taking $s = 0$ yields the binomial, while $s = -1$ gives the hypergeometric. The p.m.f. of the general case is quite straightforward to derive, and can be expressed compactly as

$$f_X(k) = \binom{n}{k} \frac{\left(\frac{b}{s}\right)^{[k]} \left(\frac{w}{s}\right)^{[n-k]}}{\left(\frac{b+w}{s}\right)^{[n]}} \mathbb{I}_{\{0,1,\dots,n\}}(k) = \binom{n}{k} \frac{B\left(\frac{b}{s}+k, \frac{w}{s}+n-k\right)}{B\left(\frac{b}{s}, \frac{w}{s}\right)} \mathbb{I}_{\{0,1,\dots,n\}}(k) ;$$

this is referred to as the Pólya–Eggenberger distribution. To confirm algebraically that this p.m.f. sums to one, use (I.9.7); see Stuart and Ord (1994, Section 5.13) or Johnson and Kotz (1977, Section 4.2), and the references therein for further detail.   ■

⊙   ***Example 7.2***   Recall that the location-zero, scale-one kernels of the normal and Laplace distributions are given by $\exp\left\{-x^2/2\right\}$ and $\exp\left\{-|x|\right\}/2$, respectively. By replacing the fixed exponent values (2 and 1) with power p, $p \in \mathbb{R}_{>0}$, the two become nested. The resulting distribution is known as the *generalized exponential distribution* (GED), with density

$$f_X(x; p) = \frac{p}{2\Gamma(p^{-1})} \exp\left\{-|x|^p\right\}, \quad p > 0 \tag{7.1}$$

(recall also Problem I.7.16). The choice of name is unfortunate, but standard; it refers to the generalization of the exponent, *not* the exponential distribution! It dates back at least to Subbotin (1923) and work by M.S. Bartlett (see Mudholkar, Freimer and Hutson, 1997) and was popularized within a Bayesian context in the classic 1973 book on Bayesian inference by George Box and George Tiao, where it is referred to as the exponential power distribution (see Box and Tiao, 1992, p. 157, and the references therein).

The GED is commonly used in applications where model residuals have excess kurtosis relative to the normal distribution (or fat tails). Another distribution which nests the Laplace and normal is the hyperbolic, discussed in Sections I.7.2 and 9.5. ■

⊖ **Example 7.3** The *generalized gamma* distribution has p.d.f.

$$f_{GGam}(x; \alpha, \beta, \gamma, c) = \frac{c(x-\gamma)^{c\alpha-1}}{\beta^{c\alpha}\Gamma(\alpha)} \exp\left[-\left(\frac{x-\gamma}{\beta}\right)^c\right] \mathbb{I}_{[\gamma,\infty)}(x), \qquad (7.2)$$

for $\alpha, \beta, c \in \mathbb{R}_{>0}$, and dates back at least to 1925. Special cases include the gamma ($c = 1$) and Weibull ($\alpha = 1$). The reader should quickly verify that $Y = [(X-\gamma)/\beta]^c$ $\sim$ Gam(α, 1). See Johnson, Kotz and Balakrishnan (1994, p. 388) for details on the history, applications, and estimation issues. ∎

The next two examples show how a constant numeric value (usually an integer) in the density can be replaced by a parameter which takes values in $\mathbb{R}$ or $\mathbb{R}_{>0}$.

⊖ **Example 7.4** Recall that the Student's t density with n degrees of freedom is given by $f_t(x; n) \propto (1 + x^2/n)^{-(n+1)/2}$. The exponent of 2 can be replaced by a parameter, say d; this gives rise to the *generalized Student's t* (GT) distribution, given by

$$f_{GT}(z; d, v) = K_{d,v}\left(1 + \frac{|z|^d}{v}\right)^{-(v+1/d)}, \qquad d, v \in \mathbb{R}_{>0}, \qquad (7.3)$$

with $K_{d,v}^{-1} = 2d^{-1}v^{1/d}B(d^{-1}, v)$, which the reader is invited to verify. This appears to have first been proposed by McDonald and Newey (1988). When a scale parameter is introduced, the Student's t becomes a special case, or is nested by, the GT (take $d = 2$ and $v = n/2$). Also, just as the Student's t p.d.f. converges pointwise to the normal as $n \to \infty$, $f_{GT} \to f_{GED}$ as $v \to \infty$. See Butler *et al.* (1990) and Bollerslev, Engle and Nelson (1994) for discussion of applications. ∎

⊖ **Example 7.5** Recall the type II Pareto distribution, with p.d.f. (I.7.29),

$$f_{Par\,II}(x; b, c) = \frac{b}{c}\left(\frac{c}{c+x}\right)^{b+1} \mathbb{I}_{(0,\infty)}, \qquad b, c \in \mathbb{R}_{>0},$$

which can be written as

$$f_{Par\,II}(x; b, c) = \frac{c^b}{1/b}\frac{x^0}{(c+x)^{1+b}}\mathbb{I}_{(0,\infty)} \quad \text{or} \quad f_{Par\,II}(x; b, c) \propto \frac{x^0}{(c+x)^{1+b}}\mathbb{I}_{(0,\infty)}. \qquad (7.4)$$

An x^0 term was introduced in the numerator of (7.4), which can be generalized to x to a positive power. In particular, the r.v. X is said to follow a *generalized (type II) Pareto distribution*, denoted $X \sim GPar\,II(a, b, c)$, if its density takes the form

$$f_X(x; a, b, c) \propto \frac{x^{a-1}}{(c+x)^{a+b}}\mathbb{I}_{(0,\infty)}, \qquad (7.5)$$

where a and b are shape parameters and c is the scale parameter. To derive the constant of integration required in (7.5), let $u = x/(c + x)$, which the reader can easily verify leads to

$$f_X(x; a, b, c) = \frac{c^b}{B(a, b)} \frac{x^{a-1}}{(c + x)^{a+b}} \mathbb{I}_{(0,\infty)}. \tag{7.6}$$

If $a = 1$, then $B(1, b) = b^{-1}$ and (7.6) reduces to (I.7.29). Using the same substitution, it is easy to see that

$$\mathbb{E}[X^k] = \frac{c^k B(a + k, b - k)}{B(a, b)}, \quad -a < k < b,$$

which implies, after a little simplification, that

$$\mathbb{E}[X] = c \frac{a}{b - 1} \quad (b > 1), \qquad \mathbb{V}(X) = c^2 a \frac{b + a - 1}{(b - 1)^2 (b - 2)} \quad (b > 2),$$

which reduce to the moments given in (I.7.32) for $a = 1$ and $c = 1$.

Remark: The extension to (7.5) actually involved a change of the functional form of density (I.7.29), and so is not as immediate as the Student's t extension above, but Example 7.18 below will demonstrate how the generalized Pareto arises in a natural way. In fact, one could have taken a different functional form which nests (I.7.29); see Example 7.6 below. ∎

The previous examples worked directly with the p.m.f. or p.d.f. to arrive at more general distributions, but use of any one-to-one function of a density can be used, most notably the c.d.f. or the characteristic function (c.f.). Both of these, however, are subject to constraints (see Section I.4.1.1 for the c.d.f., and Section 1.2.3 for the c.f.). Use of these is demonstrated in the next few examples.

⊙ ***Example 7.6*** As introduced in Section I.7.1, the type II Pareto survivor function is given by $\overline{F}(x; b) = 1 - F(x; b) = \mathbb{I}_{(-\infty,0]}(x) + (1 + x)^{-b} \mathbb{I}_{(0,\infty)}(x)$. This can be generalized to

$$\overline{F}(x; a, b, c) = \mathbb{I}_{(-\infty,0]}(x) + \frac{c^b e^{-ax}}{(c + x)^b} \mathbb{I}_{(0,\infty)}(x), \tag{7.7}$$

for $(a, b) \in \mathbb{R}_{\geq 0}^2 \setminus (0, 0)$, $c \in \mathbb{R}_{>0}$, which nests the exponential and the type II Pareto, and is referred to as the *type III Pareto distribution*; see Johnson, Kotz and Balakrishnan (1994, p. 575).[2] Observe that the tail behaviour for a and b nonzero is neither exponential nor power; one could describe the distribution as having 'semi-heavy tails'.

Figure 7.1 plots the tail probabilities (survivor function) in several ways to indicate the interesting behaviour of this c.d.f. ∎

---

[2] Johnson, Kotz and Balakrishnan (1994, p. 575) have c in the numerator of (7.7) instead of c^b, which would imply that $\lim_{x \to 0} \overline{F}(x; a, b, c) \neq 1$, so presumably the use of c^b is correct.

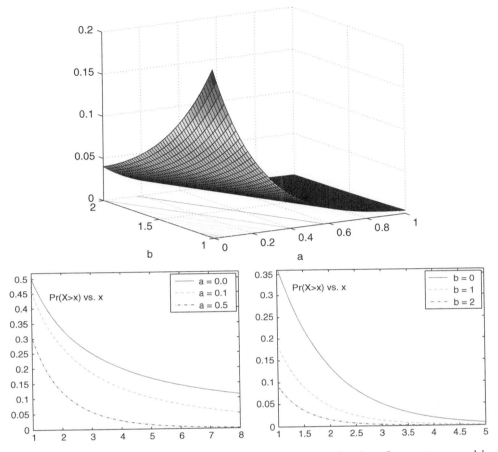

Figure 7.1   (Top) $\Pr(X > 4)$, as given by (7.7), with $c = 1$, as a function of parameters a and b. (Bottom) $\Pr(X > x)$ versus x for $b = 1$ (left) and $a = 1$ (right)

◎   ***Example 7.7***   The c.d.f. of the exponential distribution, $1 - e^{-\lambda x}$, is naturally generalized to

$$F(x; \beta, x_0, \sigma) = 1 - \exp\left\{ -\left(\frac{x - x_0}{\sigma} \right)^{\beta} \right\} \mathbb{I}_{(x_0, \infty)}(x),$$

which is a proper c.d.f. for $\beta > 0$. It is the c.d.f. of a Weibull random variable, where x_0 and $\sigma > 0$ are location and scale parameters respectively, and $\beta > 0$ is the 'power' parameter. This easily leads to the p.d.f. as given in (I.7.19), i.e.,

$$f_{\text{Weib}}(x; \beta, x_0, \sigma) = \frac{\beta}{\sigma} \left(\frac{x - x_0}{\sigma} \right)^{\beta - 1} \exp\left\{ -\left(\frac{x - x_0}{\sigma} \right)^{\beta} \right\} \mathbb{I}_{(x_0, \infty)}(x),$$

which itself is not an 'immediate' generalization of the exponential p.d.f. obtained just by introducing an exponent.   ∎

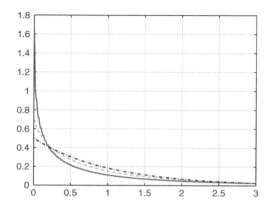

Figure 7.2 The density on the half line corresponding to $\varphi_Y(t) = (1 + |t|^\alpha)^{-1}$, $0 < \alpha \leq 2$, for $\alpha = 1$ (solid), $\alpha = 1.5$ (dashed) and $\alpha = 2$ (dash-dotted). The latter corresponds to the Laplace

⊖ **Example 7.8** The c.f. of a standard Laplace random variable X is

$$\varphi_X(t) = \mathbb{M}_X(it) = \frac{1}{1 + t^2},$$

from (5.12). A natural generalization is to relax the quadratic exponent, giving $\varphi_Y(t) = (1 + |t|^\alpha)^{-1}$. This is a valid c.f. for $0 < \alpha \leq 2$ (shown by Y. Linnik), and the corresponding p.d.f. is unimodal (shown by R. Laha); see Lukacs (1970, p. 97) for details and references. The p.d.f. must also be symmetric, because the c.f. is real (see Section 1.2.3).

Use of the p.d.f. inversion formula and a program similar to that in Listing 1.3 allows the density to be computed. Figure 7.2 shows (half of) the p.d.f. for several values of α. As α decreases from 2, the centre becomes more peaked and the tails become heavier. ∎

⊛ **Example 7.9** Recall Example 1.21, which inverted the characteristic function (c.f.) $\varphi_Z(t) = e^{-t^2/2}$ to show that $Z \sim N(0, 1)$, and Example 1.22, which inverted the c.f. $\varphi_X(t) = e^{-c|t|}$, for $c > 0$, to show that $X \sim \text{Cau}(c)$. The two c.f.s are easily nested as (omitting the scale parameter) $\varphi(t; \alpha) = \exp\{-|t|^\alpha\}$. It turns out that this is a valid c.f. for $0 < \alpha \leq 2$, and is the c.f. of a random variable which is very important both in theory and in practice, and which we will further discuss in Chapter 8. ∎

7.1.2 Asymmetric extensions

Many stochastic phenomena exhibit an asymmetric probability structure, i.e., their densities are skewed. A variety of ways of introducing asymmetry into an existing density exist, several of which are now discussed. Note that some methods are more general than others.

1. O'Hagan and Leonard (1976) and, independently and in more detail, Azzalini (1985) consider the asymmetric generalization of the normal distribution given by

$$f_{SN}(z; \lambda) = 2\phi(z)\Phi(\lambda z), \qquad \lambda \in \mathbb{R}, \tag{7.8}$$

where, adopting the notation in Azzalini (1985), SN stands for 'skew normal', and ϕ and Φ are the standard normal p.d.f. and c.d.f., respectively. Clearly, for $\lambda = 0$, f_{SN} reduces to the standard normal density, while for $\lambda \neq 0$, it is skewed. To see that (7.8) is a proper density function, we first prove a simple and intuitive result. Let X and Y be independent, continuous r.v.s, both with p.d.f.s symmetric about zero. From (I.8.41),

$$P := \Pr(X < Y) = \int_{-\infty}^{\infty} F_X(y) f_Y(y) \, dy,$$

and setting $z = -y$ and using the facts from symmetry that, for any $z \in \mathbb{R}$, $F_X(-z) = 1 - F_X(z)$ and $f_Y(y) = f_Y(-y)$,

$$P = -\int_{\infty}^{-\infty} F_X(-z) f_Y(-z) \, dz = \int_{-\infty}^{\infty} (1 - F_X(z)) f_Y(z) \, dz = 1 - P,$$

so that $P = 1/2$.

Based on this result, as in Azzalini (1985, Lemma 1), and noting that the density of λY is symmetric about zero, (I.8.41) and the previous result imply

$$\frac{1}{2} = \Pr(X < \lambda Y) = \int_{-\infty}^{\infty} F_X(\lambda y) f_Y(y) \, dy,$$

so that, from the symmetry of the standard normal distribution, the integral over $\mathbb{R}$ of $f_{SN}(z; \lambda)$ is unity.

One of the appealing properties of the SN distribution which separates it from alternative, more *ad hoc* methods of inducing skewness into a symmetric density is that, if $X \sim SN(\lambda)$, then $X^2 \sim \chi_1^2$. This is valuable for statistical inference, and is proven in Azzalini (1985).

Remark: Further aspects of the SN distribution, including more detailed derivation of the moments, methods of simulation, and the relation to truncated normal, are investigated by Henze (1986), while Azzalini (1986) proposes a skewed extension of the GED class of distributions. An extension to the multivariate normal setting, with emphasis on the bivariate case, is detailed in Azzalini and Dalla Valle (1996), and also discussed in the review article of Dalla Valle (2004). Azzalini and Capitanio (2003) develop a method of extension which gives rise to a multivariate skewed Student's t distribution. ■

2. Fernández and Steel (1998) popularized the simple method of introducing asymmetry into a symmetric, continuous density by taking

$$f(z; \theta) = \frac{2}{\theta + 1/\theta} \left\{ f\left(\frac{z}{\theta}\right) \mathbb{I}_{[0,\infty)}(z) + f(z\theta) \mathbb{I}_{(-\infty,0)}(z) \right\}, \tag{7.9}$$

where $f(z) = f(|z|)$ and $\theta > 0$. For $\theta = 1$, the density is symmetric, while for $0 < \theta < 1$, the distribution is skewed to the left and otherwise to the right. Examples include application to the Student's t distribution, resulting in, say, the Fernández–Steel4 t distribution $f_{\text{FS}-t}(z; \nu, \theta)$, given by (7.9) with $f(z; \nu)$ from (I.7.45). This was applied in Fernández and Steel (1998), and even earlier in Hansen (1994, p. 710). Another popular case is application to the GED (7.1), resulting in the p.d.f.

$$f_{\text{FS–GED}}(z; d, \theta) = K_{d,\theta} \begin{cases} \exp\left(-(-\theta z)^d\right), & \text{if } z < 0, \\ \exp\left(-(z/\theta)^d\right), & \text{if } z \geq 0, \end{cases} \tag{7.10}$$

$\theta, d \in \mathbb{R}_{>0}$, and where $K_{d,\theta}^{-1} = \left(\theta + \theta^{-1}\right) d^{-1} \Gamma\left(d^{-1}\right)$. This distribution is applied in Fernández, Osiewalski and Steel (1995) and similar constructions are examined in great detail in Ayebo and Kozubowski (2003) and Komunjer (2006).

Problem 7.7 discusses properties of a distribution which nests those given in (7.3) and (7.10) and has been found to be particularly useful in certain financial applications.

3. From the so-called generalized exponential family of distributions, Lye and Martin (1993) derive a variety of flexible distributions, one class of which is given by

$$f_{\text{LyM}'}(z; \gamma, \boldsymbol{\theta}) = K \exp\left(\theta_1 \arctan\left(\frac{z}{\gamma}\right) + \theta_2 \log\left(\gamma^2 + z^2\right) + \sum_{i=3}^{M} \theta_i z^{i-2}\right), \tag{7.11}$$

$\boldsymbol{\theta} = (\theta_1, \ldots, \theta_M)$, with integrating constant K not available in closed form. This nests the Student's t, for which $\theta_1 = 0$, $\theta_2 = -\left(1 + \gamma^2\right)/2$ and $\theta_i = 0$, $i > 2$. The highly parsimonious special case of (7.11) takes $\theta_i = 0$ for $i > 2$, i.e., with a slight change of notation,

$$f_{\text{LyM}}(z; \nu, \theta) = K \exp\left(\theta \arctan\left(\frac{z}{\sqrt{\nu}}\right) - \frac{\nu+1}{2} \log\left(1 + \frac{z^2}{\nu}\right)\right), \tag{7.12}$$

with $\theta \in \mathbb{R}$, $\nu \in \mathbb{R}_{>0}$ and the constant K obtained through numerical integration. For $\theta = 0$, density (7.12) coincides with that of a Student's t with $\nu > 0$ degrees of freedom, and is otherwise skewed to the left (right) for negative (positive) θ. As a simple extension of (7.12), consider

$$f_{\text{LyMd}}(z; d, \nu, \theta) = K \exp\left(\theta \arctan\left(\frac{z}{\nu^{1/d}}\right) - \left(\nu + \frac{1}{d}\right) \log\left(1 + \frac{|z|^d}{\nu}\right)\right), \tag{7.13}$$

$d \in \mathbb{R}_{>0}$, which is still asymmetric for $\theta \neq 0$, but now nests the GT density (7.3) as well. Figure 7.3 shows the density for a several parameter constellations.

4. Jones and Faddy (2003) suggest an asymmetric generalization of Student's t, with density

$$f_{\text{JoF}}(t; a, b) = C \left(1 + \frac{t}{\left(a+b+t^2\right)^{1/2}}\right)^{a+1/2} \left(1 - \frac{t}{\left(a+b+t^2\right)^{1/2}}\right)^{b+1/2}, \tag{7.14}$$

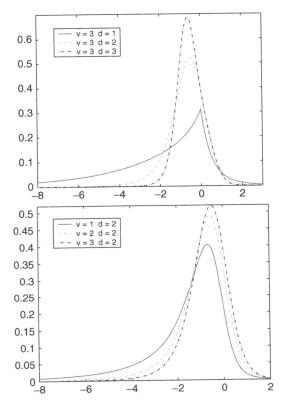

Figure 7.3 Density (7.13) with skewness parameter $\theta = -2$

where $a, b \in \mathbb{R}_{>0}$ and $C^{-1} = B(a, b)(a + b)^{1/2} 2^{a+b-1}$. If $a < b$ $(a > b)$, then S is negatively (positively) skewed, while $S \sim t(2a)$ if $a = b$. Further properties of this interesting density are explored in Problem 7.8.

5. Via continuous mixture distributions, the topic of Section 7.3.2 and Chapter 9, asymmetry can be incorporated into certain densities in a very natural way.

The generalized logistic (GL) distribution is another example of an asymmetric extension. It is discussed in Problem 7.6.

7.1.3 Extension to the real line

Let X be a continuous r.v. with p.d.f. f_X and support on $\mathbb{R}_{>0}$ and define Y to be an r.v. with p.d.f.

$$f_Y(y) = \frac{1}{2} f_X(|y|).$$

Then the distribution of Y is obtained by reflecting the p.d.f. of X onto $(-\infty, 0)$ and rescaling. The Laplace is probably the most prominent example, whereby $X \sim \text{Exp}(\lambda)$ with $f_X(x) = \lambda e^{-\lambda x}$ and $Y \sim \text{Lap}(0, \sigma)$ with $\sigma = 1/\lambda$ and $f_Y(y) = \exp\{-|y|/\sigma\}/(2\sigma)$.

⊖ **Example 7.10** The *double Weibull* distribution with location zero can also be derived from the usual Weibull distribution via this type of extension. If Y is an r.v. which follows the double Weibull distribution with shape $\beta > 0$ and scale $\sigma > 0$, we write $Y \sim \text{DWeib}(\beta, \sigma)$, where

$$f_{\text{DWeib}}(y; \beta, \sigma) = f_Y(y; \beta, \sigma) = \frac{\beta}{2\sigma}\left|\frac{y}{\sigma}\right|^{\beta-1}\exp\left(-\left|\frac{y}{\sigma}\right|^\beta\right). \tag{7.15}$$

When $\beta = 1$, Y reduces to a Laplace random variable. Dividing Y by σ removes the scale term, say $S = Y/\sigma \sim \text{DWeib}(\beta, 1)$. Its c.d.f. F_S is straightforwardly obtained; for $x \leq 0$, use $u = (-y)^\beta$ to get

$$F_S(x) = \int_{-\infty}^{x} \frac{\beta}{2}(-y)^{\beta-1}\exp\left(-(-y)^\beta\right) dy \tag{7.16}$$

$$= \frac{1}{2}\int_{(-x)^\beta}^{\infty} \exp(-u)\, du = \frac{1}{2}\exp\left(-(-x)^\beta\right), \quad x \leq 0, \tag{7.17}$$

and for $x \geq 0$, a similar calculation (or, better, use of the symmetry relation $F_S(x) = 1 - F_S(-x)$) gives $F_S(x) = 1 - \exp\left(-x^\beta\right)/2$.

From the symmetry of the density, the mean of Y is clearly zero, though a location parameter μ could also be introduced in the usual way. Estimation of the resulting three parameters and the distribution of the order statistics from an i.i.d. sample were investigated by Balakrishnan and Kocherlakota (1985).

For the variance, as σ is a scale parameter, $\mathbb{V}(Y) = \sigma^2\mathbb{V}(S) = \sigma^2\mathbb{E}[S^2]$. Similar to the derivation in (I.7.21) and using symmetry, it follows with $u = s^\beta$ that

$$\mathbb{E}[S^2] = \beta\int_0^\infty s^2 s^{\beta-1}\exp\left(-s^\beta\right) ds = \int_0^\infty u^{(1+2/\beta)-1}e^{-u}\, du = \Gamma\left(1 + \frac{2}{\beta}\right),$$

or $\mathbb{V}(Y) = \sigma^2\Gamma(1 + 2/\beta)$. As a check, when $\beta = 1$, this reduces to $2\sigma^2$, the variance of the Laplace distribution. ∎

⊖ **Example 7.11** (Example 7.10 cont.) One simple method of introducing asymmetry into (7.15) is the following. Define the *asymmetric double Weibull* density as

$$f_{\text{ADWeib}}\left(z; \beta^-, \beta^+, \sigma\right) = \begin{cases} f_{\text{DWeib}}(z; \beta^-, \sigma), & \text{if } z < 0, \\ f_{\text{DWeib}}(z; \beta^+, \sigma), & \text{if } z \geq 0, \end{cases}$$

where β^- and β^+ denote the shape parameters on the left ($z < 0$) and right support ($z \geq 0$), respectively. Clearly, if $\beta^- \neq \beta^+$, the density will be asymmetric. Letting $Z \sim$

ADWeib $(\beta^-, \beta^+, 1)$, the c.d.f. is straightforwardly seen to be $F_Z(x) = \exp(-(-x)^{\beta^-})$ /2 for $x \le 0$ and $F_Z(x) = 1 - \exp(-x^{\beta^+})/2$ for $x > 0$. Similarly,

$$\mathbb{E}[Z] = \frac{1}{2}\left[\Gamma\left(1 + \frac{1}{\beta^+}\right) - \Gamma\left(1 + \frac{1}{\beta^-}\right)\right]$$

and

$$\mathbb{E}[Z^2] = \frac{1}{2}\left[\Gamma\left(1 + \frac{2}{\beta^-}\right) + \Gamma\left(1 + \frac{2}{\beta^+}\right)\right],$$

from which the variance can be calculated. The expected shortfall (I.8.38) of Z is also easily derived: for $x < 0$,

$$\mathbb{E}[Z \mid Z \le x] = \frac{-1}{2F_Z(x)}\left[\Gamma(1 + 1/b) - \Gamma_{(-x)^b}(1 + 1/b)\right].$$

The ADWeib has been found useful for modelling asset returns, particularly extreme events; see Mittnik, Paolella and Rachev (1998). ∎

7.1.4 Transformations

Nonlinear transformations of r.v.s can give rise to more complicated (and often quite interesting) density functions. We consider two such cases.

The first is a particularly well-known one, and takes the form

$$X = \left(aU^\lambda - (1 - U)^\lambda\right)/\lambda \tag{7.18}$$

for $U \sim \text{Unif}(0, 1)$, where $a > 0$ and $\lambda \ne 0$. This results in the *Tukey lambda* class of distributions (Tukey, 1962).[3] In the limiting case of $\lambda \to 0$, $X = \ln(U^a/(1-U))$. For $a = 1$, the density is symmetric. If $\lambda \ne 0$, then

$$f_X(x; a, \lambda) = f_U(u)\left|\frac{du}{dx}\right| = \frac{\mathbb{I}_{(0,1)}(u)}{dx/du} = \frac{\mathbb{I}_{(0,1)}(u)}{au^{\lambda-1} + (1-u)^{\lambda-1}},$$

where $u = u(x)$ is given implicitly by the solution to $x = (au^\lambda - (1-u)^\lambda)/\lambda$. This will not have a closed-form solution in general. For $a = 1$ and $\lambda = 2$ it does, with $x = (u^2 - (1-u)^2)/2 = u - 1/2$, so that $f_X(x; \lambda = 2) = \mathbb{I}_{(-0.5, 0.5)}(x)$, i.e., a location-shifted uniform. Generalizations of (7.18) and other transformations of uniform deviates have been proposed; see Johnson, Kotz and Balakrishnan (1994, Ch. 12, Section 4.3).

To view the Tukey lambda density graphically, take a grid of values over the interval $(0, 1)$, say $u = 0.01, 0.02, \ldots, 0.99$, compute the corresponding vector of x-values,

---

[3] After John W. Tukey, 1915–2000. Articles honoring Tukey's accomplishments include Kafadar (2001) and Thompson (2001), as well as the set of memorial articles in Vol. 30 of *Annals of Statistics*, December 2002.

and plot x against $f_X(x)$. Except for the two cases $\lambda = 1$ and $(a = 1, \lambda = 2)$, the resulting grid of x-values will not be equally spaced. When using a tight enough grid of u-values, however, this will not be (visually) important. Matlab code for doing this is as follows:

```
a=1; lam=0.135; u=0.01:0.01:0.99;
xgrid=(a * u.^lam - (1-u).^lam)/lam;
f = 1./( a * u.^(lam-1) + (1-u).^(lam-1) ); plot(xgrid,f)
```

Clearly, a grid of (equally spaced) x-values over (a certain portion of) the support could be chosen and, for each point, $u(x)$ obtained numerically, from which $f_X(x; \lambda)$ can then be computed. This entails much more computation, though it might be desirable in certain contexts. Listing 7.1 gives a Matlab program to do this.

```
function u=tukeysolve(a,lambda,xgrid)
u=zeros(length(xgrid),1);
old=0.001; % just a guess
opts = optimset('Display','Off','tolf',1e-6);
for i=1:length(xgrid)
  x=xgrid(i);
  u(i) = fsolve(@tukeysolve_,old,opts,x,a,lambda);
  old=u(i); % use previous value as new start value
end

function f=tukeysolve_(u,x,a,lambda)
f = x - (a * u^lambda - (1-u)^lambda)/lambda;
```

Program Listing 7.1   Solves for a vector of u-values, given a vector of x-values, to be used with the Tukey lambda density

Figure 7.4 shows $f_X(x; a, \lambda)$ for three combinations of a and λ, exhibiting its ability to (i) approach normality,[4] (ii) exhibit excess kurtosis (fat tails), and (iii) be skewed.

One of the most important uses of the Tukey lambda and similar distributions arises in simulation. In particular, if details of the small-sample behaviour (i.e., distributional properties) of a statistical procedure or test statistic are desired for which analytic results are not available, one can simulate its performance using computer-generated r.v.s. As uniform transformed r.v.s such as Tukey lambda are simple to generate, they are a useful candidate. Furthermore, it is often the case that the theoretical properties of a particular statistic are known under, say, the assumption that the errors in the model are precisely normally distributed, but not otherwise (a good example being the classic t-statistic for normal means). The researcher can then choose parameter values (such as a and λ for the Tukey lambda distribution) which correspond to the properties of interest, e.g., approximately normally distributed but with 'fat tails' and negative skewness.

---

[4] In fact, values $a = 1$ and $\lambda = 0.135$ are those for which the density is closest to a (scaled) normal distribution; see Johnson, Kotz and Balakrishnan (1994, p. 41) for discussion and references.

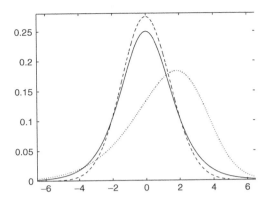

Figure 7.4 Tukey lambda density for $a = 1$ and $\lambda = 10^{-7}$ (solid), $a = 1$ and $\lambda = 0.135$ (dashed) and $a = 2$ and $\lambda = 0.1$ (dotted)

Our second example is the *inverse hyperbolic sine* (IHS) distribution, introduced by Johnson (1949), and also referred to as the S_U distribution. It is a flexible, asymmetric and leptokurtic distribution whose inverse c.d.f. is straightforwardly calculated, and has found recent use in empirical finance studies; see Brooks *et al.* (2005) and Choi, Nam and Arize (2007). The r.v. $Y \sim \text{IHS}(\lambda, \theta)$ if

$$\sinh^{-1}(Y) \sim N(\lambda, \theta^2), \quad -\infty < Y < \infty, \quad \theta > 0, \tag{7.19}$$

or, with $Z \sim N(0, 1)$, $Y = \sinh(\lambda + \theta Z)$. As $\sinh^{-1}$ is a nondecreasing function of its argument, the p.d.f. is straightforwardly obtained by transformation, giving

$$f_Y(y; \lambda, \theta) = \frac{1}{\sqrt{2\pi \left(y^2 + 1\right) \theta^2}} \exp\left\{ -\frac{\left(\sinh^{-1}(y) - \lambda\right)^2}{2\theta^2} \right\}. \tag{7.20}$$

From the relation between Z and Y, the c.d.f. is also easily derived, using the substitution

$$w = \frac{1}{\theta}\left(\sinh^{-1}(x) - \lambda\right) = \frac{1}{\theta}\left(\ln\left(x + \sqrt{1 + x^2}\right) - \lambda\right), \quad dw = \frac{dx}{\theta\sqrt{1 + x^2}},$$

to get

$$F_Y(y; \lambda, \theta) = \int_{-\infty}^{\frac{1}{\theta}\left(\ln\left(y + \sqrt{1 + y^2}\right) - \lambda\right)} f_Z(w)\, dw$$

$$= \Phi\left(\frac{1}{\theta}\left(\ln\left(y + \sqrt{1 + y^2}\right) - \lambda\right)\right) = \Phi\left(\frac{1}{\theta}\left(\sinh^{-1}(y) - \lambda\right)\right), \tag{7.21}$$

where Φ is the standard normal c.d.f., and the median of Y is $\sinh(\lambda)$. For the inverse c.d.f.,

$$\Pr\left(Y \leq y_q\right) = \Pr\left(\sinh(\lambda + \theta Z) \leq y_q\right) \Rightarrow F_Y^{-1}(q) = \sinh\left(\lambda + \theta \Phi^{-1}(q)\right).$$

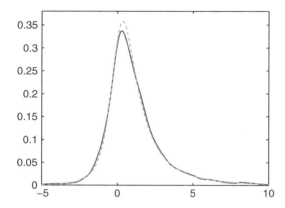

Figure 7.5   Kernel density (solid) based on 5000 simulated IHS r.v.s, for $\lambda = 2/3$ and $\theta = 1$, computed via relation (7.19) and using the Matlab `sinh` function, and the true density (7.20) (dashed)

The formulae for the mean, variance, and the third and fourth central moments were given in Johnson (1949) to be, with $\omega = \exp(\theta^2)$,

$$\mu = \omega^{1/2} \sinh(\lambda), \qquad \sigma^2 = \frac{1}{2}(\omega - 1)(\omega \cosh(2\lambda) + 1), \tag{7.22}$$

$$\mu_3 = \frac{1}{4}\omega^{1/2}(\omega - 1)^2 \{\omega(\omega + 2)\sinh(3\lambda) + 3\sinh(\lambda)\},$$

$$\mu_4 = \frac{1}{8}(\omega - 1)^2 \{\omega^2(\omega^4 + 2\omega^3 + 3\omega^2 - 3)\cosh(4\lambda)$$

$$+ 4\omega^2(\omega + 2)\cosh(2\lambda) + 3(2\omega + 1)\}.$$

Figure 7.5 shows the true and simulated density for $\lambda = 2/3$ and $\theta = 1$. To verify the c.d.f. (7.21) graphically, let Y be a vector of n simulated S_U r.v.s. The empirical c.d.f. is obtained by plotting the sorted Y values on the x-axis against the vector $(1, 2, \ldots, n)/n$. This can then be compared to an overlaid plot of (7.21).

7.1.5   Invention of flexible forms

Yet another possibility for creating density functions is 'simply' to suggest a mathematical function which possesses a desirable set of characteristics. A nice example of this is the so-called *generalized lambda distribution* (GLD), introduced by Ramberg and Schmeiser (1974). Let $X \sim \text{GLD}(\lambda_1, \lambda_2, \lambda_3, \lambda_4)$. The distribution of X is not defined by its p.d.f. f_X or its c.d.f. F_X, but rather by its quantile function

$$Q_X(p) = \lambda_1 + \frac{p^{\lambda_3} - (1 - p)^{\lambda_4}}{\lambda_2}, \qquad 0 < p < 1, \tag{7.23}$$

where this satisfies $F_X (Q_X (p)) = p$. Let $\lambda = (\lambda_1, \lambda_2, \lambda_3, \lambda_4)$. The support of X is given by the interval $\left(\lim_{p \to 0^+} f_X (Q_X (p)), \lim_{p \to 1^-} f_X (Q_X (p)) \right)$, and can have finite or infinite left or right tails, depending on λ. Regions of $\mathbb{R}^4$ for which λ gives rise to proper densities are somewhat complicated and are discussed in detail in the book-length treatment of the GLD by Karian and Dudewicz (2000).

Letting $x_p = Q_X (p)$, we have $F_X (x_p) = p$, so that $dp/dx_p = f_X (x_p)$. Thus, the density can be expressed as

$$f_X (x_p) = f_X (Q_X (p)) = \frac{dp}{dQ_X (p)} = \frac{1}{dQ_X (p) /dp} = \frac{\lambda_2}{\lambda_3 p^{\lambda_3 - 1} + \lambda_4 (1 - p)^{\lambda_4 - 1}}.$$

The value of the GLD, and the reason for its creation, is its flexibility: it can take on a large number of shapes typically associated with traditional density functions, such as uniform, beta, gamma, Weibull, normal, Student's t, F, and Pareto, and so provides the researcher with a single distribution capable of modelling data with a wide variety of possible characteristics. This is valuable because, for many statistical applications, it is not clear *a priori* what distribution (if any) is appropriate for the data at hand. Karian and Dudewicz (2000) demonstrate the GLD's ability to closely approximate the aforementioned densities (and others), as well as discussing applications to real data sets and methods of parameter estimation.

Furthermore, it is straightforward to generate GLD r.v.s, so that it can be used for simulation studies (as discussed above in Section 7.1.4). Similar to the discussion of the probability integral transform in Chapter I.7, let Q_X be the quantile function for $X \sim \text{GLD} (\lambda)$. If $U \sim \text{Unif} (0, 1)$, then $Q_X (U)$ is an r.v. such that

$$F_{Q_X(U)} (y) = \Pr (Q_X (U) \leq y)$$
$$= \Pr (F_X (Q_X (U)) \leq F_X (y)) = \Pr (U \leq F_X (y)) = F_X (y),$$

i.e., $Q_X (U)$ has the same c.d.f. as X. Thus, if $U_i \overset{\text{i.i.d.}}{\sim} \text{Unif} (0, 1)$ and $X_i = Q_X (U_i)$, then $X_i \overset{\text{i.i.d.}}{\sim} \text{GLD} (\lambda)$.

Remark: Keep in mind that the GLD and similar 'flexible forms' are parametric in nature, i.e., no matter how flexible they are, they are still functions of a finite number of parameters (e.g., four in the case of the GLD) and are limited in their variety of shapes and properties. As an important example, the GLD is unimodal, but there are situations for which bimodal (or multimodal) data exist. There do exist parametric forms which are extremely flexible and multimodal, such as the mixed normal (see Section 7.3 below), but gain this flexibility by having a potentially large number of parameters; this is a problem when they need to be estimated from a relatively small amount of data.

An alternative is to employ so-called *nonparametric methods*, which 'let the data speak for themselves' by not imposing a particular form. They too, however, require 'tuning parameters' to be specified, which embody certain assumptions about the data which might not be known. As in all situations involving uncertainty, there are no 'best' methods in all senses. ■

7.2   Weighted sums of independent random variables

Define the r.v. X as $X = \sum_{i=1}^{n} a_i X_i$, where the X_i are independent r.v.s and the a_i are known, real constants, $i = 1, \ldots, n$. Various well-known distributions are such that their r.v.s can be expressed as such a weighted sum, including the binomial, negative binomial, and gamma distributions, the properties of which should already be familiar. These cases just mentioned all require constraints on the parameters of the components, e.g., the binomial is a sum of independent Bernoulli r.v.s, each of which has the same probability parameter p. By relaxing this assumption, quite a variety of flexible and useful distributions can be obtained.

Note that, if $X_i \overset{\text{ind}}{\sim} N\left(\mu_i, \sigma_i^2\right)$, $i = 1, \ldots, n$, then X is also normally distributed with parameters $\mathbb{E}[X] = \sum_{i=1}^{n} a_i \mu_i$ and $\mathbb{V}(X) = \sum_{i=1}^{n} a_i^2 \sigma_i^2$. More generally, if $\mathbf{X} = (X_1, \ldots, X_n) \sim N(\boldsymbol{\mu}, \boldsymbol{\Sigma})$, then $X = \sum_{i=1}^{n} a_i X_i \sim N\left(\mu, \sigma^2\right)$, with expressions for μ and σ^2 given in fact 6 of Section 3.2. Such convenient results rarely hold.

⊖ **Example 7.12**   The exact p.m.f. and c.d.f. of the sum of two independent binomially distributed r.v.s with different values of p were developed using the discrete convolution formula in Section I.6.5. More generally, if $X_i \overset{\text{ind}}{\sim} \text{Bin}(n_i, p_i)$ and $X = \sum_{i=1}^{k} X_i$, then, with $q_i = 1 - p_i$, the m.g.f. of X is $\mathbb{M}_X(s) = \prod_{i=1}^{k} (p_i e^s + q_i)^{n_i}$, and the c.f. is $\varphi_X(t) = \mathbb{M}_X(it)$. The m.g.f. was used in Example 2.7 to compute the p.m.f.

Similarly, the sum of geometric (or negative binomial) r.v.s with different success probabilities was examined in Examples 2.8 and 5.3 (in the context of the occupancy distribution) and Problems I.6.7 and 5.4. ∎

The previous example considered discrete r.v.s. Sums of continuous random variables are computed similarly, examples of which were the convolution of independent normal and Laplace r.v.s (Example 2.16) and the convolution of independent normal and gamma r.v.s (Example 2.19). The important case of weighted sums of independent χ^2 r.v.s arises when studying the linear regression and time series models; this will be examined in detail in Section 10.1.

7.3   Mixtures

If you are out to describe the truth, leave elegance to the tailor.

(Albert Einstein)

Mixtures are a very general class of distributions which has enormous scope for application. We present the basic details for two special cases, namely countable and continuous mixtures, and refer the reader to other sources for more information, in particular, Johnson, Kotz and Kemp (1993, Ch. 8) for a solid introduction and overview, and the books by Everitt and Hand (1981), Titterington, Smith and Makov (1985) and McLachlan and Peel (2000). See also Bean (2001, Section 4.6).

7.3.1 Countable mixtures

There are many situations in which the random variable of interest, X, is actually the realization of one of k random variables, say $X_1, \ldots, X_k$, but from which one it came is unknown. The resulting r.v. is said to follow a *finite mixture distribution*. Denote the p.m.f. or p.d.f. of X_i as $f_{X|i}$, $i = 1, \ldots, k$, to make the dependence on i explicit. Then the p.m.f. or p.d.f. of X is given by

$$f_X(x) = \sum_{i=1}^{k} \lambda_i f_{X|i}(x), \quad \lambda_i \in (0, 1), \quad \sum_{i=1}^{k} \lambda_i = 1, \tag{7.24}$$

where the λ_i are referred to as the *mixture component weights*.

⊛ **Example 7.13** A popular and very useful model is the k-component mixed normal, $k \in \mathbb{N}$, with k in practice often very small, usually not more than 4. An r.v. X with this distribution is designated as $X \sim \text{MixN}(\boldsymbol{\mu}, \boldsymbol{\sigma}, \boldsymbol{\lambda})$, with density

$$f_{\text{MixN}}(x; \boldsymbol{\mu}, \boldsymbol{\sigma}, \boldsymbol{\lambda}) = \sum_{i=1}^{k} \lambda_i \phi(x; \mu_i, \sigma_i^2), \quad \lambda_i \in (0, 1), \quad \sum_{i=1}^{k} \lambda_i = 1. \tag{7.25}$$

As a simple illustration, imagine an electronic device which records outcomes X which are N (0, 1), but occasionally makes an error, the outcomes of which are random and distributed N (0, 5). Then, if the 'success rate' is λ, $0 < \lambda < 1$, the p.d.f. of X is $\lambda \phi(x; 0, 1) + (1 - \lambda) \phi(x; 0, 5)$, where ϕ denotes the normal p.d.f. This is sometimes referred to as a *contaminated normal distribution*. It is crucial to realize that X is *not* the weighted sum of two normal random variables (and thus itself normally distributed) but rather a random variable whose p.d.f. is a weighted sum of two p.d.f.s. The distinction lies in the fact that only one of the two random variables is realized, which one, however, not being known.

Figure 7.6 shows a mixed normal density with three components (and two modes). The right tail is relatively much fatter than the left one and, with enough components, can mimic a true fat-tailed density (with a power law) for a given range of x, but, as x increases, the tail eventually resumes its exponential behaviour and dies out quickly.[5] ∎

The form (7.24) of the p.m.f. or p.d.f. of an r.v. X with a finite mixture distribution can be further understood by introducing a discrete random variable, C, with support $\mathcal{C} = \{1, \ldots, k\}$, and p.m.f. $f_C(c) = \Pr(C = c) = \lambda_c$, $c \in \mathcal{C}$, $\lambda_c \in (0, 1)$, $\sum_{c=1}^{k} \lambda_c = 1$.

---

[5] For further details on the history of the mixed normal and its range of applications, see McLachlan and Peel (2000, Sections 1.5, 3.6 and 4.7) and the references therein. Schilling, Watkins and Watkins (2002) provide an interesting and informative discussion of the two-component case in the context of human heights and the conditions for bimodality of the distribution. For a brief survey of the use of the mixed normal distribution in finance, see Haas, Mittnik and Paolella (2004a,b), in which conditional time-varying mixed normal structures are used for modelling the volatility dynamics of financial markets.

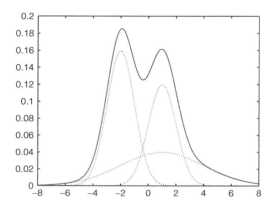

Figure 7.6 Mixed normal density (7.25) with $\mu = (-2, 1, 1)$, $\sigma = (1, 1, 3)$ and $\lambda = (0.4, 0.3, 0.3)$ (solid) and the individual weighted components (dotted)

The realization of C indicates the component p.m.f. or p.d.f. from which X is to be drawn. Thus, the p.m.f. or p.d.f. of X is, from the total probability formula (I.8.16),[6]

$$f_X(x) = \int_{\mathcal{C}} f_{X|C}(x \mid c) \, dF_C(c) = \sum_{c=1}^{k} f_{X|C}(x \mid c) f_C(c) = \sum_{c=1}^{k} \lambda_c f_{X|C}(x \mid c).$$

$$\text{(7.26)}$$

That is, X is nothing but a marginal distribution computed from the joint distribution of X and C, $f_{X,C} = f_{X|C} f_C$.

For any $k \in \mathbb{N}$, X is a finite mixture; with $k = \infty$, C becomes countably infinite (e.g., Poisson), and X is then said to follow a *countable mixture distribution*.[7] Notice that a finite mixture distribution is a special case of a countable mixture distribution.

Another consequence of X being the marginal distribution from $f_{X,C} = f_{X|C} f_C$ is that the iterated expectation (I.8.31) and conditional variance formula (I.8.37) are applicable, i.e.,

$$\mathbb{E}[X] = \mathbb{E}[\mathbb{E}[X \mid C]] \quad \text{and} \quad \mathbb{V}(X) = \mathbb{E}[\mathbb{V}(X \mid C)] + \mathbb{V}(\mathbb{E}[X \mid C]). \quad \text{(7.27)}$$

More generally, to compute the rth raw moment of X, denoted $\mu'_r(X)$ from the notation in (I.4.38), first recall (I.8.26), which states that the expected value of function $g(X)$ conditional on Y is given by

$$\mathbb{E}[g(X) \mid Y = y] = \int_{x \in \mathbb{R}} g(x) \, dF_{X|Y}(x \mid y); \quad \text{(7.28)}$$

---

[6] In the first integral in (7.26), we use the notation dF_C, introduced and defined in (I.4.31), because the same formula, and all subsequent ones which use it, will be applicable in the continuous case, discussed below in Section 7.3.2.

[7] When working with infinite sums, the question arises regarding the validity of exchanging sum and integral. This is allowed, and is a consequence of the dominated convergence theorem. See, for example, Hijab (1997, p. 171) for a basic account with applications.

in particular, for a fixed $r \in \mathbb{N}$ and $c \in \mathcal{C}$,

$$\mathbb{E}\left[X^r \mid C = c\right] = \mu'_r (X \mid C = c) = \int_{x \in \mathbb{R}} x^r \, dF_{X|C} (x \mid c) .$$

Taking expectations of both sides of (7.28) with respect to Y leads to the law of the iterated expectation $\mathbb{E}\,\mathbb{E}\left[g(X) \mid Y\right] = \mathbb{E}\left[g(X)\right]$, i.e.,

$$\mathbb{E}\left[X^r\right] = \mu'_r (X) = \int_{\mathcal{C}} \mu'_r (X \mid C = c) \, dF_C (c) = \sum_{c=1}^{k} \lambda_c \mu'_r (X \mid C = c) . \quad (7.29)$$

Similarly, the m.g.f. of X is the weighted average of the conditional m.g.f.s:

$$\mathbb{E}\left[e^{tX}\right] = \mathbb{E}_C \left[\mathbb{E}\left[e^{tX} \mid C = c\right]\right] = \int_{\mathcal{C}} \mathbb{M}_{X|C=c}(t) \, dF_C (c) = \sum_{c=1}^{k} \lambda_c \mathbb{M}_{X|C=c}(t) . \quad (7.30)$$

◎ **Example 7.14** Let $X \sim \mathrm{MixN}(\boldsymbol{\mu}, \boldsymbol{\sigma}, \boldsymbol{\lambda})$, with density (7.25). Then, from the moments of the normal distribution and (7.29),

$$\mathbb{E}[X] = \sum_{i=1}^{k} \lambda_i \mu_i, \quad \mathbb{E}\left[X^2\right] = \sum_{i=1}^{k} \lambda_i \left(\mu_i^2 + \sigma_i^2\right),$$

from which $\mathbb{V}(X) = \mathbb{E}\left[X^2\right] - (\mathbb{E}[X])^2$ can be computed. ∎

⊖ **Example 7.15** Let X be an r.v. with c.d.f. $F_X(x) = \left(2x - x^2\right)^v \mathbb{I}_{(0,1)}(x) + \mathbb{I}_{[1,\infty)}(x)$, for $v > 0$, as was considered by Topp and Leone (1955). The p.d.f. of X is

$$f_X(x; v) = 2v(1 - x)\left(2x - x^2\right)^{v-1} \mathbb{I}_{(0,1)}(x), \quad (7.31)$$

which is plotted in Figure 7.7 for several v. It is interesting because f_X can be expressed as a mixture of beta densities when $v \in \mathbb{N}$, seen by writing

$$f_X(x; v) = 2vx^{v-1}(1 - x)(1 + 1 - x)^{v-1} = 2vx^{v-1}(1 - x) \sum_{j=0}^{v-1} \binom{v-1}{j}(1 - x)^j$$

$$= 2v \sum_{j=0}^{v-1} \binom{v-1}{j} x^{v-1}(1 - x)^{j+1} =: \sum_{j=0}^{v-1} w_j(v) f_{X|j}(x; v),$$

where

$$w_j(v) = 2v \binom{v-1}{j} \frac{\Gamma(v)\Gamma(j+2)}{\Gamma(v+j+2)}, \quad f_{X|j}(x; v) = \frac{\Gamma(v+j+2)}{\Gamma(v)\Gamma(j+2)} x^{v-1}(1 - x)^{j+1},$$

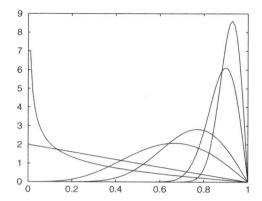

Figure 7.7 Density (7.31) shown for $v = 0.5, 1, 5, 10, 50$ and 100. As v increases, the mode of the density moves from left to right

i.e., $f_{X|j}$ is the Beta $(v, j + 2)$ density. $\sum_{j=0}^{v-1} w_j(v) = 1$, a bit of rearranging and setting $k = j + 1$ implies the nonobvious combinatoric identity

$$\sum_{k=1}^{v} \frac{k}{(v-k)!\,(v+k)!} = \frac{1}{2\,(v-1)!\,v!}$$

or, multiplying both sides by $(2v!)$ and simplifying,

$$\sum_{k=1}^{v} k \binom{2v}{v-k} = \frac{v}{2} \binom{2v}{v}, \quad v \in \mathbb{N},$$

which is proven directly in Riordan (1968, pp. 34 and 84). ∎

7.3.2 Continuous mixtures

The extension of (7.24) to the case with C a continuous random variable with support $\mathcal{C}$ follows naturally from the integral expression in (7.26). The random variable X is said to follow a *continuous mixture distribution* if its (marginal) p.d.f. can be expressed as

$$f_X(x) = \int_{\mathcal{C}} f_{X|C}(x \mid c)\, f_C(c)\, dc \qquad (7.32)$$

or, equivalently, if its c.d.f. is $F_X(x) = \int_{\mathcal{C}} F_{X|C}(x \mid c)\, f_C(c)\, dc$. The mean and variance in the continuous mixture case are again given by (7.27), while the raw moments are given by the integral expression in (7.29). Similarly, the m.g.f. of X is given by

$$\mathbb{E}\left[e^{tX}\right] = \int_{\mathcal{C}} \mathbb{M}_{X|C=c}(t)\, f_C(c)\, dc. \qquad (7.33)$$

⊚ **Example 7.16** Let $(X \mid R = r) \sim \text{Poi}(r)$, $r \in \mathbb{R}_{>0}$, and $R \sim \text{Gam}(a, b)$. Then, from (1.11),

$$\mathbb{M}_{X \mid R=r}(t) = \exp\left\{r\left(e^t - 1\right)\right\},$$

and (7.33) implies that, with $k = b + 1 - e^t$,

$$\mathbb{M}_X(t) = \int_0^\infty \exp\left\{r\left(e^t - 1\right)\right\} \frac{b^a}{\Gamma(a)} r^{a-1} \exp\left\{-br\right\} dr$$

$$= \frac{b^a}{\Gamma(a)} \int_0^\infty \exp\left\{-kr\right\} r^{a-1} dr = \frac{b^a}{\Gamma(a)} \frac{\Gamma(a)}{k^a} = \left(\frac{b}{b+1-e^t}\right)^a.$$

Setting $p = b/(b+1)$, $b = p/(1-p)$ and simplifying gives

$$\mathbb{M}_X(t) = \left(\frac{p}{1 - (1-p)e^t}\right)^a, \quad \text{or} \quad X \sim \text{NBin}\left(a, \frac{b}{b+1}\right),$$

from (1.97). Of course, we could also compute $f_X(x) = \int_0^\infty f_{X \mid R}(x \mid r) f_R(r) dr$ and use $p = b/(b+1)$ to arrive at the p.m.f. of the negative binomial, which the reader is encouraged to do.[8] ∎

⊚ **Example 7.17** (Dubey, 1968; Johnson, Kotz and Balakrishnan, 1994, p. 686) Let X be a Weibull r.v. with location zero and p.d.f.

$$f_X(x; \beta, \theta) = \beta\theta x^{\beta-1} \exp\left\{-\theta x^\beta\right\} \mathbb{I}_{(0,\infty)}(x),$$

which is a different parameterization than that used in (I.7.19). Assume that θ is not a constant, but rather a realization of a gamma r.v. Θ with shape parameter b and scale c, i.e.,

$$f_\Theta(\theta; b, c) = \frac{c^b}{\Gamma(b)} \theta^{b-1} \exp\{-c\theta\} \mathbb{I}_{(0,\infty)}(\theta). \tag{7.34}$$

Denote the conditional p.d.f. of X given $\Theta = \theta$ by $f_{X \mid \Theta}(x \mid \beta, \theta)$. From (7.34), the marginal distribution of X depends on β, b, and c, and we write its p.d.f. with three parameters as $f_X(x; \beta, b, c)$. Then, with $u = \theta\left(x^\beta + c\right)$,

$$f_X(x; \beta, b, c) = \int_{-\infty}^\infty f_{X \mid \Theta}(x \mid \beta, \theta) f_\Theta(\theta; b, c) d\theta$$

$$= \frac{c^b \beta x^{\beta-1}}{\Gamma(b)} \int_0^\infty \theta^b \exp\left\{-\theta\left(x^\beta + c\right)\right\} \mathbb{I}_{(0,\infty)}(x) d\theta$$

---

[8] A generalization of the negative binomial distribution useful for modelling a variety of data sets arises by letting R follow a type of generalized gamma distribution; see Gupta and Ong (2004) for details.

$$= \frac{c^b \beta x^{\beta-1}}{\Gamma(b)} \int_0^\infty \left(\frac{u}{x^\beta + c} \right)^b \exp\{-u\} \, \mathbb{I}_{(0,\infty)}(x) \frac{du}{x^\beta + c}$$

$$= \frac{c^b \beta}{\Gamma(b)} \frac{x^{\beta-1}}{(x^\beta + s)^{b+1}} \mathbb{I}_{(0,\infty)}(x) \int_0^\infty u^{(b+1)-1} \exp\{-u\} \, du$$

$$= bc^b \beta \frac{x^{\beta-1}}{(x^\beta + c)^{b+1}} \mathbb{I}_{(0,\infty)}(x).$$

Carrying out the univariate transformation shows that the density of $Y = X^\beta$ is $f_Y(y; b, c) = bc^b (y + c)^{-(b+1)} \mathbb{I}_{(0,\infty)}(y)$, i.e., from (I.7.29), $Y \sim \text{Par II}(b, c)$, which implies that X can be interpreted as a 'power transformation' of a type II Pareto random variable.

It is worth drawing attention to the important special case with $\beta = 1$: if $(X \mid \Theta = \theta) \sim \text{Exp}(\theta)$ and $\Theta \sim \text{Gam}(b, c)$, then, unconditionally, $X \sim \text{Par II}(b, c)$. ◼

⊚ **Example 7.18** As in the last example, let $\Theta \sim \text{Gam}(b, c)$, with p.d.f. (7.34), and let $(X \mid \Theta = \theta) \sim \text{Gam}(a, \theta)$, so that

$$f_X(x; a, b, c) = \int_{-\infty}^\infty f_{X|\Theta}(x \mid a, \theta) f_\Theta(\theta; b, c) \, d\theta$$

$$= \frac{1}{\Gamma(a) \, \Gamma(b)} c^b x^{a-1} \mathbb{I}_{(0,\infty)}(x) \int_0^\infty \exp\{-\theta(x + c)\} \theta^{a+b-1} \, d\theta$$

and, from the gamma density, the integral is just $\Gamma(a + b) / (x + c)^{a+b}$, so that

$$f_X(x; a, b, c) = \frac{c^b}{B(a, b)} \frac{x^{a-1}}{(x + c)^{a+b}} \mathbb{I}_{(0,\infty)}(x),$$

which is the type II generalized Pareto density (7.6). It is also referred to as the gamma–gamma distribution and expressed as

$$f_X(x; a, b, c) = \frac{1}{B(a, b)} \left(\frac{x}{x + c} \right)^{a-1} \left(1 - \frac{x}{x + c} \right)^{b-1} \frac{c}{(x + c)^2} \mathbb{I}_{(0,\infty)}(x);$$

see Bernardo and Smith (1994, Ch. 3). ◼

⊚ **Example 7.19** Let $(X \mid V = v) \sim N(\mu, v)$ and $V \sim \text{Exp}(\lambda)$. From (7.33) and the m.g.f. of the normal, given in (1.10), the m.g.f. of X is, with $T = \lambda - t^2/2$ and $T > 0$ (i.e., $|t| < \sqrt{2\lambda}$),

$$\mathbb{M}_X(t) = \int_0^\infty \exp\left\{ \mu t + \frac{1}{2} v t^2 \right\} \lambda \exp\{-\lambda v\} \, dv = \lambda e^{\mu t} \int_0^\infty \exp\{-v T\} \, dv$$

$$= \frac{\lambda e^{\mu t}}{\lambda - t^2/2} = e^{\mu t} \frac{1}{1 - t^2/(2\lambda)}, \quad |t| < \sqrt{2\lambda}.$$

```
n=10000; lam=1/2; unitexp = exprnd(1,n,1);
% unitexp are scale-one i.i.d. exponential r.v.s. This avoids having to
%    figure out which parameterization Matlab uses, and makes the code
%    more portable to other platforms
V=unitexp/lam; Z=randn(n,1); X=Z.*sqrt(V);
% When calculating the variance V, we divide by lam because,
%    in our notation, lambda is an 'inverse' scale parameter
[pdf,grd]=kerngau(X);
% The following code thins out the output from kerngau.
%    The plot looks the same, but takes up less memory as a graphic.
ind=kron(ones(1,floor(n/3)),[1 0 0])';
use=zeros(n,1); use(1:length(ind))=ind; use=logical(use);
thingrd=grd(use); thinpdf=pdf(use);
% Now compute the Laplace pdf for half the graph, and plot them.
xx=0:0.1:6; ff=0.5*exp(-xx);
h=plot(xx,ff,'r-',thingrd,thinpdf,'g-'); set(gca,'fontsize',16),
axis([-6 6 0 0.5])
```

Program Listing 7.2 Generates i.i.d. location-zero, scale-one Laplace r.v.s via the mixture relationship discussed in Example 7.19, and plots their kernel density and the true density (see Figure 7.8)

From (1.5), μ is just a location parameter and can be set to zero without loss of generality, and λ serves as a scale parameter (the scaling parameter is $1/\sqrt{2}\lambda$), so it suffices to examine $\mathbb{M}_X(t) = 1/(1-t^2)$ for $|t| < 1$. Recalling (5.12), we see that this is the m.g.f. of a Laplace random variable. It is an example of a *normal variance mixture*, which is discussed at length in Chapter 9. Normal variance mixtures are a very general class of distributions which are highly capable of modelling the variation in a wide variety of real data sets; Romanowski (1979) provides some intuition as to why this is true.

Simulation can easily be used to verify that $X \sim$ Lap. In particular, the mixture result has the interpretation that, if $V \sim$ Exp $(1/2)$ is drawn, followed by $(X \mid V = v) \sim$ N (μ, v), then the latter will be an observation from a standard Laplace distribution. The code in Listing 7.2 implements this, and creates Figure 7.8, which plots a kernel density estimate of 10 000 simulated draws, and, on the positive axis, the true p.d.f. They are clearly the same.

Now consider a simple generalization. Let $(X \mid \Theta = \theta) \sim$ N $(0, \theta)$ and $\Theta \sim$ Gam (b, c) with p.d.f. (7.34), so that, as before, and with $H = c - t^2/2$ and $H > 0$ (i.e., $|t| < \sqrt{2c}$),

$$\mathbb{M}_X(t) = \int_0^\infty \exp\left\{\frac{1}{2}\theta t^2\right\} \frac{c^b}{\Gamma(b)} \theta^{b-1} \exp\{-c\theta\} d\theta$$

$$= \frac{c^b}{\Gamma(b)} \int_0^\infty \theta^{b-1} \exp\{-\theta H\} d\theta = \frac{c^b}{\Gamma(b)} \frac{\Gamma(b)}{H^b}$$

$$= \left(1 - t^2/(2c)\right)^{-b}, \quad |t| < \sqrt{2c}. \tag{7.35}$$

Of course, as before, c serves as a scale parameter, and $1/\sqrt{2c}$ could be set to unity without loss of generality. This m.g.f. is not recognizable, but Example 2.10 showed

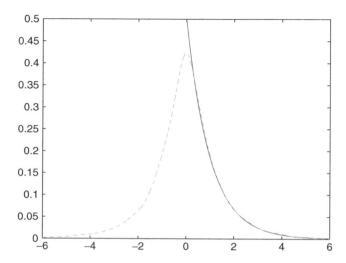

Figure 7.8 Output from Listing 7.2. The solid curve is the true Laplace p.d.f., shown just on the positive half line, and the dashed curve is the kernel density estimate of 10 000 simulated draws

that (7.35) coincides with the m.g.f. of the difference between two i.i.d. scaled gamma r.v.s.

That f_X is symmetric about zero follows immediately from its construction as a mixture of symmetric-about-zero densities, and also from the fact that t enters the m.g.f. only as a square, so that its c.f. is real (see the end of Section 1.2.3), i.e., from (7.35),

$$\varphi_X(t) = \mathbb{M}_X(it) = \left(1 - (it)^2/(2c)\right)^{-b} = \left(1 + t^2/(2c)\right)^{-b}.$$

Example 1.23 discussed the use of the inversion formulae applied to the c.f. of X to calculate its p.d.f. and c.d.f. The saddlepoint approximation offers an alternative. Basic calculation shows that

$$\mathbb{K}'_X(t) = \frac{2bt}{2c - t^2}, \quad \mathbb{K}'_X(t) = x \Longrightarrow t = \frac{-2b \pm 2\sqrt{b^2 + 2cx^2}}{2x},$$

and simplifying $t^2 < 2c$ with Maple easily shows that the value of t with the plus sign is the correct one. Thus, a closed-form s.p.a. can be computed for $x \neq 0$, which can be 'vector programmed' in Matlab for virtually instantaneous evaluation over a grid of x-values, without any numeric problems. Similarly,

$$\mathbb{K}''_X(t) = 2b\frac{2c + t^2}{\left(2c - t^2\right)^2}, \quad \mathbb{K}^{(3)}_X(t) = \frac{4bt\left(t^2 + 6c\right)}{\left(2c - t^2\right)^3}, \quad \mathbb{K}^{(4)}_X(t) = 12b\frac{t^4 + 12t^2c + 4c^2}{\left(2c - t^2\right)^4},$$

the latter two required for the second-order s.p.a.

The symmetry of f_X and the existence of the m.g.f. imply that all odd positive moments exist and are zero. For the even moments, (I.1.12) implies that

$$
\mathbb{M}_X(t) = \left(1 - \frac{t^2}{2c}\right)^{-b}
$$

$$
= 1 + \binom{b}{1}\left(\frac{t^2}{2c}\right) + \binom{b+1}{2}\left(\frac{t^2}{2c}\right)^2 + \binom{b+2}{3}\left(\frac{t^2}{2c}\right)^3 + \cdots
$$

$$
= 1 + \frac{t^2}{2!}\frac{b}{c} + \frac{3b\,(b+1)}{c^2}\frac{t^4}{4!} + \cdots,
$$

so that, from (1.3),

$$
\mu_2' = \mu_2 = \mathrm{Var}\,(X) = \frac{b}{c}, \quad \mu_4' = \mu_4 = \frac{3b\,(b+1)}{c^2}
$$

and, thus,

$$
\mathrm{kurt}\,(X) = \frac{\mu_4}{\mu_2^2} = 3\left(1 + \frac{1}{b}\right). \tag{7.36}
$$

This shows that the mixture has higher kurtosis than the normal distribution and, in the limit as $b \to \infty$, the kurtosis agrees with that of the normal. In fact, letting $c = b$ (which gives a unit variance), the asymptotic distribution is standard normal, because of

$$
\lim_{c=b\to\infty} \mathbb{M}_X(t) = \lim_{b\to\infty}\left(1 - \frac{t^2}{2b}\right)^{-b} = \exp\left\{\frac{1}{2}t^2\right\}
$$

and (1.10).

Thus, the s.p.a. will become increasingly accurate as b increases, recalling that the s.p.a. is exact for the normal density, and might be expected to break down as $b \to 0$. To verify this, the p.d.f. inversion formula could be used, but it can exhibit numerical problems for small b. It turns out, however, that an expression for the density in terms of a Bessel function is possible. In particular, working with the conditional p.d.f. via (7.32) instead of the m.g.f., setting $\theta = ux/\sqrt{2c}$ and using (I.7.61) gives, for $x > 0$,

$$
f_X(x) = \frac{c^b}{\sqrt{2\pi}\,\Gamma\,(b)} \int_0^\infty \theta^{b-3/2} \exp\left\{-c\theta - \frac{x^2}{2\theta}\right\} d\theta \tag{7.37}
$$

$$
= \frac{2c^b}{\sqrt{2\pi}\,\Gamma\,(b)}\left(\frac{x}{\sqrt{2c}}\right)^{b-1/2}\frac{1}{2}\int_0^\infty u^{b-3/2}\exp\left\{-\frac{x\sqrt{2c}}{2}\left(u + \frac{1}{u}\right)\right\} du
$$

$$
= \frac{2c^b}{\sqrt{2\pi}\,\Gamma\,(b)}\left(\frac{x}{\sqrt{2c}}\right)^{b-1/2} K_{b-1/2}\left(x\sqrt{2c}\right), \quad x > 0, \tag{7.38}
$$

as first given by Teichroew (1957), who also provides expressions for the c.d.f. An expression for $x = 0$ is easily obtained from (7.37) for $b > 1/2$; otherwise, use of the

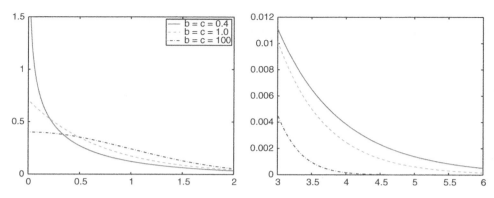

Figure 7.9 (Left) The exact p.d.f. of r.v. X, where $(X \mid \Theta = \theta) \sim N(0, \theta)$ and $\Theta \sim \text{Gam}(b, c)$. With $b = c$, the variance is unity. (Right) The 'view from the tail'

relation $K_v(z) = K_{-v}(z)$ and (I.7.62) show that $\lim_{x \downarrow 0} f_X(x) = \infty$, which the reader should verify.

Because specialized numeric routines are available for the highly accurate evaluation of the Bessel functions, the p.d.f. of X can indeed be evaluated 'exactly'. Moreover, because $K_v(x)$ is evaluated so quickly in Matlab, particularly for vector arguments, the s.p.a. is superfluous in this context (but the reader is invited to study its accuracy).

Figure 7.9 shows parts of the p.d.f. for three values of b, with $c = b$, so that the variance is unity and the p.d.f.s are comparable.   ∎

⊙ **Example 7.20** A natural way of introducing asymmetry into the previously studied normal–gamma mixture is to take

$$(X \mid \Theta = \theta) \sim N(m\theta, \theta), \qquad \Theta \sim \text{Gam}(b, c), \qquad m \in \mathbb{R}, \tag{7.39}$$

so that, for $mt/c + t^2/(2c) < 1$,

$$\mathbb{M}_X(t) = \frac{c^b}{\Gamma(b)} \int_0^\infty \theta^{b-1} \exp\left\{-\theta\left(c - mt - \frac{1}{2}t^2\right)\right\} d\theta = \left(1 - \frac{mt}{c} - \frac{t^2}{2c}\right)^{-b}.$$

The Taylor series expansion of $\mathbb{M}_X(t)$ (easily computed with Maple) is

$$1 + \left(\frac{bm}{c}\right)t + \left(\frac{b}{c} + \frac{m^2 b(b+1)}{c^2}\right)\frac{t^2}{2!} + \left(\frac{3mb(b+1)}{c^2} + \frac{m^3 b(b+1)(b+2)}{c^3}\right)\frac{t^3}{3!}$$

$$+ \left(\frac{3b(b+1)}{c^2} + \frac{6m^2 b(b+1)(b+2)}{c^3} + \frac{m^4 b(b+1)(b+2)(b+3)}{c^4}\right)\frac{t^4}{4!} + \cdots,$$

from which we get (painlessly, with the assistance of Maple),

$$\mu = \mathbb{E}[X] = \frac{bm}{c}, \qquad \mu_2 = \text{Var}(X) = \frac{b(c + m^2)}{c^2},$$

$$\text{skew}\,(X) = \frac{\mu_3}{\mu_2^{3/2}} = \frac{\mu_3' - 3\mu_2'\mu + 2\mu^3}{\mu_2^{3/2}} = \frac{m\,(3c + 2m^2)}{b^{1/2}\,(c + m^2)^{3/2}},$$

and

$$\text{kurt}\,(X) = \frac{\mu_4}{\mu_2^2} = \frac{\mu_4' - 4\mu_3'\mu + 6\mu_2'\mu^2 - 3\mu^4}{\mu_2^2}$$

$$= \frac{3\,(b+1)\,c^2 + m^2\,(2c + m^2)\,(b+2)}{b} \cdot \frac{1}{(c + m^2)^2},$$

which simplify to the expressions for the $m = 0$ case given in Example 7.19 above. The derivation of the density is similar to that in Example 7.19; with $k = (m^2/2 + c)$,

$$f_X\,(x; m, b, c) = \frac{c^b}{\sqrt{2\pi}\,\Gamma\,(b)} \int_0^\infty \theta^{b-3/2} \exp\left\{-\frac{1}{2}\frac{(x - m\theta)^2}{\theta} - c\theta\right\}\,d\theta \qquad (7.40)$$

$$= \frac{c^b e^{xm}}{\sqrt{2\pi}\,\Gamma\,(b)} \int_0^\infty \theta^{b-3/2} \exp\left\{-k\theta - \frac{x^2}{2\theta}\right\}\,d\theta,$$

which is the same form as (7.37) but with k in place of c, so that, for $x > 0$,

$$f_X\,(x; m, b, c) = \frac{2c^b e^{xm}}{\sqrt{2\pi}\,\Gamma\,(b)} \left(\frac{x}{\sqrt{m^2 + 2c}}\right)^{b-1/2} K_{b-1/2}\left(x\sqrt{m^2 + 2c}\right). \qquad (7.41)$$

For $x < 0$, inspection of (7.40) shows that

$$f_X\,(x; m, b, c) = f_X\,(-x; -m, b, c). \qquad (7.42)$$

For $x = 0$ and $b > 1/2$,

$$f_X\,(0; m, b, c) = \frac{c^b}{\sqrt{2\pi}\,\Gamma\,(b)} \int_0^\infty \theta^{b-3/2} \exp\{-k\theta\}\,d\theta = \frac{c^b}{\sqrt{2\pi}\,\Gamma\,(b)} \frac{\Gamma\,(b - 1/2)}{(m^2/2 + c)^{b-1/2}};$$

otherwise, $f_X\,(0; m, b, c) = \infty$.

Random variable X is said to follow the *variance–gamma* distribution, and was popularized by its use for modelling financial returns data by Madan and Seneta (1990). It is a special case of the generalized hyperbolic distribution, examined in great detail in Chapter 9. Setting $\lambda = b$, $\beta = m$, $\alpha = \sqrt{m^2 + 2c}$ and $\mu = 0$ yields the variance gamma density as given in (9.43).

Figure 7.10 verifies the correctness of the derivation by plotting the algebraic p.d.f. and a kernel density estimate of $n = 30\,000$ draws based on (7.39).[9] Figure 7.11 plots

---

[9] Note that the difference between the two is negligible in the right tail, where the slope of the p.d.f. is not large, but is somewhat apparent in the left part of the density, where the slope is very high. This is an artefact of kernel density estimation, and gets worse as n is reduced.

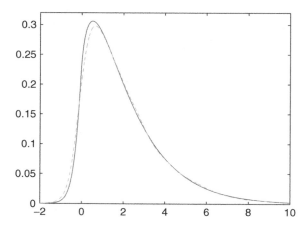

Figure 7.10 The solid curve is the true p.d.f. given in (7.41) and (7.42), based on $m = 2$ and $b = c = 1.5$. The dashed curve is the kernel density estimate based on 30 000 simulated draws from (7.39)

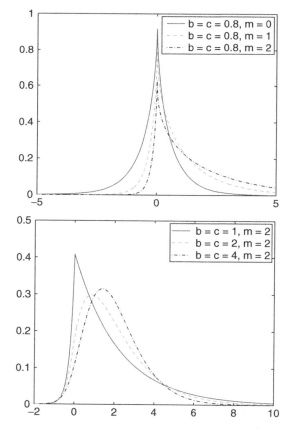

Figure 7.11 The exact p.d.f. of r.v. X, where $(X \mid \Theta = \theta) \sim N(m\theta, \theta)$ and $\Theta \sim Gam(b, c)$

the density for several parameter constellations, illustrating the great flexibility of this distribution.   ■

Instead of using a gamma distribution for the conditional normal variance, as in the previous examples, one might entertain the – arguably less natural – choice of an inverse gamma. Perhaps surprisingly, use of a special case of the inverse gamma leads to a quite common distribution: the Student's t. By taking $\mu \neq 0$, we obtain an important generalization of the t, referred to as the *noncentral Student's t*, which will be discussed in detail in Chapter 10.

⊙ **Example 7.21**   Let $(X \mid V = v) \sim N\left(\mu v^{-1/2}, v^{-1}\right)$ and $kV \sim \chi_k^2$, with

$$f_V(v; k) = \frac{k^{k/2}}{2^{k/2}\Gamma(k/2)} v^{k/2-1} e^{-vk/2} \mathbb{I}_{(0,\infty)}(v).$$ (7.43)

Then

$$f_X(x; \mu) = \int_{-\infty}^{\infty} f_{X|V}(x \mid v) f_V(v; k) \, dv$$

$$= \int_0^{\infty} \frac{v^{1/2}}{\sqrt{2\pi}} \exp\left\{-\frac{v}{2}\left(x - \mu v^{-1/2}\right)^2\right\} f_V(v; k) \, dv$$

$$= \frac{1}{\sqrt{2\pi}} \frac{k^{k/2}}{2^{k/2}\Gamma(k/2)} \int_0^{\infty} v^{(k-1)/2} \exp\left\{-\frac{1}{2}\left(\left(xv^{1/2} - \mu\right)^2 + kv\right)\right\} dv$$

or, with $z = v^{1/2}$,

$$f_X(x; \mu) = \frac{1}{\sqrt{2\pi}} \frac{k^{k/2}}{2^{k/2-1}\Gamma(k/2)} \int_0^{\infty} z^k \exp\left\{-\frac{1}{2}\left((xz - \mu)^2 + kz^2\right)\right\} dz.$$ (7.44)

The random variable X is said to follow the so-called *(singly) noncentral t* distribution with k degrees of freedom and *noncentrality parameter* μ.

Imagine how a realization x from the unconditional distribution of X would be generated. As $kV \sim \chi_k^2$, let v be drawn from a χ_k^2 r.v., then divided by k. Conditionally, $(X \mid V = v) \sim N\left(\mu v^{-1/2}, v^{-1}\right)$, so we would set

$$x = \frac{1}{v^{1/2}}(\mu + z),$$

where z is a realization from a standard normal distribution. We can write this procedure in terms of the r.v.s $kV \sim \chi_k^2$ and $Z \sim N(0, 1)$ as

$$X = \frac{\mu + Z}{\sqrt{V}},$$

and note that V is independent of Z. This is precisely how a (singly) noncentral Student's t random variable is constructed, and observe that when $\mu = 0$, this reduces to the usual Student's t distribution, as was shown via multivariate transformation in Example I.9.7. In particular, for $\mu = 0$, using the substitution $y = z^2 \left(x^2 + k \right) / 2$ in (7.44) and the fact that z is positive yields

$$f_X \left(x; 0 \right) = K \left(x^2 + k \right)^{-(k+1)/2}, \qquad K = \frac{k^{k/2}}{\sqrt{\pi}} \frac{\Gamma \left((k+1)/2 \right)}{\Gamma \left(k/2 \right)},$$

i.e., when $\mu = 0$, X is Student's t distributed with k degrees of freedom.

The moments of X can also be straightforwardly computed using (7.29). For example, with $\mu_1' \left(X \mid V = v \right) = \mathbb{E} \left[X \mid V = v \right] = \mu v^{-1/2}$ and density (7.43), a simple integration shows that

$$\mathbb{E} \left[X \right] = \int_0^\infty \mu \left(X \mid V = v \right) f_V \left(v \right) dv = \mu \sqrt{\frac{k}{2}} \frac{\Gamma \left(\frac{k-1}{2} \right)}{\Gamma \left(\frac{k}{2} \right)}.$$

Similarly, with $\mu_2' \left(X \mid V = v \right) = v^{-1} \left(1 + \mu^2 \right)$,

$$\mathbb{E} \left[X^2 \right] = \int_0^\infty \mu_2' \left(X \mid V = v \right) f_V \left(v \right) dv = \left(1 + \mu^2 \right) \frac{k}{k-2},$$

which the reader should verify. Alternatively, with $C = kV$ and from (I.7.42), (7.27) yields

$$\mathbb{E} \left[X \right] = \mathbb{E} \left[\mathbb{E} \left[X \mid V \right] \right] = \mathbb{E} \left[\mu V^{-1/2} \right] = \mu \mathbb{E} \left[\frac{k^{1/2}}{(kV)^{1/2}} \right]$$

$$= \mu k^{1/2} \mathbb{E} \left[C^{-1/2} \right] = \mu \sqrt{\frac{k}{2}} \frac{\Gamma \left(\frac{k-1}{2} \right)}{\Gamma \left(\frac{k}{2} \right)},$$

and, using the conditional variance formula (I.8.37),

$$\mathbb{V} \left(X \right) = \mathbb{E} \left[\mathbb{V} \left(X \mid V \right) \right] + \mathbb{V} \left(\mathbb{E} \left[X \mid V \right] \right) = \mathbb{E} \left[V^{-1} \right] + \mathbb{V} \left(\mu V^{-1/2} \right)$$

$$= \mathbb{E} \left[\frac{k}{kV} \right] + \mu^2 \mathbb{V} \left(\frac{k^{1/2}}{(kV)^{1/2}} \right) = k \mathbb{E} \left[C^{-1} \right] + \mu^2 k \mathbb{V} \left(C^{-1/2} \right)$$

$$= \frac{k}{k-2} + \mu^2 k \left(\frac{1}{k-2} - \frac{1}{2} \frac{\Gamma^2 \left(\frac{k-1}{2} \right)}{\Gamma^2 \left(\frac{k}{2} \right)} \right),$$

so that

$$\mathbb{E} \left[X^2 \right] = \mathbb{V} \left(X \right) + \left(\mathbb{E} \left[X \right] \right)^2 = \frac{k}{k-2} \left(1 + \mu^2 \right),$$

which agrees with the previous results. ∎

7.4 Problems

Usually when people are sad, they don't do anything. They just cry over their condition. But when they get angry, they bring about a change.   (Malcolm X)

Tears will get you sympathy; sweat will get you change.   (Jesse Jackson)

7.1. ★ Let $Z \sim$ NormLap (c) as introduced in Example 2.16; let $X \sim$ GED (p) with density given in (I.7.73) and (7.1); and let Y be a continuous random variable with density given by

$$f_Y(y; w) = w \frac{1}{\sqrt{2\pi}} \exp\left(-\frac{1}{2}y^2\right) + (1-w) \frac{1}{2} \exp\left(-|y|\right), \qquad 0 \le w \le 1,$$

i.e., the p.d.f. of Y is a weighted sum of a normal and Laplace density. This weighted sum was first proposed and used by Kanji (1985) and subsequently by Jones and McLachlan (1990). Each of these random variables nests the normal and Laplace as special cases. Show that for the kurtosis of Z, X and Y we respectively obtain

$$\frac{\mu_4}{\mu_2^2} = 3 + \frac{3}{1 + r^2 + \frac{1}{4}r^4}, \qquad r = \frac{c}{1-c}, \tag{7.45}$$

$$\frac{\mu_4}{\mu_2^2} = \frac{\Gamma(5/p)\,\Gamma(1/p)}{\Gamma^2(3/p)},$$

and

$$\frac{\mu_4}{\mu_2^2} = 3\frac{8 - 7w}{(2-w)^2}.$$

These are graphically shown in Figure 7.12. For Y, interestingly enough, the plot reveals that the kurtosis exceeds 6. Differentiating shows that the maximum occurs at $w = 2/7$, for which the kurtosis is $49/8 = 6.125$.

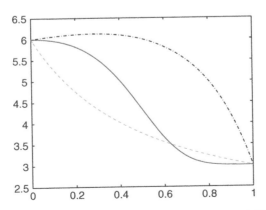

Figure 7.12 Kurtosis of Z versus c (solid); kurtosis of X versus $p - 1$ (dashed); and kurtosis of Y versus w (dash-dotted)

7.2. ★ If $X \sim \text{Beta}(\alpha, \beta)$, then

$$Y = \frac{X}{\lambda - \lambda X + X}$$

is said to follow a *generalized three-parameter beta distribution*, which is written $Y \sim \text{G3B}(\alpha, \beta, \lambda)$, for $\alpha, \beta, \lambda > 0$ (see Johnson, Kotz and Balakrishnan, 1995, p. 251 and the references therein). The p.d.f. of Y is given by

$$f_Y(y; \alpha, \beta, \lambda) = \frac{\lambda^\alpha}{B(\alpha, \beta)} \frac{y^{\alpha-1}(1-y)^{\beta-1}}{\left[1 - (1-\lambda)y\right]^{\alpha+\beta}} \mathbb{I}_{(0,1)}(y). \qquad (7.46)$$

It has, for example, been used by Lee, Lee and Wei (1991) in the context of option pricing. It simplifies to the standard beta density for $\lambda = 1$.

This can be generalized to what we call the G4B distribution; if $Z \sim \text{G4B}$ $(\alpha, \beta, \lambda, \kappa)$ with $\alpha, \beta, \lambda, \kappa > 0$, then

$$f_Z(z; \alpha, \beta, \lambda, \kappa) = C \frac{z^{\alpha-1}(1-z)^{\beta-1}}{\left[1 - (1-\lambda)z\right]^\kappa} \mathbb{I}_{(0,1)}(z), \qquad (7.47)$$

where $C^{-1} = B(\alpha, \beta) \cdot {}_2F_1(\alpha, \kappa, \alpha + \beta, 1 - \lambda)$, and ${}_2F_1$ is the hypergeometric function; see Section 5.3.

Figure 7.13(a) plots (7.46) for $\alpha = \beta = 3$ and $\lambda = 0.5, 1.0$ and 1.5. Figure 7.13(b) plots (7.47) for $\alpha = \beta = 3$, $\lambda = 0.5$, and $\kappa = 3, 6$, and 12, with $\kappa = 6 = \alpha + \beta$ corresponding to the G3B density.

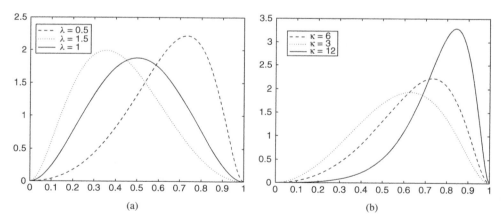

Figure 7.13 Generalized beta distributions: (a) (7.46) for $\alpha = \beta = 3$; (b) (7.47) for $\alpha = \beta = 3$ and $\lambda = 0.5$

(a) Verify via transformation the p.d.f. of Y.

(b) Give an expression for the mth moment of Y in terms of the ${}_2F_1$ function.

(c) Give an expression for the c.d.f. of Y.

Remark: It can be shown that, with $\lambda = \alpha/\beta$, the mode of the G3B density occurs at $y = 0.5$ when $\alpha > 1$ and $\beta > 1$. To see this with Maple, issue the commands

$$\texttt{assume}\,(\alpha, \texttt{positive}) \quad \text{and} \quad \texttt{assume}\,(\beta, \texttt{positive}),$$

compute

$$\frac{d}{dy} \frac{y^{\alpha-1}\,(1-y)^{\beta-1}}{\left[1 - (1 - \alpha/\beta)\,y\right]^{\alpha+\beta}},$$

set the resulting expression to zero and solve for y. This yields the two roots $y = 0.5$ and

$$y = \beta\frac{\alpha - 1}{\alpha - \beta} = \frac{1 - \alpha}{1 - \lambda}.$$

For $\alpha > 1$ and $\beta > 1$, the mode occurs at $y = 0.5$. For $0 < \beta < 1$ and $\alpha > 1$, the density has its mode at 1, a local *minimum* at 0.5, and a local maximum at $\beta\frac{\alpha-1}{\alpha-\beta}$. The same holds for $\beta > 1$ and $0 < \alpha < 1$, but the density has its mode at 0. ∎

(d) Give an expression for the mth moment of Z in terms of $_2F_1$ functions.

7.3. ★ ★ Let $Y \sim$ G3B (α, β, λ). The random variable $W = Y/(1 - Y)$ is said to follow the generalized F distribution, denoted G3F (α, β, λ) (Dyer, 1982).

(a) Express the c.d.f. of W in terms of the c.d.f. of Y.

(b) Derive the density of W.

(c) For what parameter values α, β, and λ does the density of W coincide with that of a standard F distribution $F(x; n_1, n_2)$? Denote the values (as functions of n_1 and n_2) as α', β', and λ'.

(d) Let $U = cW$ for $c > 0$. Derive the p.d.f. $f_U(u; \alpha, \beta, \lambda, c)$ of U. Next, derive values A, B, and C such that $f_U(u; \alpha, \beta, \lambda, c)$ and $f_W(u; A, B, C)$ coincide. Finally, derive that value c such that U follows a standard F random variable.

(e) Calculate the rth raw moment of $W \sim$ G3F (α, β, λ).

(f) Verify that the mean and variance of $W \sim$ G3F $(\alpha', \beta', \lambda')$ coincide with those of the standard F distribution.

7.4. ★ ★ Let $G_i \overset{\text{ind}}{\sim}$ Gam (n_i, s_i), with

$$f_{G_i}(g; n, s) = \frac{s^n}{\Gamma(n)} e^{-sg} g^{n-1} \mathbb{I}_{(0,\infty)}(g), \quad s, n > 0.$$

(Note that s is the scale and n the shape parameter.) Define $R = G_1/(G_1 + G_2)$, $Y = 1/R$, $W = G_1/G_2$ and $V = 1/W = G_2/G_1 = Y - 1$.

(a) From Example I.9.11, we know that, if $s_1 = s_2 = s$, then $R \sim \text{Beta}(n_1, n_2)$. By calculating the density of R for $s_1 \neq s_2$, show that $R \sim \text{G3B}(n_1, n_2, s_1/s_2)$ (see also Bowman, Shenton and Gailey, 1998).

(b) State an expression for the jth raw moments of R.

(c) For $s = s_1 = s_2$, derive the p.d.f. of Y, compute $\mathbb{E}[Y^j]$ and simplify $\mathbb{E}[Y]$.

(d) For $s_1 \neq s_2$, derive the p.d.f.s of Y and V. Then show that

$$W \sim \text{G3F}(n_1, n_2, s_1/s_2).$$

(e) Calculate the jth raw moment of W.

(f) Give (i) the mean, (ii) an expression for the higher *integer* raw moments, and (iii) the variance of Y.

(g) Derive an expression for $\mathbb{E}[Y^j]$ for $j \in \mathbb{R}$ and specify when it is valid.

(h) Derive the p.d.f. of $R = \sum_{i=1}^{h} G_i / \sum_{i=1}^{k} G_i$ for known $1 \leq h \leq k$ and assuming $s = s_i \ \forall i$.

7.5. ★ ★ Let $X \sim \text{Log}(x; \alpha, \beta)$ with p.d.f. given in (I.7.51) by

$$f_X(x; \alpha, \beta) = \beta^{-1} \exp\left\{-\frac{x - \alpha}{\beta}\right\} \left(1 + \exp\left\{-\frac{x - \alpha}{\beta}\right\}\right)^{-2}, \tag{7.48}$$

for $\alpha \in \mathbb{R}$ and $\beta \in \mathbb{R}_{>0}$, and support on the whole real line.

(a) First verify that

$$\text{(i)} \int \ln x \, dx = x \ln x - x \quad \text{and} \quad \text{(ii)} \int_0^1 \ln \frac{u \, du}{1 - u} = 0.$$

Then evaluate $\mathbb{E}[X] = \int_{-\infty}^{\infty} x f_X(x) \, dx$ by using the transformation

$$u = \frac{e^{-\frac{x-\alpha}{\beta}}}{1 + e^{-\frac{x-\alpha}{\beta}}}.$$

(b) From $\mathbb{M}_Y(t)$, the m.g.f. of $Y = \beta^{-1}(X - \alpha)$, calculate $\mathbb{E}[Y]$ and $\mathbb{V}(Y)$ and, hence, $\mathbb{E}[X]$ and $\mathbb{V}(X)$.

7.6. ★ ★ The so-called type IV generalized logistic distribution is a useful generalization of (7.48), with (location zero, scale one) density

$$f_{\text{GLog}}(x; p, q, 0, 1) = \frac{1}{B(p, q)} \frac{e^{-qx}}{(1 + e^{-x})^{p+q}}, \quad p, q \in \mathbb{R}_{>0}, \tag{7.49}$$

with support on the whole real line. It first appeared in Fisher (1921) and has been studied by Ahuja and Nash (1967), Prentice (1975), Barndorff-Nielsen, Kent and Sørensen (1982), and McDonald (1991, 1997).

Let $X \sim \text{GLog}(x; p, q, 0, 1)$.

(a) Show that $\int_{-\infty}^{\infty} f_X(x)\, dx = 1$.

(b) More generally, derive an expression for the c.d.f. of X in terms of the incomplete beta ratio.

(c) Calculate the m.g.f. of X.

(d) Calculate $\mathbb{E}[X]$ and $\mathbb{V}(X)$ in terms of the digamma function (see Problem 2.19). Hint:

$$\frac{d^2}{ds^2}\Gamma(s) = \Gamma(s)\left[\psi^2(s) + \psi'(s)\right],$$

where $\psi'(s) = d^2 \ln \Gamma(s)\, / ds^2$ is the *trigamma* function.

7.7. ★ ★ The generalized asymmetric t (GAt) density is given by

$$f_{\text{GAt}}(z; d, v, \theta) = K \times \begin{cases} \left(1 + \dfrac{(-z \cdot \theta)^d}{v}\right)^{-\left(v+\frac{1}{d}\right)}, & \text{if } z < 0, \\[4mm] \left(1 + \dfrac{(z/\theta)^d}{v}\right)^{-\left(v+\frac{1}{d}\right)}, & \text{if } z \geq 0, \end{cases} \quad (7.50)$$

$d, v, \theta \in \mathbb{R}_{>0}$. It is noteworthy because limiting cases include the GED (7.1), and hence the Laplace and normal, while the Student's t (and, thus, the Cauchy) distributions are special cases. For $\theta > 1$ ($\theta < 1$) the distribution is skewed to the right (left), while for $\theta = 1$ it is symmetric. It has been shown by Mittnik and Paolella (2000) and Giot and Laurent (2004) to be of great use in the context of modelling the conditional returns on financial assets, and Paolella (2004) for so-called partially adaptive estimation of a macroeconomic model of money demand. Let $Z \sim \text{GAt}(d, v, \theta)$.

(a) Calculate the rth raw integer moment ($0 < r < vd$) and, in doing so, derive the constant K. Hint: One possible substitution for the integral over $z < 0$ is $u = 1 + (-z\theta)^d v^{-1}$ followed by $x = (u-1)u^{-1}$. As a special case, the mean of Z is

$$\mathbb{E}[Z] = \frac{\theta^2 - \theta^{-2}}{\theta^{-1} + \theta} \frac{B(2/d, v - 1/d)}{B(1/d, v)} v^{1/d} \quad (7.51)$$

when $vd > 1$.

(b) Derive an expression for $\mathbb{E}\left[(|Z| - \gamma Z)^r\right]$ for $|\gamma| < 1$. Hint: Use the previous result.

(c) Show that the c.d.f. of Z is

$$
F_Z(z) = \begin{cases} \dfrac{\overline{B}_L(v, 1/d)}{1 + \theta^2}, & \text{if } z \leq 0, \\[3mm] \dfrac{\overline{B}_U(1/d, v)}{1 + \theta^{-2}} + \left(1 + \theta^2\right)^{-1}, & \text{if } z > 0, \end{cases}
$$

where $\overline{B}$ is the incomplete beta ratio,

$$
L = \frac{v}{v + (-z\theta)^d}, \quad \text{and} \quad U = \frac{(z/\theta)^d}{v + (z/\theta)^d}.
$$

Relate this to the c.d.f. of a Student's t random variable. Program the c.d.f. in Matlab.

(d) Show that $S_r(c) = \mathbb{E}[Z^r \mid Z < c]$ for $c < 0$ is given by

$$
S_r(c) = (-1)^r v^{r/d} \frac{\left(1 + \theta^2\right)}{\left(\theta^r + \theta^{r+2}\right)} \frac{\overline{B}_L(v - r/d, (r+1)/d)}{\overline{B}_L(v, 1/d)}, \quad L = \frac{v}{v + (-c\theta)^d}.
$$

7.8. ★ ★ Let $T \sim \text{JoF}(a, b)$, $a, b \in \mathbb{R}_{>0}$, with p.d.f. (7.14).

(a) Verify that, for $a = b$, the distribution of T reduces to that of a Student's t r.v. with $2a$ degrees of freedom.

(b) Let $B \sim \text{Beta}(a, b)$, $k = a + b$, and define S to be the r.v. given by

$$
S = g(B), \quad \text{where } g : (0, 1) \to \mathbb{R}, \quad g(B) = \frac{\sqrt{k}(2B - 1)}{2\sqrt{B(1 - B)}}. \tag{7.52}
$$

Show that
(i) $B = g^{-1}(S)$, where $g^{-1}(s) = 1/2 + s\left(s^2 + k\right)^{-1/2}/2$,
(ii) $dg^{-1}(s)/ds = k\left(s^2 + k\right)^{-3/2}/2$,
(iii) $S \sim \text{JoF}(a, b)$ and
(iv) $F_S(t) = \overline{B}_y(a, b)$, where $y = 1/2 + t\left(t^2 + k\right)^{-1/2}/2$ and $\overline{B}$ is the incomplete beta ratio.

(c) Let $U \sim \chi^2(2a)$ independent of $V \sim \chi^2(2b)$ and define

$$
R = \frac{\sqrt{k}(U - V)}{2\sqrt{UV}}, \quad k = a + b. \tag{7.53}
$$

Show that $R \sim \text{JoF}(a, b)$. Similarly, let $W = aF/b$, where $F \sim \text{F}(2a, 2b)$. Show that

$$
Z = \frac{\sqrt{k}}{2}\left(W^{1/2} - W^{-1/2}\right) \sim \text{JoF}(a, b).
$$

Hint: State and use well-known relationships between beta, gamma, χ^2, and F random variables.

(d) Derive a general expression for the integer moments of S and show that, as a special case,

$$\mathbb{E}[S] = (a - b) \frac{(a+b)^{1/2}}{2} \frac{\Gamma(a - 1/2)\,\Gamma(b - 1/2)}{\Gamma(a)\,\Gamma(b)}, \qquad (7.54)$$

from which it is clear that, for $a = b$, $\mathbb{E}[S] = 0$.

(e) Compute the expected shortfall of $S \sim \mathrm{JoF}(a, b)$, i.e., $E = \mathbb{E}[S \mid S < c]$.

7.9. ★ Let $(X \mid P = p) \sim \mathrm{Bin}(n, p)$ independent of $P \sim \mathrm{Beta}(a, b)$. The unconditional distribution of X is often referred to as the *beta-binomial distribution*, denoted $\mathrm{BetaBin}(n, a, b)$, with $n \in \mathbb{N}$ and $a, b \in \mathbb{R}_{>0}$.

(a) Calculate $f_X(x)$.

(b) Calculate $\mathbb{E}[X]$ and $\mathbb{V}(X)$.

(c) For $n = 15$, compute values a and b such that $\mathbb{E}[X] = 6$ and $\mathbb{V}(X) = 9$ and plot the resulting mass function.

8

The stable Paretian distribution

For tests of significance and maximum likelihood estimation, the comprehensive assumption of normally-distributed variables seems ubiquitous, even to the extent of being unstated in many articles and research reports!

(Hamouda and Rowley, 1996, p. 135)

The stable distribution exhibits heavy tails and possible skewness, making it a useful candidate for modelling a variety of data which exhibit such characteristics. However, there is another property of the stable which separates its use from that of other, *ad hoc* distributions which also possess fat tails and skewness. This is the result of the *generalized central limit theorem*, which, as its name suggests, generalizes the standard central limit theorem (Section 4.4). In particular, it relaxes the finite-mean and finite-variance constraints, allowing the sequence of r.v.s in the sum to be *any* set of i.i.d. r.v.s.

This chapter provides a very basic introduction to the univariate stable Paretian distribution. The emphasis is on its computation and basic properties. Detailed monographs and collections on various aspects of stable Paretian distributions include Zolotarev (1986), Samorodnitsky and Taqqu (1994), Janicki and Weron (1994), Nikias and Shao (1995), Adler, Feldman and Taqqu (1998), Uchaikin and Zolotarev (1999) and Nolan (2007). Those which concentrate more on the applications of stable Paretian r.v.s in finance include Rachev and Mittnik (2000), Rachev (2003) and the contributions in the special issues of *Mathematical and Computer Modelling*, volumes 29 (1999) and 34 (2001).

8.1 Symmetric stable

Example 7.9 demonstrated how the Cauchy and normal r.v.s could be easily nested via use of their characteristic functions, i.e., by taking

$$\varphi_X(t; \alpha) = \exp\left\{-|t|^\alpha\right\}, \quad 0 < \alpha \leq 2. \tag{8.1}$$

Intermediate Probability: A Computational Approach M. Paolella
© 2007 John Wiley & Sons, Ltd

For $\alpha = 2$, $\varphi_X(t; \alpha) = \exp\{-t^2\}$, which is the same as $\exp\{-t^2\sigma^2/2\}$ with $\sigma^2 = 2$, i.e., as $\alpha \to 2$, X approaches the $N(0, 2)$ distribution.

Listing 8.1 uses the inversion formulae from Chapter 1 to compute the p.d.f. and c.d.f. corresponding to $\varphi_X(t; \alpha)$. Equivalently, as in Example 1.22 for the Cauchy c.f. but using (8.1),

$$2\pi f_X(x) = \int_{-\infty}^{\infty} e^{-ixt} e^{-|t|^\alpha} dt$$

$$= \int_{-\infty}^{0} \cos(tx) e^{-(-t)^\alpha} dt - i \int_{-\infty}^{0} \sin(tx) e^{-(-t)^\alpha} dt$$

$$+ \int_{0}^{\infty} \cos(tx) e^{-t^\alpha} dt - i \int_{0}^{\infty} \sin(tx) e^{-t^\alpha} dt$$

or

$$f_X(x) = \frac{1}{\pi} \int_{0}^{\infty} \cos(tx) e^{-t^\alpha} dt. \tag{8.2}$$

Location and scale parameters are incorporated as usual by setting $Y = cX + \mu, c > 0$, and using $f_Y(y; \alpha, \mu, c) = c^{-1} f_X((y - \mu)/c)$. For numerically computing (8.2), use of the substitution $u = 1/(1 + t)$ as in (1.60) leads to

$$f_X(x) = \frac{1}{\pi} \int_{0}^{1} \cos\left(x\frac{1-u}{u}\right) \exp\left(-\left(\frac{1-u}{u}\right)^\alpha\right) u^{-2} du, \tag{8.3}$$

while use of $u = t/(1 + t)$ leads to

$$f_X(x) = \frac{1}{\pi} \int_{0}^{1} \cos\left(x\frac{u}{1-u}\right) \exp\left(-\left(\frac{u}{1-u}\right)^\alpha\right) (1 - u)^{-2} du. \tag{8.4}$$

Either one can be used; the integrand in (8.4) is the same as that in (8.3), but flipped about $u = 1/2$. When a large number of p.d.f. points is desired, use of the FFT and linear interpolation will be (much) faster, as discussed in Section 1.3. The c.d.f. is computed via (1.71).

It can be shown (see Lukacs, 1970, Section 5.7, and the references therein) that only values of α such that $0 < \alpha \le 2$ give rise to a valid c.f. For illustration, the 'p.d.f.' corresponding to $\alpha = 25$ is shown in Figure 8.1; it is negative for intervals of x. While no density can be negative, multimodality is, in general, certainly possible. However, for $0 < \alpha \le 2$, it can be shown (Lukacs, 1970, Section 5.10) that the p.d.f. is unimodal (this also holds for the asymmetric case discussed below).

The characteristic function (8.1) with $0 < \alpha \le 2$ defines a class of distributions which nests the Cauchy and normal; it is called the *symmetric stable Paretian distribution with tail index* α, or the α-stable distribution for short. Note that, as $\varphi_X(t; \alpha)$ is real, f_X must be symmetric about zero. The study of this, and the more general asymmetric case shown below, goes back to the work of Paul Lévy in the 1920s and 1930s.

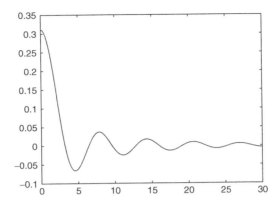

Figure 8.1 Inversion of the c.f. $\exp\{-|t|^{25}\}$, which does not lead to a proper p.d.f.

We will write $S_\alpha(\mu, c)$, where μ and $c > 0$ are location and scale parameters, though the notation SαS ('symmetric alpha stable') is also popular and emphasizes the symmetry. Note that, if $Z \sim S_\alpha(0, 1)$ and $X = \mu + cZ$, then $X \sim S_\alpha(\mu, c)$ and $\varphi_X(t) = \exp(i\mu t - c^\alpha |t|^\alpha)$.

One of the most important properties of the stable distribution is summability (or *stability*, hence part of the name): with $X_i \overset{\text{ind}}{\sim} S_\alpha(\mu_i, c_i)$ and $S = \sum_{i=1}^n X_i$,

$$\varphi_S(t) = \prod_{i=1}^n \varphi_{X_i}(t) = \exp\left(i\mu_1 t - c_1^\alpha |t|^\alpha\right) \cdots \exp\left(i\mu_n t - c_n^\alpha |t|^\alpha\right)$$

$$= \exp\left(i\mu t - c^\alpha |t|^\alpha\right),$$

i.e.,

$$S \sim S_\alpha(\mu, c), \quad \text{where} \quad \mu = \sum_{t=1}^n \mu_i \quad \text{and} \quad c = \left(c_1^\alpha + \cdots + c_n^\alpha\right)^{1/\alpha}.$$

The word 'Paretian' in the name reflects the fact that the asymptotic tail behaviour of the S_α distribution is the same as that of the Pareto distribution, i.e., S_α has power tails, for $0 < \alpha < 2$. In particular, it can be shown that, for $X \sim S_\alpha(0, 1)$, $0 < \alpha < 2$, as $x \to \infty$,

$$\overline{F}_X(x) = \Pr(X > x) \approx k(\alpha) x^{-\alpha}, \quad \text{where} \quad k(\alpha) = \pi^{-1} \sin(\pi\alpha/2) \Gamma(\alpha),$$

$$(8.5)$$

and $a \approx b$ means that a/b converges to unity as $x \to \infty$. Informally, differentiating the limiting value of $1 - \overline{F}_X(x)$ gives the asymptotic density in the right tail,

$$f_X(x) \approx \alpha k(\alpha) x^{-\alpha-1}. \tag{8.6}$$

Expressions for the asymptotic left-tail behaviour follow from the symmetry of the density about zero, i.e., $f_X(x) = f_X(-x)$ and $F_X(x) = \overline{F}_X(-x)$. The term Paretian goes back to Mandelbrot (1960), who pointed out that Vilfredo Pareto had suggested the use of distributions with power tails at the end of the nineteenth century (far before the stable class was characterized) for modelling a variety of economic phenomena.

Recall the discussion in Chapter I.7 on power tails and how the thickness of the tails dictates the maximally existing moment. From the Pareto moments (I.7.25) and result (8.5), it follows that, for $0 < \alpha < 2$, the (fractional absolute) moments of $X \sim S_\alpha$ of order α and higher do not exist, i.e., $\mathbb{E}\left[|X|^r\right]$ is finite for $r < \alpha$, and infinite otherwise. (We will detail this below in Section 8.3.) For $\alpha = 2$, all positive moments exist. This indeed coincides with the special cases Cauchy ($\alpha = 1$) and normal ($\alpha = 2$). If $\alpha > 1$, then the mean exists,[1] and, from the symmetry of the density, it is clearly zero. The variance does not exist unless $\alpha = 2$.

Figure 8.2 shows the densities of $X \sim S_\alpha (0, 1)$ (on the half line) corresponding to several values of α. The plot on the right better illustrates the power-tail behaviour of the density, and the increasing weight of the tails as α moves from 2 to 0.

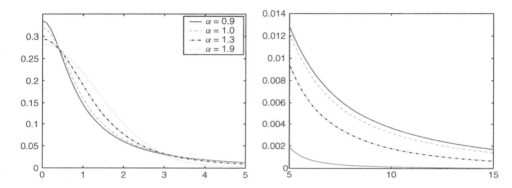

Figure 8.2 Density functions of symmetric stable Paretian r.v.s

Numeric calculation of the density is unfortunately not as simple as it appears. As α moves from 2 towards 0, and as $|x|$ increases, the integrand in (1.60) begins to exhibit increasingly problematic oscillatory behaviour. Thus, in the worst case, the tail of the density for small values of α is practically impossible to determine numerically via (1.60). The use of the FFT to obtain the density will also suffer accordingly; see Menn and Rachev (2006) and Mittnik, Doganoglu and Chenyao (1999) for further details on the use of the FFT in this context. Fortunately, for most applied work, interest centres on values of α larger than 1, and observations (x-values) 'excessively' far into the tail do not occur.[2] Moreover, the simple limiting approximation for the p.d.f. from (8.6)

---

[1] Distributions with $\alpha \leq 1$ might appear to be of only academic interest, but there exist data sets, such as file sizes of data downloaded from the World Wide Web, whose tails are so heavy that the mean does not exist; see Resnick and Rootzén (2000) and the references therein.

[2] Of course, all values on $\mathbb{R}$ are possible, and for small α, values 'extremely' far from zero have non-negligible probability. In practice, use of S_α is only an approximation to the distribution of real data, which are often restricted to a finite-length interval. For example, the lowest return on a financial asset (say, a stock price) is -100%, while an upper limit could be approximated based on structural characteristics of the relevant financial and monetary institutions. Note that such constraints will ultimately apply to all conceivable situations in the universe, so that *any* distribution with infinite support can only be an approximation.

becomes increasingly accurate as one moves further into the tails, obviating the need to grapple with the numeric complexities of exact evaluation.

8.2 Asymmetric stable

The symmetric stable Paretian distribution can be extended to allow for asymmetry. To help motivate the (much less intuitive) c.f. in this case, let X be Cauchy, or $S_1(0, 1)$, and consider introducing a parameter, say β, into the c.f. such that, for $\beta \neq 0$, X is skewed. If we take the form $\varphi(t) = \exp\{-|t|(1 + \beta h(t))\}$, where $h(t)$ is some function, then summability still holds. Let $S = \sum_{i=1}^{n} X_i$, where X_i are independent r.v.s with $\varphi_{X_i}(t) = \exp\{-|t|(1 + \beta_i h(t))\}$. Then, with $\overline{\beta} = n^{-1} \sum_{i=1}^{n} \beta_i$,

$$\ln \varphi_S(t) = \ln \varphi_{X_1}(t) + \cdots + \ln \varphi_{X_n}(t) = -n |t| \left[1 + \overline{\beta} h(t) \right],$$

i.e., the c.f. of S is still of the form $\varphi(t) = \exp\{-|t|(1 + \beta h(t))\}$, but with the introduction of a scale parameter n.

The c.f. of the *asymmetric stable Paretian distribution* with tail index $\alpha = 1$ and skewness parameter β, $-1 \leq \beta \leq 1$, takes $h(t)$ to be $(2i/\pi)\, \mathrm{sgn}(t) \ln |t|$, where $\mathrm{sgn}(\cdot)$ is the sign (or signum) function, so that

$$\ln \varphi(t) = -c|t| \left[1 + i\beta \frac{2}{\pi} \mathrm{sgn}(t) \ln |t| \right] + i\mu t, \tag{8.7}$$

designated $S_{1,\beta}(\mu, c)$. Thus, with $X_i \overset{\mathrm{ind}}{\sim} S_{1,\beta_i}(0, 1)$, $S = \sum_{i=1}^{n} X_i$, and $\overline{\beta} = n^{-1} \sum_{i=1}^{n} \beta_i$,

$$\ln \varphi_S(t) = -n |t| \left[1 + i\overline{\beta} \frac{2}{\pi} \mathrm{sgn}(t) \ln |t| \right],$$

i.e., $S \sim S_{1,\overline{\beta}}(0, n)$. The density is clearly symmetric for $\beta = 0$ and skewed to the right (left) for $\beta > 0$ ($\beta < 0$). When $|\beta| = 1$, the density is said to be maximally or totally skewed. The p.d.f. and c.d.f. can be computed by computing the usual inversion formulae applied to (8.7). Figure 8.3 plots the density corresponding to $\beta = 0$ (Cauchy) and $\beta = 1$. Numerically speaking, in this case, the integral is quite 'stable' (pun intended) for the x-range shown, and all $|\beta| \leq 1$.

The most general case is the *asymmetric stable Paretian distribution*, with tail index α, $0 < \alpha \leq 2$, and skewness parameter β, $-1 \leq \beta \leq 1$. For brevity, we refer to this as just the stable distribution, and write $X \sim S_{\alpha,\beta}(\mu, c)$. The stable c.f. is given by

$$\ln \varphi_X(t) = \begin{cases} -c^\alpha |t|^\alpha \left[1 - i\beta\, \mathrm{sgn}(t) \tan \dfrac{\pi \alpha}{2} \right] + i\mu t, & \text{if } \alpha \neq 1, \\[2ex] -c|t| \left[1 + i\beta \dfrac{2}{\pi} \mathrm{sgn}(t) \ln |t| \right] + i\mu t, & \text{if } \alpha = 1. \end{cases} \tag{8.8}$$

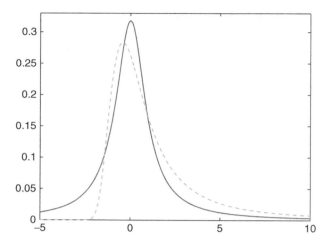

Figure 8.3 Cauchy (solid) and 'totally skewed Cauchy' (dashed) computed by inverting (8.7) with $\beta = 1$

As $\alpha \to 2$, the effect of β diminishes because $\tan(\pi\alpha/2) \to 0$; and when $\alpha = 2$, $\tan(\pi) = 0$, and β has no effect. Thus, there is no 'skewed normal' distribution within the stable family.

The form of (8.8) does not lend itself to much intuition. It arises from characterizing the class of distributions for which summability holds, and is quite an involved derivation well beyond the scope of this text; see Lukacs (1970, Section 5.7) and the references therein for details. As in the symmetric case, summability means that the sum of independent stable r.v.s, *each with the same tail index* α, also follows a stable distribution. In particular,

$$\text{if} \quad X_i \overset{\text{ind}}{\sim} S_{\alpha,\beta_i}(\mu_i, c_i), \quad \text{then} \quad S = \sum_{i=1}^{n} X_i \sim S_{\alpha,\beta}(\mu, c),$$

where

$$\mu = \sum_{t=1}^{n} \mu_i, \quad c = \left(c_1^\alpha + \cdots + c_n^\alpha\right)^{1/\alpha}, \quad \text{and} \quad \beta = \frac{\beta_1 c_1^\alpha + \cdots + \beta_n c_n^\alpha}{c_1^\alpha + \cdots + c_n^\alpha}.$$

The class of stable distributions is the only one which has the property of summability (or stability). Of course, it also holds for the normal and Cauchy, which are nested in the stable family. (Recall that sums of independent gamma r.v.s are also gamma, but only if their scale parameters are the same.)

Using the program in Listing 8.1, Figure 8.4 shows the p.d.f. for various α and $\beta = 0.9$ (and location zero, scale one). It appears somewhat strange but it is correct: for positive β, as $\alpha \uparrow 1$, the mode drifts to $+\infty$, while for $\alpha \downarrow 1$, the mode drifts to $-\infty$.

```
function [f,F] = asymstab(xvec,a,b)

bordertol=1e-8; lo=bordertol; hi=1-bordertol; tol=1e-7;
x1=length(xvec); F=zeros(x1,1); f=F;
for loop=1:length(xvec)
  x=xvec(loop); dopdf=1;
  f(loop)= quadl(@fff,lo,hi,tol,[],x,a,b,1) / pi;
  if nargout>1
    F(loop)=0.5-(1/pi)* quadl(@fff,lo,hi,tol,[],x,a,b,0);
  end
end;

function I=fff(uvec,x,a,b,dopdf);
for ii=1:length(uvec)
  u=uvec(ii);   t  = (1-u)/u;
  if a==1
    cf = exp( -abs(t)*( 1 + i*b*(2/pi)*sign(t) * log(t) ) );
  else
    cf = exp( - ((abs(t))^a) *( 1 - i*b*sign(t) * tan(pi*a/2) ) );
  end
  z  = exp(-i*t*x) .* cf;
  if dopdf==1, g=real(z); else g=imag(z)./t; end
  I(ii)  = g / u^2;
end
```

Program Listing 8.1 Computes the p.d.f. and c.d.f. of the asymmetric stable Paretian distribution. The value of `bordertol` dictates how close the range of integration comes to the edges of [0, 1]; a value of 10^{-8} was found to work well in practice. The tolerance on the integration error is passed as `tol` to the function `quadl`; its optimal value depends on x, α, and β, with too small a value resulting in 'chaotic' behaviour and gross errors. Values between 10^{-7} and 10^{-9} appear to work best

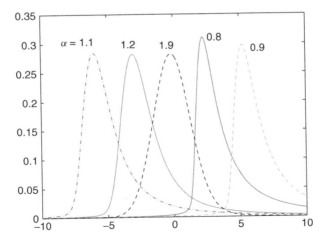

Figure 8.4 Densities of $S_{\alpha,0.9}(0, 1)$ for various α

At $\alpha = 1$, the mode is near zero (and depends on β). This phenomenon is of course less pronounced when β is not so extreme; Figure 8.5 shows p.d.f.s using the same values of α as in Figure 8.4, but with $\beta = 0.2$. There are other parameterizations of the $S_{\alpha,\beta}$ c.f. which are more advantageous, depending on the purpose; Zolotarev's (M) parameterization (Zolotarev, 1986, p. 11) avoids the 'erratic mode' problem. Nolan (2007) offers a detailed discussion of the various parameterizations and their respective advantages.

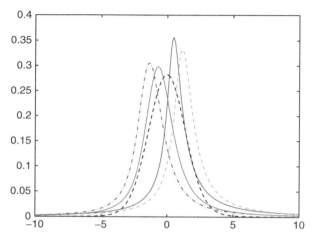

Figure 8.5   Same as Figure 8.4, but with $\beta = 0.2$

The same power law (rate of decay) applies to the tail behaviour of the asymmetric stable Paretian distribution, but the parameter β dictates the 'relative heights' of the two tails. In particular, for $X \sim S_{\alpha,\beta}(0, 1)$, as $x \to \infty$,

$$\overline{F}_X(x) = \Pr(X > x) \approx k(\alpha)(1 + \beta) x^{-\alpha} \qquad (8.9)$$

and

$$f_X(x) \approx \alpha k(\alpha)(1 + \beta) x^{-\alpha-1}, \qquad (8.10)$$

where $k(\alpha)$ is given in (8.5). For the left tail, the symmetry relations $f_X(x; \alpha, \beta) = f_X(-x; \alpha, -\beta)$ and $F_X(x; \alpha, \beta) = \overline{F}_X(-x; \alpha, -\beta)$ can be used. Figure 8.6 shows an example of the quality of these asymptotic results by plotting the RPE of using (8.10) to approximate the density, with $\alpha = 1.5$ and $\beta = 0.5$. The magnitude of the ordinate x in each tail was chosen such that, for larger values, the numeric integration (8.2) began to break down.[3]

---

[3] Obviously, this is not an absolute statement on the use of (8.2) (via (1.60)) for these parameter values, but rather an artefact of the simple, implementation we used, based on Matlab's general integration routine quadl with convergence tolerance 10^{-9} and range of integration $0 + \epsilon$ to $1 - \epsilon$, with $\epsilon = 10^{-8}$, in (1.60). More sophisticated methods of integration exist which are designed for oscillatory integrands, though, for our purposes, this is adequate. Of course, use of better routines would also prevent the glitches in accuracy which are apparent in Figure 8.6.

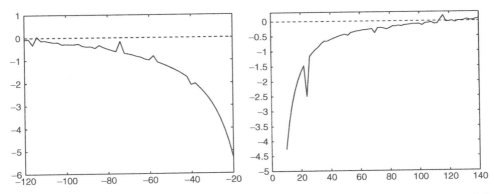

Figure 8.6 Relative percentage error of using (8.10) to approximate the $S_{1.5,0.5}$ density in the left and right tails

From the asymptotic tail behaviour, we can ascribe a particular meaning to parameter β. Let $\alpha < 2$ and $X \sim S_{\alpha,\beta}(0, 1)$. Then, from (8.9) and the symmetry relation $F_X(x; \alpha, \beta) = \bar{F}_X(-x; \alpha, -\beta)$, a calculation devoid of any mathematical rigour suggests (correctly), as $x \to \infty$,

$$\frac{P(X > x) - P(X < -x)}{P(X > x) + P(X < -x)} \approx \frac{k(\alpha)(1 + \beta)x^{-\alpha} - k(\alpha)(1 - \beta)x^{-\alpha}}{k(\alpha)(1 + \beta)x^{-\alpha} + k(\alpha)(1 - \beta)x^{-\alpha}}$$

$$= \frac{2k(\alpha)\beta x^{-\alpha}}{2k(\alpha)x^{-\alpha}} = \beta,$$

i.e., β measures the asymptotic difference in the two tail masses, scaled by the sum of the two tail areas.

There is one more special case of the stable distribution for which a closed-form expression for the density exists, namely when $\alpha = 1/2$ and $\beta = 1$, which is often referred to as the Lévy, or Smirnov, distribution. If $X \sim S_{1/2,1}(0, 1)$, then

$$f_X(x) = (2\pi)^{-1/2} x^{-3/2} e^{-1/(2x)} \mathbb{I}_{(0,\infty)}(x). \tag{8.11}$$

The Lévy distribution arises in the context of hitting times for Brownian motion (see, for example, Feller, 1971, pp. 52, 173). Example I.7.6 showed that it arises as a limiting case of the inverse Gaussian distribution, and Example I.7.15 showed that, if $Z \sim N(0, 1)$, then Z^{-2} follows a Lévy distribution. Example 1.19 discussed the relationship between its m.g.f. and its c.f.

Two applications of l'Hôpital's rule show that $f_X(x) \downarrow 0$ as $x \downarrow 0$. The exact density (8.11) is plotted in Figures 8.7 and 8.8. (Numeric inversion of the c.f. also works in this rather extreme case, but only for about $x < 10$.) The tail approximation (8.10) simplifies in this case to $(2\pi)^{-1/2} x^{-3/2}$, which should be obvious from the density (8.11). It is also shown in Figure 8.8.

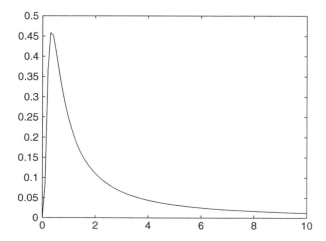

Figure 8.7   The p.d.f. of the Lévy distribution (8.11)

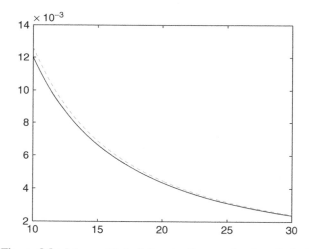

Figure 8.8   Lévy p.d.f. (solid) and tail approximation (dashed)

Other than the normal, Cauchy and Lévy distributions, no other cases exist with closed-form p.d.f.s, although in 1954 Zolotarev expressed the p.d.f. of a handful of other cases in terms of higher transcendental functions, these being $(\alpha = 2/3, \beta = 1)$, $(\alpha = 2/3, \beta = 0)$, $(\alpha = 3/2, \beta = 1)$, $(\alpha = 1/3, \beta = 1)$, and $(\alpha = 1/2, -1 \leq \beta \leq 1)$ (see Lukacs, 1970, p. 143).

8.3 Moments

8.3.1 Mean

As in the symmetric case, it is clear from (8.10) that moments of $X \sim S_{\alpha,\beta}$ of order α and higher do not exist when $\alpha < 2$. When $1 < \alpha \leq 2$, the mean of $X \sim S_{\alpha,\beta}(\mu, c)$ is μ, irrespective of β. This appears implausible for $\beta \neq 0$, but recall Figure 8.4, which shows that, for $\beta > 0$, the mode of the density is negative, and continues to move left as $\alpha \downarrow 1$, so that the mass in the thicker right tail is balanced out by the mass around the mode. Consider how this would be numerically verified: using (8.10), the p.d.f. can be approximated by $K_- (-x)^{-(\alpha+1)}$ as $x \rightarrow -\infty$, and by $K_+ (x)^{-(\alpha+1)}$ as $x \rightarrow +\infty$, where

$$K_- = \pi^{-1}\alpha\Gamma(\alpha) \sin(\pi\alpha/2)(1-\beta), \qquad K_+ = \pi^{-1}\alpha\Gamma(\alpha)\sin(\pi\alpha/2)(1+\beta),$$

so that the mean (for $\alpha > 1$) can be approximated by

$$\mathbb{E}[X] \approx K_- \int_{-\infty}^{\ell} x(-x)^{-(\alpha+1)} dx + \int_{\ell}^{h} xf_X(x) dx + K_+ \int_{h}^{\infty} x(x)^{-(\alpha+1)} dx,$$

$$(8.12)$$

where the interval $\ell < x < h$ denotes the region for which the c.f. inversion formula for the p.d.f. f_X can be accurately evaluated. Writing the first integral as

$$\int_{-\infty}^{\ell} x(-x)^{-(\alpha+1)} dx = \int_{-\infty}^{\ell} (-1)(-x)(-x)^{-(\alpha+1)} dx$$

$$= -\int_{-\infty}^{\ell} (-x)^{1-(\alpha+1)} dx = \int_{\infty}^{-\ell} u^{1-(\alpha+1)} du$$

$$= -\int_{-\ell}^{\infty} u^{1-(\alpha+1)} du = \frac{(-\ell)^{1-\alpha}}{1-\alpha},$$

and the third as

$$\int_{h}^{\infty} x^{1-(\alpha+1)} dx = -\frac{h^{1-\alpha}}{1-\alpha},$$

(8.12) reduces to

$$\mathbb{E}[X] \approx \frac{K_-(-\ell)^{1-\alpha}}{1-\alpha} - \frac{K_+ h^{1-\alpha}}{1-\alpha} + \int_{\ell}^{h} xf_X(x) dx. \qquad (8.13)$$

Evaluating (8.13) for $\alpha = 1.5$ and $\beta = 0.5$, with $\ell = -120$ and $h = 140$ (these low and high cutoff values were found during the exercise which produced Figure 8.6) and an integration error tolerance of 10^{-6}, we get -8.5×10^{-6}, which seems quite good, given the use of the tail approximation.[4]

8.3.2 Fractional absolute moment proof I

In empirical applications, the quantity $\mathbb{E}\left[\,|X|^r\,\right]$, i.e., the rth fractional absolute moment, $r < \alpha$, is often of value.[5] We derive an expression for $\mathbb{E}\left[\,|X|^r\,\right]$ via two methods, restricting attention to the case of $\alpha \neq 1$, $\mu = 0$, and $c = 1$. The derivations are somewhat long, but just involve a variety of basic concepts from calculus which we give in great detail, so that they are both readable and instructive. They can, of course, be skipped; the final result is given below in (8.26).

The first proof we illustrate follows that given in Kuruoglu (2001) and the references therein, and involves a nonobvious trick, namely working with a particular function times $\mathbb{E}\left[|X|^r\right]$. The interested reader could also pursue the references given in Samorodnitsky and Taqqu (1994, p. 18).

First we consider the case of $r > 0$. Let

$$\rho(r) := \int_0^\infty u^{-(r+1)} \sin^2(u)\, du. \tag{8.14}$$

A closed-form expression for (8.14) is given in Hoffmann-Jørgensen (1994, p. 128) and can be easily derived from the general result in Gradshteyn and Rynhik (2007, formula 3.823, p. 460), which states that

$$\int_0^\infty u^{(\mu-1)} \sin^2(bu)\, du = -\frac{\Gamma(\mu)\cos(\mu\pi/2)}{2^{\mu+1}\, b^\mu}, \qquad b > 0,\ \mathrm{Re}(\mu) \in (-2, 0) \setminus \{-1\}. \tag{8.15}$$

By setting $b = 1$ and $\mu = -r$, $r \in (0, 2) \setminus \{1\}$, in (8.15), we have

$$\rho(r) = -2^{r-1}\Gamma(-r)\cos\left(-\frac{r\pi}{2}\right) = 2^{r-1}\frac{1}{r}\Gamma(1-r)\cos\left(\frac{r\pi}{2}\right).$$

It can be shown (e.g., using the dominated convergence theorem) that $\rho(r)$ depends continuously on r, and the expression for $\rho(1)$ is then obtained as follows. From Euler's reflection formula (1.14) and the well-known relation

$$\sin a = 2\sin(a/2)\cos(a/2), \tag{8.16}$$

---

[4] To illustrate just how heavy the tails are, integrating 'only' from $-10\,000$ to $10\,000$ (replacing the $\pm\infty$ with $\pm10\,000$ in (8.12)) results in -0.0060. This accuracy of this result is questionable, however, depending on numerical issues inherent in the procedures and tolerances used for c.f. inversion and numeric integration. The interested reader is encouraged to investigate this in further detail, possibly using more sophisticated software for its evaluation.

[5] For example, in empirical finance, it plays a central role in the so-called stable GARCH process; see Mittnik, Paolella and Rachev (2002) for details.

we have

$$\rho\,(r) = \frac{2^{r-1}}{r}\,\frac{\pi}{\sin(r\pi)\Gamma(r)}\cos\left(\frac{\pi r}{2}\right) = \frac{2^{r-1}}{r}\,\frac{\pi}{2\sin(r\pi/2)\cos(r\pi/2)\Gamma(r)}\cos\left(\frac{\pi r}{2}\right),$$

so that, for $r = 1$,

$$\rho\,(1) = \frac{\pi}{\Gamma(1)2\sin(\pi/2)} = \frac{\pi}{2}.$$

Thus,

$$\rho\,(r) = \begin{cases} r^{-1}2^{r-1}\Gamma(1-r)\cos(\pi r/2), & \text{if } 0 < r < 2, r \neq 1, \\ \pi/2, & \text{if } r = 1. \end{cases}$$

Then

$$\rho\,(r)\,\mathbb{E}\big[\,|X|^r\,\big] = \int_0^\infty u^{-(r+1)}\sin^2(u)\,du \int_{-\infty}^\infty |x|^r f_X(x)\,dx$$

$$= \int_0^\infty u^{-(r+1)}\sin^2(u)\,du \int_0^\infty x^r\big(f_X(x) + f_X(-x)\big)\,dx$$

$$= \int_0^\infty x^r\big(f_X(x) + f_X(-x)\big)\left[\int_0^\infty u^{-(r+1)}\sin^2(u)\,du\right]dx$$

or, with $t = u/x$,

$$\rho\,(r)\,\mathbb{E}\big[\,|X|^r\,\big] = \int_0^\infty x^r\big(f_X(x) + f_X(-x)\big)\left[\int_0^\infty t^{-(r+1)}x^{-r}\sin^2(tx)\,dt\right]dx$$

$$= \int_0^\infty \int_0^\infty t^{-(r+1)}\sin^2(tx)\big(f_X(x) + f_X(-x)\big)\,dt\,dx.$$

The integral $\int_0^\infty \sin^2(tx)f_X(x)\,dx$ is clearly convergent for $t \in \mathbb{R}$, and as

$$\int_0^\infty \sin^2(tx)\big(f_X(x) + f_X(-x)\big)\,dx = \int_{-\infty}^0 \sin^2(tx)\big(f_X(x) + f_X(-x)\big)\,dx,$$

we have

$$\rho\,(r)\,\mathbb{E}\big[\,|X|^r\,\big] = \int_0^\infty t^{-(r+1)}\left[\int_0^\infty \sin^2(tx)\big(f_X(x) + f_X(-x)\big)\,dx\right]dt$$

$$= \int_0^\infty t^{-(r+1)}\left[\frac{1}{2}\int_{-\infty}^\infty \sin^2(tx)\big(f_X(x) + f_X(-x)\big)\,dx\right]dt,$$

where the exchange of integrals can be validated by Fubini's theorem.

Next, as $\sin(z) = \left(e^{iz} - e^{-iz}\right)/(2i)$,

$$\rho(r)\,\mathbb{E}\big[\,|X|^r\,\big] = \frac{1}{2}\int_0^\infty t^{-(r+1)}\left[\int_{-\infty}^\infty \frac{\left(e^{itx} - e^{-itx}\right)^2}{-4}\left(f_X(x) + f_X(-x)\right)dx\right]dt$$

$$= -\frac{1}{8}\int_0^\infty t^{-(r+1)}\left[\int_{-\infty}^\infty \left(e^{i2tx} - 2 + e^{-i2tx}\right)\left(f_X(x) + f_X(-x)\right)dx\right]dt$$

$$= -\frac{1}{4}\int_0^\infty t^{-(r+1)}\left[\varphi_X(2t) + \varphi_X(-2t) - 2\right]dt, \tag{8.17}$$

where φ_X is the stable c.f. (8.8), for $\mu = 0$ and $c = 1$.

Observe that, up to this point, neither the particular form of the p.d.f. nor of the c.f. of the stable distribution has been used, so that (8.17) is actually a general result, perhaps of use (algebraic or computational) with other distributions.

It is now convenient to write the c.f. as

$$\varphi_X(t) = \exp\left\{-z_p t^\alpha\right\} \quad \text{and} \quad \varphi_X(-t) = \exp\left\{-z_n t^\alpha\right\}, \quad t \geq 0, \tag{8.18}$$

where

$$z_p := 1 - i\beta\tan(\pi\alpha/2) \quad \text{and} \quad z_n := 1 + i\beta\tan(\pi\alpha/2). \tag{8.19}$$

Use of the transformation $w = (2t)^\alpha$ then gives

$$\rho(r)\,\mathbb{E}\big[\,|X|^r\,\big] = -\frac{1}{4}\int_0^\infty \left(\frac{w^{1/\alpha}}{2}\right)^{-(r+1)}\left[\varphi_X(w^{1/\alpha}) + \varphi_X(-w^{1/\alpha}) - 2\right]\frac{1}{2\alpha}w^{1/\alpha-1}\,dw$$

$$= -\frac{2^{r-2}}{\alpha}\int_0^\infty w^{-r/\alpha-1}\left[\varphi_X(w^{1/\alpha}) + \varphi_X(-w^{1/\alpha}) - 2\right]dw$$

$$= -\frac{2^{r-2}}{\alpha}\int_0^\infty w^{-r/\alpha-1}\left[\exp\left(-z_p w\right) + \exp\left(-z_n w\right) - 2\right]dw. \tag{8.20}$$

Now let

$$u = \exp\left(-z_p w\right) + \exp\left(-z_n w\right) - 2 \quad \text{and} \quad dv = w^{-r/\alpha-1}\,dw,$$

so that

$$du = -\left[z_p\exp\left(-z_p w\right) + z_n\exp\left(-z_n w\right)\right]dw \quad \text{and} \quad v = -\frac{\alpha}{r}w^{-r/\alpha}.$$

Integration by parts then gives

$$\rho(r)\,\mathbb{E}\big[\,|X|^r\,\big] = -\frac{2^{r-2}}{\alpha}\left\{uv\Big|_0^\infty - \int_0^\infty v\,du\right\}$$

$$= \left(\frac{2^{r-2}}{r}\right)\left[\exp\left(-z_p w\right) + \exp\left(-z_n w\right) - 2\right]\left(w^{-r/\alpha}\right)\Big|_0^\infty \tag{8.21}$$

$$+ \frac{2^{r-2}}{r}\int_0^\infty w^{-r/\alpha}\left[z_p\exp\left(-z_p w\right) + z_n\exp\left(-z_n w\right)\right]dw. \tag{8.22}$$

Term (8.21) evaluated at $w = \infty$ is clearly zero. For $w = 0$,

$$\lim_{w \to 0} \left[\left(e^{-z_p w} + e^{-z_n w} - 2 \right) w^{-r/\alpha} \right] = \lim_{w \to 0} \left[\frac{e^{-z_p w} + e^{-z_n w} - 2}{w^{r/\alpha}} \right],$$

and l'Hôpital's rule is applicable.[6] Using it gives

$$\lim_{w \to 0} \left[\frac{-z_p e^{-z_p w} - z_n e^{-z_n w}}{w^{r/\alpha - 1} (r/\alpha)} \right] = \lim_{w \to 0} \left[\frac{\alpha}{r} \left(-z_p e^{-z_p w} - z_n e^{-z_n w} \right) w^{1 - r/\alpha} \right]$$

$$= 0, \quad 0 < r < \alpha.$$

This explicitly shows why moments for $r > \alpha$ do not exist. For the term (8.22), use of the transformations $m = z_p w$ and $m = z_n w$ easily leads to

$$\rho(r) \, \mathbb{E} \left[\, |X|^r \, \right] = \frac{1}{r} 2^{r-2} \Gamma \left(1 - \frac{r}{\alpha} \right) \left(z_p^{r/\alpha} + z_n^{r/\alpha} \right), \quad 0 < r < \alpha. \qquad (8.23)$$

This is essentially the result: together with (8.19), (8.23) can be used to evaluate $\mathbb{E} \left[\, |X|^r \, \right]$ numerically if complex arithmetic is supported.

If desired, the term $z_p^{r/\alpha} + z_n^{r/\alpha}$ can be simplified so that complex numbers are not needed. Let $\theta := \arctan \{ \beta \tan(\pi \alpha / 2) \}$. Then, as

$$\tan z = \frac{\sin z}{\cos z} = \frac{\left(e^{iz} - e^{-iz} \right) / (2i)}{\left(e^{iz} + e^{-iz} \right) / 2} = \frac{e^{iz} - e^{-iz}}{i \left(e^{iz} + e^{-iz} \right)},$$

we have

$$z_p^{r/\alpha} + z_n^{r/\alpha} = \left(1 - i \beta \tan \left(\frac{\pi \alpha}{2} \right) \right)^{r/\alpha} + \left(1 + i \beta \tan \left(\frac{\pi \alpha}{2} \right) \right)^{r/\alpha}$$

$$= (1 - i \tan \theta)^{r/\alpha} + (1 + i \tan \theta)^{r/\alpha}$$

$$= \left(1 - \frac{e^{i\theta} - e^{-i\theta}}{e^{i\theta} + e^{-i\theta}} \right)^{r/\alpha} + \left(1 + \frac{e^{i\theta} - e^{-i\theta}}{e^{i\theta} + e^{-i\theta}} \right)^{r/\alpha}$$

$$= \frac{2^{r/\alpha}}{\left(e^{i\theta} + e^{-i\theta} \right)^{r/\alpha}} \left[\left(e^{-i\theta} \right)^{r/\alpha} + \left(e^{i\theta} \right)^{r/\alpha} \right].$$

But, as $\cos z = \left(e^{iz} + e^{-iz} \right) / 2$,

$$z_p^{r/\alpha} + z_n^{r/\alpha} = \frac{1}{(\cos \theta)^{r/\alpha}} \times 2 \cos \left(\frac{r\theta}{\alpha} \right) = 2 \left(1 + \tan^2 \theta \right)^{r/2\alpha} \cos \left(\frac{r\theta}{\alpha} \right),$$

where the latter equality follows by dividing $\cos^2 x + \sin^2 x = 1$ by $\cos^2 x$ and taking square roots.

---

[6] Notice that w is real, and only the numerator contains complex numbers. The numerator approaches zero iff its real and imaginary parts do, and is differentiable iff its real and imaginary parts are (real) differentiable. A version of l'Hôpital's rule applicable when both the numerator and denominator are complex analytic functions of complex argument w, nonconstant in a disc around zero, is given in Palka (1991, p. 304).

Finally, let $\tau := \tan \theta$, so that

$$z_p^{r/\alpha} + z_n^{r/\alpha} = 2 \left(1 + \tau^2\right)^{r/2\alpha} \cos \left(\frac{r}{\alpha} \arctan \tau\right) \tag{8.24}$$

and, thus,

$$\rho \left(r\right) \mathbb{E}\left[\,|X|^r\,\right] = \frac{1}{r} 2^{r-1} \Gamma \left(1 - \frac{r}{\alpha}\right) \left(1 + \tau^2\right)^{r/2\alpha} \cos \left(\frac{r}{\alpha} \arctan \tau\right) \tag{8.25}$$

or (writing the final result with the wider range $-1 < r < \alpha$ instead of $0 < r < \alpha$, as will be verified below)

$$\boxed{\mathbb{E}\left[\,|X|^r\,\right] = \kappa^{-1} \Gamma \left(1 - \frac{r}{\alpha}\right) \left(1 + \tau^2\right)^{r/2\alpha} \cos \left(\frac{r}{\alpha} \arctan \tau\right), \quad -1 < r < \alpha,} \tag{8.26}$$

where

$$\kappa = \begin{cases} \Gamma(1 - r) \cos(\pi r / 2), & \text{if } r \neq 1, \\ \pi/2, & \text{if } r = 1. \end{cases}$$

It is easy to see that (8.26) is also valid for $r = 0$, and the extension in (8.26) to $-1 < r < \alpha$ instead $0 < r < \alpha$ is now shown.

Let $-1 < r < 0$. From the inversion formula (1.54),

$$
\begin{aligned}
\mathbb{E}\left[\,|X|^r\,\right] &= \frac{1}{2\pi} \int_{-\infty}^{\infty} \int_{-\infty}^{\infty} |x|^r e^{-iux} \varphi_X(u) \, du \, dx \\
&= \frac{1}{2\pi} \int_{-\infty}^{\infty} \int_{0}^{\infty} |x|^r \left[e^{-iux} \varphi_X(u) + e^{iux} \varphi_X(-u)\right] du \, dx \\
&= \frac{1}{2\pi} \int_{0}^{\infty} \int_{0}^{\infty} x^r \left(e^{-iux} + e^{iux}\right) \left(\varphi_X(u) + \varphi_X(-u)\right) du \, dx \\
&= \frac{1}{\pi} \int_{0}^{\infty} \left[\int_{0}^{\infty} x^r \cos(ux) \, dx\right] \left(\varphi_X(u) + \varphi_X(-u)\right) du.
\end{aligned}
$$

Gradshteyn and Ryzhik (2007, formula 3.761(9), p. 437) give the result

$$\int_{0}^{\infty} x^{\mu-1} \cos(ax) \, dx = \frac{\Gamma(\mu)}{a^{\mu}} \cos\left(\frac{\mu\pi}{2}\right), \quad a > 0, \quad 0 < \mathrm{Re}(\mu) < 1,$$

from which, for $a = u$ $(0 < u < \infty)$ and $\mu = r + 1$ $(-1 < r < 0)$, we obtain

$$\int_{0}^{\infty} x^r \cos(ux) \, dx = -\frac{\Gamma(1 + r)}{u^{r+1}} \sin\left(\frac{r\pi}{2}\right),$$

so that

$$\mathbb{E}\left[\,|X|^r\,\right] = -\frac{1}{\pi} \Gamma(1 + r) \sin\left(\frac{r\pi}{2}\right) \int_{0}^{\infty} u^{-1-r} \left(\varphi_X(u) + \varphi_X(-u)\right) du.$$

Now, from (8.18), the transformations $w = z_p u^\alpha$ and $w = z_n u^\alpha$, and (8.24),

$$
\mathbb{E}[|X|^r] = -\frac{1}{\pi}\Gamma(1+r)\sin\left(\frac{r\pi}{2}\right)\int_0^\infty u^{-1-r}\left(\exp(-z_p u^\alpha) + \exp(-z_n u^\alpha)\right) du
$$

$$
= -\frac{1}{\pi}\Gamma(1+r)\sin\left(\frac{r\pi}{2}\right)\frac{1}{\alpha}\Gamma\left(-\frac{r}{\alpha}\right)\left[z_p^{r/\alpha} + z_n^{r/\alpha}\right]
$$

$$
= \frac{\Gamma(1+r)}{r\pi}\sin\left(\frac{r\pi}{2}\right)\Gamma\left(1-\frac{r}{\alpha}\right)2(1+\tau^2)^{r/2\alpha}\cos\left(\frac{r}{\alpha}\arctan\tau\right).
$$

From Euler's reflection formula (1.14) and basic properties of the gamma and sine functions,

$$
\Gamma(1+z) = \frac{\pi}{\Gamma(-z)\sin(\pi+z\pi)} = -\frac{z\pi}{\Gamma(1-z)\sin(\pi+z\pi)} = \frac{z\pi}{\Gamma(1-z)\sin(z\pi)},
$$
$$
\tag{8.27}
$$

giving

$$
\mathbb{E}[|X|^r] = \frac{2\sin(r\pi/2)}{\Gamma(1-r)\sin(r\pi)}\Gamma\left(1-\frac{r}{\alpha}\right)(1+\tau^2)^{r/2\alpha}\cos\left(\frac{r}{\alpha}\arctan\tau\right)
$$

or, via (8.16),

$$
\mathbb{E}[|X|^r] = \frac{\Gamma(1-r/\alpha)}{\Gamma(1-r)}\frac{1}{\cos(r\pi/2)}(1+\tau^2)^{r/2\alpha}\cos\left(\frac{r}{\alpha}\arctan\tau\right), \quad -1 < r < 0,
$$
$$
\tag{8.28}
$$

which justifies (8.26).

8.3.3 Fractional absolute moment proof II

The second method of proving (8.26) is outlined in Zolotarev (1986, Sections 2.1 and 2.2), and also uses a nonobvious trick. One starts with the general relation

$$
|x|^r = \frac{2\sin(\pi/2)\Gamma(r+1)}{\pi}\int_0^\infty t^{-(r+1)}(1-\cos(xt))\,dt, \quad 0 < r < 2, \tag{8.29}
$$

which we first prove, using (8.15).[7] Using the identity $2\sin^2 z = 1 - \cos(2z)$, we can write (8.15) as

$$
\int_0^\infty \frac{1}{2}t^{(\mu-1)}(1-\cos(2bt))\,dt
$$
$$
= -\frac{\Gamma(\mu)\cos(\mu\pi/2)}{2^{\mu+1}b^\mu}, \quad b > 0,\ \mathrm{Re}(\mu) \in (-2,0)\setminus\{-1\},
$$

---

[7] This is from Zolotarev (1986, p. 63, eq. 2.1.10), but his equation is erroneously missing the π in the denominator.

or, because $\cos(z) = \cos(-z)$, as

$$\int_0^\infty \frac{1}{2} t^{(\mu-1)} (1 - \cos(2bt)) \, dt$$

$$= -\frac{\Gamma(\mu) \cos(\mu\pi/2)}{2^{\mu+1} |b|^\mu}, \quad b \in \mathbb{R}, \ \mathrm{Re}(\mu) \in (-2, 0) \setminus \{-1\}.$$

Setting $b = x/2$ and $\mu = -r$, $0 < r < 2$, leads to

$$\int_0^\infty t^{-(r+1)} (1 - \cos(xt)) \, dt = -\frac{\Gamma(-r) \cos(-r\pi/2)}{|x|^{-r}}, \quad x \in \mathbb{R}, \ r \in (0, 2) \setminus \{1\},$$

or

$$|x|^r = -\frac{1}{\Gamma(-r) \cos(r\pi/2)} \int_0^\infty t^{-(r+1)} (1 - \cos(xt)) \, dt, \quad r \in (0, 2) \setminus \{1\}$$

Finally, using (8.16) and the relation $\Gamma(z)\Gamma(-z) = -\pi/[z \sin(z\pi)]$, which easily follows from (8.27) and $\Gamma(z+1) = z\Gamma(z)$, we get

$$|x|^r = \frac{r \sin(r\pi) \Gamma(r)}{\pi \cos(r\pi/2)} \int_0^\infty t^{-(r+1)} (1 - \cos(xt)) \, dt$$

$$= \frac{2 \sin(\pi/2) \Gamma(r+1)}{\pi} \int_0^\infty t^{-(r+1)} (1 - \cos(xt)) \, dt,$$

which is (8.29).

Thus, for any random variable X, taking expectations on both sides gives

$$\mathbb{E}[|X|^r] = \frac{2 \sin(\pi/2) \Gamma(r+1)}{\pi} \int_{-\infty}^\infty \left[\int_0^\infty t^{-(r+1)} (1 - \cos(xt)) \, dt \right] f_X(x) \, dx$$

$$= \frac{2 \sin(\pi/2) \Gamma(r+1)}{\pi} \int_{-\infty}^\infty \left[\int_0^\infty t^{-(r+1)} \left(1 - \frac{e^{itx} + e^{-itx}}{2}\right) dt \right] f_X(x) \, dx$$

$$= \frac{\sin(\pi/2) \Gamma(r+1)}{\pi} \int_{-\infty}^\infty \left[\int_0^\infty t^{-(r+1)} \left(2 - e^{itx} - e^{-itx}\right) dt \right] f_X(x) \, dx$$

or, switching the order of integration,

$$\mathbb{E}[|X|^r] = \frac{\sin(\pi/2) \Gamma(r+1)}{\pi} \int_0^\infty t^{-(r+1)} \left[\int_{-\infty}^\infty \left(2 - e^{itx} - e^{-itx}\right) f_X(x) \, dx \right] dt$$

$$= \frac{\sin(\pi/2) \Gamma(r+1)}{\pi} \int_0^\infty t^{-(r+1)} [2 - \varphi(t) - \varphi(-t)] \, dt.$$

Using (1.45), we arrive at

$$E[|X|^r] = \frac{2\sin(\pi/2)\Gamma(r+1)}{\pi} \int_0^\infty t^{-(r+1)} [1 - \text{Re}(\varphi(t))]\,dt, \quad 0 < r < 2,$$

(8.30)

which is of interest in itself.[8]

Now let X be a stable Paretian random variable. Using (8.18),

$$\mathbb{E}[|X|^r] = \frac{\sin(\pi/2)\,\Gamma(r+1)}{\pi} \int_0^\infty t^{-(r+1)} [2 - \varphi(t) - \varphi(-t)]\,dt$$

$$= -\frac{\sin(\pi/2)\,\Gamma(r+1)}{\pi} \int_0^\infty t^{-(r+1)} \left[\left(\exp\{-z_p t^\alpha\} + \exp\{-z_n t^\alpha\}\right) - 2\right]dt.$$

Using the transformation $w = t^\alpha$,

$$\mathbb{E}[|X|^r] = -\frac{\sin(\pi/2)\,\Gamma(r+1)}{\pi}$$

$$\times \int_0^\infty \left(w^{1/\alpha}\right)^{-(r+1)} \left[\exp\left(-z_p w\right) + \exp\left(-z_n w\right) - 2\right] \frac{1}{\alpha} w^{1/\alpha - 1}\,dw$$

$$= -\frac{\sin(\pi/2)\,\Gamma(r+1)}{\alpha\pi} \int_0^\infty \left[\exp\left(-z_p w\right) + \exp\left(-z_n w\right) - 2\right] w^{-r/\alpha - 1}\,dw.$$

This integral is the same as that in (8.20), so, following the same derivation which led to (8.25) and again using $\tau = \tan\theta$,

$$\mathbb{E}[|X|^r] = -\frac{\sin(\pi/2)\Gamma(r+1)}{\alpha\pi} \left[-\frac{2\alpha}{r}\Gamma\left(1 - \frac{r}{\alpha}\right)(1+\tau^2)^{r/2\alpha}\cos\left(\frac{r}{\alpha}\arctan\tau\right)\right]$$

$$= \frac{2\sin(\pi/2)\Gamma(r+1)}{r\pi}\Gamma\left(1 - \frac{r}{\alpha}\right)(1+\tau^2)^{r/2\alpha}\cos\left(\frac{r}{\alpha}\arctan\tau\right)$$

$$= \frac{2\sin(\pi/2)\Gamma(r)}{\pi}\Gamma\left(1 - \frac{r}{\alpha}\right)(1+\tau^2)^{r/2\alpha}\cos\left(\frac{r}{\alpha}\arctan\tau\right).$$

Using (8.27) with $z = r - 1$ now gives

$$\mathbb{E}[|X|^r] = \frac{2\sin(\pi/2)}{\Gamma(1-r)\sin(r\pi)}\Gamma\left(1 - \frac{r}{\alpha}\right)(1+\tau^2)^{r/2\alpha}\cos\left(\frac{r}{\alpha}\arctan\tau\right)$$

$$= \frac{\Gamma(1-r/\alpha)}{\Gamma(1-r)}\frac{1}{\cos(r\pi/2)}(1+\tau^2)^{r/2\alpha}\cos\left(\frac{r}{\alpha}\arctan\tau\right),$$

which is (8.26).

---

[8] Equation (8.30) differs from Zolotarev (1986, p. 63, eq. 2.1.9) because the latter contains two errors: the π in the denominator is missing, and the exponent of t in the integral is erroneously $r - 1$.

⊖ **Example 8.1**   With $\alpha = 1.5$ and $\beta = 0.5$, (8.26) with $r = 1$ gives $\mathbb{E}[|X|] = 1.75009$. For numeric evaluation, take

$$\mathbb{E}[|X|^r] \approx K_- \int_{-\infty}^{\ell} (-x)^r (-x)^{-(\alpha+1)} \, dx$$
$$+ \int_{\ell}^{h} |x|^r f_X(x) \, dx + K_+ \int_{h}^{\infty} x^r (x)^{-(\alpha+1)} \, dx$$

or, via a similar calculation which led to (8.13),

$$\mathbb{E}[|X|^r] \approx -K_- \frac{(-\ell)^{r-\alpha}}{r-\alpha} - K_+ \frac{h^{r-\alpha}}{r-\alpha} + \int_{\ell}^{h} |x|^r f_X(x) \, dx. \tag{8.31}$$

Computing this for $\alpha = 1.5$, $\beta = 0.5$ and $r = 1$ with the same low and high cutoff values as before yields $\mathbb{E}[|X|] \approx 1.749996$, which is virtually the same as the exact answer.

A program to compute the exact and numeric values is given in Listing 8.2.   ■

```
function m=stabmom(r,a,b,l,h)

if r==1, k=pi/2; else, k=gamma(1-r)*cos(pi*r/2); end
m=gamma(1-r/a)/k;
if (b~=0)
   tau=b * tan(a*pi/2);   f1 = (1 + tau^2)^( r/(2*a) );
   f2 = cos( (r/a) * atan( tau ) );   m=m*f1*f2;
end
mexact=m
if nargin>3 % compute numerically also
   tmp = a*gamma(a)*sin(pi*a/2)/pi; Km=tmp*(1-b); Kp=tmp*(1+b);
   d=r-a; I=-(Km/d)*((-1)^d); III=-(Kp/d)*(h^d);
   II=quadl(@ff,l,h,1e-6,1,r,a,b); m=I+II+III;
end

function I=ff(xvec,r,a,b) % compute |x|^r * f(x)
I=zeros(size(xvec));
for ii=1:length(xvec)
   x=xvec(ii); f = asymstab(x,a,b); I(ii) = ((abs(x))^r) .* f;
end
```

Program Listing 8.2   Computes (8.26) and, optionally, (8.31)

8.4   Simulation

A method for simulating stable Paretian r.v.s was developed in Chambers, Mallows and Stuck (1976); see also Janicki and Weron (1994) and Weron (1996). Let $U \sim$

Unif $(-\pi/2, \pi/2)$ independent of $E \sim \text{Exp}(1)$. Then $Z \sim S_{\alpha,\beta}(0, 1)$, where

$$
Z = \begin{cases}
\dfrac{\sin \alpha (\theta + U)}{(\cos \alpha\theta \, \cos U)^{1/\alpha}} \left[\dfrac{\cos (\alpha\theta + (\alpha - 1) U)}{E} \right]^{(1-\alpha)/\alpha}, & \text{if } \alpha \neq 1, \\[4mm]
\dfrac{2}{\pi} \left[\left(\dfrac{\pi}{2} + \beta U \right) \tan U - \beta \ln \left(\dfrac{(\pi/2) E \cos U}{\pi/2 + \beta U} \right) \right], & \text{if } \alpha = 1,
\end{cases}
$$

and $\theta = \arctan(\beta \tan(\pi\alpha/2))/\alpha$ for $\alpha \neq 1$. Incorporation of location and scale changes is done as usual when $\alpha \neq 1$, i.e., $X = \mu + cZ \sim S_{\alpha,\beta}(\mu, c)$, but for $\alpha = 1$, $X = \mu + cZ + (2\beta c \ln c)/\pi \sim S_{1,\beta}(\mu, c)$ (note that the latter term is zero for $\beta = 0$ or $c = 1$).

For example, Figure 8.9 plots a kernel density estimate (see Section I.7.4.2), truncated to $(-10, 10)$, of a simulated i.i.d. data set of 20 000 observations for $\alpha = 1.2$ and $\beta = 0.9$. Overlaid is the true density, which is indeed very close to the kernel density, even in the tails.

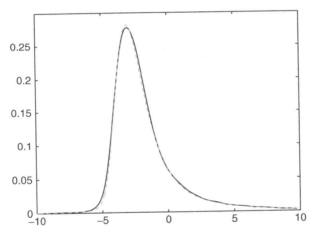

Figure 8.9 Kernel density estimate (solid) of the $S_{1.2,0.9}(0, 1)$ p.d.f. based on an i.i.d. sample of size 20 000, and the true density (dashed)

8.5 Generalized central limit theorem

Let $X_1, X_2, \ldots$ be an i.i.d. sequence of r.v.s (with or without finite means and variances). The generalized central limit theorem states that there exist real constants $a_n > 0$ and b_n, and a nondegenerate r.v. Z such that, as $n \to \infty$,

$$
a_n \sum_{j=1}^{n} X_j - b_n \xrightarrow{d} Z,
$$

iff $Z \sim S_{\alpha,\beta}(\mu, c)$. In the standard CLT, $a_n = n^{-1/2}$. In the generalized CLT, a_n is of the form $n^{-1/\alpha}$, which nests the $\alpha = 2$ case.

This can be invoked in practice in the same way that the CLT is used. The normal distribution is justified as the appropriate error distribution in many applications with the reasoning that the discrepancy between the model and the observations is the cumulative result of many independent random factors (too numerous and/or difficult to capture in the model). If these factors have finite variance, then the CLT states that their sum will approach a normal distribution (and no other). Based on the same reasoning and assumptions, except allowing for infinite variances of the random factors, the generalized CLT states that the asymmetric stable Paretian distribution (with special case the normal) is the only possible candidate for modelling the error term.

It is indeed the case that there are many data sets, from a variety of disciplines, for which the data points consist of a set of independent observations[9] but exhibit much fatter tails than the normal, and sometimes considerable skewness as well. Such data are much better modelled by use of a stable distribution instead of the normal. The most notable example is in the field of finance, where many observed quantities, such as the daily returns on financial assets (stock prices, exchange rates, etc.) deviate remarkably from the normal distribution, but are quite well fitted by a stable distribution. This also holds for the returns on real estate (see Young and Graff, 1995). Even Hollywood is far from 'normal': De Vany and Walls (2004) demonstrate the outstanding ability of the asymmetric stable distribution to model the profit in the motion picture business, and provide reasons for its success (see also Walls, 2005, for use of the skewed t distribution for modelling box-office revenue). Other fields include telecommunications, biology, and geology; see, for example, Uchaikin and Zolotarev (1999) for a variety of examples.

---

[9] The extent of the observations being identically distributed is more of a modelling assumption, and depends on what other r.v.s are being conditioned on.

9

Generalized inverse Gaussian and generalized hyperbolic distributions

The significant problems we face cannot be solved at the same level of thinking we were at when we created them.   (Albert Einstein)

This chapter, and its problems and solutions, were written by Walther Paravicini.

9.1 Introduction

Chapter 7 introduced a variety of flexible distributions, with particular attention paid to ones with skewness and fatter tails than the normal, resulting in, for example, numerous *ad hoc* generalizations of the Student's t. Chapter 8 introduced the asymmetric stable Paretian, which is theoretically well motivated, but has the drawback that the variance does not exist, an assumption which might be deemed too extreme for certain applications.

This chapter discusses the family of generalized inverse Gaussian distributions, which generalize the inverse Gaussian distribution presented in Section I.7.2, and the family of generalized hyperbolic distributions, which generalize the hyperbolic distribution, also introduced in Section I.7.2. The GHyp are fat-tailed and skewed, but with a finite variance, and also possess other interesting and desirable properties. Also recall that the logarithm of the hyperbolic density function is a (possibly skewed) hyperbola, while that of the normal density is a parabola.

The GHyp was introduced in Barndorff-Nielsen (1977) and has since found a great variety of applications. For a recent discussion of its multivariate extension, see Mencía and Sentana (2004). A standard reference for the GIG is still Jørgensen (1982).

Intermediate Probability: A Computational Approach   M. Paolella
© 2007 John Wiley & Sons, Ltd

In discussing the properties of the GIG and GHyp families, use of continuous mixture distributions (introduced in Section 7.3.2) and the modified Bessel function of the third kind (mentioned in the context of the hyperbolic distribution in Section I.7.2) will be omnipresent. We begin with a more detailed discussion of the latter.

9.2 The modified Bessel function of the third kind

The integral representation of the Bessel function is the one most suitable for our purposes, so we simply take (I.7.61) as its definition: for every $\nu \in \mathbb{R}$, the modified Bessel function of the third kind, here simply called the Bessel function, with index ν is defined as

$$K_\nu(x) := \frac{1}{2} \int_0^\infty t^{\nu-1} e^{-\frac{1}{2}x(t+t^{-1})} \, dt, \quad x > 0. \tag{9.1}$$

This choice of integral representation is interesting, because it is similar to the gamma function. Below, we will discuss how the two functions are related. As with the gamma function, values of the Bessel function can sometimes be computed explicitly, but in general, numerical methods are necessary. In Matlab, $K_\nu(x)$ is computed with the built-in function `besselk(v,x)`.

We can deduce some simple computational rules that enable us to treat the Bessel function analytically. For example, we have

$$K_\nu(x) = K_{-\nu}(x), \quad \nu \in \mathbb{R}, \ x \in \mathbb{R}_{>0}. \tag{9.2}$$

To derive this, substitute $s = t^{-1}$ in the integral expression (9.1) to get

$$K_\nu(x) = \frac{1}{2} \int_\infty^0 s^{1-\nu} e^{-\frac{1}{2}x(s+s^{-1})} (-s^{-2}) \, ds$$

$$= \frac{1}{2} \int_0^\infty s^{-\nu-1} e^{-\frac{1}{2}x(s+s^{-1})} \, ds = K_{-\nu}(x).$$

Also, the Bessel function can be stated explicitly for $\nu = 1/2$, with

$$K_{1/2}(x) = K_{-1/2}(x) = \sqrt{\frac{\pi}{2x}} e^{-x}, \quad x \in \mathbb{R}_{>0}. \tag{9.3}$$

In addition, a recursive formula that can be used to derive explicit expressions for the Bessel function is given by

$$K_{\nu+1}(x) = \frac{2\nu}{x} K_\nu(x) + K_{\nu-1}(x), \quad \nu \in \mathbb{R}, \ x \in \mathbb{R}_{>0}. \tag{9.4}$$

This can be derived as follows. If $v = 0$ we just have to show that $K_{-1}(x) = K_1(x)$, a fact we already know. So let $v \neq 0$. Let $0 < a < b < \infty$. Then, using integration by parts,

$$\frac{1}{2} \int_a^b t^{v-1} e^{-\frac{1}{2}x(t+t^{-1})} \, dt = \frac{t^v}{v} e^{-\frac{1}{2}x(t+t^{-1})} \Big|_a^b - \frac{1}{2} \int_a^b \frac{t^v}{v} \left(-\frac{1}{2}x(1 - t^{-2}) \right) e^{-\frac{1}{2}x(t+t^{-1})} \, dt$$

$$= \frac{t^v}{v} e^{-\frac{1}{2}x(t+t^{-1})} \Big|_a^b + \frac{x}{4v} \int_a^b \left(t^v - t^{v-2} \right) e^{-\frac{1}{2}x(t+t^{-1})} \, dt.$$

Now if $a \to 0$ and $b \to \infty$, the l.h.s. approaches $K_v(x)$, whereas the second term on the r.h.s. converges to $\frac{x}{2v}(K_{v+1}(x) - K_{v-1}(x))$. The first term on the r.h.s. vanishes. By rearranging the equation, we get (9.4). See also Problem 9.2.

For $x = 0$, the Bessel function has a singularity. The following formulae explain how the function behaves for small values of x. For $v = 0$,

$$K_0(x) \simeq -\ln(x) \quad \text{for } x \downarrow 0, \ v \doteq 0, \tag{9.5}$$

whereas for $v \neq 0$,

$$K_v(x) \simeq \Gamma(v) 2^{|v|-1} x^{-|v|} \quad \text{for } x \downarrow 0, \ v \neq 0. \tag{9.6}$$

Further information on Bessel functions can be found in Lebedev (1972), Andrews, Askey and Roy (1999), and also Appendix C of Mencía and Sentana (2004).

Later in this chapter, we will frequently meet the following integral which is closely related to the Bessel function:

$$k_\lambda (\chi, \psi) := \int_0^\infty x^{\lambda-1} \exp\left[-\frac{1}{2}(\chi x^{-1} + \psi x) \right] dx. \tag{9.7}$$

This integral converges for arbitrary $\lambda \in \mathbb{R}$ and $\chi, \psi > 0$. In this case we have, using the notation $\eta := \sqrt{\chi/\psi}$ and $\omega := \sqrt{\chi \psi}$ and the substitution $x = \eta y$,

$$\int_0^\infty x^{\lambda-1} \exp\left[-\frac{1}{2}(\chi x^{-1} + \psi x) \right] dx = \int_0^\infty x^{\lambda-1} \exp\left[-\frac{1}{2}\omega((x/\eta)^{-1} + x/\eta) \right] dx$$

$$= \int_0^\infty (\eta y)^{\lambda-1} \exp\left[-\frac{1}{2}\omega(y^{-1} + y) \right] \eta \, dy$$

$$= 2\eta^\lambda \frac{1}{2} \int_0^\infty y^{\lambda-1} \exp\left[-\frac{1}{2}\omega(y^{-1} + y) \right] dy$$

$$= 2\eta^\lambda K_\lambda(\omega),$$

so that

$$
k_\lambda\,(\chi,\,\psi) = 2\eta^\lambda K_\lambda(\omega) = 2\left(\frac{\chi}{\psi}\right)^{\lambda/2} K_\lambda\left(\sqrt{\chi\psi}\right). \tag{9.8}
$$

The integral in (9.7) also converges in two 'boundary cases'. If $\chi = 0$ and $\psi > 0$, it converges if and only if (iff) $\lambda > 0$. On the other hand, if $\chi > 0$ and $\psi = 0$, it converges iff $\lambda < 0$. In the first boundary case one gets

$$
\int_0^\infty x^{\lambda-1} e^{-\frac{1}{2}\psi x}\,dx = \left(\frac{\psi}{2}\right)^{-\lambda}\int_0^\infty x^{\lambda-1}e^{-x}\,dx = \left(\frac{\psi}{2}\right)^{-\lambda}\Gamma(\lambda),
$$

whereas, in the second case, we use the substitution $y = \psi x^{-1}/2$ to compute

$$
\int_0^\infty x^{\lambda-1} e^{-\frac{1}{2}\chi x^{-1}}\,dx = \left(\frac{\chi}{2}\right)^{\lambda}\int_0^\infty y^{-\lambda-1}e^{-y}\,dx = \left(\frac{\chi}{2}\right)^{\lambda}\Gamma(-\lambda).
$$

Hence we have

$$
k_\lambda\,(0,\,\psi) = \left(\frac{\psi}{2}\right)^{-\lambda}\Gamma(\lambda) \tag{9.9}
$$

and

$$
k_\lambda\,(\chi,\,0) = \left(\frac{\chi}{2}\right)^{\lambda}\Gamma(-\lambda). \tag{9.10}
$$

The function $k_\lambda\,(\chi,\,\psi)$ possesses some symmetries, namely

$$
k_\lambda\,(\chi,\,\psi) = k_{-\lambda}\,(\psi,\,\chi) \tag{9.11}
$$

and

$$
k_\lambda\,(\chi,\,\psi) = r^\lambda\,k_\lambda\left(r^{-1}\chi,\,r\psi\right), \tag{9.12}
$$

for all $r > 0$. To prove the latter formula, one could either check that it is true for the three explicit expressions for $k_\lambda\,(\chi,\,\psi)$ that we have just derived, or proceed as follows: substitute x with ry in the definition of $k_\lambda\,(\chi,\,\psi)$ to get

$$
\begin{aligned}
k_\lambda\,(\chi,\,\psi) &= \int_0^\infty x^{\lambda-1}\exp\left[-\frac{1}{2}(\chi x^{-1} + \psi x)\right]dx \\
&= \int_0^\infty (ry)^{\lambda-1}\exp\left[-\frac{1}{2}(\chi(ry)^{-1} + \psi ry)\right]r\,dy \\
&= r^\lambda\int_0^\infty y^{\lambda-1}\exp\left[-\frac{1}{2}((r^{-1}\chi)y^{-1} + (r\psi)y)\right]dy \\
&= r^\lambda\,k_\lambda\left(r^{-1}\chi,\,r\psi\right).
\end{aligned}
$$

9.3 Mixtures of normal distributions

9.3.1 Mixture mechanics

The technique of constructing (countable or continuous) mixture distributions was introduced in Section 7.3, and will be used heavily in what follows. Here we consider a special type of mixture, called the *variance–mean mixture of normals*, which will be seen to give rise to a variety of interesting examples. As the name suggests, it is a mixture of normal distributions such that the mean and variance of the mixed distributions are in a special (affine-linear) relation. But let us be more precise.

Suppose Z is a positive random variable and μ and β are constants. If X is a random variable satisfying

$$X \mid Z \sim N(\mu + \beta Z, Z),$$

then we can calculate the density function f_X of X as follows. If Z is a continuous random variable with p.d.f. f_Z, then

$$f_X(x) = \int_0^\infty f_N(x; \mu + \beta z, z) f_Z(z)\, dz,$$

and if Z is a discrete r.v. with mass function f_Z, then

$$f_X(x) = \sum_{\substack{z > 0 \\ f_Z(z) \neq 0}} f_N(x; \mu + \beta z, z) f_Z(z).$$

The distribution of X is thus constructed by taking normal distributions and mixing them together using the random variable Z. The density of X depends on Z only via the p.d.f. or p.m.f. of Z, hence only through the distribution of Z. To streamline the notation, we will denote the distribution of X by $\text{Mix}_\pi(\mu, \beta)$, where π is the distribution of Z, subsequently called the *weight*, and $X \mid Z \sim N(\mu + \beta Z, Z)$. The distribution of X has μ as a location parameter, seen by setting $Y = X - \mu$, $y = x - \mu$, and, in the continuous case, using the usual univariate transformation (I.7.65) to get

$$f_Y(y) = f_X(y + \mu) = \int_0^\infty f_N(y + \mu; \mu + \beta z, z) f_Z(z)\, dz = \int_0^\infty f_N(y; \beta z, z) f_Z(z)\, dz,$$

which does not involve μ. If the parameter β is nonzero, then the distribution will usually be skewed. In the simplest case, the random variable Z has finitely many possible values, i.e., the weight π has finite support, and we are mixing a finite number of normal distributions, as in Example 7.13.

The situation is far more interesting with continuous weights, as the next example shows.

⊙ **Example 9.1** Let $Z \sim \mathrm{Exp}(\lambda)$ and $\mu = \beta = 0$. Then

$$
\begin{aligned}
f_{\mathrm{Mix}_{\mathrm{Exp}(\lambda)}(0,0)}(x) &= \int_0^\infty f_{\mathrm{N}}(x; 0, z) f_{\mathrm{Exp}}(z; \lambda)\, dz \\
&= \int_0^\infty \frac{1}{\sqrt{2\pi z}} \exp\left[-\frac{x^2}{2z}\right] \lambda e^{-\lambda z}\, dz \\
&= \frac{\lambda}{\sqrt{2\pi}} \int_0^\infty z^{-1/2} \exp\left[-\frac{1}{2}\left(x^2 z^{-1} + 2\lambda z\right)\right] dz \\
&\overset{(9.7)}{=} \frac{\lambda}{\sqrt{2\pi}}\, k_{1/2}\left(x^2,\, 2\lambda\right) \\
&\overset{(9.8)}{=} \frac{\lambda}{\sqrt{2\pi}}\, 2 \left(\frac{|x|}{\sqrt{2\lambda}}\right)^{1/2} K_{1/2}\left(\sqrt{2\lambda}|x|\right) \\
&\overset{(9.3)}{=} \frac{\lambda}{\sqrt{2\pi}}\, 2 \left(\frac{|x|}{\sqrt{2\lambda}}\right)^{1/2} \sqrt{\frac{\pi}{2\sqrt{2\lambda}|x|}} e^{-\sqrt{2\lambda}|x|} = \frac{\sqrt{2\lambda}}{2} e^{-\sqrt{2\lambda}|x|}.
\end{aligned}
$$

This is the Laplace density with scale parameter $1/\sqrt{2\lambda}$, as was obtained in Example 7.19 via use of the m.g.f. ∎

Now the question arises as to which distributional class is obtained when taking a more general distribution family as mixing weights, such as gamma. Before treating this problem, a rather general distribution family is introduced which nests the gamma. Plugging this family into the mixing procedure will provide the vast class of distributions referred to as generalized hyperbolic.

9.3.2 Moments and generating functions

One of the features of constructing a distribution by mixing is that one can essentially read off the properties of the distribution given the properties of the chosen weight. The central tools we require for the moments and generating functions are as follows. Suppose $Z \sim \pi$ and $X \sim \mathrm{Mix}_\pi(\mu, \beta)$. Then we have expected value

$$
\boxed{\mathbb{E}[X] = \mu + \beta\, \mathbb{E}[Z],} \tag{9.13}
$$

variance

$$
\boxed{\mathbb{V}(X) = \mathbb{E}[Z] + \beta^2 \mathbb{V}(Z),} \tag{9.14}
$$

third central moment

$$\mu_3(X) = 3\beta \mathbb{V}(Z) + \beta^3 \mu_3(Z),$$

(9.15)

moment generating function

$$\mathbb{M}_X(t) = e^{\mu t} \mathbb{M}_Z(\beta t + t^2/2),$$

(9.16)

and characteristic function

$$\varphi_X(v) = e^{i\mu v} \varphi_Z(\beta v + iv^2/2).$$

(9.17)

In these formulae, if the right-hand side exists then so does the left-hand side.
Furthermore, the mixed distribution $\mathrm{Mix}_\pi(\mu, \beta)$ depends on π, μ, and β in a continuous fashion. This makes it easier to determine the limiting cases of certain families of mixed distributions by analysing the limiting cases of the family of mixing weights involved.

⊙ ***Example 9.2*** To prove (9.13) when Z is a continuous random variable, first note that, for every $z > 0$, the integral

$$\int_{-\infty}^{\infty} x f_N(x; \mu + \beta z, z) \, dx$$

is just the mean of a normal distribution with parameters $\mu + \beta z$ and z, so an application of Fubini's theorem gives

$$\mathbb{E}[X] = \int_{-\infty}^{\infty} x f_X(x) \, dx = \int_{-\infty}^{\infty} x \int_0^{\infty} f_N(x; \mu + \beta z, z) f_Z(z) \, dz \, dx$$

$$= \int_0^{\infty} \int_{-\infty}^{\infty} x f_N(x; \mu + \beta z, z) \, dx f_Z(z) \, dz = \int_0^{\infty} (\mu + \beta z) f_Z(z) \, dz$$

$$= \int_0^{\infty} \mu f_Z(z) \, dz + \beta \int_0^{\infty} z f_Z(z) \, dz = \mu + \beta \, \mathbb{E}[Z],$$

which is (9.13). ∎

⊙ ***Example 9.3*** Analogously, to prove (9.16) when Z is a continuous random variable, we use the equation

$$\int_{-\infty}^{\infty} e^{tx} f_N(x; \mu + \beta z, z) \, dx = \exp\left[(\mu + \beta z)t + \frac{z}{2}t^2\right],$$

which is just the moment generating function for a normal distribution with parameters $\mu + \beta z$ and z, where $z > 0$ is arbitrary (see Example 1.3 or Table I.C.7). Thus

$$\mathbb{M}_X(t) = \int_{-\infty}^{+\infty} e^{tx} f_X(x) \, dx = \int_{-\infty}^{+\infty} e^{tx} \int_0^\infty f_N(x; \mu + \beta z, z) f_Z(z) \, dz \, dx$$

$$= \int_0^\infty \int_{-\infty}^{+\infty} e^{tx} f_N(x; \mu + \beta z, z) \, dx f_Z(z) \, dz$$

$$= \int_0^\infty \exp\left[(\mu + \beta z)t + \frac{z}{2}t^2\right] f_Z(z) \, dz = e^{\mu t} \mathbb{M}_Z(\beta t + t^2/2),$$

which is (9.16). ∎

One can derive the other equations in a very similar fashion (see Problem 9.3).

9.4 The generalized inverse Gaussian distribution

9.4.1 Definition and general formulae

The generalized inverse Gaussian (GIG) p.d.f. is given by

$$f_{\text{GIG}}(x; \lambda, \chi, \psi) = \frac{1}{k_\lambda(\chi, \psi)} x^{\lambda-1} \exp\left[-\frac{1}{2}(\chi x^{-1} + \psi x)\right] \mathbb{I}_{(0,\infty)}(x),$$

where $k_\lambda(\chi, \psi)$ is the function defined in (9.7) depending on the parameters λ, χ, and ψ. We denote the parameter space by Θ_{GIG}, which consists of three cases given by

$$\Theta_{\text{GIG}} := \{(\lambda, \chi, \psi) \in \mathbb{R}^3 : \quad \lambda \in \mathbb{R}, \ \chi > 0, \ \psi > 0 \quad \text{(normal case)},$$

$$\text{or} \ \lambda > 0, \ \chi = 0, \ \psi > 0 \quad \text{(boundary case I)},$$

$$\text{or} \ \lambda < 0, \ \chi > 0, \ \psi = 0\} \quad \text{(boundary case II)}.$$

One can show that the GIG distribution depends continuously on its parameters λ, χ, and ψ. The GIG distribution generalizes quite a number of distribution families; for example, the GIG reduces to the gamma distribution in boundary case I. The normal case of the GIG distribution, i.e., if the parameters have values $\lambda \in \mathbb{R}$ and $\chi, \psi > 0$, is usually considered to be the GIG distribution in the strict sense. The boundary cases are included in the definition of GIG because it helps organize all the interesting subfamilies and limit cases that we will detail in the next section.

The simple program in Listing 9.1 computes the GIG density in the normal case. Numerically integrating the density via `quadl(@gigpdf,lo,hi,1e-8,0,lambda,chi,psi)` (where `lo` and `hi` specify the range of integration) indeed yields 1.0 to (over) eight significant digits. Figure 9.1 shows the GIG density for several parameter values.

```
function f=gigpdf(x,lambda,chi,psi)
eta = sqrt(chi/psi); omega = sqrt(chi*psi);
c = 1 / (2 * eta^lambda * besselk(lambda,omega));
p1 = c * x.^(lambda-1); p2 = -0.5 * (chi ./ x + psi * x);
f = p1 .* exp(p2);
```

Program Listing 9.1   Computes the GIG p.d.f. in the normal case ($\lambda \in \mathbb{R}$, $\chi > 0$, $\psi > 0$)

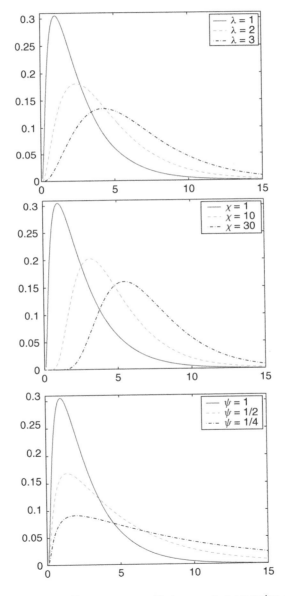

Figure 9.1   The GIG p.d.f.: in all cases, nonspecified parameters are unity, e.g., in the top graph, $\chi = \psi = 1$

We will now compute general expressions for the moments and the m.g.f. of the GIG distribution using the function $k_\lambda(\chi, \psi)$. Afterwards, we will apply these results to the subclasses of GIG and discuss the actual *existence* of the moments – in general, a somewhat precarious[1] endeavour.

Assume that $(\lambda, \chi, \psi) \in \Theta_{GIG}$ and $X \sim GIG(\lambda, \chi, \psi)$. Then the rth raw moment of X is

$$\mathbb{E}[X^r] = \int_0^\infty x^r f_{GIG}(x; \lambda, \chi, \psi)\, dx$$

$$= \frac{1}{k_\lambda(\chi, \psi)} \int_0^\infty x^{\lambda+r-1} \exp\left[-\frac{1}{2}(\chi x^{-1} + \psi x)\right] dx$$

$$= \frac{k_{\lambda+r}(\chi, \psi)}{k_\lambda(\chi, \psi)}. \tag{9.18}$$

It follows that

$$\mathbb{E}[X] = \frac{k_{\lambda+1}(\chi, \psi)}{k_\lambda(\chi, \psi)} \tag{9.19}$$

and

$$\mathbb{V}(X) = \mathbb{E}[X^2] - \mathbb{E}[X]^2 = \frac{k_\lambda(\chi, \psi)k_{\lambda+2}(\chi, \psi) - (k_{\lambda+1}(\chi, \psi))^2}{(k_\lambda(\chi, \psi))^2}. \tag{9.20}$$

The moment generating function of X can also be determined:

$$\mathbb{M}_{GIG}(t; \lambda, \chi, \psi) = \int_0^\infty e^{tx} f_{GIG}(x; \lambda, \chi, \psi)\, dx$$

$$= \frac{1}{k_\lambda(\chi, \psi)} \int_0^\infty x^{\lambda-1} \exp\left[-\frac{1}{2}(\chi x^{-1} + (\psi - 2t)x)\right] dx$$

$$= \frac{k_\lambda(\chi, \psi - 2t)}{k_\lambda(\chi, \psi)}. \tag{9.21}$$

9.4.2 The subfamilies of the GIG distribution family

For an overview consult Table 9.1 and Figure 9.2.

9.4.2.1 *The proper* GIG *distribution (if* $\lambda \in \mathbb{R}$, $\chi > 0$, $\psi > 0$)

Here, in the normal case, an interesting alternative parameterization of the GIG distribution is available: set $\eta := \sqrt{\chi/\psi}$ and $\omega := \sqrt{\chi\psi}$. Then the parameters λ and ω are

---

[1] *Precarious* is a word that contains all vowels exactly once, like equation, Austin Powers, Julia Roberts, sequoia, and behaviour (at least in the UK). There are words in foreign languages as well: Alufolie (German, 'tin foil'), vino spumante (Italian, 'sparkling wine') and luómàndìkè (Chinese, 'romantic'), and, among the shortest, oiseau (French, 'bird') and aiuole (Italian, 'flower bed').

Table 9.1 Special cases of GIG

Name	Parameter range		
normal case	$\lambda \in \mathbb{R}$	$\chi > 0$	$\psi > 0$
gamma (Gam)	$\lambda > 0$	$\chi = 0$	$\psi > 0$
inverse gamma (IGam)	$\lambda < 0$	$\chi > 0$	$\psi = 0$
exponential (Exp)	$\lambda = 1$	$\chi = 0$	$\psi > 0$
positive hyperbolic (pHyp)	$\lambda = 1$	$\chi > 0$	$\psi > 0$
Lévy	$\lambda = -1/2$	$\chi > 0$	$\psi = 0$
inverse Gaussian (IG)	$\lambda = -1/2$	$\chi > 0$	$\psi > 0$
limiting case Dirac Δ_x, $x \geq 0$	$\lambda \in \mathbb{R}$	$\eta \to x$	$\omega \to \infty$

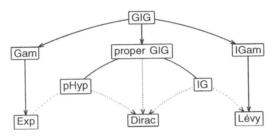

Figure 9.2 The subclasses of the GIG distribution: solid arrows point to subclasses, dashed arrows indicate limits

scale-invariant, and η is a scale parameter (see Problem 9.5). We will switch between the parameterizations throughout this chapter whenever it seems useful. Note that, in the normal case, the p.d.f. of the GIG distribution can be expressed as

$$f_{\mathrm{GIG}}(x; \lambda, \chi, \psi) = \frac{1}{2\eta K_\lambda(\omega)} \left(\frac{x}{\eta}\right)^{\lambda-1} \exp\left[-\frac{1}{2}\omega\left(\left(\frac{x}{\eta}\right)^{-1} + \frac{x}{\eta}\right)\right] \mathbb{I}_{(0,\infty)}(x),$$

which, recalling the location–scale form (I.7.1), confirms that $\eta = \sqrt{\chi/\psi}$ is a scale parameter. This distribution possesses all moments, and a moment generating function on a neighbourhood of zero: if $r \in \mathbb{R}$, then the rth raw moment of $X \sim \mathrm{GIG}(\lambda, \chi, \psi)$ is

$$\mathbb{E}[X^r] = \int_0^\infty x^r f_{\mathrm{GIG}}(x; \lambda, \chi, \psi)\,dx = \eta^r \frac{K_{\lambda+r}(\omega)}{K_\lambda(\omega)}. \tag{9.22}$$

This equation can be derived by plugging (9.8) into (9.18). In Problem 9.8 we use this equation to prove

$$\mathbb{E}[\ln X] = \ln \eta + \frac{\frac{\partial}{\partial\lambda} K_\lambda(\omega)}{K_\lambda(\omega)}. \tag{9.23}$$

By making use of (9.19), (9.20), and (9.21), the following can be shown:

$$\mathbb{E}[X] = \eta \frac{K_{\lambda+1}(\omega)}{K_\lambda(\omega)}, \quad \mathbb{V}(X) = \eta^2 \frac{K_\lambda(\omega)K_{\lambda+2}(\omega) - K_{\lambda+1}(\omega)^2}{K_\lambda(\omega)^2}, \tag{9.24}$$

and

$$\mathbb{M}_{\mathrm{GIG}}(t; \lambda, \chi, \psi) = \frac{K_\lambda\left(\omega\sqrt{1 - 2t/\psi}\right)}{K_\lambda(\omega)\,(1 - 2t/\psi)^{\lambda/2}}, \quad -\infty < t < \psi/2. \tag{9.25}$$

9.4.2.2   *The gamma distribution* Gam *(if* $\lambda > 0$, $\chi = 0$, $\psi > 0$)

In boundary case I, the GIG distribution reduces to the gamma distribution

$$\mathrm{Gam}(\lambda, \psi/2) = \mathrm{GIG}(\lambda, 0, \psi).$$

It possesses a moment generating function on a neighbourhood of zero, and all positive, but not all negative moments. Let $r \in \mathbb{R}$. Then the rth raw moment of $X \sim \mathrm{GIG}(\lambda, 0, \psi)$ exists iff $r > -\lambda$, and then

$$\mathbb{E}[X^r] = \int_0^\infty x^r f_{\mathrm{GIG}}(x; \lambda, 0, \psi)\,\mathrm{d}x = \frac{\Gamma(\lambda + r)}{\Gamma(\lambda)} \left(\frac{\psi}{2}\right)^{-r}. \tag{9.26}$$

This formula is well known, and we can easily read it off equation (9.18) using (9.9). Similarly,

$$\mathbb{E}[X] = \frac{\lambda}{\psi/2}, \quad \mathbb{V}(X) = \frac{\lambda}{(\psi/2)^2}, \tag{9.27}$$

and

$$\mathbb{M}_{\mathrm{GIG}}(t; \lambda, 0, \psi) = \left(1 - \frac{2t}{\psi}\right)^{-\lambda}, \quad -\infty < t < \psi/2. \tag{9.28}$$

9.4.2.3   *The inverse gamma distribution* IGam *(if* $\lambda < 0$, $\chi > 0$, $\psi = 0$)

In boundary case II, the GIG distribution reduces to the inverse gamma distribution

$$\mathrm{IGam}(-\lambda, \chi/2) = \mathrm{GIG}(\lambda, \chi, 0),$$

with density

$$f_{\mathrm{IGam}}(x; -\lambda, \chi/2) = f_{\mathrm{GIG}}(x; \lambda, \chi, 0)$$

$$= \left(\frac{\chi}{2}\right)^{-\lambda} \Gamma(-\lambda)^{-1} x^{\lambda-1} \exp\left[-\frac{\chi}{2} x^{-1}\right] \mathbb{I}_{(0,\infty)}(x)$$

(see Problem I.7.9). It possesses all negative, but not all positive moments. Its m.g.f. is only defined on the negative half-line. Let $r \in \mathbb{R}$. Then the rth raw moment of $X \sim \text{GIG}(\lambda, \chi, 0)$ exists iff $r < -\lambda$, and

$$\mathbb{E}[X^r] = \int_0^\infty x^r f_{\text{GIG}}(x; \lambda, \chi, 0) \, dx = \frac{\Gamma(-\lambda - r)}{\Gamma(-\lambda)} \left(\frac{\chi}{2}\right)^r. \tag{9.29}$$

The mean (and variance) exist if $\lambda < -1$ (and $\lambda < -2$); these are

$$\mathbb{E}[X] = -\frac{\chi}{2} \frac{1}{\lambda + 1} \quad \text{and} \quad \mathbb{V}(X) = -\left(\frac{\chi}{2}\right)^2 \frac{1}{(\lambda + 1)^2(\lambda + 2)}. \tag{9.30}$$

The preceding equations follow from (9.18), (9.19), and (9.20), respectively, by an application of (9.10). The moment generating function can be computed from (9.21), but we have to be somewhat careful. The numerator of (9.21) has to be computed using (9.8), whereas (9.10) is needed for the denominator:

$$\mathbb{M}_{\text{GIG}}(t; \lambda, \chi, 0) = \frac{2K_\lambda\left(\sqrt{-2\chi t}\right)}{\Gamma(-\lambda)\left(-\chi t/2\right)^{\lambda/2}}, \quad -\infty < t \le 0, \tag{9.31}$$

where $\mathbb{M}_{\text{GIG}}(0; \lambda, \chi, 0) = 1$.

9.4.2.4 The exponential distribution Exp (if $\lambda = 1$, $\chi = 0$, $\psi > 0$)

If $\lambda = 1$, $\chi = 0$ and $\psi > 0$, then the p.d.f. of the GIG distribution reduces to

$$f_{\text{GIG}}(x; 1, 0, \psi) = \frac{\psi}{2} \exp\left[-\frac{\psi}{2}x\right] \mathbb{I}_{(0,\infty)}(x) = f_{\text{Exp}}(\psi/2),$$

i.e., an exponential distribution with parameter $\psi/2$,

$$\text{Exp}(\psi/2) = \text{GIG}(1, 0, \psi),$$

this being a special case of the gamma distribution mentioned above.

9.4.2.5 The positive hyperbolic distribution pHyp [if $\lambda = 1$, $\chi > 0$, $\psi > 0$]

The log-density of this distribution is a hyperbola and, like all GIG distributions, it is positive (i.e., it is supported on the positive half-line), hence the name. Below, we detail the hyperbolic distribution family Hyp, the log-density of which is also a hyperbola, but living on the whole real line. To avoid possible confusion, one could alternatively deem the GIG distribution with $\lambda = 1$ a generalized exponential distribution, because we have just seen that $\text{GIG}(1, 0, \psi)$ is an exponential distribution. However, we have already given this name to the GED distribution we met in (I.7.73) and (7.1). Moreover, the name *positive hyperbolic distribution* is widespread in the literature, where it is, for instance, used to model particle size data collected by the Viking Mars landers (see Garvin, 1997; and Christiansen and Hartmann, 1991, pp. 237–248, and the references therein).

We parameterize it as follows:

$$\text{pHyp}(\chi, \psi) := \text{GIG}(1, \chi, \psi), \quad \chi > 0, \psi > 0.$$

It is a matter of taste whether to include the case $\chi = 0$ in the pHyp family; to be consistent with the literature, we treat $\text{GIG}(1, 0, \psi)$, i.e., the exponential distribution $\text{Exp}(\psi/2)$, as a limiting case $(\chi \to 0)$ rather than a subcase of the positive hyperbolic distribution. The p.d.f. of pHyp is given by

$$f_{\text{pHyp}}(x; \chi, \psi) = f_{\text{GIG}}(x; 1, \chi, \psi)$$

$$= \frac{1}{2\eta K_1(\omega)} \exp\left[-\frac{1}{2}(\chi x^{-1} + \psi x)\right] \mathbb{I}_{(0,\infty)}(x). \tag{9.32}$$

The moments and the m.g.f. of the pHyp distribution can be obtained by evaluating the formulae for the normal case stated above for $\lambda = 1$.

Note that the parameter space of the exponential distribution, considered as a subfamily of GIG, is located precisely where the parameter spaces of the gamma distribution $(\chi = 0)$ and the positive hyperbolic distribution $(\lambda = 1)$ meet.

9.4.2.6   *The Lévy distribution (if $\lambda = -1/2$, $\chi > 0$, $\psi = 0$)*

The Lévy distribution is a very interesting subfamily of the inverse gamma distribution,

$$\text{Lévy}(\chi/2) = \text{IGam}(-1/2, \chi/2) = \text{GIG}(-1/2, \chi, 0),$$

which we have already encountered in (I.7.56) and (8.11). Its p.d.f. is

$$f_{\text{Lévy}}(x; \chi/2) = \left(\frac{\chi}{2\pi}\right)^{1/2} x^{-3/2} \exp\left[-\frac{\chi}{2x}\right] \mathbb{I}_{(0,\infty)}(x). \tag{9.33}$$

The moment generating function of this distribution can be calculated as follows for $t \leq 0$:

$$M_{\text{Lévy}}(t, \chi/2) = \frac{2K_{1/2}\left(\sqrt{-2\chi t}\right)}{\Gamma(1/2)(-\chi t/2)^{-1/4}} = \frac{2\sqrt{\pi/2}\sqrt{-2\chi t}^{-1/2}e^{-\sqrt{-2\chi t}}}{\sqrt{\pi}(-\chi t/2)^{-1/4}}$$

$$= 2^{1-1/2-1/4-1/4}e^{-\sqrt{-2\chi t}} = \exp\left[-\sqrt{\chi}\sqrt{-2t}\right], \qquad t \leq 0. \tag{9.34}$$

This formula implies that the sum of independent Lévy-distributed random variables again follows a Lévy distribution. We will show how to derive this fact when we discuss the analogous result for the IG distribution. As we saw in Example I.7.6, the Lévy distribution is a limiting case of the inverse Gaussian distribution, which in turn can be found to be a subfamily of GIG, as detailed next.

9.4.2.7 The inverse Gaussian distribution IG (if $\lambda = -1/2$, $\chi > 0$, $\psi > 0$)

If $\lambda = -1/2$, then the GIG distribution in the normal case reduces to the inverse Gaussian distribution:

$$\text{IG}_1(\chi, \psi) := \text{GIG}(-1/2, \chi, \psi), \quad \chi, \psi > 0.$$

Note that we introduced an alternative parameterization, $\text{IG}_2(\lambda, \mu)$, in (I.7.54). The p.d.f. of a $\text{IG}_1(\chi, \psi)$ random variable is

$$f_{\text{IG}_1}(x; \chi, \psi) = f_{\text{GIG}}(x; -1/2, \chi, \psi)$$

$$= \frac{1}{k_{-1/2}(\chi, \psi)} x^{-3/2} \exp\left[-\frac{1}{2}(\chi x^{-1} + \psi x)\right] \mathbb{I}_{(0,\infty)}(x),$$

with

$$k_{-1/2}(\chi, \psi) = 2\eta^{-1/2} K_{-1/2}(\omega) = 2\eta^{-1/2} \sqrt{\frac{\pi}{2\omega}} e^{-\omega} = \frac{\sqrt{2\pi}}{\sqrt{\chi}} e^{-\omega} = \left(\frac{2\pi}{\chi}\right)^{1/2} e^{-\omega},$$

where $\omega := \sqrt{\chi\psi}$ and $\eta := \sqrt{\chi/\psi}$. That is,

$$f_{\text{IG}_1}(x; \chi, \psi) = \left(\frac{\chi}{2\pi}\right)^{1/2} e^{\sqrt{\chi\psi}} x^{-3/2} \exp\left[-\frac{1}{2}(\chi x^{-1} + \psi x)\right] \mathbb{I}_{(0,\infty)}(x), \qquad (9.35)$$

as in (I.7.55).

Remarks:
(a) Expressing (9.35) as

$$f_{\text{IG}_1}(x; \chi, \psi) = \left(\frac{\chi}{2\pi x^3}\right)^{1/2} \exp\left[\omega - \frac{1}{2}(\chi x^{-1} + \psi x)\right] \mathbb{I}_{(0,\infty)}(x)$$

$$= \left(\frac{\chi}{2\pi x^3}\right)^{1/2} \exp\left[-\frac{1}{2}(\chi x^{-1} - 2\sqrt{\chi\psi} + \psi x)\right] \mathbb{I}_{(0,\infty)}(x),$$

using the relation

$$\chi x^{-1} - 2\sqrt{\chi\psi} + \psi x = \psi \frac{x^2 - 2\eta x + \eta^2}{x} = \psi \frac{(x - \eta)^2}{x}$$

and $\psi = \chi/\eta^2$ (from the definition of η above), the IG density can also be written as

$$\left(\frac{\chi}{2\pi x^3}\right)^{1/2} \exp\left[-\frac{\chi}{2\eta^2} \frac{(x - \eta)^2}{x}\right] \mathbb{I}_{(0,\infty)}(x),$$

which is (the same form as) (I.7.54).

(b) As in the case of pHyp and Exp, it is again a matter of taste whether one should consider the limiting case $\lambda = -1/2$, $\chi > 0$, $\psi = 0$, i.e., the Lévy distribution, to be a subclass of IG. We decided, for the sake of consistency with the literature, to exclude the Lévy distribution from IG, so it is just a limiting case. However, there are some conceptual properties that both distribution families share, indicating that Lévy and IG could also be put into a single family.   ∎

The moments of the IG distribution are calculated in Problem 9.12. The m.g.f. of the IG distribution can be expressed in the following simple form:

$$\mathbb{M}_{\mathrm{IG}_1}(t, \chi, \psi) = \exp\left[\sqrt{\chi}\left(\sqrt{\psi} - \sqrt{\psi - 2t}\right)\right], \quad -\infty < t \le \psi/2. \tag{9.36}$$

Indeed, from (9.25) and (9.3), the m.g.f. of the IG distribution reduces as follows:

$$\mathbb{M}_{\mathrm{IG}_1}(t, \chi, \psi) = \mathbb{M}_{\mathrm{GIG}}(t; -1/2, \chi, \psi) = \frac{K_{1/2}\left(\omega\sqrt{1 - \frac{2t}{\psi}}\right)}{K_{1/2}(\omega)\left(1 - \frac{2t}{\psi}\right)^{-1/4}}$$

$$= \frac{\left(\omega\sqrt{1 - \frac{2t}{\psi}}\right)^{-1/2} e^{-\left(\omega\sqrt{1 - \frac{2t}{\psi}}\right)}}{\omega^{-1/2} e^{-\omega}\left(1 - \frac{2t}{\psi}\right)^{-1/4}}$$

$$= \exp\left[\omega\left(1 - \sqrt{1 - \frac{2t}{\psi}}\right)\right] = \exp\left[\sqrt{\chi}\left(\sqrt{\psi} - \sqrt{\psi - 2t}\right)\right].$$

This makes sense whenever $t \le \psi/2$. Note that this formula also makes sense for $\psi = 0$, and, in this case, we get the m.g.f. of the Lévy distribution. We now discuss the distribution of the sum of independent IG r.v.s by analysing the m.g.f. The following argument also holds for the Lévy distribution by plugging in $\psi = 0$.

Recalling from (2.2) that, for independent r.v.s X and Y, $\mathbb{M}_{X+Y}(t) = \mathbb{M}_X(t)\,\mathbb{M}_Y(t)$, it follows that, if $X_i \overset{\text{ind}}{\sim} \mathrm{IG}_1(\chi_i, \psi)$, then the m.g.f. of $S = X_1 + X_2 + \cdots + X_n$ is given by

$$\mathbb{M}_S(t) = \prod_{i=1}^{n} \exp\left[\sqrt{\chi_i}\left(\sqrt{\psi} - \sqrt{\psi - 2t}\right)\right] \tag{9.37}$$

$$= \exp\left[\left(\sum_{i=1}^{n}\sqrt{\chi_i}\right)\left(\sqrt{\psi} - \sqrt{\psi - 2t}\right)\right] = \exp\left[\sqrt{\chi}\left(\sqrt{\psi} - \sqrt{\psi - 2t}\right)\right],$$

for all $t \in (-\infty, \psi/2)$, where $\chi = \left(\chi_1^{1/2} + \chi_2^{1/2} + \cdots + \chi_n^{1/2}\right)^2$. Because the m.g.f. of a nonnegative r.v. uniquely determines its distribution, as discussed in Section 1.2.4, we can conclude that $S \sim \mathrm{IG}_1(\chi, \psi)$. If we let $\psi = 0$, then the same argument shows that the sum of independent Lévy r.v.s also follows a Lévy distribution.

9.4.2.8 The degenerate or Dirac distribution as a limiting case

Suppose that X is a *constant* r.v. having some constant value $x \in \mathbb{R}$. Then X has a distribution which we call the *degenerate* or *Dirac distribution* with value x, and denote it by Δ_x.

Let $\lambda \in \mathbb{R}$ and $x \in [0, \infty)$. Then, with $\omega = \sqrt{\chi \psi}$ and $\eta = \sqrt{\chi / \psi}$,

$$\text{GIG}(\lambda, \chi, \psi) \to \Delta_x \quad \text{for } \eta \to x \text{ and } \omega \to \infty \text{ where } \chi, \psi > 0. \tag{9.38}$$

These special cases are presented in Table 9.1, and a synopsis of the connections between the various special cases is given in Figure 9.2.

9.5 The generalized hyperbolic distribution

The family of generalized hyperbolic (GHyp) distributions is obtained by mixing normals using the GIG family as weights. This approach would give us a natural parameterization of the GHyp family, taking the three parameters λ, χ, and ψ of the GIG distribution plus the parameters β and μ appearing in the mixing procedure. But historically, another parameterization has dominated the literature, which we will also use. As such, we show how to convert between the two parameterizations. This enables us to work with either of them, switching from one to the other whenever it is convenient.

9.5.1 Definition, parameters and general formulae

For given parameters α, β, and δ, set $\chi := \delta^2$ and $\psi := \alpha^2 - \beta^2$. Now define the generalized hyperbolic distribution (GHyp) as follows. Let $\alpha, \delta \in \mathbb{R}_{\geq 0}$, $\lambda, \mu \in \mathbb{R}$ and $\beta \in [-\alpha, \alpha]$. For $(\lambda, \chi, \psi) \in \Theta_{\text{GIG}}$, we define the generalized hyperbolic distribution $\text{GHyp}(\lambda, \alpha, \beta, \delta, \mu)$ as a $\text{GIG}(\lambda, \chi, \psi)$–variance–mean mixture of normal distributions, i.e.,

$$\text{GHyp}(\lambda, \alpha, \beta, \delta, \mu) := \text{Mix}_{\text{GIG}(\lambda, \delta^2, \alpha^2 - \beta^2)}(\mu, \beta). \tag{9.39}$$

The domain of variation of the parameters of the GHyp distribution is $\lambda, \mu \in \mathbb{R}$, $\alpha, \delta \geq 0$, $\beta \in [-\alpha, \alpha]$ such that the following conditions are satisfied:

$$\begin{array}{llllll}
|\beta| < \alpha & \text{and} & \delta > 0 & \text{if} & \lambda = 0, \\
|\beta| < \alpha & \text{and} & \delta \geq 0 & \text{if} & \lambda > 0, \\
|\beta| = \alpha & \text{and} & \delta > 0 & \text{if} & \lambda < 0.
\end{array}$$

In order to calculate the p.d.f. of the GHyp distribution, we have to solve the integrals appearing in the mixing procedure. At first glance, this does not appear to be an attractive activity! In order to save time and effort, we do the calculation just once, in its most general form, giving us a somewhat mysterious formula for the p.d.f. involving the function $k_\lambda (\chi, \psi)$ (which has already appeared as the normalizing constant of the GIG distribution). When discussing the subfamilies of GHyp, we will deduce more explicit formulae for the p.d.f.

Let $\lambda, \mu \in \mathbb{R}$, $\alpha, \delta \geq 0$, $\beta \in [-\alpha, \alpha]$ such that $(\lambda, \delta^2, \alpha^2 - \beta^2) \in \Theta_{\text{GIG}}$. Then from (9.39),

$$
f_{\text{GHyp}}(x; \lambda, \alpha, \beta, \delta, \mu) = f_{\text{Mix}_{\text{GIG}(\lambda, \delta^2 \alpha^2 - \beta^2)}}(x; \mu, \beta)
$$

$$
= \int_0^\infty f_N(x; \mu + \beta z, z) f_{\text{GIG}}(z; \lambda, \delta^2, \alpha^2 - \beta^2)\, dz
$$

$$
= \int_0^\infty \frac{1}{\sqrt{2\pi z}} \exp\left\{ -\frac{1}{2} \frac{(x - (\mu + \beta z))^2}{z} \right\}
$$

$$
\times \frac{1}{k_\lambda (\delta^2, \alpha^2 - \beta^2)} z^{\lambda - 1} \exp\left\{ -\frac{1}{2}\left(\delta^2 z^{-1} + (\alpha^2 - \beta^2) z \right) \right\} dz
$$

$$
= \frac{1}{\sqrt{2\pi}\, k_\lambda (\delta^2, \alpha^2 - \beta^2)} \int_0^\infty z^{\lambda - 1 - \frac{1}{2}} e^{-\frac{1}{2}\left(z^{-1}[(x-\mu)^2 + \delta^2] + z[\alpha^2 - \beta^2 + \beta^2] - 2\beta x + 2\beta\mu \right)} dz
$$

$$
= \frac{1}{\sqrt{2\pi}\, k_\lambda (\delta^2, \alpha^2 - \beta^2)} e^{\beta(x-\mu)} \underbrace{\int_0^\infty z^{[\lambda - \frac{1}{2}] - 1} e^{-\frac{1}{2}\left(z^{-1}[(x-\mu)^2 + \delta^2] + z\alpha^2 \right)} dz}_{= k_{\lambda - \frac{1}{2}}\left((x-\mu)^2 + \delta^2,\, \alpha^2\right)}
$$

or

$$
\boxed{f_{\text{GHyp}}(x; \lambda, \alpha, \beta, \delta, \mu) = \frac{k_{\lambda - \frac{1}{2}}\left((x - \mu)^2 + \delta^2,\, \alpha^2\right)}{\sqrt{2\pi}\, k_\lambda (\delta^2, \alpha^2 - \beta^2)} e^{\beta(x-\mu)}.}
\tag{9.40}
$$

Before we arrive at more explicit, but less general, expressions for the density, we derive the general formulae for the mean, variance, and m.g.f. of GHyp. To calculate the expected value, we use (9.13) for mixtures of normals, and the expression calculated for the mean of the GIG distribution. Suppose $X \sim \text{GHyp}(\lambda, \alpha, \beta, \delta, \mu)$, and define $\chi := \delta^2$ and $\psi := \alpha^2 - \beta^2$. For $Z \sim \text{GIG}(\lambda, \chi, \psi)$, we proved above that

$$
\mathbb{E}[Z] = \frac{k_{\lambda+1} (\chi, \psi)}{k_\lambda (\chi, \psi)},
$$

so, by the mean equation (9.13), it follows that

$$
\mathbb{E}[X] = \mu + \beta \mathbb{E}[Z] = \mu + \beta \frac{k_{\lambda+1} (\chi, \psi)}{k_\lambda (\chi, \psi)}.
$$

Similarly, it follows from

$$
\mathbb{V}(Z) = \frac{k_\lambda (\chi, \psi)\, k_{\lambda+2} (\chi, \psi) - (k_{\lambda+1} (\chi, \psi))^2}{(k_\lambda (\chi, \psi))^2}
$$

and (9.14) that

$$V(X) = \mathbb{E}[Z] + \beta^2 \mathbb{V}(Z) = \frac{k_{\lambda+1}(\chi, \psi)}{k_\lambda(\chi, \psi)} + \beta^2 \frac{k_\lambda(\chi, \psi)k_{\lambda+2}(\chi, \psi) - (k_{\lambda+1}(\chi, \psi))^2}{(k_\lambda(\chi, \psi))^2}.$$

Admittedly, this latter expression is somewhat complicated and not particularly handy.[2] But this complexity precisely illustrates the utility of the mixture approach, via the GIG distribution. Attempting to calculate the variance of the GHyp distribution directly would be clumsy and unnecessarily difficult.

Analogously to the above computations, we use (9.16) and the general formula (9.21) for the m.g.f. of the GIG distribution to get

$$\mathbb{M}_X(t) = e^{\mu t}\mathbb{M}_Z(\beta t + t^2/2) = e^{\mu t}\frac{k_\lambda(\chi, \psi - 2(\beta t + t^2/2))}{k_\lambda(\chi, \psi)}.$$

Now

$$\psi - 2(\beta t + t^2/2) = \alpha^2 - \beta^2 - 2\beta t - t^2 = \alpha^2 - (\beta + t)^2,$$

so

$$\mathbb{M}_X(t) = e^{\mu t}\frac{k_\lambda(\delta^2, \alpha^2 - (\beta + t)^2)}{k_\lambda(\delta^2, \alpha^2 - \beta^2)}.$$

This holds for all t such that $-\alpha - \beta < t < \alpha - \beta$.

9.5.2 The subfamilies of the GHyp distribution family

The GHyp distribution family is quite flexible, but this comes at the price of complexity. However, we can use the mixing mechanics as a tool to get an overview of the vast number of important subfamilies. As GHyp is constructed as a GIG mixture of normals, it follows that to every subfamily of GIG there corresponds a subfamily of GHyp, and, moreover, this GHyp subfamily inherits properties from the mixing weights used. This is the reason for having discussed the subfamilies of GIG in such detail.

9.5.2.1 *The proper* GHyp *distribution (if $\lambda \in \mathbb{R}$, $\alpha > 0$, $\beta \in (-\alpha, \alpha)$, $\delta > 0$, $\mu \in \mathbb{R}$)*

It follows that $\chi := \delta^2 > 0$ and $\psi := \alpha^2 - \beta^2 > 0$, so we have a mixture of proper GIG distributions. The normal case ($\chi > 0$, $\psi > 0$) can be considered to be a GHyp distribution in the strict sense, and we can use (9.8) to derive its p.d.f. from the general formula (9.40),

---

[2] Mencía and Sentana (2004) give impressive expressions even for the skewness and kurtosis of GHyp in the multivariate case.

$$f_{\text{GHyp}}(x; \lambda, \alpha, \beta, \delta, \mu)$$

$$\overset{(9.40)}{=} \frac{k_{\lambda-\frac{1}{2}}\left((x-\mu)^2 + \delta^2, \alpha^2\right)}{\sqrt{2\pi}\, k_\lambda \left(\delta^2, \alpha^2 - \beta^2\right)}\, e^{\beta(x-\mu)}$$

$$\overset{(9.8)}{=} \frac{(\alpha^2 - \beta^2)^{\frac{\lambda}{2}}\, K_{\lambda-\frac{1}{2}}\left(\alpha\sqrt{(x-\mu)^2 + \delta^2}\right)}{\sqrt{2\pi}\,\delta^\lambda \alpha^{\lambda-\frac{1}{2}} K_\lambda\left(\delta\sqrt{\alpha^2 - \beta^2}\right)}\, \sqrt{(x-\mu)^2 + \delta^2}^{\lambda-\frac{1}{2}}\, e^{\beta(x-\mu)}$$

or, with $\boxed{y_x = \sqrt{\delta^2 + (x-\mu)^2},}$

$$\boxed{f_{\text{GHyp}}(x; \lambda, \alpha, \beta, \delta, \mu) = \frac{(\alpha^2 - \beta^2)^{\frac{\lambda}{2}}\, y_x^{\lambda-\frac{1}{2}}}{\sqrt{2\pi}\,\alpha^{\lambda-\frac{1}{2}}\,\delta^\lambda\, K_\lambda(\delta\sqrt{\alpha^2 - \beta^2})}\, K_{\lambda-\frac{1}{2}}(\alpha y_x)\, e^{\beta(x-\mu)}.}$$

$$(9.41)$$

Remark: There is a considerable amount of confusion in the literature about the suitable domain of variation for the parameters of the GHyp distribution. Because we have introduced GHyp as a GIG mixture of normals, we decided to give GHyp the rather complicated parameter space which is described on page 315, reflecting the fact that the natural domain of variation for the parameters of the GIG distribution comprises two boundary cases. Various authors give GHyp this larger parameter space but introduce it having the p.d.f. (9.41), which does not make sense on the boundary, e.g., for $\alpha = \beta = 0$. In fact, the general formula (9.40) represents the p.d.f. of GHyp on the whole parameter space and (9.41) is only true for the 'normal case' which we termed the *proper* GHyp distribution. ∎

Listing 9.2 gives a program which computes the GHyp p.d.f. in the normal case. Numerically integrating the density via `quadl(@ghyppdf,lo,hi,1e-8,0,lambda,alpha,beta)` (where `lo` and `hi` specify the range of integration) verifies that the density is proper. Numeric integration can also be used to calculate the c.d.f. Figure 9.3 shows the GHyp density for several parameter values.

```
function f=ghyppdf(x,lambda,alpha,beta,delta,mu)
if nargin<6, mu=0; end
if nargin<5, delta=1; end
a=alpha; b=beta; lam=lambda;   % just easier to work with
c=sqrt(a^2 - b^2); y = sqrt(delta^2 + (x-mu).^2);
t1=c^lam * y.^(lam - 0.5);
t2=sqrt(2*pi) * a^(lam-0.5) * delta^lam * besselk(lam,delta*c);
d1=besselk(lam-0.5,a*y) .* exp(b*(x-mu));
f = (t1./t2) .* d1;
```

Program Listing 9.2 Computes the GHyp p.d.f. in the normal case

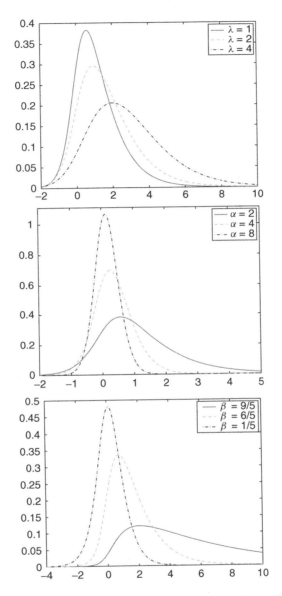

Figure 9.3 The GHyp p.d.f. for $\delta = 1$ and $\mu = 0$: (top) $\alpha = 2$ and $\beta = 1$; (middle) $\lambda = 1$ and $\beta = 1$; (bottom) $\lambda = 1$ and $\alpha = 2$

In the normal case, we use the abbreviations

$$\eta := \sqrt{\frac{\chi}{\psi}} = \sqrt{\frac{\delta^2}{\alpha^2 - \beta^2}} \quad \text{and} \quad \omega := \sqrt{\chi\psi} = \sqrt{\delta^2\left(\alpha^2 - \beta^2\right)} = \delta\sqrt{\alpha^2 - \beta^2}.$$

Suppose $X \sim \mathrm{GHyp}(\lambda, \alpha, \beta, \delta, \mu)$ and $Z \sim \mathrm{GIG}(\lambda, \chi, \psi)$. To deduce an expression for the mean, we could either use the general formula for the mean of GHyp derived above, or apply equation (9.13) for mixtures of normals and the expression calculated for the mean of the GIG distribution in the normal case,

$$\mathbb{E}[Z] = \eta \frac{K_{\lambda+1}(\omega)}{K_\lambda(\omega)},$$

implying

$$\mathbb{E}[X] = \mu + \beta \mathbb{E}[Z] = \mu + \beta \eta \frac{K_{\lambda+1}(\omega)}{K_\lambda(\omega)}.$$

By a similar argument,

$$\mathbb{V}(X) = \mathbb{E}[Z] + \beta^2 \mathbb{V}(Z) = \eta \frac{K_{\lambda+1}(\omega)}{K_\lambda(\omega)} + \beta^2 \eta^2 \frac{K_\lambda(\omega) K_{\lambda+2}(\omega) - K_{\lambda+1}(\omega)^2}{K_\lambda(\omega)^2}.$$

An expression for the m.g.f. of the GHyp distribution in the normal case can be derived in the same way using (9.16) and (9.25), but we can also directly deduce it from the general formula. In any case,

$$\mathbb{M}_X(t) = e^{\mu t} \frac{K_\lambda \left(\delta \sqrt{\alpha^2 - (\beta + t)^2} \right)}{K_\lambda \left(\delta \sqrt{\alpha^2 - \beta^2} \right) \left(\frac{\alpha^2 - (\beta+t)^2}{\alpha^2 - \beta^2} \right)^{\lambda/2}}, \tag{9.42}$$

with convergence strip given by those values of t such that $-\alpha - \beta < t < \alpha - \beta$. Using the abbreviation $\psi_t := \alpha^2 - (\beta + t)^2$, and recalling that $\chi = \delta^2 > 0$ and $\psi = \alpha^2 - \beta^2 > 0$, this can also be expressed as

$$\mathbb{M}_X(t) = e^{\mu t} \frac{K_\lambda \left(\sqrt{\chi \psi_t} \right)}{K_\lambda \left(\sqrt{\chi \psi} \right) (\psi_t/\psi)^{\lambda/2}}.$$

9.5.2.2 The variance–gamma distribution VG (if $\lambda > 0$, $\alpha > 0$, $\beta \in (-\alpha, \alpha)$, $\delta = 0$, $\mu \in \mathbb{R}$)

Then $\chi = \delta^2 = 0$ and $\psi = \alpha^2 - \beta^2 > 0$, *resulting in a gamma mixture of normals. If we want to apply the general formula* (9.40) *to compute the p.d.f. in this case, we have to be careful. As $\delta = 0$, we have to use* (9.9) *rather than* (9.8) *to compute the denominator in* (9.40)*, whereas the numerator can be calculated as in the normal case:*

$$f_{\mathrm{GHyp}}(x; \lambda, \alpha, \beta, 0, \mu) = \frac{2 \left(\frac{\alpha^2 - \beta^2}{2} \right)^\lambda}{\sqrt{2\pi} \Gamma(\lambda)} \left(\frac{|x - \mu|}{\alpha} \right)^{\lambda - \frac{1}{2}} K_{\lambda - \frac{1}{2}}(\alpha |x - \mu|) e^{\beta(x - \mu)}. \tag{9.43}$$

We will refer to this as the *variance–gamma* distribution (VG), so that

$$VG(\lambda, \alpha, \beta, \mu) := GHyp(\lambda, \alpha, \beta, 0, \mu).$$

It was derived directly in Example 7.20, with $\mu = 0$, and substituting $b = \lambda$, $m = \beta$, and $c = (\alpha^2 - \beta^2)/2$ yields (7.41).

This distribution was popularized by Madan and Seneta (1990) in their study of financial returns data,[3] and continues to receive attention in this context; see Seneta (2004) and the references therein. Bibby and Sørensen (2003) also propose the name normal–gamma (NG) distribution. See Problem 9.9 for expressions for mean, variance, and m.g.f. of the GHyp distribution in boundary case I, i.e., of the variance–gamma distribution.

9.5.2.3 The hyperbolic asymmetric (Student's) t distribution HAt (if $\lambda < 0$, $\beta \in \mathbb{R}$, $\alpha = |\beta|$, $\delta > 0$, $\mu \in \mathbb{R}$)

Now, $\chi = \delta^2 > 0$ *and* $\psi = \alpha^2 - \beta^2 = 0$, *and we have an inverse gamma mixture of normals. There are two cases to distinguish,* $\alpha = |\beta| > 0$ *and* $\alpha = \beta = 0$.

For the former, note that the denominator of (9.40) has to be computed using (9.10), and the numerator can be treated as above to get

$$f_{GHyp}(x; \lambda, |\beta|, \beta, \delta, \mu) = \frac{2\left(\delta^2/2\right)^{-\lambda}}{\sqrt{2\pi}\,\Gamma(-\lambda)} \left(\frac{y}{|\beta|}\right)^{\lambda - \frac{1}{2}} K_{\lambda - \frac{1}{2}}(|\beta|y)\, e^{\beta(x - \mu)}, \qquad (9.44)$$

where

$$y = \sqrt{\delta^2 + (x - \mu)^2}.$$

In the second case, where $\alpha = \beta = 0$, the equation changes quite dramatically. Both the numerator and the denominator of (9.40) have to be calculated by the use of (9.10), yielding

$$
\begin{aligned}
f_{GHyp}(x; \lambda, 0, 0, \delta, \mu) &= \int_0^\infty f_N(x; \mu, z)\, f_{GIG}(z; \chi, \delta^2, 0)\, dz \\[2mm]
&\overset{(9.40)}{=} \frac{k_{\lambda - \frac{1}{2}}\left((x - \mu)^2 + \delta^2,\, 0\right)}{\sqrt{2\pi}\, k_\lambda\left(\delta^2,\, 0\right)}\, e^{0 \cdot (x - \mu)} \\[2mm]
&\overset{(9.10)}{=} \frac{\left((x - \mu)^2 + \delta^2\right)^{\lambda - \frac{1}{2}} \Gamma\left(-\lambda + \frac{1}{2}\right)}{\sqrt{2\pi}\,\left(\delta^2\right)^{\lambda} \Gamma(-\lambda)} \\[2mm]
&= \frac{\Gamma\left(\frac{-2\lambda + 1}{2}\right)}{\Gamma\left(\frac{-2\lambda}{2}\right)}\, \frac{1}{\sqrt{\delta^2 \pi}} \left(1 + \frac{(x - \mu)^2}{\delta^2}\right)^{-\frac{-2\lambda + 1}{2}}, \qquad (9.45)
\end{aligned}
$$

---

[3] Madan and Seneta (1990) only consider the symmetric case, and use a different parameterization than the one we adopt. To convert *to* their notation, set $v = 1/\lambda$, $\sigma^2 = (2\lambda)/\alpha^2$, and $m = 1$. To convert *from* their notation, set $\beta = 0$, $\lambda = 1/v$, and $\alpha^2 = 2/(\sigma^2 v)$.

which (as $\beta = 0$) is symmetric about μ. If $\delta^2 = -2\lambda =: n$, then this is a Student's t density with n degrees of freedom. The parameter δ is just a scale parameter of the distribution $\mathrm{GHyp}(\lambda, 0, 0, \delta, \mu)$, and μ is a location parameter.

In light of (9.45), we see that (9.44) can be interpreted as (yet another) skewed t distribution. Running out of names for generalized t distributions, we term this the *hyperbolic asymmetric* (Student's) t, or HAt, given by

$$\mathrm{HA}t(n, \beta, \mu, \delta) := \mathrm{GHyp}(\lambda, |\beta|, \beta, \delta, \mu),$$

where $n = -2\lambda$, $n > 0$, $\beta, \mu \in \mathbb{R}$, $\delta > 0$, and $\lambda = -n/2$. Bibby and Sørensen (2003) refer to it as the *asymmetric scaled t distribution*, whereas Aas and Haff (2005, 2006), who have also analysed its applications to skewed financial data, name it the *(generalized hyperbolic) skew Student's t*. For $\beta \neq 0$, this distribution has p.d.f.

$$f_{\mathrm{HA}t}(x; n, \beta, \mu, \delta) = \frac{2^{\frac{-n+1}{2}} \delta^n}{\sqrt{\pi}\Gamma(n/2)} \left(\frac{y_x}{|\beta|}\right)^{-\frac{n+1}{2}} K_{-\frac{n+1}{2}}(|\beta|y_x)\, e^{\beta(x-\mu)}, \qquad (9.46)$$

where, as before, $y_x = \sqrt{\delta^2 + (x - \mu)^2}$. As in (9.45), if $\beta = 0$, it reduces to the Student's t p.d.f.

$$f_{\mathrm{HA}t}(x; n, 0, \mu, \delta) = \frac{\Gamma\left(\frac{n+1}{2}\right)}{\Gamma\left(\frac{n}{2}\right)} \frac{1}{\sqrt{\delta^2\pi}} \left(1 + \left(\frac{x-\mu}{\delta}\right)^2\right)^{-\frac{n+1}{2}}.$$

More precisely, this is the usual t distribution with n degrees of freedom if we set $\delta^2 = n$. Figure 9.4 illustrates the density of HAt for three values of β.

For a discussion of the mean, variance, and m.g.f. of GHyp in boundary case II, the reader is referred to Problem 9.10.

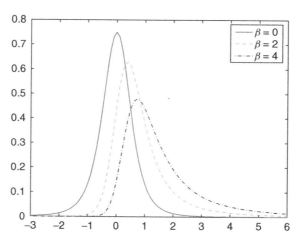

Figure 9.4 Hyperbolic asymmetric t density (9.46) for $\mu = 0$ and $\delta = 1$

9.5.2.4 The asymmetric Laplace distribution ALap *(if* $\lambda = 1$, $\alpha > 0$, $\beta \in (-\alpha, \alpha)$,
$\delta = 0$, $\mu \in \mathbb{R}$)

Then $\chi = 0$, $\psi > 0$, *and* $\lambda = 1$, *i.e., we have an exponential mixture of normals.*
This is a subfamily of the variance–gamma distribution, the resulting density being

$$f_{\mathrm{ALap}}(x; \alpha, \beta, \mu) = f_{\mathrm{GHyp}}(x; 1, \alpha, \beta, 0, \mu) = \frac{\alpha^2 - \beta^2}{2\alpha} \, \mathrm{e}^{-\alpha|x-\mu|+\beta(x-\mu)}, \qquad (9.47)$$

with distributional notation

$$\mathrm{ALap}(\alpha, \beta, \mu) := \mathrm{GHyp}(1, \alpha, \beta, 0, \mu).$$

Note that, with $\beta = 0$, this simplifies to

$$\frac{\alpha}{2} \, \mathrm{e}^{-\alpha|x-\mu|},$$

which is the usual location–scale Laplace p.d.f. (I.7.17); more precisely,

$$\mathrm{GHyp}(1, \alpha, 0, 0, \mu) = \mathrm{Lap}\left(\mu, \alpha^{-1}\right).$$

This agrees with Example 9.1, which showed that the exponential mixture of normals results in a Laplace distribution. So the question arises as to what we get if we take the positive hyperbolic distribution, which generalizes the exponential distribution, as the mixing weight. The answer is Hyp, the hyperbolic distribution, which we discuss next.

Remark: The density of the asymmetric Laplace distribution can be reparameterized in the case $\mu = 0$ as follows. Define $\beta^+ := \alpha - \beta$ and $\beta^- := \alpha + \beta$, then

$$\mathrm{e}^{-\alpha|x|+\beta x} = \begin{cases} \mathrm{e}^{-(\alpha-\beta)|x|} = \mathrm{e}^{-\beta^+|x|}, & \text{if } x > 0, \\ \mathrm{e}^{-(\alpha+\beta)|x|} = \mathrm{e}^{-\beta^-|x|}, & \text{if } x \leq 0. \end{cases}$$

So the asymmetric Laplace can be constructed by taking two Laplace densities with parameters β^+ and β^-, one for the positive and one for the negative half-line. We have met this way of producing asymmetric from symmetric distributions in Chapter 7, and indeed, if we apply the formula (7.9) of Fernández and Steel to the Laplace distribution, we get ALap. More precisely, remembering that the density of Laplace is $f_{\mathrm{Lap}}(x; \sigma) = 1/(2\sigma)\mathrm{e}^{|x|/\sigma}$, we are looking for θ and σ such that $\beta^+ = 1/(\sigma\theta)$ and $\beta^- = \theta/\sigma$. It is straightforward to check that

$$\sigma = \frac{1}{\sqrt{\alpha^2 - \beta^2}} \quad \text{and} \quad \theta = \frac{\sqrt{\alpha^2 - \beta^2}}{\alpha - \beta}$$

give the solution. ∎

9.5.2.5 *The hyperbolic distribution* Hyp *(if* $\lambda = 1$, $\alpha > 0$, $\beta \in (-\alpha, \alpha)$, $\delta > 0$, $\mu \in \mathbb{R}$)

Here $\lambda = 1$, $\chi > 0$, $\psi > 0$, *so this is a* pHyp *mixture of normals.* The hyperbolic distribution, Hyp, is defined as the distribution[4]

$$\text{Hyp}_3(\alpha, \beta, \delta, \mu) := \text{GHyp}(1, \alpha, \beta, \delta, \mu).$$

The density, as derived in Problem 9.11, is

$$f_{\text{Hyp}_3}(x; \alpha, \beta, \delta, \mu) = \frac{\sqrt{\alpha^2 - \beta^2}}{2\alpha\delta K_1(\delta\sqrt{\alpha^2 - \beta^2})} \exp\left[-\alpha\sqrt{\delta^2 + (x - \mu)^2} + \beta(x - \mu)\right].$$

$$(9.48)$$

In the boundary case $\delta \to 0$, we have $\sqrt{\delta^2 + (x - \mu)^2} \to |x - \mu|$ and we get the skewed Laplace distribution we met in (9.47), this being an Exp mixture of normals as we have seen above.

As already mentioned, the name of this family originates from the shape of the log-density. The log-densities of the hyperbolic distribution are hyperbolae, even in the skewed case ($\beta \neq 0$), while the log-density of the normal distribution is a parabola. Exactly as parabolae are limits of hyperbolae, the normal distribution is a limit of hyperbolic distributions.

9.5.2.6 *An asymmetric Cauchy distribution (if* $\lambda = -1/2$, $\beta \in \mathbb{R}$, $\alpha = |\beta|$, $\delta > 0$, $\mu \in \mathbb{R}$)

This is a special case of the hyperbolic asymmetric t distribution HAt, discussed above. If we use the Lévy distribution as a mixing weight, then we get a distribution having the following p.d.f. (just take the p.d.f. of HAt for $\lambda = -1/2$):

$$f_{\text{HA}t}(x; -1/2, \beta, \delta, \mu) = \frac{2(\delta^2/2)^{1/2}}{\sqrt{2\pi}\Gamma(1/2)}\left(\frac{y_x}{|\beta|}\right)^{-1} K_{-1}(|\beta|y_x) e^{\beta(x-\mu)}$$

$$= \frac{\delta|\beta|}{\pi y_x} K_{-1}(|\beta|y_x) e^{\beta(x-\mu)},$$

for $\beta \neq 0$ and letting

$$y_x = \sqrt{\delta^2 + (x - \mu)^2}.$$

In the symmetric case, i.e., with $\beta = 0$, we get

$$f_{\text{HA}t}(x; -1/2, 0, \delta, \mu) = \frac{\delta}{\pi\left(\delta^2 + (x - \mu)^2\right)},$$

i.e., HA$t(-1/2, 0, \delta, \mu)$ is a Cauchy distribution. The Cauchy distribution does not possess a moment generating function, but we can prove that, if X follows

---

[4] We write Hyp$_3$ for this parameterization because, in Chapter I.7, we have already introduced two alternative parameterizations, called Hyp$_1$ and Hyp$_2$.

$f_{\mathrm{HAt}}(x; -1/2, \beta, \delta, \mu)$ with $\beta \neq 0$ and $-\alpha - \beta \leq t \leq \alpha + \beta$, then

$$M_X(t) = e^{\mu t} e^{-\delta \sqrt{\alpha^2 - (\beta + t)^2}}. \tag{9.49}$$

Note that, as $\alpha = |\beta|$, the m.g.f. is not defined on a neighbourhood of zero but rather on a closed interval such that one of the borders is zero.

We do not bother to give $\mathrm{HAt}(-1/2, \beta, \delta, \mu)$ a name of its own. It is a subfamily of the hyperbolic asymmetric t distribution HAt and a limiting case of the following, rather general subfamily of GHyp, which is of much greater importance in our context, and the calculations concerning the m.g.f. can be done in this more general situation.

9.5.2.7 The normal inverse Gaussian distribution NIG (if $\lambda = -1/2$, $\alpha > 0$, $\beta \in (-\alpha, \alpha)$, $\delta > 0$, $\mu \in \mathbb{R}$)

The normal inverse Gaussian (NIG) distribution is defined as

$$\mathrm{NIG}(\alpha, \beta, \delta, \mu) := \mathrm{GHyp}(-1/2, \alpha, \beta, \delta, \mu),$$

and has density

$$f_{\mathrm{NIG}}(x; \alpha, \beta, \delta, \mu) = e^{\delta \sqrt{\alpha^2 - \beta^2}} \frac{\alpha \delta}{\pi \sqrt{\delta^2 + (x - \mu)^2}} K_1 \left(\alpha \sqrt{\delta^2 + (x - \mu)^2} \right) e^{\beta (x - \mu)}. \tag{9.50}$$

The preceding formula still makes sense if $\alpha = |\beta| > 0$ and it is then just the p.d.f. of the asymmetric Cauchy distribution that we have just met. So the asymmetric Cauchy distribution is a limiting case of NIG, and if $\alpha, \beta \to 0$, we get the usual Cauchy distribution.

The formulae for the mean, variance, and skewness for the NIG distribution take a much simpler form than for the general case of GHyp. If $X \sim \mathrm{NIG}(\alpha, \beta, \delta, \mu)$, then

$$\mathbb{E}[X] = \mu + \beta \eta, \quad \mathbb{V}(X) = \eta + \beta^2 \frac{\eta^2}{\omega}, \quad \mu_3(X) = 3\beta \frac{\eta^2}{\omega} + 3\beta^3 \frac{\eta^3}{\omega^2}, \tag{9.51}$$

where $\eta = \delta / \sqrt{\alpha^2 - \beta^2}$ and $\omega = \delta \sqrt{\alpha^2 - \beta^2}$. For the m.g.f., we obtain

$$M_X(t) = e^{\mu t} e^{\delta \left(\sqrt{\alpha^2 - \beta^2} - \sqrt{\alpha^2 - (\beta + t)^2} \right)}, \quad -\alpha - \beta \leq t \leq \alpha + \beta. \tag{9.52}$$

Problems 9.11 and 9.12 derive these results.

By definition, the NIG distribution is an IG mixture of normals, and we have already seen that the family of inverse Gaussian distributions has a stability property in that the sum of independent IG r.v.s is again IG distributed. This carries over to the NIG distribution, as we shall see in the last section of this chapter.

Remark: Above we have seen that the Lévy distribution is a limit of IG distributions and could even be considered as a subcase. In the same way, the (asymmetric) Cauchy distribution could be considered a subfamily of the NIG distribution with $\alpha = |\beta|$. That this would make sense can be underpinned by the fact that the Cauchy distribution and the NIG distribution share important properties such as stability under sums of independent r.v.s. However, we decided to exclude the Cauchy distribution from NIG to be consistent with the literature, and also to ensure that it is easier to reparameterize the NIG, as we will see below. ∎

Tables 9.2 and 9.3 and Figure 9.5 provide an overview of the results pertaining to the GHyp distributional class.

Table 9.2 Special cases of GHyp

Name	Abbrev.	Parameter range				
variance–gamma	VG	$\lambda > 0$	$\alpha > 0$	$\beta \in (-\alpha, \alpha)$	$\delta = 0$	$\mu \in \mathbb{R}$
asymmetric Laplace	ALap	$\lambda = 1$	$\alpha > 0$	$\beta \in (-\alpha, \alpha)$	$\delta = 0$	$\mu \in \mathbb{R}$
Laplace	Lap	$\lambda = 1$	$\alpha > 0$	$\beta = 0$	$\delta = 0$	$\mu \in \mathbb{R}$
hyperbolic	Hyp	$\lambda = 1$	$\alpha > 0$	$\beta \in (-\alpha, \alpha)$	$\delta > 0$	$\mu \in \mathbb{R}$
hyperbolic asymmetric t	HAt	$\lambda < 0$	$\alpha = |\beta|$	$\beta \geq 0$	$\delta > 0$	$\mu \in \mathbb{R}$
Student's t	t	$\lambda < 0$	$\alpha = 0$	$\beta = 0$	$\delta > 0$	$\mu \in \mathbb{R}$
Cauchy	Cau	$\lambda = -1/2$	$\alpha = 0$	$\beta = 0$	$\delta > 0$	$\mu \in \mathbb{R}$
normal inverse Gaussian	NIG	$\lambda = -1/2$	$\alpha > 0$	$\beta \in (-\alpha, \alpha)$	$\delta > 0$	$\mu \in \mathbb{R}$
normal (as a limit)	N	$\lambda \in \mathbb{R}$	$\alpha \to \infty$	$\beta \to \beta_0$	$\frac{\delta}{\alpha} \to \sigma^2$	$\mu \in \mathbb{R}$

Table 9.3 Connections between GIG and GHyp

Mixing weight	Resulting distribution
proper GIG	proper GHyp
gamma (Gam)	variance–gamma (VG)
exponential (Exp)	asymmetric Laplace (ALap)
exponential (Exp) with $\beta = 0$	Laplace (Lap)
positive hyperbolic (pHyp)	hyperbolic (Hyp)
inverse gamma (IGam)	hyperbolic asymmetric t (HAt)
inverse gamma with $\beta = 0$	Student's t
Lévy	an asymmetric Cauchy
Lévy with $\beta = 0$	Cauchy
inverse Gaussian (IG)	normal inverse Gaussian (NIG)
Dirac Δ_x with $x > 0$	normal distribution

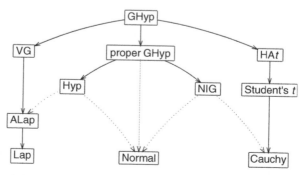

Figure 9.5 The subclasses of the GHyp distribution: solid arrows point to subclasses, dashed arrows indicate limits

9.5.3   Limiting cases of GHyp

There are two limiting cases of the GHyp distribution which are noteworthy: the normal distribution and the GIG distribution. The first is of obvious practical and theoretical interest, whereas the second limit just shows what happens as the GHyp distribution becomes maximally skewed, and is primarily of 'academic interest'.

As mentioned in (9.38), the one-point distribution is a limiting case of the GIG distribution family. Taking the one-point distribution as a mixing weight results in a degenerate mixture, i.e., just a normal distribution. To be more precise, let $\mu, \beta_0 \in \mathbb{R}$ and $\sigma^2 \in \mathbb{R}_{>0}$, so that

$$\mathrm{Mix}_{\Delta_{\sigma^2}}(\mu, \beta_0) = \mathrm{N}(\mu + \beta_0\sigma^2, \sigma^2).$$

As the mixture depends continuously[5] on the weight and the mixing parameters, we can conclude that, if GIG converges to a Dirac distribution, then GHyp converges to a normal distribution; hence the following convergence behaviour of GHyp.

Normal distribution as a limiting case: Let $\sigma_0^2 > 0$ and $\mu_0, \beta_0 \in \mathbb{R}$. Then, for all $\lambda \in \mathbb{R}$,

$$\mathrm{GHyp}(\lambda, \alpha, \beta, \delta, \mu) \to \mathrm{N}(\mu_0 + \beta_0\sigma_0^2, \sigma_0^2)$$

as $\alpha \to \infty$, $\delta \to \infty$, $\mu \to \mu_0$, and $\beta \to \beta_0$ with $\delta/\alpha \to \sigma_0^2$.

Similarly, as is shown in Eberlein and Hammerstein (2004) for the proper GIG distribution, direct calculation reveals the following.

GIG distribution as a limiting case: Let $\lambda_0 \in \mathbb{R}$ and $\psi_0, \chi_0 \geq 0$ such that $(\lambda_0, \chi_0, \psi_0)$ is in the parameter space Θ_{GIG} of the GIG distribution. Then

$$\mathrm{GHyp}\,(\lambda_0, \alpha, \beta, \delta, 0) \to \mathrm{GIG}(\lambda_0, \chi_0, \psi_0)$$

as $\alpha \to \infty$, $\beta \to \infty$, and $\delta \to 0$ such that $\alpha - \beta \to \psi_0/2$ and $\alpha\delta^2 \to \chi_0$.

Note that the condition $\alpha - \beta \to \psi_0/2$ means that β and α converge 'at the same speed' to infinity, so the GHyp distribution with these parameter values becomes more and more skewed to the right until there is no mass remaining on the negative half-line; this results in a positive GIG distribution. Doing the same thing, but bending the GHyp distribution more and more to the left, we get a symmetric picture, namely the reflected GIG distribution on the negative half-line, which we denote by $-\mathrm{GIG}$:

The reflected GIG distribution as a limiting case: Let $\lambda_0 \in \mathbb{R}$ and $\psi_0, \chi_0 \geq 0$ such that $(\lambda_0, \chi_0, \psi_0) \in \Theta_{\mathrm{GIG}}$. Then

$$\mathrm{GHyp}\,(\lambda_0, \alpha, \beta, \delta, 0) \to -\mathrm{GIG}(\lambda_0, \chi_0, \psi_0)$$

as $\alpha \to \infty$, $\beta \to -\infty$, and $\delta \to 0$ such that $\alpha + \beta \to \psi_0/2$ and $\alpha\delta^2 \to \chi_0$.

---

[5] The notion of continuity (or convergence) that has to be used at this point is the notion of *weak continuity* (or *weak convergence*) which also gives rise to the term *convergence in distribution* for random variables (see, for example, Fristedt and Gray, 1997). As we do not want to say too much about this topic here, we prefer not to give the precise definitions and proofs. If desired, the reader may consider checking that the density functions converge pointwise in the two cases of convergence that we mention in this paragraph.

In both cases, if $\mu \neq 0$, we have to shift the resulting distribution by μ (yielding a shifted GIG or $-$ GIG distribution).

9.6 Properties of the GHyp distribution family

9.6.1 Location–scale behaviour of GHyp

If $X \sim \text{GHyp}(\lambda, \alpha, \beta, \delta, \mu)$ and $a, b \in \mathbb{R}$ with $a \neq 0$, then

$$aX + b \sim \text{GHyp}(\lambda,\ \alpha/|a|,\ \beta/a,\ \delta|a|,\ a\mu + b). \tag{9.53}$$

This fact can be proved in several ways. We choose the brute force method and calculate the p.d.f. of $aX + b$ using the general formula (9.40) for the p.d.f. of GHyp (we use the abbreviations $\alpha' = \alpha/|a|$, $\beta' = \beta/a$, and $\delta' = \delta|a|$ as well as $\mu' = a\mu + b$):

$$f_{aX+b}(x) = a^{-1} f_{\text{GHyp}}((x-b)/a;\ \lambda, \alpha, \beta, \delta, \mu)$$

$$= a^{-1} \frac{k_{\lambda-\frac{1}{2}}\left(((x-b)/a - \mu)^2 + \delta^2,\ \alpha^2\right)}{\sqrt{2\pi}\, k_\lambda\left(\delta^2,\ \alpha^2 - \beta^2\right)}\, e^{\beta((x-b)/a - \mu)}$$

$$= a^{-1} \frac{k_{\lambda-\frac{1}{2}}\left((x-b-a\mu)^2/a^2 + (\delta|a|)^2/a^2,\ a^2(\alpha/|a|)^2\right)}{\sqrt{2\pi}\, k_\lambda\left((\delta|a|)^2/a^2,\ a^2\left((\alpha/|a|)^2 - (\beta/a)^2\right)\right)}\, e^{(\beta/a)(x-b-a\mu)}$$

$$= a^{-1} \frac{k_{\lambda-\frac{1}{2}}\left(\left((x-\mu')^2 + \delta'^2\right)/a^2,\ a^2\alpha'^2\right)}{\sqrt{2\pi}\, k_\lambda\left(\delta'^2/a^2,\ a^2\left(\alpha'^2 - \beta'^2\right)\right)}\, e^{\beta'(x-\mu')} = (*).$$

Now use formula (9.12) in the numerator and denominator:

$$(*) = a^{-1} \frac{(a^2)^{-(\lambda-\frac{1}{2})} k_{\lambda-\frac{1}{2}}\left((x-\mu')^2 + \delta'^2,\ \alpha'^2\right)}{\sqrt{2\pi}\, (a^2)^{-\lambda} k_\lambda\left(\delta'^2,\ \alpha'^2 - \beta'^2\right)}\, e^{\beta'(x-\mu')}$$

$$= \underbrace{a^{-1-2\lambda+1-2\lambda}}_{=1}\, \frac{k_{\lambda-\frac{1}{2}}\left((x-\mu')^2 + \delta'^2,\ \alpha'^2\right)}{\sqrt{2\pi}\, k_\lambda\left(\delta'^2,\ \alpha'^2 - \beta'^2\right)}\, e^{\beta'(x-\mu')}$$

$$= f_{\text{GHyp}}(x;\ \lambda, \alpha', \beta', \delta', \mu').$$

Remark: There is another way to analyse the location–scale behaviour of GHyp which gives us more insight than the direct calculation. By our construction, GHyp is a mixture with mixing weight GIG, and there is close relation between the scale behaviour of the weight and the location–scale behaviour of the resulting mixed distribution that we will now outline.

First, an obvious bit of notation. If π is a distribution on $\mathbb{R}$ and $a, b \in \mathbb{R}$, then we write $a\pi + b$ for the distribution with the property that, if $Z \sim \pi$, then $aZ + b \sim a\pi + b$. Now the following theorem is true.

If π is a positive (continuous) distribution and $a, b \in \mathbb{R}$ such that $a \neq 0$, then for all $\beta, \mu \in \mathbb{R}$:

$$a \left(\mathrm{Mix}_\pi(\mu, \beta)\right) + b = \mathrm{Mix}_{a^2\pi}(a\mu + b, \beta/a).$$

To see this, let X and Z be random variables such that $Z \sim \pi$ and $(X \mid Z) \sim N(\mu + \beta Z, Z)$. Then, by definition, $X \sim \mathrm{Mix}_\pi(\mu, \beta)$. To derive the distribution of $aX + b$ we use

$$(aX + b \mid Z) \sim a\,N(\mu + \beta Z, Z) + b = N\left(a\mu + a\beta Z + b, a^2 Z\right).$$

Because $aX + b \mid Z$ and $aX + b \mid a^2 Z$ describe the same distribution, we can rewrite the formula for it as

$$(aX + b \mid a^2 Z) \sim N\left((a\mu + b) + \frac{\beta}{a}(a^2 Z), a^2 Z\right),$$

making it possible to use the definition of the mixed distribution again, which says that

$$aX + b \sim \mathrm{Mix}_{a^2\pi}(a\mu + b, \beta/a).$$

If we use $\pi = \mathrm{GIG}(\lambda, \delta^2, \alpha^2 - \beta^2)$ as a mixing weight, then we can use the formula

$$r\,\mathrm{GIG}(\lambda, \chi, \psi) = \mathrm{GIG}(\lambda, r\chi, r^{-1}\psi),$$

proved in Problem 9.5, to get

$$\begin{aligned}
a\,\mathrm{GHyp}(\lambda, \alpha, \beta, \delta, \mu) + b &= a\left(\mathrm{Mix}_{\mathrm{GIG}(\lambda, \delta^2, \alpha^2 - \beta^2)}(\mu, \beta)\right) + b \\
&= \mathrm{Mix}_{a^2\,\mathrm{GIG}(\lambda, \delta^2, \alpha^2 - \beta^2)}(a\mu + b, \beta/a) \\
&= \mathrm{Mix}_{\mathrm{GIG}(\lambda, (a\delta)^2, (\alpha/|a|)^2 - (\beta/a)^2)}(a\mu + b, \beta/a) \\
&= \mathrm{GHyp}(\lambda, \alpha/|a|, \beta/a, a\delta, a\mu + b),
\end{aligned}$$

providing a conceptual proof of the above result. ■

9.6.2 The parameters of GHyp

The GHyp parameters admit the following interpretations.

- α is the tail parameter, in the sense that it dictates their fatness. The larger α is, the lighter are the tails of the GHyp distribution. When $\alpha = 0$ and $-2\lambda = \delta =: n$, the GHyp reduces to a t distribution with n degrees of freedom.

- $\beta \in [-\alpha, \alpha]$ is the skewness parameter. As $|\beta|$ grows (compared to α), so does the amount of skewness, and the distribution is symmetric when $\beta = 0$.

Note that β also influences the fatness of the tails of the GHyp distribution in that it can shift mass from one tail to the other. More precisely,

$$f_{\text{GHyp}}(x; \lambda, \alpha, \beta, \delta, \mu) \propto |x|^{\lambda-1} e^{(\mp\alpha+\beta)x} \quad \text{as } x \to \pm\infty. \tag{9.54}$$

Observe how the limiting form of the p.d.f. is the product of the function $|x|^{\lambda-1}$ and an exponential-tail law (at least if $|\beta| < \alpha$); hence, the tails are referred to as 'semi-heavy' (see Barndorff-Nielsen, 1998; Prause, 1999, eq. 1.19).

If $|\beta| = \alpha > 0$, then one of the tails is a (heavy) power-tail and the other is semi-heavy, while, if $\alpha = \beta = 0$, then both tails are heavy. Note that $|\beta| = \alpha$ means that we are in boundary case II (with $\lambda < 0$), so we are talking about the hyperbolic asymmetric t distribution HAt. Aas and Haff (2005, 2006) discuss the unbalanced tails of the HAt distribution and demonstrate its usefulness for modelling financial data.

- μ is a location parameter. When $\beta = 0$, the distribution is symmetric, and μ coincides with the mean, if the first moment of the distribution exists.

- δ is a 'peakedness' parameter, and controls the shape of the p.d.f. near its mode. The larger δ is, the flatter the peak of the density becomes. If δ assumes its minimum allowed value of zero, then the GHyp reduces to a (generalized) Laplace distribution, with its characteristic pointed peak.

- λ is a shape parameter which influences the p.d.f. in a nonspecific way.

9.6.3 Alternative parameterizations of GHyp

There is one obvious parameterization for GHyp that we have already used in Section 9.5.1 when we first introduced GHyp and calculated its moments:

$$(\lambda, \psi, \beta, \chi, \mu) \quad \text{with } \chi := \delta^2, \ \psi := \alpha^2 - \beta^2.$$

In our context of representing GHyp as a GIG mixture of normals, this is the *natural parameterization*. It is obviously quite close to our *standard parameterization* $(\lambda, \alpha, \beta, \delta, \mu)$ and we only use the variables χ and ψ as abbreviations to streamline our formulae.

There are at least four alternative parameterizations, three of which can be found throughout the literature. They differ from the parameterizations given above in that they parameterize *just the proper GHyp distribution* with $|\beta| < \alpha$ and $\delta > 0$.

1. The proper GHyp distribution has the feature that $\chi > 0$, $\psi > 0$, so the mixing weights are proper GIG distributions. So we can use the alternative parameterization for proper GIG to get an alternative parameterization for proper GHyp:

$$(\lambda, \omega, \beta, \eta, \mu) \quad \text{with } \omega = \sqrt{\chi\psi} = \delta\sqrt{\alpha^2 - \beta^2}, \ \eta := \sqrt{\chi/\psi} = \frac{\delta}{\sqrt{\alpha^2 - \beta^2}}.$$

This parameterization was introduced on page 319 and facilitates calculations for the proper GHyp distribution.

2. As we have seen in (9.53), the standard parameters of GHyp cannot be divided up into a location parameter, a scale parameter, and three location- and scale-invariant parameters. However, quite obviously, μ is a location parameter and a direct consequence of (9.53) is that

$$(\lambda, \overline{\alpha}, \overline{\beta}, \delta, \mu) \quad \text{with } \overline{\alpha} := \alpha\delta, \ \overline{\beta} := \beta\delta$$

is a parameterization such that δ is a scale parameter and $\lambda, \overline{\alpha}, \overline{\beta}$ are location- and scale-invariant.

3. Another common parameterization of proper GHyp is

$$(\lambda, \zeta, \rho, \delta, \mu) \quad \text{with } \zeta := \delta\sqrt{\alpha^2 - \beta^2} = \omega, \ \rho := \frac{\beta}{\alpha}.$$

Again, μ and δ are location and scale parameters, respectively, and λ, ζ, ρ are location- and scale-invariant. This can be deduced from the fact that ζ and ρ can be expressed in terms of $\overline{\alpha}$ and $\overline{\beta}$, namely $\zeta = (\overline{\alpha}^2 - \overline{\beta}^2)^{(1/2)}$ and $\rho = \overline{\beta}/\overline{\alpha}$. As $\overline{\alpha}$ and $\overline{\beta}$ are both location- and scale-invariant, so are ζ and ρ. The parameter ρ is already a good measure for the skewness of the distribution: if $\rho \to 1$, then the right tail of the distribution becomes heavy, whereas the left tail stays light, and vice versa if $\rho \to -1$.

4. A rather instructive reparameterization of the proper GHyp distribution is

$$(\lambda, p, q, \delta, \mu) \quad \text{with } p := (1 + \zeta)^{-1/2}, \ q := \rho p,$$

with ζ and ρ defined in case 3 above. Because p and q are formed out of the location- and scale-invariant parameters ζ and ρ, they (and λ as well) are again location- and scale-invariant; μ and δ are a location and a scale parameter, respectively. Let us describe the domain of variation of these parameters. Obviously, $\lambda, \mu \in \mathbb{R}$, $\delta > 0$. The parameter ζ is allowed to be any positive number, so $0 < p < 1$. As $-1 < \rho < 1$, we get $-p < q < p$. So the parameters p and q vary in the so-called *shape triangle* defined by

$$\{(q, p) \in \mathbb{R}^2 \mid 0 \leq |q| < p < 1\},$$

which we analyse in the next section.

Note that, whenever $\beta = \overline{\beta} = \rho = q = 0$, the GHyp distribution is symmetric. Thus, in the respective parameterizations, these parameters measure the asymmetry of GHyp.

9.6.4   The shape triangle

The shape triangle was introduced by Barndorff-Nielsen, Blæsild, Jensen and Sørensen (1985), who realized that the skewness (γ_1) and the kurtosis (γ_2) of the GHyp distribution behave like

$$(\gamma_1, \gamma_2) \sim (3q, 3p^2)$$

for small values of q and p. So it seems natural to take the parameters q and p to measure asymmetry and kurtosis of the (proper) GHyp distributions. As mentioned above, the parameters (q, p) vary in the triangle $\{(q, p) \in \mathbb{R}^2 \mid 0 \le |q| < p < 1\}$, referred to as the shape triangle of the (proper) GHyp distribution.

What is it good for? Imagine you are working with a particular data set which lends itself to modelling with a GHyp distribution, or, better, several such data sets. To get an impression of the shapes of the estimated GHyp distributions, represent the estimated parameters q and p corresponding to each data set as points in the shape triangle. If the point is in the middle, the density is symmetric, while the more to the left or right the point is, the more the distribution is skewed. The further down the point is located, the less excess kurtosis is present in the distribution, and the lower value ($q = 0$, $p = 0$) represents the normal distribution. The upper boundary ($p = 1$) is, for $\lambda > 0$, formed by the variance–gamma distribution; in particular, we find the Laplace distribution at the upper boundary.

Let us be more precise. We have already analysed the various limiting cases of the generalized hyperbolic distribution for our standard parameterization. Now we want to find the limiting cases again in the shape triangle parameterization ($\lambda, p, q, \delta, \mu$); this is done in great detail in Eberlein and Hammerstein (2004), and an instructive illustration can be found in Bibby and Sørensen (2003, p. 217).

For convergence considerations, it is convenient to look at the GHyp distribution for fixed $\lambda, \mu \in \mathbb{R}$, to let the parameters q and p vary in the shape triangle, and to use the parameter δ to rescale the distributions if necessary to achieve convergence when approaching the boundary of the triangle.

Before we discuss the behaviour of the GHyp distribution at the boundary of the shape triangle, have a look at Figure 9.6. This depicts the domain of variation of the parameters p and q for various choices of λ. The upper edge corresponds to $p = 1$, the dashed line in the middle corresponds to the case $q = 0$, i.e., distributions with parameters on this line will be symmetric. On the right-hand side, they will be skewed to the right; on the left-hand side, to the left.

Note that, for $\lambda < 0$, there are many different distributions that are concentrated in the upper left-hand and and upper right-hand corners. For $\delta > 0$, we get hyperbolic asymmetric t distributions, depending on δ. For $\delta \to 0$, the distribution gets more skewed, and, for $\delta = 0$, we have an IGam distribution in the upper right-hand corner, and the reflected distribution, $-$ IGam, in the upper left-hand corner.

To get a better idea of the boundary behaviour, consider Figure 9.7. This displays the domain of variation not only of the parameters p and q, but also of the scale parameter $\delta > 0$. In this diagram, which might be called the *shape prism* of the GHyp distribution, we have the shape triangle as the top view of the prism. Note that distributions

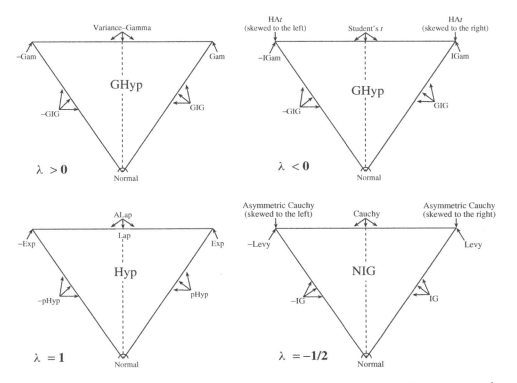

Figure 9.6 The shape triangle for $\lambda > 0$, $\lambda < 0$ and the special cases $\lambda = 1$ (Hyp) and $\lambda = -\frac{1}{2}$ (NIG)

which are at opposite parts of the prism, such as the normal distribution and the GIG distribution, are very close to each other in the triangle.

• **The lower corner:** If $p \to 0$, $\zeta = \delta\sqrt{\alpha^2 - \beta^2} \to \infty$, so we need to inspect the cases $\alpha \to \infty$ and $\delta \to \infty$. A way (and presumably the only nontrivial way) to achieve convergence of the distributions is in fact

$$\alpha \to \infty, \quad \beta \to \beta_0, \quad \delta \to \infty, \quad \delta/\alpha \to \sigma_0^2,$$

with $\beta_0 \in \mathbb{R}$ and $\sigma_0 > 0$. As we have seen in Section 9.5.3, this implies *convergence to the normal distribution* $N(\mu + \beta_0\sigma_0^2, \sigma_0^2)$.

• **The upper boundary:**

– If $\lambda > 0$, we can achieve convergence at the upper boundary by considering

$$\alpha \to \alpha_0, \quad \beta \to \beta_0, \quad \delta \to 0, \quad -\alpha_0 < \beta_0 < \alpha_0.$$

Note that $\zeta = \delta\sqrt{\alpha^2 - \beta^2} \to 0$ and, hence, $p \to 1$, as well as $q \to \beta_0/\alpha_0$, so (q, p) indeed converges to some point in the upper boundary. In this case we get as a limit distribution simply GHyp$(\lambda, \alpha_0, \beta_0, 0, \mu)$, i.e., a *variance–gamma distribution*.

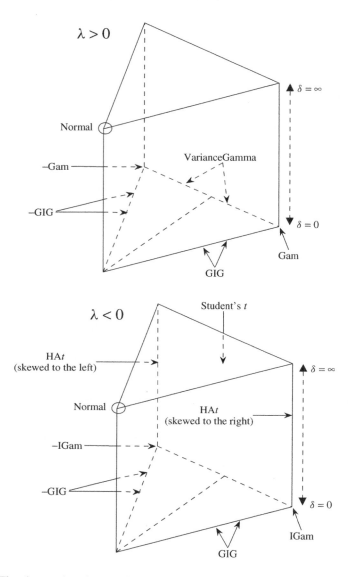

Figure 9.7 The shape prism for $\lambda > 0$ (top) and $\lambda < 0$ (bottom). The top view yields the shape triangle

- If $\lambda = 0$, there seems to be no interesting convergence behaviour at the upper boundary.

- If $\lambda < 0$, we can consider the case

$$\alpha \to |\beta_0|, \quad \beta \to \beta_0, \quad \delta \to \delta_0, \quad \beta/\alpha \to \rho_0,$$

with $-1 \le \rho_0 \le 1$, $\delta_0 > 0$, $\beta_0 \in \mathbb{R}$. If $\beta_0 \ne 0$, then ρ_0 is automatically either -1 or 1. On the other hand, if $\beta_0 = 0$, then this means that both α and β converge to zero, but their quotient, β/α, converges to some fixed number $\rho_0 \in [-1, 1]$. Now $\zeta = \delta\sqrt{\alpha^2 - \beta^2} \to 0$ and $p \to 1$. On the other hand, $q \to \rho_0$, so we are at the point $(\rho_0, 1)$ of the upper boundary. The resulting distribution is GHyp$(\lambda, |\beta_0|, \beta_0, \delta_0, \mu)$, i.e., a *hyperbolic asymmetric Student's t distribution*. If $\beta_0 = 0$, we get the (symmetric) *Student's t distribution* as a special case. Note that, if $\beta_0 = 0$, then the distribution does not depend on ρ_0, so at the whole upper boundary we get the same, symmetric distribution depending only on λ, δ and μ.

If $\beta_0 \neq 0$, then we are either in the upper left-hand corner ($\beta_0 < 0$), and the distribution is skewed (heavily) to the left, or we are in the upper right-hand corner ($\beta_0 > 0$), and the distribution is skewed to the right.

- **The right boundary:** To achieve convergence in this case, we want to ensure that $\zeta = \delta\sqrt{\alpha^2 - \beta^2}$ converges to some $\zeta_0 \geq 0$. We do this by letting

$$\alpha \to \infty, \quad \beta \to \infty, \quad \delta \to 0, \quad \alpha - \beta \to \frac{\psi_0}{2}, \quad \alpha\delta^2 \to \chi_0,$$

where $\chi_0, \psi_0 \geq 0$ are such that $(\lambda, \chi_0, \psi_0)$ is in the parameter space of the GIG distribution. Then

$$\beta\delta^2 = \alpha\delta^2 - \frac{\psi_0}{2}\delta^2 \to \chi_0$$

and, hence,

$$\delta^2(\alpha^2 - \beta^2) = (\delta^2\alpha + \delta^2\beta)(\alpha - \beta) \to 2\chi_0\frac{\psi_0}{2} = \chi_0\psi_0,$$

so that

$$\zeta \to \sqrt{\chi_0\psi_0} \quad \text{and} \quad p \to p_0 := \left(1 + \sqrt{\chi_0\psi_0}\right)^{-1/2}.$$

On the other hand,

$$\frac{\beta}{\alpha} = \frac{\alpha - \psi_0/2}{\alpha} = 1 - \frac{\psi_0}{2\alpha} \to 1,$$

so that $\rho \to 1$ and hence $q \to p_0$. So we get something (namely (p_0, p_0)) on the right boundary, the limiting distribution being a *shifted generalized inverse Gaussian distribution* GIG$(\lambda, \chi_0, \psi_0) + \mu$; see also Section 9.5.3.

Pay particular attention to the upper right-hand corner: $p_0 = 1$ is equivalent to $\chi_0\psi_0 = 0$. If $\lambda > 0$, this can be achieved if $\chi_0 = 0$, resulting in a *gamma distribution* in the upper right-hand corner. If $\lambda < 0$, this is possible for $\psi_0 = 0$, giving us an *inverse gamma distribution* in the upper right-hand corner.

- **The left boundary:** By an argument completely symmetric to the one for the right-hand boundary, we can achieve convergence to $- GIG(\lambda, \chi_0, \psi_0) + \mu$ at the left-hand boundary.

9.6.5   Convolution and infinite divisibility

Suppose X and Y are independent random variables with distributions π and ρ, respectively. The distribution of $X + Y$ is obtained using the convolution formula, as introduced in Section I.8.2.6 and used subsequently in many chapters. Note that the distribution of $X + Y$ depends on X and Y only through their distributions π and ρ. As such, a popular notation to designate the convolution of two r.v.s is $\pi \star \rho$, read as the *convolution of π and ρ*. If π and ρ have probability density functions f_π and f_ρ, then the p.d.f. of $\pi \star \rho$ is, from (I.8.42),

$$f_{\pi \star \rho}(t) = \int_{-\infty}^{+\infty} f_\pi(t - s) f_\rho(s) \, ds.$$

If π, ρ and σ are distributions, then the following rules easily carry over from the addition of random variables:

$$\pi \star \rho = \rho \star \pi \qquad \text{and} \qquad (\pi \star \rho) \star \sigma = \pi \star (\rho \star \sigma).$$

The n-fold convolution of π with itself is abbreviated by $\pi^{\star n}$, i.e.,

$$\pi^{\star n} = \underbrace{\pi \star \pi \star \cdots \star \pi}_{n \text{ times}}.$$

A distribution ρ is called an nth *convolution root of* π if $\rho^{\star n} = \pi$. If π is a distribution on $(0, \infty)$, i.e., π is positive with probability one, then we want ρ to be positive as well. There are distributions which do not possess convolution roots. A distribution π is called *infinitely divisible* if, for every n, an nth convolution root of π exists (see also Problem 1.13).

◎ ***Example 9.4***   Suppose $\mu \in \mathbb{R}$ and $\sigma > 0$. Then the normal distribution $N(\mu, \sigma^2)$ is infinitely divisible. To see this, let $n \in \mathbb{N}$. If $X_i \overset{\text{ind}}{\sim} N(\mu/n, \sigma^2/n)$, then

$$S = X_1 + \cdots + X_n \sim N(n\mu/n, n\sigma^2/n) = N(\mu, \sigma^2).$$

So $N(\mu/n, \sigma^2/n)$ is an nth convolution root of $N(\mu, \sigma^2)$.   ■

A mixture of normals behaves well under convolution. In particular, suppose π and ρ are distributions on $(0, \infty)$ and $\mu_\pi, \mu_\rho, \beta \in \mathbb{R}$. Then

$$\text{Mix}_{\pi \star \rho}(\mu_\pi + \mu_\rho, \beta) = \text{Mix}_\pi(\mu_\pi, \beta) \star \text{Mix}_\rho(\mu_\rho, \beta). \tag{9.55}$$

In words, it does not matter whether you take the convolution product first and then mix, or mix first and subsequently build the convolution product.

To show (9.55) if π and ρ have p.d.f.s f_π and f_ρ, respectively, write

$$f_{\mathrm{Mix}_{\pi \star \rho}(\mu_\pi + \mu_\rho, \beta)}(x) = \int_0^\infty f_N(x; \mu_\pi + \mu_\rho + \beta z, z) f_{\pi \star \rho}(z)\, dz$$

$$= \int_0^\infty f_N(x; \mu_\pi + \mu_\rho + \beta z, z) \int_{-\infty}^{+\infty} f_\pi(z - s) f_\rho(s)\, ds\, dz$$

$$= \int_{-\infty}^{+\infty} \int_0^\infty f_N(x; \mu_\pi + \mu_\rho + \beta z, z) f_\pi(z - s) f_\rho(s)\, dz\, ds$$

$$= \int_{-\infty}^{+\infty} \int_0^\infty \int_{-\infty}^{+\infty} f_N(x - r; \mu_\pi + \beta(z - s), z - s)$$
$$\times f_N(r; \mu_\rho + \beta s, s)\, dr f_\pi(z - s) f_\rho(s)\, dz\, ds$$

$$= \int_{-\infty}^{+\infty} \int_{-\infty}^{+\infty} \int_0^\infty f_N(x - r; \mu_\pi + \beta(z - s), z - s)$$
$$\times f_\pi(z - s)\, dz\, f_N(r; \mu_\rho + \beta s, s) f_\rho(s)\, ds\, dr$$

$$= \int_{-\infty}^{+\infty} \int_{-\infty}^{+\infty} f_{\mathrm{Mix}_\pi(\mu_\pi, \beta)}(x - r) f_N(r; \mu_\rho + \beta s, s) f_\rho(s)\, ds\, dr$$

$$= \int_{-\infty}^{+\infty} f_{\mathrm{Mix}_\pi(\mu_\pi, \beta)}(x - r) \int_{-\infty}^{+\infty} f_N(r; \mu_\rho + \beta s, s) f_\rho(s)\, ds\, dr$$

$$= \int_{-\infty}^{+\infty} f_{\mathrm{Mix}_\pi(\mu_\pi, \beta)}(x - r) f_{\mathrm{Mix}_\rho(\mu_\rho, \beta)}(r)\, dr$$

$$= f_{\mathrm{Mix}_\pi(\mu_\pi, \beta) \star \mathrm{Mix}_\rho(\mu_\rho, \beta)}(x).$$

Barndorff-Nielsen and Halgreen (1977) show that the GIG distribution is indeed infinitely divisible. In the special case of the IG distribution, we have derived above the explicit formula (9.37) which shows that the convolution of two IG distributions is again IG:

$$\mathrm{IG}_1(\chi_1, \psi) \star \mathrm{IG}_1(\chi_2, \psi) = \mathrm{IG}_1((\sqrt{\chi_1} + \sqrt{\chi_2})^2, \psi). \tag{9.56}$$

This applies also to the limiting case $\psi \to 0$, i.e., the Lévy distribution is also invariant under convolution. Analogous statements were shown to be true for the Cauchy (Example I.8.16) and normal (Examples I.9.6 and 2.5). All three cases are nested in the stable Paretian family, which is invariant under convolution; see Chapter 8.

From the result of Barndorff-Nielsen and Halgreen (1977) mentioned above, it follows directly that the GHyp distribution is infinitely divisible (see Problem 9.13). Again, we have an explicit formula if $\lambda = -1/2$, i.e., for the NIG distribution, which is derived in Problem 9.14. This indicates why the NIG is an important subfamily:

$$\mathrm{NIG}(\alpha, \beta, \delta_1, \mu_1) \star \mathrm{NIG}(\alpha, \beta, \delta_2, \mu_2) = \mathrm{NIG}(\alpha, \beta, \delta_1 + \delta_2, \mu_1 + \mu_2). \tag{9.57}$$

This result also applies to the limiting case $\alpha = |\beta|$, i.e., for the (asymmetric) Cauchy distribution.

9.7 Problems

Success is a matter of luck. Just ask any failure. (Earl Wilson)

Some succeed because they are destined to, but most succeed because they are determined to. (Roscoe Dunjee)

Das Wort Schwierigkeit muß gar nicht für einen Menschen von Geist als existent gedacht werden. Weg damit! (Georg Christoph Lichtenberg)

9.1. ★ Prove the following convenient expression for the derivative of the Bessel function:

$$-2K_\nu'(x) = K_{\nu-1}(x) + K_{\nu+1}(x), \quad \nu \in \mathbb{R}, \ x \in \mathbb{R}_{>0}. \tag{9.58}$$

Hint: Use Leibniz' rule for differentiating under the integral sign.

9.2. ★ Show the equations

$$K_{3/2}(x) = \left(\frac{1}{x} + 1\right)\sqrt{\frac{\pi}{2x}}e^{-x} \quad \text{and} \quad K_{5/2}(x) = \left(\frac{3}{x^2} + \frac{3}{x} + 1\right)\sqrt{\frac{\pi}{2x}}e^{-x}.$$

Hint: Use equations (9.3) and (9.4).

9.3. ★ ★ Proceed as in Section 9.3.2 to prove the equations for the variance and the third moment, when Z is a continuous random variable.

9.4. Show that, if $X \sim \text{GIG}(\lambda, \chi, \psi)$, then

$$X^{-1} \sim \text{GIG}(-\lambda, \psi, \chi).$$

This means that X follows boundary case I iff X^{-1} follows boundary case II. In this sense, they are inverses to each other. Because of this, the GIG distribution in boundary case II is the inverse gamma distribution (IGam).

9.5. Show that, if $X \sim \text{GIG}(\lambda, \chi, \psi)$ and $r > 0$, then

$$rX \sim \text{GIG}(\lambda, r\chi, r^{-1}\psi).$$

9.6. Verify the formulae for the raw moments and the m.g.f. of the GIG distribution for boundary case I and II, i.e., for the gamma and the inverse gamma distribution.

9.7. ★ Assume $\lambda \in \mathbb{R}_{>0}$ and $r \in \mathbb{R}$ with $r > -\lambda$. Let $\psi > 0$. Show that, as $\chi \downarrow 0$, the rth raw moment of $\text{GIG}(\lambda, \chi, \psi)$ as given in formula (9.22) converges to the rth raw moment of the gamma distribution $\text{GIG}(\lambda, 0, \psi)$ given in (9.26). Use the relation given in formula (9.6).

9.8. ★ ★ Assume $X \sim \text{GIG}(\lambda, \chi, \psi)$ with $\lambda \in \mathbb{R}$ and $\chi, \psi > 0$.

    (a) Let $Z := \ln X$. Give a formula for $\mathbb{M}_Z(t)$ for all $t \in \mathbb{R}$. Hint: Equation (9.22).

    (b) Prove formula (9.23) for $\mathbb{E}[Z] = \mathbb{E}[\ln X]$ using the cumulant generating function of Z.

    (c) Use the same trick to determine $\mathbb{E}[\ln X]$ when X is a Gamma r.v. Compare your result with Example 1.8, where the special case $X \sim \chi_\nu^2$ is treated.

    (d) Do the same for an inverse gamma r.v.

9.9. Show that the mean, variance, and m.g.f. of the GHyp distribution for boundary case I, i.e., for the variance–gamma distribution, are given by

$$\mathbb{E}[X] = \mu + \beta \frac{\lambda}{\psi/2}, \quad \mathbb{V}(X) = 2\lambda \frac{\alpha^2 + \beta^2}{\psi^2},$$

and

$$\mathbb{M}_X(t) = e^{\mu t} \left(\frac{\alpha^2 - \beta^2}{\alpha^2 - (\beta + t)^2} \right)^\lambda,$$

where $X \sim \text{GHyp}(\lambda, \alpha, \beta, 0, \mu)$, $-\alpha - \beta < t < \alpha - \beta$, and $\psi = \alpha^2 - \beta^2$.

9.10. Show that the mean and variance of the GHyp distribution for boundary case II, i.e., for the HAt distribution, are given by

$$\mathbb{E}[X] = \mu - \beta \frac{\chi}{2} \frac{1}{\lambda + 1} \quad \text{(if } \lambda < -1\text{),}$$

$$\mathbb{V}(X) = \frac{\chi}{2} \frac{-1}{\lambda + 1} + \beta^2 \left(\frac{\chi}{2} \right)^2 \frac{-1}{(\lambda + 1)^2(\lambda + 2)} \quad \text{(if } \lambda < -2\text{),}$$

where $X \sim \text{GHyp}(\lambda, \alpha, \beta, \delta, \mu)$, writing $\chi = \delta^2$ and $\alpha = |\beta|$.

Note that substituting $\lambda = -n/2$ yields the expressions for the mean and variance found in Aas and Haff (2005, 2006); they also give expressions for the skewness and kurtosis of HAt.

Show that the m.g.f. is

$$\mathbb{M}_X(t) = e^{\mu t} \frac{2 K_\lambda \left(\sqrt{-2\chi \left(\beta t + t^2/2 \right)} \right)}{\Gamma(-\lambda) \left(\frac{-\chi(\beta t + t^2/2)}{2} \right)^{\lambda/2}},$$

where $-2\beta < t \leq 0$.

Discuss what happens if $\beta = 0$, i.e., if X is symmetric.

9.11. Verify the formulae (9.48) and (9.50) for the densities of the Hyp and the NIG distribution.

9.12. (a) ★ Suppose that $X \sim \mathrm{IG}_1(\chi, \psi)$, where $\chi > 0$ and $\psi > 0$. Show that

$$\mathbb{E}[X] = \eta, \quad \mathbb{V}(X) = \frac{\eta^2}{\omega}, \quad \mu_3(X) = 3\eta^3 \omega^{-2},$$

where $\eta = \sqrt{\chi/\psi}$ and $\omega = \sqrt{\chi\psi}$. Hint: Use the explicit formula (9.3) for the Bessel function.

(b) ★ Calculate the mean, variance, skewness, and m.g.f. of the NIG distribution as given in (9.51) and (9.52). Hint: Use the equations from Section 9.3.2 and the results of part **(a)**.

9.13. (a) Show that, if ρ is a distribution on $(0, \infty)$ and $\beta, \mu \in \mathbb{R}$, then

$$\mathrm{Mix}_\rho(\mu, \beta)^{*n} = \mathrm{Mix}_{\rho^{*n}}(n\mu, \beta).$$

(b) Argue that, if π is an infinitely divisible distribution on $(0, \infty)$ and $\mu, \beta \in \mathbb{R}$, then $\mathrm{Mix}_\pi(\mu, \beta)$ is also infinitely divisible.

9.14. (a) Show the convolution (9.57) formula for the NIG distribution using the convolution formula (9.56) for the IG distributions and our knowledge of mixtures.

(b) Show (9.57) using the formula (9.52) for the m.g.f. of the NIG distribution.

(c) ★ ★ Show the following formula for the variance–gamma distribution:

$$\mathrm{VG}(\lambda_1, \alpha, \beta, \mu_1) \star \mathrm{VG}(\lambda_2, \alpha, \beta, \mu_2) = \mathrm{VG}(\lambda_1 + \lambda_2, \alpha, \beta, \mu_1 + \mu_2)$$

for all $\lambda_1, \lambda_2 > 0$, $\alpha > 0$, $\beta \in (-\alpha, \alpha)$, $\mu_1, \mu_2 \in \mathbb{R}$.

10

Noncentral distributions

The intelligent man finds almost everything ridiculous, the sensible man hardly
anything.   (Johann Wolfgang von Goethe)

Seriousness is the only refuge of the shallow.   (Oscar Wilde)

The derivations of the Student's t and F densities in Examples I.9.7 and I.9.8, along
with result (2.4) for the χ^2 distribution, hinge on (i) independence and (ii) zero means
of the relevant normal random variables. The *noncentral* χ^2, F and t distributions relax
the mean-zero assumption. They arise in a variety of statistical contexts, most notably
for determining the power function of certain popular test statistics, though they also
crop up in numerous other areas, such as engineering and finance applications. We will
see that accurate calculation of their density and distribution functions entails much
higher computation cost than that associated with their central counterparts, though
in each noncentral case a saddlepoint approximation is applicable, underscoring once
again their value. The p.d.f.s, c.d.f.s, and moments of the noncentral distributions are
summarized in Tables A.9 and A.10.

10.1 Noncentral chi-square

The noncentral χ^2 distribution arises directly in goodness-of-fit tests for contingency
tables and likelihood ratio tests, and is fundamental for constructing the noncentral
F distribution, which is ubiquitous in statistics. Perhaps somewhat unexpectedly, the
noncentral χ^2 also has a number of uses in finance; see Johnson, Kotz and Balakrishnan
(1994, pp. 467–470) for an overview.

10.1.1 Derivation

Let $X_i \overset{\text{ind}}{\sim} N(\mu_i, 1)$, $\mu_i \in \mathbb{R}$, $i = 1, \ldots, n$, or $\mathbf{X} = (X_1, \ldots, X_n) \sim N_n(\boldsymbol{\mu}, \mathbf{I})$, with
$\boldsymbol{\mu} = (\mu_1, \ldots, \mu_n)' \in \mathbb{R}^n$, so that $f_{\mathbf{X}}(\mathbf{x}) = (2\pi)^{-n/2} \exp\{-\sum_{i=1}^{n} (x_i - \mu_i)^2 / 2\}$.

Intermediate Probability: A Computational Approach M. Paolella
© 2007 John Wiley & Sons, Ltd

Interest centres on the distribution of $X = \sum_{i=1}^{n} X_i^2$. Define $\theta = \boldsymbol{\mu}'\boldsymbol{\mu} = \sum_{i=1}^{n} \mu_i^2$, which is referred to as the *noncentrality parameter*. Starting as in Rao (1973, p. 181) and Stuart, Ord and Arnold (1999, Section 22.4), let $\mathbf{B}$ be an orthogonal $n \times n$ matrix (i.e., $\mathbf{B}'\mathbf{B} = \mathbf{B}\mathbf{B}' = \mathbf{I}$) with first row given by $\boldsymbol{\mu}'\theta^{-1/2}$ if $\theta > 0$, and take $\mathbf{B} = \mathbf{I}_n$ if $\theta = 0$. These conditions imply that $\mathbf{B}\boldsymbol{\mu} = (\theta^{1/2}, 0, \ldots, 0)'$.[1] For example, with $n = 2$ and $n = 3$, and assuming $\theta > 0$,

$$\mathbf{B}_2 = \theta^{-1/2} \begin{pmatrix} \mu_1 & \mu_2 \\ \mu_2 & -\mu_1 \end{pmatrix}, \qquad \mathbf{B}_3 = \theta^{-1/2} \begin{pmatrix} \mu_1 & \mu_2 & \mu_3 \\ -\mu_1\mu_2/U & U & -\mu_2\mu_3/U \\ \mu_3\theta^{1/2}/U & 0 & -\mu_1\theta^{1/2}/U \end{pmatrix},$$

$$\tag{10.1}$$

where $U = \sqrt{\mu_1^2 + \mu_3^2}$, satisfy the constraints (though are not unique); see Problem 10.4. Let $\mathbf{Y} = \mathbf{B}\mathbf{X}$, with $\mathbb{E}[\mathbf{Y}] = \mathbf{B}\boldsymbol{\mu}$ and $(\mathbf{B}\boldsymbol{\mu})'\mathbf{B}\boldsymbol{\mu} = \boldsymbol{\mu}'\boldsymbol{\mu} = \theta$. From the results in Section 3.2, $\mathbf{Y} \sim \mathrm{N}(\mathbf{B}\boldsymbol{\mu}, \mathbf{I})$, so that the Y_i are independent, unit variance, normal r.v.s with $\mathbb{E}[Y_1] = \theta^{1/2}$, and $\mathbb{E}[Y_i] = 0$, $i = 2, \ldots, n$. They have joint density

$$f_{\mathbf{Y}}(\mathbf{y}) = (2\pi)^{-n/2} \exp\left\{ -\frac{1}{2}(y_1 - \theta^{1/2})^2 - \frac{1}{2}\sum_{i=2}^{n} y_i^2 \right\}.$$

Thus,

$$X = \mathbf{X}'\mathbf{X} = \mathbf{X}\mathbf{B}'\mathbf{B}\mathbf{X} = \mathbf{Y}'\mathbf{Y} = Y_1^2 + (Y_2^2 + \cdots + Y_n^2) = Y_1^2 + Z,$$

where $Z := (Y_2^2 + \cdots + Y_n^2) \sim \chi_{n-1}^2$ (see Example 2.3) with density

$$f_Z(z) = \frac{1}{2^{(n-1)/2}\Gamma((n-1)/2)} z^{(n-1)/2 - 1} e^{-z/2} \mathbb{I}_{(0,\infty)}(z)$$

and, from Problem I.7.6,

$$f_{Y_1^2}(y) = \frac{1}{2\sqrt{y}} \frac{1}{\sqrt{2\pi}} \exp\left(-\frac{y+\theta}{2}\right) \left(\exp\left(\theta^{1/2}\sqrt{y}\right) + \exp\left(-\theta^{1/2}\sqrt{y}\right)\right) \mathbb{I}_{(0,\infty)}(y)$$

$$= \frac{1}{\sqrt{y}} \frac{1}{\sqrt{2\pi}} \exp\left(-\frac{y+\theta}{2}\right) \sum_{i=0}^{\infty} \frac{(y\theta)^i}{(2i)!} \mathbb{I}_{(0,\infty)}(y),$$

noting that

$$\frac{e^z + e^{-z}}{2} = 1 + \frac{z^2}{2!} + \frac{z^4}{4!} + \frac{z^6}{6!} + \cdots = \cosh(z).$$

From (2.7) with

$$K = \frac{1}{2^{(n-1)/2}\Gamma((n-1)/2)\sqrt{2\pi}}$$

---

[1] To see this, let $\mathbf{b}_j$ denote the jth row of $\mathbf{B}$. Then, from the condition on the first row of $\mathbf{B}$, $\boldsymbol{\mu}' = \mathbf{b}_1'\theta^{1/2}$ and, from the orthogonality of $\mathbf{B}$, $\mathbf{b}_1\boldsymbol{\mu} = \mathbf{b}_1\mathbf{b}_1'\theta^{1/2} = \theta^{1/2}$ and, for $j = 2, \ldots, n$, $\mathbf{b}_j\boldsymbol{\mu} = \mathbf{b}_j\mathbf{b}_1'\theta^{1/2} = 0$, i.e., $\mathbf{B}\boldsymbol{\mu} = (\theta^{1/2}, 0, \ldots, 0)'$.

and substituting $u = (x - y)/x$,

$$
\begin{aligned}
f_X(x) &= \int_{-\infty}^{\infty} f_{Y_1^2}(y) \, f_Z(x - y) \, dy \\
&= K \int_{-\infty}^{\infty} y^{-1/2} \exp\left(-\frac{y + \theta}{2}\right) \\
&\quad \times \sum_{i=0}^{\infty} \frac{(y\theta)^i}{(2i)!} \mathbb{I}_{(0,\infty)}(y)(x - y)^{(n-1)/2-1} \, e^{-(x-y)/2} \mathbb{I}_{(0,\infty)}(x - y) \, dy
\end{aligned}
$$

or

$$
\begin{aligned}
f_X(x) &= K e^{-\theta/2} e^{-x/2} \sum_{i=0}^{\infty} \frac{\theta^i}{(2i)!} \int_0^x y^{i-1/2}(x - y)^{(n-1)/2-1} \, dy \, \mathbb{I}_{(0,\infty)}(x) \\
&= K e^{-\theta/2} e^{-x/2} \sum_{i=0}^{\infty} \frac{\theta^i}{(2i)!} x^{i-1/2+(n-1)/2} \int_0^1 (1 - u)^{i-1/2} u^{(n-1)/2-1} \, du \, \mathbb{I}_{(0,\infty)}(x) \\
&= \frac{1}{2^{(n-1)/2}\sqrt{2\pi}} e^{-x/2} \sum_{i=0}^{\infty} \frac{e^{-\theta/2}\theta^i}{(2i)!} x^{n/2+i-1} \frac{\Gamma(i + 1/2)}{\Gamma(i + n/2)} \mathbb{I}_{(0,\infty)}(x) \\
&= e^{-x/2} \sum_{i=0}^{\infty} \frac{e^{-\theta/2}(\theta/2)^i}{(2i)!} x^{n/2+i-1} \frac{1}{\sqrt{\pi}} 2^{i-n/2} \frac{\Gamma(i + 1/2)}{\Gamma(i + n/2)} \mathbb{I}_{(0,\infty)}(x) .
\end{aligned}
$$

Then, using the relations

$$
\frac{1 \cdot 3 \cdot 5 \cdots (2i - 1)}{(2i)!} = \frac{2^{-i}}{i!} \qquad \text{and} \qquad \Gamma(i + 1/2) = \frac{1 \cdot 3 \cdot 5 \cdots (2i - 1)}{2^i} \sqrt{\pi}
$$

$$\tag{10.2}$$

from (I.1.3) and (I.1.51) respectively, we obtain

$$
\begin{aligned}
f_X(x; n, \theta) &= e^{-x/2} \sum_{i=0}^{\infty} \frac{e^{-\theta/2}(\theta/2)^i}{(2i)!} x^{n/2+i-1} 2^{-n/2} \frac{1 \cdot 3 \cdot 5 \cdots (2i - 1)}{\Gamma(i + n/2)} \mathbb{I}_{(0,\infty)}(x) \\
&= e^{-x/2} \sum_{i=0}^{\infty} \frac{e^{-\theta/2}(\theta/2)^i}{i!} \frac{x^{n/2+i-1}}{2^{n/2+i}\Gamma(i + n/2)} \mathbb{I}_{(0,\infty)}(x) \\
&= \sum_{i=0}^{\infty} \omega_{i,\theta} \, g_{n+2i}(x) ,
\end{aligned}
$$

$$\tag{10.3}$$

where g_ν denotes the χ_ν^2 density and $\omega_{i,\theta} = e^{-\theta/2}(\theta/2)^i/i!$ are weights corresponding to a Poisson distribution. Thus, X is a countable mixture distribution, as introduced in Section 7.3.1.

The c.d.f. F_X can thus be expressed as

$$\Pr(X \le x) = \int_0^x f_X(x)\,dx = \sum_{i=0}^{\infty} \omega_{i,\theta} \int_0^x g_{n+2i}(x)\,dx = \sum_{i=0}^{\infty} \omega_{i,\theta}\, G_{n+2i}(x),$$

(10.4)

where G_ν is the c.d.f. of a χ_ν^2 random variable. We will use the notation $X \sim \chi^2(n,\theta)$ to denote a χ^2 r.v. with n degrees of freedom and noncentrality parameter θ. Thus, the notation $X \sim \chi_n^2$ is equivalent to $X \sim \chi^2(n,0)$.

Starting from (10.3), simple manipulations show that the p.d.f. of $X \sim \chi^2(n,\theta)$ can be represented as

$$f_X(x) = \frac{1}{2} e^{-(x+\theta)/2} x^{(n-2)/4} \theta^{-(n-2)/4} \sum_{i=0}^{\infty} \frac{(\sqrt{\theta x}/2)^{(n-2)/2+2i}}{i!\,\Gamma(i+(n-2)/2+1)} \mathbb{I}_{(0,\infty)}(x)$$

$$= \frac{1}{2} e^{-(x+\theta)/2} x^{(n-2)/4} \theta^{-(n-2)/4} I_{(n-2)/2}(\sqrt{\theta x}) \mathbb{I}_{(0,\infty)}(x),$$

(10.5)

where

$$I_\nu(z) = \sum_{i=0}^{\infty} \frac{(z/2)^{\nu+2i}}{i!\,\Gamma(\nu+i+1)}$$

is the modified Bessel function of the first kind (which also arose in Example I.6.11 for the difference of two independent Poisson r.v.s, and in Example I.8.9 for Moran's bivariate exponential).

Analogous to the central χ^2 case, it should be clear that, if $X_i \overset{\text{ind}}{\sim} \chi^2(\nu_i, \theta_i)$, then $\sum_{i=1}^k X_i \sim \chi^2(\nu, \theta)$, where $\nu = \sum_{i=1}^k \nu_i$ and $\theta = \sum_{i=1}^k \theta_i$.

10.1.2 Moments

The expected value of $X \sim \chi^2(n,\theta)$ is, with $P \sim \text{Poi}(\theta/2)$,

$$\mathbb{E}[X] = \sum_{i=0}^{\infty} \omega_{i,\theta} \int_0^\infty x g_{n+2i}(x)\,dx$$

$$= \sum_{i=0}^{\infty} \frac{e^{-\theta/2}(\theta/2)^i}{i!}(n+2i) = \mathbb{E}[n+2P] = n+\theta,$$

(10.6)

using (I.7.42) and (I.4.34).[2] Similarly,

---

[2] Note that, if $W \sim N(\mu, \sigma^2)$, then $W/\sigma \sim N(\mu/\sigma, 1)$ and $(W/\sigma)^2 \sim \chi^2(1, \mu^2/\sigma^2)$. Thus, from (10.6),

$$\mathbb{E}[(W/\sigma)^2] = 1 + \frac{\mu^2}{\sigma^2} \quad \text{and} \quad \mathbb{E}[W^2] = \sigma^2 \mathbb{E}[(W/\sigma)^2] = \sigma^2 + \mu^2,$$

which agrees with the former expression in (I.7.67).

$$\mathbb{E}\left[X^2\right] = \sum_{i=0}^{\infty} \frac{e^{-\theta/2}\,(\theta/2)^i}{i!}\,(n+2k)\,(n+2k+2)$$

$$= n^2 + 2n + 4\mathbb{E}\left[P^2 + P\,(n+1)\right]$$

$$= n^2 + 2n + 4\left[\theta/2 + (\theta/2)^2 + (n+1)\,\theta/2\right]$$

$$= n^2 + 2n + 4\theta + \theta^2 + 2\theta n, \tag{10.7}$$

so that

$$\mathbb{V}\,(X) = n^2 + 2n + 4\theta + \theta^2 + 2\theta n - (n+\theta)^2 = 2n + 4\theta.$$

More generally, for $s \in \mathbb{R}$ with $s > -n/2$, $\mathbb{E}\,[X^s] = \int_0^\infty x^s f_X\,(x)\,dx$ is given by

$$\sum_{i=0}^{\infty} \frac{\omega_{i,\theta}}{2^{n/2+i}\Gamma\,(n/2+i)} \int_0^\infty x^{s+n/2+i-1}\exp\left(-\frac{x}{2}\right) dx = 2^s \sum_{i=0}^{\infty} \omega_{i,\theta}\,\frac{\Gamma\,(n/2+i+s)}{\Gamma\,(n/2+i)}. \tag{10.8}$$

From the relation $a^{[j]}\Gamma\,(a) = \Gamma\,(a+j)$ and (5.25), this can be written as

$$\mathbb{E}\left[X^s\right] = \frac{2^s}{e^{\theta/2}}\,\frac{\Gamma\,(n/2+s)}{\Gamma\,(n/2)}\sum_{i=0}^{\infty} \frac{(\theta/2)^i}{i!}\,\frac{(n/2+s)^{[i]}}{(n/2)^{[i]}}$$

$$= \frac{2^s}{e^{\theta/2}}\,\frac{\Gamma\,(n/2+s)}{\Gamma\,(n/2)}\,{}_1F_1\,(n/2+s, n/2; \theta/2), \tag{10.9}$$

where ${}_1F_1$ is the confluent hypergeometric function (see Section 5.3 for details). For $s = -1$, (10.8) simplifies to

$$\mathbb{E}\left[X^{-1}\right] = \sum_{i=0}^{\infty} \omega_{i,\theta}\,\frac{1}{n+2i-2},$$

i.e., $\mathbb{E}[X^{-1}] = \mathbb{E}[(n+2P-2)^{-1}]$, where P is a Poisson random variable. Furthermore, using the integral form of ${}_1F_1$ from (5.27),

$$\mathbb{E}\left[X^{-1}\right] = 2^{-1}e^{-\theta/2}\int_0^1 y^{n/2-2}e^{\theta y/2}\,dy,$$

which can be repeatedly simplified by integration by parts for n even. For example, with $n = 4$, this simplifies to $\left(1 - e^{-\theta/2}\right)/\theta$ for $\theta > 0$. For $\theta = 0$ and $n \geq 3$, it simplifies to $1/(n-2)$.

For $s \in \mathbb{N}$, it can be shown (see Johnson, Kotz and Balakrishnan, 1994, p. 448) that

$$\mathbb{E}\left[X^s\right] = 2^s\Gamma\left(s+\frac{n}{2}\right)\sum_{i=0}^{s}\binom{s}{i}\frac{(\theta/2)^i}{\Gamma\,(i+n/2)}. \tag{10.10}$$

This yields $n + \theta$ for $s = 1$ while, for $s = 2$, (10.10) indeed simplifies to (10.7). Problem 10.6 shows that the m.g.f. of X is[3]

$$\mathbb{M}_X(t) = (1 - 2t)^{-n/2} \exp\left\{\frac{t\theta}{1 - 2t}\right\}, \quad t < 1/2. \tag{10.11}$$

With c.g.f. $\mathbb{K}_X(t) = \ln \mathbb{M}_X(t)$, it follows that $\mathbb{K}'_X(t) = n(1 - 2t)^{-1} + \theta(1 - 2t)^{-2}$, with higher-order terms easily computed, from which we obtain

$$\kappa_i = \mathbb{K}_X^{(i)}(0) = 2^{i-1}(i - 1)!\,(n + i\theta).$$

For $i = 1$ and $i = 2$, this immediately yields the mean and variance. For the third central moment, $\mu_3 = \kappa_3 = 8(n + 3\theta)$ while, from (10.10),

$$\mathbb{E}\left[X^3\right] = 2^3 \Gamma\left(3 + \frac{n}{2}\right) \left(\frac{1}{\Gamma(n/2)} + 3\frac{\theta/2}{\Gamma(1 + n/2)} + 3\frac{(\theta/2)^2}{\Gamma(2 + n/2)} + \frac{(\theta/2)^3}{\Gamma(3 + n/2)}\right)$$

$$= 8\left(2 + \frac{n}{2}\right)\left(1 + \frac{n}{2}\right)\left(\frac{n}{2}\right) + 12\theta\left(2 + \frac{n}{2}\right)\left(1 + \frac{n}{2}\right) + 6\theta^2\left(2 + \frac{n}{2}\right) + \theta^3$$

$$= (n + 4)(n + 2)(n + 3\theta) + 3\theta^2(n + 4) + \theta^3. \tag{10.12}$$

Of course, using this in conjunction with (I.4.49) yields $\mu_3 = \mu'_3 - 3\mu'_2\mu + 2\mu^3 = 8(n + 3\theta)$ after simplifying.

10.1.3 Computation

Both f_X and F_X can be computed to any desired degree of accuracy by directly evaluating (10.3) and (10.4) using a finite number of terms in the sum. Alternatively, the expression in (1.74), obtained by inverting the characteristic function of X, can be evaluated with numerical integration. Other methods exist, along with a variety of approximations (see Johnson, Kotz and Balakrishnan, 1994, pp. 458–467 and the references therein). The simplest approximation is to use a multiple of a central χ^2; equating the first two moments of $X \sim \chi^2(n, \theta)$ and $Y \sim c\chi^2(k)$ implies $n + \theta = ck$ and $2n + 4\theta = 2c^2k$, or

$$c = \frac{n + 2\theta}{n + \theta}, \quad k = \frac{(n + \theta)^2}{n + 2\theta}.$$

Thus, $F_X(x) = \Pr(X \le x)$ is approximated by $G_k(x/c)$, where G_k is the c.d.f. of a $\chi^2(k)$ random variable.

The simple form of the cumulative generating function obtained from (10.11) implies that the saddlepoint approximation can be easily implemented. In particular, we require

---

[3] This was also derived for $n = 1$ in Problem I.7.18, which directly computed $\mathbb{E}\left[\exp\left(t(Z + b)^2\right)\right]$, where $Z \sim N(0, 1)$ and $Y = (Z + b)^2 \sim \chi^2(1, b^2)$.

the solution to $x = \mathbb{K}'_X(t)$, or the zeros of $4xt^2 - 2t(2x - n) - n - \theta + x$, given by

$$t_\pm = \frac{1}{4x}\left(2x - n \pm \sqrt{n^2 + 4x\theta}\right).$$

Rearranging and using the facts that (i) the constraint $t < 1/2$ from (10.11), (ii) $\theta \geq 0$, and (iii) the interior of the support of X is $\mathbb{R}_{>0}$ (i.e., $x > 0$), easily shows that t_- is always the correct solution. This is another example of when a closed-form solution to the saddlepoint equation and, thus, the approximate p.d.f. exists. It appears that this application of the s.p.a. was first observed by Hougaard (1988).

10.1.4 Weighted sums of independent central χ^2 random variables

There are many examples of test statistics (in particular, when working with linear regression models and time series analysis) in which the distribution of a weighted sum of independent χ^2 r.v.s arises. The special case with central χ^2 r.v.s is of most importance, so we consider it separately now; the general case is discussed below.

Let $X_i \overset{\text{ind}}{\sim} \chi^2(n_i)$ and define $X = \sum_{i=1}^k a_i X_i$, $a_i \neq 0$. No simple expression exists for f_X or F_X, although several numerical methods exist for their approximation. First observe that the m.g.f. takes on a very amenable form: with $\vartheta_i = \vartheta_i(s) = (1 - 2sa_i)^{-1}$, the m.g.f. of X is

$$\mathbb{M}_X(s) = \prod_{i=1}^k \mathbb{M}_{a_i X_i}(s) = \prod_{i=1}^k \mathbb{M}_{X_i}(a_i s) = \prod_{i=1}^k (1 - 2a_i s)^{-n_i/2} = \prod_{i=1}^k \vartheta_i^{n_i/2}, \quad (10.13)$$

which is valid for s such that $1 - 2a_i s > 0$, $i = 1, \ldots, k$. The saddlepoint approximation is clearly applicable. To use it, it is necessary to consider the following three cases, depending on the a_i. Let $\underline{a} = 2\min a_i$ and $\overline{a} = 2\max a_i$. If $\underline{a} > 0$, i.e., all a_i are positive, then $s < \overline{a}^{-1}$. If $\overline{a} < 0$, i.e., all a_i are negative, then $s > \underline{a}^{-1}$. Otherwise, $\underline{a}^{-1} < s < \overline{a}^{-1}$. It is easy to verify that

$$\mathbb{K}_X(s) = \frac{1}{2}\sum_{i=1}^k n_i \ln \vartheta_i, \quad \mathbb{K}'_X(s) = \sum_{i=1}^k n_i a_i \vartheta_i, \quad \mathbb{K}''_X(s) = 2\sum_{i=1}^k n_i a_i^2 \vartheta_i^2,$$

so that the s.p.a. can be straightforwardly applied. For $k = 2$, an explicit expression for the solution to the saddlepoint equation $\mathbb{K}'_X(\hat{s}) = x$ can be obtained. Otherwise, it needs to be solved numerically. The most important applications are such that $\underline{a} < 0$ and $\overline{a} > 0$, in which case $\underline{a}^{-1} < \hat{s} < \overline{a}^{-1}$. Knowing bounds for $\hat{s}$ (along with the fact that a unique value of $\hat{s}$ which satisfies the saddlepoint equation exists in that range) greatly simplifies the numerical search for the saddlepoint.

The c.f. of X is given by

$$\varphi_X(t) = \mathbb{M}_X(it) = \prod_{j=1}^{k} \left(1 - 2a_j it\right)^{-n_j/2} = \exp\left\{ -\frac{1}{2} \sum_{j=1}^{k} n_j \ln\left(1 - 2a_j it\right) \right\} \quad (10.14)$$

(the r.h.s. is better suited for numerical purposes), so that the inversion formulae for the p.d.f. and c.d.f. can be evaluated.

To illustrate, let $k = 5$, $\mathbf{a} = (a_1, \ldots, a_5) = (-3, -2, -1, 1, 2)$ and each $n_i = 1$. The density over a large range of x is shown in Figure 10.1(a) (solid line) along with the first- and second-order saddlepoint approximations. Figure 10.1(b) shows the RPE of the latter approximation along with the asymptote of -2.843, obtained by evaluating the RPE for $x = -450$.

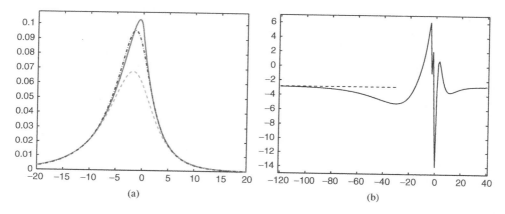

(a)　　　　　　　　　　　　(b)

Figure 10.1 (a) Density of $\sum_{i=1}^{5} a_i X_i$, where $\mathbf{a} = (-3, -2, -1, 1, 2)$ and $X_i \overset{\text{i.i.d.}}{\sim} \chi^2(1)$. Solid line computed as $(F_X(x + \delta) - F_X(x - \delta))/(2\delta)$ for $\delta = 10^{-7}$ and F_X computed using Pan's procedure. Dashed and dash-dotted lines are the first- and second-order saddlepoint approximations computed with Program 10.3 given in Section 10.1.5. (b) Relative percentage error of the second-order s.p.a.

A specific method for the c.d.f. has been developed by Grad and Solomon (1955) and Pan (1968) and is valid for this case (i.e., a weighted sum of independent central χ^2 r.v.s), but also with the restriction that all the degrees of freedom are one, and that the weights are all distinct, i.e., $X = \sum_{i=1}^{k} a_i X_i$ for $X_i \overset{\text{i.i.d.}}{\sim} \chi^2(1)$ and $a_i \neq a_j$. The derivation involves contour integration applied to the inversion formula, and we omit it and just state the result (see also Durbin and Watson, 1971, for a proof, and an improvement, which we implement below). Let the a_i be in descending order, i.e., $a_1 > a_2 > \cdots > a_k$, and let $\delta_v = (1 - (-1)^v)/2$ and v such that $a_{v+1} < 0 < a_v$. Then

$$F_X(x) = 1 + \frac{1}{\pi} \sum_{j=1}^{\lfloor v/2 \rfloor} (-1)^j \int_{a_{2j}}^{a_{2j-1}} \frac{y^{k/2-1}}{\prod_{i=1}^{k} |y - a_i|^{1/2}} \exp\left(-\frac{x}{2y}\right) dy$$

$$+ \frac{\delta_v}{\pi} (-1)^{\frac{v+1}{2}} \int_{0}^{a_v} \frac{y^{k/2-1}}{\prod_{i=1}^{k} |y - a_i|^{1/2}} \exp\left(-\frac{x}{2y}\right) dy. \quad (10.15)$$

Because the integrals in (10.15) can also be expressed as

$$F_X(x) = 1 + \frac{1}{\pi} \sum_{j=1}^{\lfloor v/2 \rfloor} (-1)^j \int_{-1}^1 \left| \frac{y_{2j}^{k-2}}{P_{1,j}(t)} \right|^{1/2} \exp\left(-\frac{x}{2y_{2j}}\right) (1-t^2)^{-1/2} \, dt$$

$$+ \frac{\delta_v}{\pi} (-1)^{\frac{v+1}{2}} \int_{-1}^1 \left| \frac{y_v^{k-1}}{P_2(t)} \right|^{1/2} \exp\left(-\frac{x}{2y_v}\right) (1-t^2)^{-1/2} \, dt, \tag{10.16}$$

where

$$P_{1,j}(t) = \prod_{\substack{i=1 \\ i \neq 2j, 2j-1}}^k (y_{2j} - a_i), \qquad P_2(t) = \prod_{\substack{i=1 \\ i \neq v}}^k (y_v - a_i)$$

and

$$y_{2j} = \frac{1}{2}(a_{2j-1} + a_{2j}) - \frac{1}{2}(a_{2j-1} - a_{2j})t, \qquad y_v = \frac{1}{2}a_v - \frac{1}{2}a_v t,$$

they can be approximated via (see Farebrother, 1980a)

$$\int_{-1}^1 \frac{f(t)}{(1-t^2)^{1/2}} \, dt = \lim_{N \to \infty} \frac{\pi}{N} \sum_{j=1}^N f\left(y_j^{(N)}\right), \qquad y_j^{(N)} = \cos\frac{(2j-1)\pi}{2N}, \tag{10.17}$$

where f denotes the functional form in (10.16) without the term $(1-t^2)^{-1/2}$. A value of only $N = 20$ has been found to yield quite high accuracy in typical cases.

Durbin and Watson (1971) recommend using (10.15) if $x > 0$, or if $x = 0$ and $v \leq k - v$. Otherwise, one should use

$$F_X(x) = \frac{1}{\pi} \sum_{j=1}^{\lfloor (k-v)/2 \rfloor} (-1)^j \int_{a_{k-2j+1}}^{a_{k-2j-2}} \frac{|y|^{k/2-1}}{\prod_{i=1}^k |y - a_i|^{1/2}} \exp\left(-\frac{x}{2y}\right) dy$$

$$+ \frac{\delta_{(k-v)}}{\pi} (-1)^{\frac{k-v+1}{2}} \int_0^{a_{v+1}} \frac{|y|^{k/2-1}}{\prod_{i=1}^k |y - a_i|^{1/2}} \exp\left(-\frac{x}{2y}\right) dy. \tag{10.18}$$

The integrals in this expression can also be transformed so that (10.17) can be applied. The program in Listing 10.1 implements this method.

```
function F=pan(xvec,a,N)
if nargin<3, N=20; end;
a=reshape(a,length(a),1); F=zeros(size(xvec));
yin=cos((2*cumsum(ones(N,1))-1)*pi/(2*N));
for loop=1:length(xvec)
  x=xvec(loop);
  if all(a<0),    F(loop)=1-AS_R52(-x,-a,yin);
  else,    F(loop)=AS_R52(x,a,yin); end;
end
```

Program Listing 10.1 Computes the c.d.f. of $X = \sum_{i=1}^k a_i X_i$ at x, for $X_i \overset{\text{i.i.d.}}{\sim} \chi^2(1)$ and $a_i \neq a_j$. The a_i are input as vector a, and vector x indicates the points at which to evaluate the c.d.f. Input N dictates the number of terms in the sum (10.17), with default value 20. The program is continued in Listing 10.2

```
function F = AS_R52(c,lambda,yin)
lambda=-sort(-lambda); m=length(lambda);n=length(yin);
v=min(find(lambda<=0))-1; if isempty(v),v=m; end;
deltav=0.5*(1-(-1)^v); deltamv=0.5*(1-(-1)^(m-v)); F=0;
if c>=0 | (c==0 & v<=m-v)
  for j=1:floor(0.5*v)
    F=F+((-1)^j) * mean(panintegrand(yin,c,lambda,[2*j,2*j-1]));
  end
  F=1+F;
  if deltav>0
    F=F+(deltav)*(-1)^(0.5*(v+1))*mean(panintegrand(yin,c,lambda,v));
  end
else
  for j=1:floor(0.5*(m-v))
    F = F + ((-1)^j) * ...
      mean(panintegrand(yin,c,lambda,[m-2*j+1,m-2*j+2]));
  end;
  if deltamv>0
    F=F+(deltamv)*(-1)^(0.5*(m-v+1)) ...
      *mean(panintegrand(yin,c,lambda,v+1));
  end
  F=-F;
end;

function I=panintegrand(tvec,c,lambda,outind)
m=length(lambda); outnumber=length(outind); inind=1:m;
for i=1:outnumber;
  inind=inind(find(inind~=outind(i)));
end
sv=1; if outnumber==2, sv=[1 ;-1]; end
I=zeros(size(tvec));
for tloop=1:length(tvec)
  t=tvec(tloop);
  yv=0.5*(sum(lambda(outind)))-0.5*(lambda(outind)'*sv)*t;
  if yv==0
    helper=0;
  else
    helper=(exp(-c/(2*(yv))));
  end
  I(tloop)=helper ...
    *sqrt(abs(((yv)^(m-outnumber))/prod(yv-lambda(inind))));
end;
```

Program Listing 10.2 Continuation of the program in Listing 10.1

The following code provides an example of Pan's procedure. The output can be compared to the call to myimhof, which is a more general (but far slower) routine developed below. The last line computes the saddlepoint approximation, also discussed below.

```
lambda=[-3,-1,3,4]; xvec=-10:4:20;
df=ones(1,length(lambda)); noncen=zeros(1,length(lambda));
pan(xvec,lambda)'
myimhof (xvec,lambda,df,noncen,500)
[pdf,cdf] = spaweightedsum(2,xvec,lambda, df, noncen); cdf
```

10.1.5 Weighted sums of independent $\chi^2(n_i, \theta_i)$ random variables

Generalizing the central case, let $X_i \overset{\text{ind}}{\sim} \chi^2(n_i, \theta_i)$ and define $X = \sum_{i=1}^{k} a_i X_i$, $a_i \neq 0$. Notice that, because of the independence, the moments of X are just straightforward functions of those of the X_i. However, as before, tractable expressions for f_X and F_X are not available. From (10.11),

$$\mathbb{M}_{X_i}(a_i s) = (1 - 2a_i s)^{-n_i/2} \exp\left\{\frac{a_i s \theta_i}{1 - 2a_i s}\right\}, \quad 1 - 2a_i s > 0, \tag{10.19}$$

so that

$$\mathbb{M}_X(s) = \prod_{i=1}^{k} \mathbb{M}_{X_i}(a_i s) \tag{10.20}$$

for s in a sufficiently small neighbourhood of zero – the conditions for which are stated below equation (10.13). For convenience, let $\vartheta_i = \vartheta_i(s) = (1 - 2sa_i)^{-1}$. Then straightforward calculation yields

$$\mathbb{K}_X(s) = \frac{1}{2}\sum_{i=1}^{k} n_i \ln \vartheta_i + s \sum_{i=1}^{k} a_i \theta_i \vartheta_i, \quad \mathbb{K}'_X(s) = \sum_{i=1}^{k} a_i \vartheta_i (n_i + \theta_i \vartheta_i)$$

and

$$\mathbb{K}''_X(s) = 2\sum_{i=1}^{k} a_i^2 \vartheta_i^2 (n_i + 2\theta_i \vartheta_i), \quad \mathbb{K}'''_X(s) = 8\sum_{i=1}^{k} a_i^3 \vartheta_i^3 (n_i + 3\theta_i \vartheta_i),$$

and $\mathbb{K}_X^{(4)}(s) = 48\sum_{i=1}^{k} a_i^4 \vartheta_i^4 (n_i + 4\theta_i \vartheta_i)$, from which the s.p.a. can be calculated once $\hat{s}$ is determined.

Observe that, as $\theta_i \geq 0$, $i = 1, \ldots, k$, $\mathbb{K}''_X(s) > 0$, and $\mathbb{K}'_X(s)$ is monotone increasing. Thus, the saddlepoint equation $\mathbb{K}'_X(\hat{s}) = x$ always has a unique solution for x in the interior of the support of X. The situation with $\min a_i < 0$ and $\max a_i > 0$ arises often and so is of particular importance. In that case, with $\underline{a} = 2\min a_i$ and $\bar{a} = 2\max a_i$, $\underline{a}^{-1} < s < \bar{a}^{-1}$, i.e., $\hat{s}$ lies in a known, finite-length region, which can greatly assist its numerical determination.

A program for calculating the saddlepoint approximation to the p.d.f. and c.d.f. of X is given in Listings 10.3–10.5. The saddlepoint is obtained using Newton's (or the Newton–Raphson) method.[4]

The s.p.a. in this context is computationally fast and yields high relative accuracy which is usually enough for practical work. However, there are situations in which

---

[4] That is, if $f : [a, b] \mapsto \mathbb{R}$ is a continuously differentiable function on the closed interval $[a, b]$ such that $\exists z \in [a, b]$ such that $f(z) = 0$ and $f'(z) \neq 0$, and $\{x_n\}$ is a sequence in $[a, b]$ defined by the recursion $x_{n+1} = x_n - f(x_n)/f'(x_n)$ (where x_1 is an arbitrary starting value in $[a, b]$), then x_n will converge (quadratically) to z.

```
function [pdf,cdf,svec] = spaweightedsum(daniels,xvec,a,df,q,s);

if nargin<6, s=0; end
if nargin<5, q = zeros(length(a),1); end
if nargin<4, df = ones(length(a),1); end

alo=2*min(a) ; ahi=2*max(a);
if alo>0
  lower=-Inf;    upper = 1/ahi;  boundtype=1;
elseif ahi<0
  lower=1/alo;   upper=Inf;      boundtype=2;
else
  lower=1/alo;   upper=1/ahi;    boundtype=3;
end

sstart=s;
pdf=zeros(length(xvec),1); cdf=pdf; svec=pdf;
for i=1:length(xvec)
  x=xvec(i);
  [s,report]=spaweightedsumsadroot(sstart,x,a,df,q,lower,upper);
  if (report>0) & (boundtype==3) % try again, using a grid of s values
    ss=lower:(upper-lower)/200:upper; x
    for j=2:length(ss)-1
      [s,report]=spaweightedsumsadroot(ss(j),x,a,df,q,lower,upper);
      if report==0, disp('valid saddlepoint found'), break, end
    end
  end
  svec(i)=s;
  %sstart=s; % turn on to use shat from xvec(i-1) as starting value

  if ((report==0) | (report==5))
    v = 1./(1-2*s*a);
    K = 0.5 * sum(df .* log(v)) + s*sum(a .* q .* v);
    Kpp  = 2 * sum(a.^2 .* v.^2 .* (df + 2 * q .* v));
    pdf(i) = exp(K - x*s) / sqrt(2*pi*Kpp);
    fac = 2*s*x-2*K;
    ww=sign(s)*sqrt(fac); u=s*sqrt(Kpp);
```

Program Listing 10.3   The s.p.a. of the p.d.f. and c.d.f. of $\sum a_i X_i$, $a_i \neq 0$, for $X_i \overset{\text{ind}}{\sim} \chi^2(df_i, q_i)$, where df_i and q_i denote the degrees of freedom and noncentrality parameters, respectively. Pass a, df and q as column vectors. Optionally pass scalar s as a starting value for $\hat{s}$. The program is continued in Listing 10.4. Function spaweightedsumsadroot is given in Listing 10.5

more precise calculations are required (such as assessing the accuracy of the s.p.a.). As was discussed in Section 10.1.4 for the central χ^2 case, the p.d.f. and c.d.f. inversion formulae (1.59) and (1.70) can be straightforwardly evaluated numerically using software which supports complex numbers and offers built-in numerical integration routines. Recall Example 1.25 in which the characteristic function inversion formula (1.70) was used to obtain a noncomplex integral expression for the c.d.f. of a noncentral χ^2 random variable. The generalization to the weighted sum X is straightforward and is useful for computing environments which do not support complex arithmetic.

```
if ((report==5) | (abs(ww)<1e-7))
    disp(['x = ',num2str(x),' is close to the expected value'])
    Kpp0  = 2 * sum(a.^2 .* (df + 2 * q));
    Kppp0 = 8 * sum(a.^3 .* (df + 3 * q));
    cdf(i)=0.5+Kppp0/6/sqrt(2*pi)/Kpp0^(3/2);
else
    npdf=normpdf(ww); tempcdf=normcdf(ww)+npdf*(1/ww - 1/u);
    if daniels==2
        K3= 8*sum(a.^3 .* v.^3 .* (df + 3*q.*v));
        K4=48*sum(a.^4 .* v.^4 .* (df+4*q.*v));
        kap3= K3/(Kpp)^(3/2);  kap4=K4/(Kpp)^(4/2);
        forboth = (kap4/8 - 5*kap3^2/24);

        % the cdf correction can be problematic!
        bigterm = forboth/u - 1/u^3 - kap3/2/u^2 + 1/ww^3;
        correction=npdf * bigterm;
        if abs(correction)/tempcdf > 0.1
            disp('The 2nd order CDF correction term is being ditched!')
            correction=0;
        end
        cdf(i) = tempcdf - correction;

        if ((cdf(i)<0) | (cdf(i)>1))
            disp('The 2nd order CDF correction term is being ditched!')
            cdf(i) = tempcdf;
        end
    else
        cdf(i) = tempcdf;
    end
end
if daniels==2
    pdf(i) = pdf(i) * (1+forboth);
end
else
    pdf(i)=0; cdf(i)=0;
end
end
```

Program Listing 10.4   Continuation of Listing 10.3

In particular, taking $\varphi_X(t) = \mathbb{M}_X(it)$,

$$\varphi_X(t) = \prod_{j=1}^{k} (1 - 2a_j it)^{-n_j/2} \exp\left\{ \sum_{j=1}^{k} \frac{a_j \theta_j it}{1 - 2a_j it} \right\}, \quad t \in \mathbb{R}, \tag{10.21}$$

and letting $z_x(t) := e^{-itx} \varphi_X(t)$, extending the results in Example 1.25 yields

$$|z_x(t)| = |\varphi_X(t)| = \prod_{j=1}^{k} (1 + 4a_j^2 t^2)^{-n_j/4} \exp\left(-\sum_{j=1}^{k} \frac{2\theta_j a_j^2 t^2}{1 + 4a_j^2 t^2} \right)$$

```
function [s,report]=spaweightedsumsadroot(s0,x,a,df,q,lower,upper)

toll = 1e-8;   % how close to the edge of the support
tol2 = 1e-12; % how close to zero Kpp can be in the Newton step
tol3 = 1e-12; % when to stop the root search
tol4 = 1e-8;   % how close to zero shat may be
MAXIT = 200; s=0; iter=0; derivOK=1; disc=10; report=0;

while ( (iter<=MAXIT) & (derivOK==1) & (disc>=tol3) )
   iter=iter+1; v = 1./(1-2*s0*a);
   kp = sum(a .* v .* (df + q .* v));
   kpp= 2 * sum(a.^2 .* v.^2 .* (df + 2 * q .* v));
   derivOK = (kpp > tol2);
   if derivOK
      s = s0 - (kp-x) / kpp;  disc = abs(s-s0);  s0 = s;
   else, report=6; break, end
end
if report==0
   if      s<lower+toll, report = 11;
   elseif s>upper-toll, report = 12;
   elseif iter==MAXIT, report = 3;
   elseif abs(s) < tol4, report = 5; end
end
```

Program Listing 10.5   Called by program `spaweightedsum` to get $\hat{s}$

and

$$\arg z_x(t) = \arg e^{-itx} + \arg \varphi_X(t) = -tx + \sum_{j=1}^{k} \frac{n_j}{2} \arctan\left(2a_j t\right) + \sum_{j=1}^{k} \frac{t a_j \theta_j}{1 + 4a_j^2 t^2}.$$

Letting $u = 2t$ simplifies things a little. Then (1.70) gives

$$F_X(x) = \frac{1}{2} - \frac{1}{\pi} \int_0^\infty \frac{\sin \beta(u, x)}{u \gamma(u)} \, du, \tag{10.22}$$

where, incorporating the product term in $|z_x(t)|$ into the exponent to improve numerical accuracy,

$$\beta(u, x) = \frac{1}{2} \sum_{j=1}^{k} \left(n_j \arctan p_j + \frac{\theta_j p_j}{c_j} \right) - \frac{1}{2} xu$$

and

$$\gamma(u) = \frac{1}{|z_x(t)|} = \exp\left\{ \frac{1}{2} \sum_{j=1}^{k} \frac{\theta_j b_j}{c_j} + \frac{1}{4} \sum_{j=1}^{k} n_j \ln c_j \right\},$$

with $p_j = a_j u$, $b_j = p_j^2$ and $c_j = 1 + b_j$. Expression (10.22) is due to Imhof (1961), which is still a very fequently cited article because of the vast number of statistical applications which require evaluation of the c.d.f. of X. The method was particularly popularized by the econometric examples and Fortran programs given in Koerts and Abrahamse (1969).

As discussed in Section 1.2.5, one way of evaluating (10.22) is just to replace the indefinite integral with a definite one, by choosing an appropriate upper bound for the integral, while the second way is to transform the integrand so that the range of integration is over a finite interval, say $(0, 1)$, as in (1.71). For certain parameter constellations, such as small k, this transformation could induce high oscillatory behavior in the integrand, making numerical integration difficult. The program in Listing 10.6 allows both ways, depending on how parameter `uplim` is passed. To illustrate, we use the sum of k independent χ_1^2 r.v.s, each with unit weight (and noncentrality parameter zero), so that the resulting distribution is just χ_k^2, with a c.d.f. available to machine precision. Running

```
x=1; k=1; true=chi2cdf(x,k)
m=myimhof(x,ones(k,1),ones(k,1),zeros(k,1),-1), m-true
m=myimhof(x,ones(k,1),ones(k,1),zeros(k,1),2000), m-true
```

illustrates the problem with transforming to $(0, 1)$ when k is small. With $k = 5$ instead of 1, the transformation is successful, and obviates the need to determine an appropriate upper limit.

```
function cdf = myimhof (xvec,wgt,df,nc,uplim)
if nargin<5, uplim=-1; end
k=length(wgt); tol=1e-8;
if nargin < 3, df=ones(k,1); end
if nargin < 4, nc=zeros(k,1); end

lx=length(xvec); cdf=zeros(lx,1);
for i=1:lx, x=xvec(i);
  if uplim<0
    cdf(i)=0.5-(1/pi)*quadl(@ff,tol,1-tol,tol,0,x,wgt,df,nc,uplim);
  else
    cdf(i)=0.5-(1/pi)*quadl(@ff,tol,uplim,tol,0,x,wgt,df,nc,uplim);
  end
end

function I = ff(tvec,x,wgt,df,nc,uplim)
vlen=length(tvec); I=zeros(1,vlen);
usetransform=(uplim<0);
for ii=1:vlen,  t=tvec(ii);
  if usetransform, u=(1-t)/t; else  u=t; end
  p = u*wgt; b=p.^2; c=1+b;
  ss0=sum(df.*atan(p) + nc.*p./c); beta = (ss0 - x*u)/2;
  ss1=sum(nc.*b./c); ss2=sum(df.*log(c)); gam=exp(0.5*ss1 + 0.25*ss2);
  I(ii) = sin(beta) / (u*gam);
  if usetransform, I(ii) = I(ii) / t^2; end % don't forget du/dt
end
```

Program Listing 10.6 Implementation of (10.22). Pass `uplim` as a negative value to transform to interval $(0, 1)$, otherwise it calculates the integral as in (10.22), with upper limit `uplim`. Inputs `wgt`, `df` and `nc` refer to the weights a_i, the degrees of freedom n_i, and the noncentrality parameters θ_i in (10.21), respectively. Pass them as column vectors

Remark: Another approach for dealing with the upper limit in the indefinite integral (10.22) is to analytically derive an upper bound to $\int_U^\infty f(u)\,du$, which then allows determination of U as a function of the desired accuracy. This was first given by Imhof (1961). Using this upper bound in conjunction with a chosen tolerance on the numerical integration will, in theory, ensure an upper bound on the error. However, implementations of this method which are commonly used in practice can still fail to deliver the c.d.f. with the desired accuracy. In particular, the popular Pascal implementation given in Farebrother (1990), and direct Matlab interpretations of it available from the internet, can perform so poorly as to return c.d.f. values outside $[0, 1]$. See Farebrother (1980, 1990) and Davies (1973, 1980) for further issues regarding the computation method.■

⊖ **Example 10.1** To illustrate the accuracy of the s.p.a. method, let $\mathbf{a} = (-5, -4, \ldots,$ $4, 5)$ and $X_i \overset{\text{i.i.d.}}{\sim} \chi_1^2$. (The zero weight on X_6 is not problematic for the s.p.a. or c.f. inversion.) The density of $X = \sum_{j=1}^{11} a_i X_i$ is clearly symmetric about zero with variance $2 \sum_{j=1}^{11} a_i^2 = 220$.

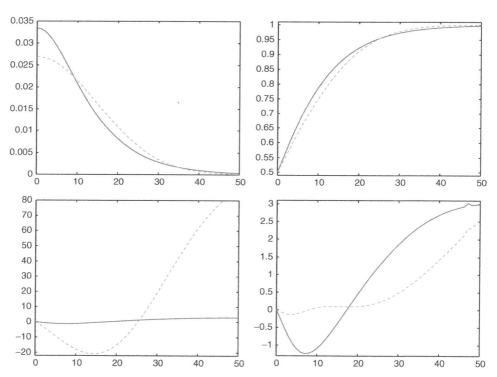

Figure 10.2 Accuracy of s.p.a. and normal approximation for the symmetric (about zero) r.v. $X = \sum_{j=1}^{11} a_i X_i$, as given in Example 10.1. (Top left) The p.d.f. of X using second-order s.p.a. (solid) and normal approximation (dashed). (Top right) The exact c.d.f. (solid) computed using (10.22) and the normal approximation (dashed). The s.p.a. to the c.d.f. is graphically indistinguishable from the true values and is not shown. (Bottom left) The RPE, defined as $100(A - E)/\min(E, 1 - E)$, of the first-order s.p.a. (solid) and normal (dashed) approximation, where A denotes the approximation and E the exact c.d.f. values. (Bottom right) The RPE for the first-order (solid) and second-order (dashed) c.d.f. s.p.a.

Figure 10.2 shows the quality of the s.p.a. and normal approximation. In terms of relative error, note how poorly the normal approximation performs, even though (i) X is symmetric, (ii) X consists of the sum of 10 independent r.v.s, each with finite variance, and (iii) the X_i have exponential (as opposed to power) tails. ∎

10.2 Singly and doubly noncentral F

> The price one pays for pursuing any profession or calling is an intimate know-ledge of its ugly side.   (James Arthur Baldwin)

The singly noncentral F distribution plays a central role in the analysis of variance, where it determines the power for tests of linear hypotheses. The doubly noncentral F generalizes the singly noncentral case, and also arises in such contexts; see Scheffé (1959, pp. 134–135). In econometrics, the doubly noncentral F arises naturally in simultaneous equation models, in testing linear models with proxy variables (Kuru-mai and Ohtani, 1998) and in Ramsey's (1969) popular regression specification error (RESET) test (see also DeBenedictis and Giles, 1998). It is also of use in signal pro-cessing and pattern recognition applications (Price, 1964; Helstrom and Ritcey, 1985).

10.2.1 Derivation

Let $X_i \overset{\text{ind}}{\sim} \chi^2(n_i, \theta_i)$, $n_i > 0$ (though usually $n_i \in \mathbb{N}$), $\theta_i \geq 0$, $i = 1, 2$, and define

$$F = \frac{X_1/n_1}{X_2/n_2} \quad \text{and} \quad \omega_{i,\theta} = \frac{e^{-\theta/2}(\theta/2)^i}{i!} \quad \text{for} \quad \theta \geq 0, \quad i \in \mathbb{N}_0 = \{0, 1, 2, \ldots\}.$$

It is straightforward to show (Problem 10.8) that the density of F, $f_F(x; n_1, n_2, \theta_1, \theta_2)$ $= f_F(x)$, is given by

$$f_F(x) = \sum_{i=0}^{\infty} \sum_{j=0}^{\infty} \omega_{i,\theta_1} \omega_{j,\theta_2} \frac{n_1^{n_1/2+i}}{n_2^{-n_2/2-j}} \frac{x^{n_1/2+i-1}(xn_1+n_2)^{-(n_1+n_2)/2-i-j}}{B(i+n_1/2, j+n_2/2)}, \tag{10.23}$$

and F is said to follow a *doubly noncentral F distribution*, $\mathrm{F}(n_1, n_2, \theta_1, \theta_2)$. If $\theta_2 = 0$, then this reduces to $f_F(x) = f_F(x; n_1, n_2, \theta_1, 0) = f_F(x; n_1, n_2, \theta_1)$, where

$$f_F(x) = \sum_{i=0}^{\infty} \omega_{i,\theta_1} \frac{n_1^{n_1/2+i}}{n_2^{-n_2/2}} \frac{x^{n_1/2+i-1}(xn_1+n_2)^{-(n_1+n_2)/2-i}}{B(i+n_1/2, n_2/2)}, \tag{10.24}$$

(taking $(\theta_2/2)^0 = 1$), which is the p.d.f. associated with the *(singly) noncentral F distribution*.[5] Of course, with $\theta_1 = 0$, the singly noncentral F p.d.f. reduces to the

---

[5] If the noncentrality is desired only in the denominator, recall that $\Pr(F < x) = \Pr(1/F > 1/x)$ and use the ratio construction of F given above, i.e., $F = (X_1/n_1)/(X_2/n_2)$, so that the noncentrality parameter appears in the numerator term.

usual (central) F distribution (I.7.48), i.e.,

$$f_F(x) = \frac{n_1^{n_1/2} n_2^{n_2/2}}{B(n_1/2, n_2/2)} x^{n_1/2-1} (xn_1 + n_2)^{-(n_1+n_2)/2}$$

$$= \frac{n_1/n_2}{B(n_1/2, n_2/2)} \frac{((n_1/n_2)x)^{n_1/2-1}}{(1 + (n_1/n_2)x)^{(n_1+n_2)/2}}.$$

The c.d.f. $F_F(x; n_1, n_2, \theta_1, \theta_2) = F_F(x)$ can be expressed as

$$F_F(x) = \sum_{i=0}^{\infty} \sum_{j=0}^{\infty} \omega_{i,\theta_1} \omega_{j,\theta_2} \overline{B}_y(i + n_1/2, \ j + n_2/2), \qquad y = \frac{n_1 x}{n_1 x + n_2}, \qquad (10.25)$$

where $\overline{B}$ is the incomplete beta function ratio. For $\theta_2 = 0$,

$$F_F(x) = \sum_{i=0}^{\infty} \omega_{i,\theta_1} \overline{B}_y(i + n_1/2, \ n_2/2), \qquad y = \frac{n_1 x}{n_1 x + n_2}, \qquad (10.26)$$

which reduces to (I.7.49) in the central case.

10.2.2 Moments

We will denote the rth raw (central) moment of a doubly noncentral F r.v. as $_2\mu_r'$ $(_2\mu_r)$, and that of a singly noncentral F as $_1\mu_r'$ $(_1\mu_r)$.

From the independence of X_1 and X_2, the rth raw moment of $X \sim F(n_1, n_2, \theta_1, \theta_2)$, is,[6] for $r < n_2/2$,

$$_2\mu_r' = \mathbb{E}[X^r] = \left(\frac{n_2}{n_1}\right)^r \mathbb{E}[X_1^r] \mathbb{E}[X_2^{-r}]$$

$$= \left(\frac{n_2}{n_1}\right)^r \sum_{i=0}^{\infty} \sum_{j=0}^{\infty} \omega_{i,\theta_1} \omega_{j,\theta_2} \frac{\Gamma(n_1/2 + i + r) \Gamma(n_2/2 + j - r)}{\Gamma(n_1/2 + i)} \frac{\Gamma(n_2/2 + j - r)}{\Gamma(n_2/2 + j)}, \qquad r < \frac{n_2}{2}.$$

(10.27)

If $X \sim F(n_1, n_2, \theta_1, 0)$, i.e., only singly noncentral, then its rth raw moment is

$$_1\mu_r' = \mathbb{E}[X^r] = \left(\frac{n_2}{n_1}\right)^r \mathbb{E}[X_1^r] \mathbb{E}[X_2^{-r}]$$

$$= \left(\frac{n_2}{n_1}\right)^r \frac{\Gamma(n_2/2 - r)}{\Gamma(n_2/2)} \sum_{i=0}^{\infty} \omega_{i,\theta_1} \frac{\Gamma(n_1/2 + i + r)}{\Gamma(n_1/2 + i)}, \qquad r < \frac{n_2}{2}, \qquad (10.28)$$

---

[6] Note the minor typo in the bottom equation on p. 500 of Johnson, Kotz and Balakrishnan (1994).

so that, recalling (10.8) and the simple expressions for the first two moments of $\chi^2 (n, \theta)$,

$$_1\mu_1' = \mathbb{E}[X] = \frac{n_2}{n_1} \frac{\Gamma (n_2/2 - 1)}{\Gamma (n_2/2)} 2^{-1} \mathbb{E}[X_1] = \frac{n_2}{n_1} \frac{n_1 + \theta_1}{n_2 - 2}, \quad n_2 > 2, \qquad (10.29)$$

and

$$_1\mu_2' = \mathbb{E}[X^2] = \left(\frac{n_2}{n_1}\right)^2 \frac{1}{(n_2/2 - 1)(n_2/2 - 2)} 2^{-2} \mathbb{E}[X_1^2]$$

$$= \left(\frac{n_2}{n_1}\right)^2 \frac{n_1^2 + 2n_1 + 4\theta_1 + \theta_1^2 + 2\theta_1 n_1}{(n_2 - 2)(n_2 - 4)}, \quad n_2 > 4,$$

or

$$_1\mu_2 = \mathbb{V}(X) = 2\frac{n_2^2}{n_1^2} \frac{(n_1 + \theta_1)^2 + (n_1 + 2\theta_1)(n_2 - 2)}{(n_2 - 2)^2 (n_2 - 4)}, \quad n_2 > 4. \qquad (10.30)$$

Also, from (10.28) and (10.12),

$$_1\mu_3' = \mathbb{E}[X^3] = \frac{n_2^3}{n_1^3} \frac{\Gamma (n_2/2 - 3)}{\Gamma (n_2/2)} 2^{-3} \mathbb{E}[X_1^3]$$

$$= \frac{n_2^3}{n_1^3} \frac{(n_1 + 4)(n_1 + 2)(n_1 + 3\theta_1) + 3\theta_1^2 (n_1 + 4) + \theta_1^3}{(n_2 - 2)(n_2 - 4)(n_2 - 6)}$$

so that, from (I.4.49)[7]

$$_1\mu_3 = \mathbb{E}[(X - \mu)^3] = \mu_3' - 3\mu_2'\mu + 2\mu^3$$

$$= \frac{8n_2^3 (n_1 + n_2 - 2) k}{n_1^2 (n_2 - 2)^3 (n_2 - 4)(n_2 - 6)} \left(1 + 3\ell + \frac{6n_1}{k}\ell^2 + \frac{2n_1^2}{k (n_1 + n_2 - 2)}\ell^3\right), \qquad (10.31)$$

where $\ell = \theta_1/n_1$ and $k = (2n_1 + n_2 - 2)$. Finally, it can be shown that

$$_1\mu_4 = \frac{12n_2^4 (n_1 + n_2 - 2)}{n_1^3 (n_2 - 2)^4 (n_2 - 4)(n_2 - 6)(n_2 - 8)} (k_1 (1 + 4\ell) + k_2\ell^2 + k_3\ell^3 + k_4\ell^4),$$

where

$$k_1 = 2 (3n_1 + n_2 - 2)(2n_1 + n_2 - 2) + (n_1 + n_2 - 2)(n_2 - 2)(n_1 + 2),$$

$$k_2 = 2n_1 (3n_1 + 2n_2 - 4)(n_2 + 10),$$

$$k_3 = 4n_1^2 (n_2 + 10),$$

$$k_4 = n_1^3 (n_2 + 10) / (n_1 + n_2 - 2).$$

---

[7] This equation is also given in the middle of p. 482 of Johnson, Kotz and Balakrishnan (1994). However, the top equation for μ_3 on that same page is erroneous.

Now returning to the doubly noncentral case, from (10.27),

$$2\mu'_r = \left(\frac{n_2}{n_1}\right)^r \frac{\Gamma(n_2/2 - r)}{\Gamma(n_2/2)}$$

$$\times \sum_{i=0}^{\infty} \omega_{i,\theta_1} \frac{\Gamma(n_1/2 + i + r)}{\Gamma(n_1/2 + i)} \frac{\Gamma(n_2/2)}{\Gamma(n_2/2 - r)} \sum_{j=0}^{\infty} \omega_{j,\theta_2} \frac{\Gamma(n_2/2 + j - r)}{\Gamma(n_2/2 + j)}$$

or, using (10.28),

$$2\mu'_r = {}_1\mu'_r e^{-\theta_2/2} \sum_{j=0}^{\infty} \frac{(\theta_2/2)^j}{j!} \frac{(n_2/2 - r)^{[j]}}{(n_2/2)^{[j]}}$$

$$= {}_1\mu'_r e^{-\theta_2/2} {}_1F_1(n_2/2 - r, n_2/2, \theta_2/2)$$

$$= {}_1\mu'_r {}_1F_1(r, n_2/2, -\theta_2/2), \tag{10.32}$$

as $a^{[i]} = \Gamma(a+i)/\Gamma(a)$ and using the Kummer transformation (5.29), i.e., ${}_1F_1(a, b, x) = e^x {}_1F_1(b - a, b, -x)$. Given the simple expressions for the lower-order moments in the singly noncentral case, (10.32) is clearly more useful for computational purposes than the r.h.s. of (10.27). Furthermore, Section 5.3 gives a highly accurate and easily computed approximation to the ${}_1F_1$ function, so that, unless high accuracy is required, the moments of the doubly noncentral F are easily computed. Unfortunately, using (I.4.47), it is easy to see that (10.32) does not hold for the central moments.

10.2.3 Exact computation

Both (10.25) and (10.26) can be directly evaluated, terminating the infinite sum when the desired degree of accuracy is reached. Consider first the singly noncentral case. Schader and Schmid (1986), Lenth (1987), and Randall (1994) made use of recursions for the incomplete beta ratio to help speed up the 'bottleneck' part of the evaluation of (10.26) (see also Knüsel and Bablok, 1996, and the references therein). The main difficulty nevertheless is that the summands, say c_i, are not necessarily monotonically decreasing, implying that a finite upper limit on i cannot be determined simply by checking for a small enough relative c_i-value. As an illustration, Figure 10.3 plots the c_i for values

$$n_1 = 1, \quad n_2 = 12, \quad \theta_1 = 2316, \quad x = 990, \tag{10.33}$$

which correspond to an application in Chow and Shao (1990). The summand c_i reaches its maximum at $i = 1149 \approx \theta_1/2$. A somewhat obvious strategy then suggests itself: start the sum in (10.26) at $i = \theta_1/2$ and sum in both directions (increasing and decreasing i) until the desired degree of accuracy is obtained. Particularly for large θ_1, this 'summing outwards' is both time-saving and numerically more stable.

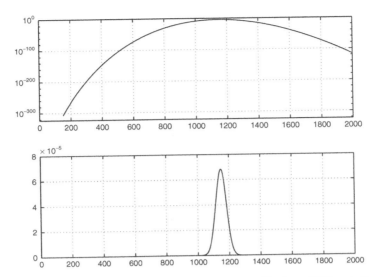

Figure 10.3 The summands c_i in (10.26) for $n_1 = 1$, $n_2 = 12$, $\theta_1 = 2316$ and $f = 990$ as in (10.33). The top panel uses $\log_{10}$ scale

Similar results also hold for the p.d.f. Denoting the summands in (10.24) by d_i, empirical evidence suggests that

$$\arg\max_i (d_i) \approx \arg\max_i (c_i) \approx \theta_1/2 \qquad (10.34)$$

for $x \geq \mathbb{E}[F]$, with both $\arg\max_i (d_i)$ and $\arg\max_i (c_i)$ decreasing as x moves to the left of $\mathbb{E}[F]$. To increase numerical precision, d_i should be evaluated as $\exp\{\ln d_i\}$.

Remark: The calculation of the singly noncentral F c.d.f. provides a good illustration of the errors and oversights in popular statistical software.

The relevance of the nonmonotonic behaviour of summands d_i and c_i is seen in the implementation in function ncfcdf of Matlab, version 5. Summing is stopped when $c_i/(C_i + \epsilon^{1/4}) < \epsilon^{1/2}$, where $\epsilon = 2.2 \times 10^{-16}$ represents machine tolerance and $C_i = \sum_{j=0}^i c_j$. For the example in (10.33), the cutoff value for i would be set to zero and the algorithm fails. (It now works correctly in versions 6 and 7.)

Knüsel (1995) reported erroneous results in Gauss-386 (version 3.2.6) using parameters $x = 100$, $n_1 = 10$, $n_2 = 1$, and $\theta_1 = 38$. The true c.d.f. is 0.828266; the s.p.a., detailed below, returns 0.825842.

McCullough (2000) noted that, after 2.5 minutes of computing, Mathematica version 4 failed with an error message for $x = 1$, $n_1 = n_2 = 1000$, and $\theta_1 = 200$. The true c.d.f. is 0.001860337908; the s.p.a. returns 0.001860337910.

With these errors in mind, one can only speculate what could go wrong with the calculations in the doubly noncentral F case. This emphasizes that one should never put complete trust in even well-established computational and statistical software. ∎

For the doubly noncentral case, a similar result to (10.34) applies, so that each sum in (10.23) and (10.25) may be evaluated by summing outwards, i.e., j would start at $\theta_2/2$ and work outwards and, for each j, i would work outwards from $\theta_1/2$.

This method of 'summing outwards' works well in terms of accuracy and stability and can readily be programmed in 'low-level' languages which are compiled, such as C++ or Pascal. However, in an interpreted language such as Matlab or S-Plus, it will be very slow, because it does not take advantage of the pre-compiled vectorized routines native to these languages. As such, it is much faster in Matlab to compute the p.d.f. and c.d.f. in the singly noncentral case by summing all terms from zero to, say, $3 \lceil \theta_1/2 \rceil$ (which should catch most of the 'mass' in the summands) and then, if the last term is still larger than some tolerance (say, 10^{-14}), add the terms up to, say, $10 \lceil \theta_1/2 \rceil$. This is implemented in Listing 10.7 for the p.d.f. This approach is somewhat rudimentary and could easily be improved.[8]

Evaluation in the doubly noncentral case just involves repeated evaluation of the singly noncentral case. This is done in the program given in Listing 10.8, though in a simple way. (A more sophisticated implementation would generalize the vectorized elements to matrices and avoid the FOR loop over k.)

```
function pdf = ncf1pdf(xvec,nu1,nu2,theta1,k)
if nargin<5, k=0; end
if (theta1==0), pdf=fpdf(xvec,nu1,nu2); return, end
xlen=length(xvec); pdf=zeros(xlen,1);
n12=nu1/2; n22=nu2/2; n1n22=(nu1+nu2)/2; rndtheta=round(theta1+1);
for xloop=1:xlen,  x=xvec(xloop);
   tmp=nu1*x/nu2; tmp1=tmp+1; ltmp=log(tmp); ltmp1=log(tmp1);
   up=3*rndtheta; c=0:up;   % This upper limit is often adequate...
   ppln=ppdfln(c,theta1/2);
   bbln=gammaln(n12+c)+gammaln(n22+k)-gammaln(n1n22+c+k);
   dln =  ppln - bbln + (n12+c-1)*ltmp - (n1n22+c+k)*ltmp1;
   pdf(xloop) = sum(exp(dln));
   if exp(dln(end))>1e-14 % ...but if NOT enough, use a large value.
      c=(up+1):(10*rndtheta);  ppln=ppdfln(c,theta1/2);
      bbln=gammaln(n12+c)+gammaln(n22)-gammaln(n12+c+n22);
      dln =  ppln - bbln + (n12+c-1)*ltmp - (n1n22+c)*ltmp1;
      extra = sum(exp(dln)); pdf(xloop) = pdf(xloop) + extra;
   end
end
pdf=pdf*nu1/nu2;

function y=ppdfln(x,lambda)
y = -lambda + x .* log(lambda) - gammaln(x + 1);
```

Program Listing 10.7　Computes the exact p.d.f. of the singly noncentral F distribution. Argument k should be ignored; it is used by program `ncfpdf` in Listing 10.8

---

[8] Moreover, without knowledge of the speed of decay of the terms in the sum, so that an upper bound on the sum of the infinite number of neglected terms could be computed, it is not apparent when summing should stop. It *appears* that the algorithm employed here is adequate, but this should only be seen as a rule of thumb. To do things correctly, the skills of numerical analysis should be invoked.

```
function pdf = ncfpdf(xvec,nu1,nu2,theta1,theta2)
if nargin<5, theta2=0; end
xlen=length(xvec); pdf=zeros(xlen,1);
if (theta2==0)
  pdf = ncf1pdf(xvec,nu1,nu2,theta1); return
end
rndtheta2=round(theta2+1);
for xloop=1:xlen
  x=xvec(xloop);
  for k=0:rndtheta2
    poiswgt=exp(ppdfln(k,theta2/2));
    single=ncf1pdf(x,nu1,nu2,theta1,k);
    val=poiswgt*single; pdf(xloop)=pdf(xloop)+val;
  end
  while val>1e-14
    k=k+1;
    poiswgt=exp(ppdfln(k,theta2/2));
    single=ncf1pdf(x,nu1,nu2,theta1,k);
    val=poiswgt*single; pdf(xloop)=pdf(xloop)+val;
  end
end

function y=ppdfln(x,lambda)
y = -lambda + x .* log(lambda) - gammaln(x + 1);
```

Program Listing 10.8 Computes the exact p.d.f. of the doubly noncentral F distribution. Uses program `ncf1pdf` given in Listing 10.7

The c.d.f. of $F = (X_1/n_1)/(X_2/n_2)$ can also be computed by writing

$$\Pr(F \le x) = \Pr\left(\frac{n_2}{n_1}X_1 - xX_2 \le 0\right) = \Pr(Y_x \le 0), \qquad (10.35)$$

where Y_x is the so-defined linear combination of $k = 2$ independent noncentral χ^2 r.v.s, and then using the inversion method for weighted sums, given in (10.22). This has the advantage that the computation time involved is not a function of the noncentrality terms as it is in the outward summing method, and is thus of particular use in the doubly noncentral case with large noncentrality parameters.[9]

10.2.4 Approximate computation methods

A popular and straightforward method based on matching moments has been given by Tiku (1972) and is discussed first. The use of the saddlepoint approximation in this context proves to be far more accurate and is developed subsequently.

---

[9] Based on this computation of the c.d.f., the p.d.f. could be approximated by numerical differentiation, though this can be problematic and is not recommended.

10.2.4.1   Matching the first three moments

Tiku (1972) proposed approximating the distribution of $X \sim F(n_1, n_2, \theta_1, \theta_2)$ with that of a location- and scale-shifted central F distribution, i.e., with $F_{k_1, k_2} \sim F(k_1, k_2, 0, 0)$,

$$\Pr(X < x) \approx \Pr\left(F_{\hat{v}_1, n_2} < (x + \hat{a})/\hat{h}\right), \tag{10.36}$$

by matching the mean, variance, and third central moment of X and $Y := h F_{v_1, n_2} - a$. From (10.29)–(10.31), the first three moments of Y are given by

$$\mu_Y = \mathbb{E}[Y] = h\frac{n_2}{(n_2 - 2)} - a, \qquad \mathbb{V}(Y) = h^2 \frac{2n_2^2(v_1 + n_2 - 2)}{v_1(n_2 - 2)^2(n_2 - 4)},$$

$$\mathbb{E}\left[(Y - \mu_Y)^3\right] = h^3 \frac{8n_2^3(v_1 + n_2 - 2)(2v_1 + n_2 - 2)}{v_1^2(n_2 - 2)^3(n_2 - 4)(n_2 - 6)},$$

provided $n_2 > 6$. Denote the rth central moment of X as $_2\mu_r$ and define $\beta_1 = {_2\mu_3^2}/{_2\mu_2^3}$ and $\mu_X = \mathbb{E}[X]$. Then, with the convenient help of Maple, we obtain[10]

$$\hat{a} = \frac{\hat{h}n_2}{n_2 - 2} - \mu_X, \qquad \hat{h} = \frac{\hat{v}_1(n_2 - 2)(n_2 - 6)\,{_2\mu_3}}{4n_2(2\hat{v}_1 + n_2 - 2)\,{_2\mu_2}},$$

and

$$\hat{v}_1 = \frac{n_2 - 2}{2}\left(\left(1 - \frac{32(n_2 - 4)}{(n_2 - 6)^2 \beta_1}\right)^{-1/2} - 1\right).$$

To make this operational without having to compute the $_1F_1$ function, Tiku (1972) used the approximation (5.32), i.e.,

$$_1F_1\left(r, \frac{n_2}{2}, \frac{-\theta_2}{2}\right) \approx \left(1 + \frac{\theta_2}{n_2}\right)^{-r}, \tag{10.37}$$

which increases in accuracy as $n_2 \to \infty$, so that, recalling (10.32),

$$_2\mu_r \approx {_1\mu_r} \cdot \left(1 + \frac{\theta_2}{n_2}\right)^{-r}. \tag{10.38}$$

Notice that (10.38) contains two approximations – one is (10.37) and the other is the fact that (10.32) does not hold for the central moments. Naturally, the accuracy of the method will increase if exact values for $_2\mu_r$ are used instead (computed via (10.32) and (I.4.47)), which, however, somewhat defeats the object of computational simplicity. The reader may verify, however, that the increase turns out to be small, except in cases when $\theta_2/n_2 > 1/2$, in which case the method is still not particularly accurate, most notably so as n_2 approaches 6.

---

[10] Although his calculations are correct, the formula for $\hat{v}_1$ in Tiku (1972, eq. 8) contains a misprint. Also, the expressions for $\hat{v}_1$ and $\hat{h}$ given in Johnson, Kotz and Balakrishnan (1994, p. 501) both contain misprints, and the M function in equation (30.52)' on the same page refers to the $_1F_1$ function, and not to $_2F_0$.

Tiku and Yip (1978) extended the method to the use of four moments, which does increase the accuracy in some, but not all cases. A detailed study of the accuracy of these moment-based methods is given in Butler and Paolella (1999).

10.2.4.2 Saddlepoint approximation to the c.d.f.

By (approximately) inverting the moment generating function, the s.p.a. takes *all* the moments into account, not just the first several. As such, we expect it to perform better than the previous method. This is indeed true, by several orders of magnitude in fact. Also, the previous method using three (four) moments is only valid for $n_2 > 6$ ($n_2 > 8$), while the s.p.a. can be used for any $n_2 > 0$.

As throughout this section, let $X_i \overset{\text{ind}}{\sim} \chi^2(n_i, \theta_i)$, so that $F = (X_1/n_1)/(X_2/n_2) \sim F(n_1, n_2, \theta_1, \theta_2)$. Recall from (10.35) that the c.d.f. of F can be expressed as $\Pr(F \leq x) = \Pr(Y_x \leq 0)$, where Y_x is defined in (10.35). The m.g.f. of Y_x is given by the tractable expression

$$\mathbb{M}_{Y_x}(s) = \left(1 - 2s\frac{n_2}{n_1}\right)^{-n_1/2} (1 + 2sx)^{-n_2/2} \exp\left(\frac{s\theta_1 \frac{n_2}{n_1}}{1 - 2s\frac{n_2}{n_1}} - \frac{s\theta_2 x}{1 + 2sx}\right) \tag{10.39}$$

from (2.1), so that the s.p.a. of Y_x is readily calculated. As $\Pr(F \leq x) = \Pr(Y_x \leq 0)$, the saddlepoint equation is $\mathbb{K}'_{Y_x}(s) = 0$, and

$$\mathbb{K}'_{Y_x}(s) = \frac{n_2}{n_1}\frac{1}{1 - 2s\frac{n_2}{n_1}}\left(n_1 + \frac{\theta_1}{1 - 2s\frac{n_2}{n_1}}\right) - \frac{x}{1 + 2sx}\left(n_2 + \frac{\theta_2}{1 + 2sx}\right). \tag{10.40}$$

By defining $\ell_1 = n_2/n_1$, $\ell_2 = -x$ and $\vartheta_i = \vartheta_i(s) = (1 - 2s\ell_i)^{-1}$, the dth-order derivative of the c.g.f. of Y_x can be conveniently expressed as

$$\mathbb{K}_{Y_x}^{(d)}(s) = k_d \sum_{i=1}^{2} \ell_i^d \vartheta_i^d (n_i + d\theta_i \vartheta_i), \quad d \geq 1, \tag{10.41}$$

where $k_1 = 1$ and, for $d \geq 2$, $k_d = 2(d-1)k_{d-1}$, i.e.,

$$k_d = (d-1)! \, 2^{d-1}, \quad d \in \mathbb{N}.$$

Note that $\mathbb{M}_{Y_x}(s)$ is convergent on the neighbourhood of zero given by $-(2x)^{-1} < s < n_1/(2n_2)$, which does not depend upon the noncentrality parameters. The saddlepoint $\hat{s} = 0$ occurs for that value of x such that $\mathbb{K}'_{Y_x}(0) = 0$, easily seen from (10.40) to be

$$x = \frac{1 + \theta_1/n_1}{1 + \theta_2/n_2} \neq \mathbb{E}[F].$$

For this value of x (and, for numerical reasons, a small neighbourhood around it), the limiting approximation (5.8) could be used, or, more accurately, linear interpolation based on adjacent points $x - \epsilon$ and $x + \epsilon$, where ϵ is a small positive value such that the s.p.a. still delivers numerically stable results.

Conveniently, even for the doubly noncentral case, an explicit form of $\hat{s}$ is available. Using the notation of Abramowitz and Stegun (1972, p. 17), the solution to $\mathbb{K}'_{Y_x}(s) = 0$ can be expressed as the root of the cubic $s^3 + a_2 s^2 + a_1 s + a_0$, where

$$a \cdot a_2 = 8x\,(1-x)\,n_1 n_2^2 + 4x\left(n_2^3 + \theta_2 n_2^2 - n_1^2 n_2 x - n_1 n_2 \theta_1 x\right),$$

$$a \cdot a_1 = 2\left(n_2^3 n_1 + n_1^2 n_2 x^2\right) - 4x n_1 n_2\,(n_1 + n_2 + \theta_1 + \theta_2),$$

$$a \cdot a_0 = x\theta_2 n_1^2 - (1-x)\,n_1^2 n_2 - n_1 n_2 \theta_1,$$

$$a = 8x^2 n_2^2\,(n_1 + n_2).$$

Upon further defining

$$q = \frac{1}{3}a_1 - \frac{1}{9}a_2^2, \quad r = \frac{1}{6}(a_1 a_2 - 3a_0) - \frac{1}{27}a_2^3, \quad m = q^3 + r^2, \tag{10.42}$$

and the two values $s_{1,2} = \sqrt[3]{r \pm m^{1/2}}$, the three roots to the cubic equation are

$$z_1 = (s_1 + s_2) - \frac{a_2}{3}, \quad z_{2,3} = -\frac{1}{2}(s_1 + s_2) - \frac{a_2}{3} \pm \frac{i\sqrt{3}}{2}(s_1 - s_2). \tag{10.43}$$

It turns out that, for any $x, n_1, n_2, \theta_1, \theta_2 \in \mathbb{R}_{>0}$, the three roots are real and ordered according to $z_2 < z_3 \le z_1$. Furthermore, the saddlepoint solution is always

$$\hat{s} = z_3 = -\frac{1}{2}(s_1 + s_2) - \frac{a_2}{3} - \frac{i\sqrt{3}}{2}(s_1 - s_2). \tag{10.44}$$

z_3 is the only root in the convergence strip of $\mathbb{M}_{Y_x}(s)$, $-1/(2x) < s < n_1/(2n_2)$, and the only root for which $\mathbb{K}''_{Y_x}(z_i) > 0$. The proof of this result is given in Butler and Paolella (2002) and is repeated in Section 10.5 below. An alternative expression for the saddlepoint that is useful with software not supporting complex arithmetic is

$$\hat{s} = z_3 = \sqrt{-q}\left\{\sqrt{3}\sin(\phi) - \cos(\phi)\right\} - \frac{a_2}{3}, \tag{10.45}$$

where

$$3\phi = \arg\left(r + i\sqrt{-m}\right) = \begin{cases} \tan^{-1}\left(\sqrt{-m}/r\right), & \text{if } r \ge 0, \\ \pi + \tan^{-1}\left(\sqrt{-m}/r\right), & \text{if } r < 0, \end{cases}$$

and m and q are always negative. For the singly noncentral case in which $\theta_2 = 0$, this simplifies to

$$\hat{s} = \frac{x n_1\,(n_1 + 2n_2 + \theta_1) - n_1 n_2 - \sqrt{n_1 y}}{4 n_2 x\,(n_1 + n_2)}, \tag{10.46}$$

where

$$y = x^2 n_1^3 + 2x^2 n_1^2 \theta_1 + 2n_1^2 x n_2 + 4x^2 n_1 n_2 \theta_1 + n_1 \theta_1^2 x^2 + 2n_1 \theta_1 x n_2 + n_2^2 n_1 + 4x n_2^2 \theta_1.$$

Finally, for the central F case with $\theta_1 = 0 = \theta_2$, we obtain

$$\hat{s} = \frac{n_1 (x - 1)}{2x (n_1 + n_2)}, \tag{10.47}$$

which was also noted by Marsh (1998).

This is straightforward to implement in Matlab, and is very similar to the program shown below which computes the s.p.a. to the p.d.f.

10.2.4.3 Saddlepoint approximation to the density

The s.p.a. to the density of $F \sim \mathrm{F}(n_1, n_2, \theta_1, \theta_2)$ is not as straightforward as the c.d.f. because the 'trick' used via Y_x is not available. Using (1.63) and (1.64), it can be shown that

$$\hat{f}_F (x) = \frac{(n_2 + \theta_2) M (\hat{s})}{\sqrt{2\pi (\ln M^*)''(\hat{s})}}, \tag{10.48}$$

where

$$M (s) = \left(1 - 2s\frac{n_2}{n_1}\right)^{-n_1/2} (1 + 2sx)^{-n_2/2} \exp\left(\frac{s\theta_1 \frac{n_2}{n_1}}{1 - 2s\frac{n_2}{n_1}} - \frac{s\theta_2 x}{1 + 2sx}\right) \tag{10.49}$$

$$\times (n_2 + \theta_2)^{-1} \left\{\frac{\theta_2}{(1 + 2sx)^2} + \frac{n_2}{1 + 2sx}\right\}, \tag{10.50}$$

function M^* is M in (10.49), ignoring the term in (10.50), and the density saddlepoint $\hat{s}$ is the same as the c.d.f. saddlepoint given above; see Marsh (1998) and Butler and Paolella (2002) for further details. Listing 10.9 implements (10.48), and also the extension to the higher-order s.p.a. based on (5.10).

It is noteworthy that the approximate density for both the singly and doubly noncentral cases is a closed-form expression which is trivial to evaluate on a computer, in stark comparison to the exact expressions, particularly in the doubly noncentral case. Furthermore, in the central case with $\theta_1 = \theta_2 = 0$ and using the saddlepoint in (10.47), (10.48) reduces to

$$\hat{f}_F (x) = \left(\frac{n_2}{n_1}\right)^{n_2/2} \frac{1}{\hat{B} (n_1/2, n_2/2)} \frac{x^{n_1/2-1}}{(1 + (n_1/n_2)x)^{(n_1+n_2)/2}}, \tag{10.51}$$

where $\hat{B} (n_1/2, n_2/2)$ is Stirling's approximation to the beta function. Thus, the normalized density s.p.a. is exact for the central F distribution. From this result, one might expect the s.p.a. to exhibit high accuracy in the noncentral cases as well. This is indeed the case. As an example, let $F \sim \mathrm{F}(1, 12, 2316)$, which are the parameters in (10.33). Figure 10.4 shows the normalized density $\hat{f}_F$ (with constant of integration 1.00161574, so that normalization is hardly necessary), which is graphically indistinguishable from the exact density f_F. Superimposed on the graph is the RPE, which is seen to be between 0.0005 and -0.003 for x-values for which $f_F (x) > 10^{-10}$.

```
function f=SPncfpdf(xords,n1,n2,theta1,theta2,acclevel)
if nargin<6, acclevel=2; end
if nargin<5, theta2=0; end
xlen=length(xords); f=zeros(1,xlen);
for xloop=1:xlen
  x=xords(xloop);
  if x<1e-30, f(xloop)=0;
  else
    l1= n2/n1; l2=-x;
    if (theta2==0)
      x2=x^2; a=n1; a2=n1^2; a3=a*a2; b=n2; b2=n2^2;
      t=theta1; t2=t^2;
      y=x2 * a3 + 2*x2*a2*t + 2*a2*x*b + 4*x2*a*b*t ...
        + a*t2*x2 + 2*a*t*x*b + b2*a + 4*x*b2*t;
      num=x*a*(a+2*b+t) - a*b - sqrt(a*y);
      den=4*b*x*(a+b);
      s=num/den; roots=s;
    else
      x2=x^2; n12=n1^2; n22=n2^2; n13=n1^3; n23=n2^3;
      t1=theta1; t2=theta2;
      a=8*(n23*x2+n22*n1*x2);
      b=4*(-2*n22*n1*x2 + x*n23 + 2*x*n1*n22 ...
        + x*t2*n22 - n12*n2*x2 - n1*n2*t1*x2);
      c=2*(n22*n1 + n12*n2*x2) - 4*(x*n12*n2 ...
        + n1*n2*t1*x + x*t2*n1*n2 + x*n1*n22);
      d=-n12*n2 + x*n12*n2 + x*t2*n12 - n1*n2*t1;

      % Abramowitz and Stegun notation
      a2=b/a; a1=c/a; a0=d/a;
      q=a1/3 -a2^2/9; r=(a1*a2-3*a0)/6 - a2^3/27;
      m=q^3 + r^2;
      if m >= 0, error ('this should not happen!'), end
      s1=(r+sqrt(m))^(1/3); s2=(r-sqrt(m))^(1/3);
      sps=s1+s2; sms=s1-s2; z3=-sps/2 - a2/3 - sqrt(-3)*sms/2;
      roots=z3;
    end
```

Program Listing 10.9 Computes the s.p.a. of the p.d.f. of the doubly noncentral F distribution. Continued in Listing 10.10

```
    s=roots; v1 = 1/(1-2*s*l1); v2 = 1/(1-2*s*l2);
    K=0.5*(n1*log(v1) + n2*log(v2)) + s*(l1*theta1*v1 + l2*theta2*v2);
    kpp=2*(l1^2*v1^2*(n1+2*theta1*v1) + l2^2*v2^2*(n2+2*theta2*v2) );
    M=exp(K) * ( theta2/(1+2*s*x)^2 + n2/(1+2*s*x) );
    f(xloop)=M/sqrt(2*pi*kpp);
    if acclevel==2
      K3= 8*(l1^3*v1^3*(n1+3*theta1*v1) + l2^3*v2^3*(n2+3*theta2*v2) );
      K4=48*(l1^4*v1^4*(n1+4*theta1*v1) + l2^4*v2^4*(n2+4*theta2*v2) );
      kap3= K3/(kpp)^(3/2); kap4=K4/(kpp)^(4/2);
      O1= 1 + kap4/8 - 5*kap3^2/24;
    else, O1=1;
    end
    f(xloop)=f(xloop)*O1;
  end
end
```

Program Listing 10.10 Continued from Listing 10.9

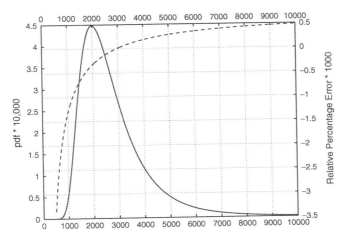

Figure 10.4 Density function (solid) and 1000 times the RPE of the saddlepoint approximation (dashed) for the singly noncentral F with parameters $n_1 = 1$, $n_2 = 12$, $\theta_1 = 2316$, and $f = 990$ as in (10.33), where RPE $= 100\,(\text{s.p.a.} - \text{exact}) / \min(\text{exact}, 1 - \text{exact})$

10.3 Noncentral beta

Problem I.7.20 showed that, if $X \sim F(n_1, n_2)$, then

$$B = \frac{nX}{1 + nX} \sim \text{Beta}(n_1/2, n_2/2),$$

where $n = n_1/n_2$. This relationship can be used as the basis for the definition of the noncentral beta distribution: if $X \sim F(2n_1, 2n_2, \theta_1, \theta_2)$, then

$$B = \frac{nX}{1 + nX}, \qquad n = n_1/n_2,$$

is said to follow a *doubly noncentral beta distribution*, Beta$(n_1, n_2, \theta_1, \theta_2)$. Going the other way, if $B \sim \text{Beta}(n_1, n_2, \theta_1, \theta_2)$, then

$$X = \frac{B}{n(1 - B)} \sim F(2n_1, 2n_2, \theta_1, \theta_2), \qquad n = n_1/n_2.$$

With

$$x = \frac{b}{n(1 - b)}, \qquad \frac{dx}{db} = \frac{1}{n(1 - b)^2}, \tag{10.52}$$

it follows from the univariate transformation formula (I.7.65) and the density of the doubly noncentral F distribution (10.23), after simplifying, that

$$f_B(b) = f_B(b; n_1, n_2, \theta_1, \theta_2) = f_X(x) \frac{dx}{db} \tag{10.53}$$

$$= \sum_{i=0}^{\infty} \sum_{j=0}^{\infty} \frac{\omega_{i,\theta_1} \omega_{j,\theta_2}}{B(i+n_1, j+n/2)} b^{n_1+i-1} (1-b)^{n_2+j-1} \mathbb{I}_{(0,1)}(b)$$

$$= \sum_{i=0}^{\infty} \sum_{j=0}^{\infty} \omega_{i,\theta_1} \omega_{j,\theta_2} f_{\text{Beta}}(b; n_1+i, n_2+j),$$

where $\omega_{i,\theta} = e^{-\theta/2} (\theta/2)^i / i!$ and $f_{\text{Beta}}(b; p, q)$ denotes the central beta density with parameters p and q. Via relation (10.53), computation of f_B can be done via that of f_X, using either exact methods or with the s.p.a. Similarly, for the c.d.f.,

$$F_B(b; n_1, n_2, \theta_1, \theta_2) = \Pr\left(\frac{nX}{1+nX} < b\right)$$

$$= \Pr\left(X < \frac{b}{n(1-b)}\right) = F_X\left(\frac{b}{n(1-b)}; 2n_1, 2n_2, \theta_1, \theta_2\right).$$

In particular, if `SPncfpdf` and `SPncf` denote the programs which compute the s.p.a. of the doubly noncentral F density and distribution, respectively, then the program in Listing 10.11 will compute the p.d.f. and c.d.f. of the noncentral beta.

```
function [pdf,cdf] = ncbeta(bvec,n1,n2,theta1,theta2)
n=n1/n2; xvec=( bvec./(1-bvec) ) / n; dxdb = (1-bvec).^(-2) / n;
pdf = SPncfpdf(xvec,2*n1,2*n2,theta1,theta2) .* dxdb;
if nargout>1
  cdf = SPncf(xvec,2*n1,2*n2,theta1,theta2);
end
```

Program Listing 10.11 Computes the s.p.a. to the p.d.f. and, optionally, the c.d.f. of the non-central beta distribution using programs for the s.p.a. of the noncentral F distribution

10.4 Singly and doubly noncentral *t*

Similar to the noncentral F, the noncentral *t* distribution arises in power calculations for statistical hypothesis testing and also in construction of certain confidence intervals. It has also found use in a large variety of other contexts; see the examples and references in Johnson, Kotz and Balakrishnan (1994, p. 512).

10.4.1 Derivation

Let $X \sim N(\mu, 1)$ independent of $Y \sim \chi^2(k, \theta)$. The random variable $T = X/\sqrt{Y/k}$ is said to follow a *doubly noncentral t distribution* with k degrees of freedom, numerator noncentrality parameter μ and denominator noncentrality parameter θ. We write $T \sim t''(k, \mu, \theta)$. If $\theta = 0$, then T is singly noncentral t with noncentrality parameter μ, and we write $T \sim t'(k, \mu)$.

10.4.1.1 Singly noncentral t

Consider first the singly noncentral t distribution. From Problem I.7.5, the density of $Z = \sqrt{Y/k}$ is given by

$$f_Z(z; k) = \frac{2^{-k/2+1} k^{k/2}}{\Gamma(k/2)} z^{k-1} e^{-\left(kz^2\right)/2} \mathbb{I}_{(0,\infty)}(z).$$

From (I.8.41) with X and Z independent, the c.d.f. of $T = X/\sqrt{Y/k} = X/Z$ is therefore

$$
\begin{aligned}
F_T(t; k, \mu) &= \Pr(X \leq tZ) \\
&= \int_{-\infty}^{\infty} F_{X|Z}(tz) f_Z(z) \, \mathrm{d}z \\
&= \int_0^{\infty} F_X(tz) f_Z(z) \, \mathrm{d}z \\
&= \frac{2^{-k/2+1} k^{k/2}}{\Gamma(k/2)} \int_0^{\infty} \Phi(tz; \mu, 1) z^{k-1} \exp\left\{-\frac{1}{2} kz^2\right\} \mathrm{d}z, \quad (10.54)
\end{aligned}
$$

where

$$\Phi(tz; \mu, 1) = (2\pi)^{-1/2} \int_{-\infty}^{tz} \exp\left\{-\frac{1}{2}(x - \mu)^2\right\} \mathrm{d}x = \Phi(tz - \mu; 0, 1) \equiv \Phi(tz - \mu).$$

As the standard normal c.d.f. $\Phi(\cdot)$ is a pre-programmed function (in Matlab, S-Plus, R, etc.) which is evaluated quickly and to machine precision, the integral in (10.54) can be computed numerically to a high degree of accuracy. The program in Listing 10.12 can be used to compute (10.54). However, Matlab's implementation in its function nctcdf appears not to suffer from faulty convergence criteria (as does its p.d.f. implementation; see footnote 11 below) and is both more accurate than our program in Listing 10.12 (because it uses the infinite-sum representation) and faster, and so is preferred.

```
function cdf = tcdfsing(tvec,k,mu)
cdf=zeros(length(tvec),1);
for tloop=1:length(tvec)
  t=tvec(tloop);
  intsum=0; lo=0; up=1; done=0;
  while ~done
    s = quadl(@ff,lo,up,1e-10,0,t,k,mu); intsum=intsum+s;
    if (s<1e-12) & (ff(up,t,k,mu) <= ff(up,t,k,mu)), done=1; end
    % this piece of the integrand is extremely small and decreasing
    lo=up; up=2*up;
  end
  cdf(tloop) = intsum * 2^(1-k/2) * k^(k/2) / gamma(k/2);
end

function s=ff(z,t,k,mu)
s=normcdf(t*z,mu,1) .* z.^(k-1) .* exp(-0.5 * k * z.^2);
```

Program Listing 10.12 Computes the singly noncentral t c.d.f. (10.54). The practical upper limit on the integral is a function of the parameters, so a `while` loop is used to sum over disjoint parts of it (which double in size, starting with (0, 1)) until convergence (when the integrand is decreasing and close enough to zero). Alternatively, the transformation $t = (1+z)^{-1}$ could be used to map the integrand to (0, 1)

Differentiating (10.54) using Leibniz' rule (see Section I.A.3.4.3) yields an expression for the p.d.f. as

$$f_T(t; k, \mu) = \frac{d}{dt} F_T(t)$$

$$= K \int_0^\infty \frac{d}{dt} \int_{-\infty}^{tz} \exp\left\{-\frac{1}{2}(x-\mu)^2\right\} z^{k-1} \exp\left\{-\frac{1}{2}kz^2\right\} dx \, dz$$

$$= K \int_0^\infty z^k \exp\left\{-\frac{1}{2}\left[(tz-\mu)^2 + kz^2\right]\right\} dz, \quad K = \frac{2^{-k/2+1}k^{k/2}}{\Gamma(k/2)\sqrt{2\pi}},$$

$$\tag{10.55}$$

which can also be readily evaluated using numerical integration. Note that, in the central case with $\mu = 0$, this integral expression reduces to precisely that given in (2.17), which can be further simplified. With $\mu \neq 0$, matters are more complicated. We can, however, write

$$\exp\left\{-\frac{1}{2}\left[(tz-\mu)^2 + kz^2\right]\right\} = \exp\left\{-\frac{1}{2}\left[(tz)^2 + kz^2\right]\right\} \exp\left\{-\frac{1}{2}\left[\mu^2 - 2tz\mu\right]\right\}$$

$$= \exp\left\{-\frac{1}{2}\left[(tz)^2 + kz^2\right]\right\} \exp\left\{-\mu^2/2\right\} \exp(tz\mu)$$

$$= \exp\left\{-\frac{1}{2}\left[(tz)^2 + kz^2\right]\right\} \exp\left\{-\mu^2/2\right\} \left(\sum_{i=0}^\infty \frac{(zt\mu)^i}{i!}\right), \quad \tag{10.56}$$

so that

$$f_T(t; \mu, k) = K e^{-\mu^2/2} \sum_{i=0}^{\infty} \frac{(t\mu)^i}{i!} \int_0^{\infty} z^{k+i} \exp\left\{-\frac{1}{2}\left[(t^2 + k) z^2\right]\right\} dz. \qquad (10.57)$$

Standard manipulations (Problem 10.10) show that

$$f_T(t; k, \mu) = e^{-\mu^2/2} \frac{\Gamma((k+1)/2) k^{k/2}}{\sqrt{\pi} \Gamma(k/2)} \left(\frac{1}{k + t^2}\right)^{\frac{k+1}{2}}$$
$$\times \left(\sum_{i=0}^{\infty} \frac{(t\mu)^i}{i!} \left(\frac{2}{t^2 + k}\right)^{i/2} \frac{\Gamma((k+i+1)/2)}{\Gamma((k+1)/2)}\right). \qquad (10.58)$$

This expression immediately yields the central t density when $\mu = 0$. It also lends itself well to numerical computation, given that the gamma function can be quickly and accurately evaluated in numerical software packages.[11] As an illustration, Figure 10.5 shows the singly noncentral t density for $\mu = 1$ and $k = 1, 4$ and 40. The $k = 1$ ($k = \infty$) case can be seen as a skewed generalization of the Cauchy (normal) distribution.

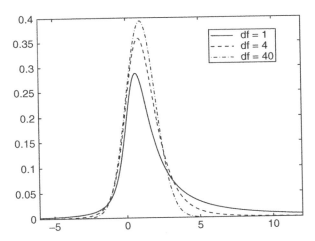

Figure 10.5   The singly noncentral t density for $\mu = 1$ and three different values for the degrees of freedom

The numerical integration associated with evaluation of the c.d.f. expression (10.54) can also be avoided. From (10.58), $\Pr(0 \leq T \leq t)$ for $t > 0$ can be expressed as

---

[11] Expression (10.58) is used, for example, by the program `nctpdf` in Matlab 6, where summing starts from $i = 0$ and is terminated when the absolute value of the ith term divided by the current total is 'small' relative to machine precision. This can fail, however, when the initial summands (and, hence, the current total) are negative. For example, take $t = -3$, $k = 4$ and $\mu = 1$, which results in a negative p.d.f. value. The patch is easy: replace the convergence test `if all(abs(newterm(:))./infsum(:) < eps^(3/4))` with `if all(abs(newterm(:)./infsum(:)) < eps^(3/4))`.

$$Pr(0 \leq T \leq t) = e^{-\mu^2/2} \sum_{i=0}^{\infty} \frac{\Gamma((k+i+1)/2)}{\sqrt{\pi}\Gamma(k/2)i!}$$
$$\times k^{k/2}2^{i/2}\mu^i \int_0^t x^i(k+x^2)^{-(k+1+i)/2} dx. \tag{10.59}$$

From this, we obtain (Problem 10.11)

$$Pr(0 \leq T \leq t) = \frac{1}{2}e^{-\mu^2/2} \sum_{i=0}^{\infty} \frac{2^{i/2}\mu^i \Gamma\left(\frac{1}{2}(1+i)\right)}{\sqrt{\pi}i!} \overline{B}_{m(t)}\left(\frac{1+i}{2}, \frac{k}{2}\right) \tag{10.60}$$

$$= \frac{1}{2}e^{-\mu^2/2} \sum_{i=0}^{\infty} \mu^i \frac{(1/2)^{i/2}}{\Gamma(i/2+1)} \overline{B}_{m(t)}\left(\frac{1+i}{2}, \frac{k}{2}\right), \quad t > 0, \tag{10.61}$$

where $\overline{B}$ is the incomplete beta ratio and $m(t) = t^2/(k+t^2)$. Similar to (10.58), this expression lends itself well to numerical computation.

Recall the construction of the r.v. $T \sim t'(k, \mu)$ as $T = X/\sqrt{Y/k}$, where $X \sim N(\mu, 1)$, independent of $Y \sim \chi^2(k, 0)$. Then

$$Pr(T \leq 0) = Pr(X \leq 0) = \Phi(-\mu), \tag{10.62}$$

so that, for $t > 0$, we can compute the c.d.f. using (10.61) and (10.62), i.e.,

$$Pr(T \leq t) = Pr(T \leq 0) + Pr(0 \leq T \leq t), \quad t > 0.$$

For $t < 0$, start with the relationship between the $t'(k, \mu)$ and F distributions

$$Pr(-s \leq T \leq s) = Pr(|T| \leq s) = Pr(T^2 \leq s^2) = F_F(s^2; 1, k, \mu^2, 0), \quad s > 0,$$

so that $Pr(-s \leq T \leq s)$ can be computed using the methods for the singly noncentral F distribution. Thus, for $t < 0$, we can evaluate

$$F_T(t; k, \mu, 0) = Pr(T \leq t)$$
$$= Pr(T \leq 0) + Pr(0 \leq T \leq |t|) - Pr(-|t| \leq T \leq |t|), \quad t < 0.$$

Remark: An interesting result emerges by using the fact that $Pr(T > 0) = 1 - \Phi(-\mu) = \Phi(\mu)$ and taking $t \to \infty$ in (10.61), so that $m(t) \to 1$, $\overline{B}_{m(t)}(\cdot, \cdot) \to 1$, and

$$\Phi(\mu) = \frac{1}{2}e^{-\mu^2/2} \sum_{i=0}^{\infty} \mu^i \frac{(1/2)^{i/2}}{\Gamma(i/2+1)}, \tag{10.63}$$

as noted by Hawkins (1975). As

$$\sum_{i=0}^{\infty} z^{2i} \frac{(1/2)^i}{\Gamma(i+1)} = \sum_{i=0}^{\infty} \frac{(z^2/2)^i}{i!} = \exp\left(\frac{1}{2}z^2\right),$$

separating the even and odd terms in (10.63) and replacing μ with z yields

$$\Phi(z) = \frac{1}{2}e^{-z^2/2}\left(\sum_{i=0}^{\infty} z^{2i}\frac{(1/2)^i}{\Gamma(i+1)} + \sum_{i=0}^{\infty} z^{2i+1}\frac{(1/2)^{(2i+1)/2}}{\Gamma(i+3/2)}\right)$$

$$= \frac{1}{2} + \frac{1}{2}e^{-z^2/2}\sum_{i=0}^{\infty} \frac{\left(z/\sqrt{2}\right)^{(2i+1)}}{\Gamma(i+3/2)}. \tag{10.64}$$

Similar to (I.7.34), besides being of theoretical interest, (10.64) can also be used to numerically evaluate the normal c.d.f., without requiring a routine for the gamma function, recalling its recursive property and the l.h.s. of (10.2). For example, with $z = -1.64485362695147$, the exact value of $\Phi(z)$ is 0.05 to 14 significant digits. With only 6 terms in the sum, (10.64) yields 0.0506; with 11 terms, we obtain 0.05000003; and, to 14 significant digits, 0.05 using 18 terms. The expression is obviously exact for $z = 0$ and, to obtain a given accuracy level, we expect more terms in the sum to be necessary as $|z|$ increases. ∎

10.4.1.2 Doubly noncentral t

We now turn to the doubly noncentral t distribution, in which case $T = X/\sqrt{Y/k}$, where $X \sim N(\mu, 1)$, independent of $Y \sim \chi^2(k, \theta)$, so that $T \sim t''(k, \mu, \theta)$. Using the representations in (10.5) and (10.3), the density of $W = \sqrt{Y}$ is

$$f_W(w) = 2wf_Y\left(w^2\right)$$

$$= e^{-\left(w^2+\theta\right)/2}w^{k/2}\theta^{-(k-2)/4}I_{(k-2)/2}\left(w\sqrt{\theta}\right)\mathbb{I}_{(0,\infty)}(w)$$

$$= e^{-\left(w^2\right)/2}\sum_{i=0}^{\infty}\frac{e^{-\theta/2}(\theta/2)^i}{i!}\frac{w^{k+2i-1}}{2^{k/2+i-1}\Gamma(i+k/2)}\mathbb{I}_{(0,\infty)}(w).$$

Then, similar calculations to those for deriving (10.58) yield

$$f_T(t; k, \mu, \theta) = \frac{e^{-\left(\theta+\mu^2\right)/2}}{\sqrt{\pi k}}\sum_{i=0}^{\infty}\sum_{j=0}^{\infty}\frac{(\theta/2)^i}{i!j!}\frac{\Gamma\left((k+2i+j+1)/2\right)}{\Gamma(i+k/2)}$$

$$\times \left(t\mu\sqrt{2/k}\right)^j\left(1+t^2/k\right)^{-(k+2i+j+1)/2} \tag{10.65}$$

and, numerically more useful,

$$f_T(t; k, \mu, \theta) = \frac{e^{-\left(\theta+\mu^2\right)/2}}{\sqrt{\pi k}}\sum_{j=0}^{\infty}A_j(t), \tag{10.66}$$

where

$$A_j(t) = \frac{1}{j!} \frac{\left(t\mu\sqrt{2/k}\right)^j}{\left(1+t^2/k\right)^{(k+j+1)/2}} \frac{\Gamma\left(\frac{k+j+1}{2}\right)}{\Gamma\left(\frac{k}{2}\right)} {}_1F_1\left(\frac{k+j+1}{2}, \frac{k}{2}; \frac{\theta}{2\left(1+t^2/k\right)}\right)$$

(10.67)

(see Problem 10.12). If $\mu = 0$, then, as $0^0 = 1$ by convention, (10.66) reduces to

$$f_T(t; k, 0, \theta) = \frac{e^{-\theta/2}}{\sqrt{\pi k}} \frac{1}{\left(1+t^2/k\right)^{(k+1)/2}} \frac{\Gamma\left(\frac{k+1}{2}\right)}{\Gamma\left(\frac{k}{2}\right)} {}_1F_1\left(\frac{k+1}{2}, \frac{k}{2}; \frac{\theta}{2\left(1+t^2/k\right)}\right).$$

(10.68)

For $t < 0$, the terms (10.67) oscillate. Figure 10.6 shows these terms (as a function of j) for $t = -2$, $\mu = 3$, $k = 10$, and $\theta = 5$, computed using the approximation to ${}_1F_1$ given in Section 5.3 (see Programs 10.13 and 5.6). Notice that the integral representation of the ${}_1F_1$ function is not valid for any j in (10.66), which otherwise could have be used to evaluate it numerically to a high degree of precision. Thus, the approximation is quite convenient in this case. It is used via Program 10.14 to compute the doubly noncentral t densities illustrated in Figure 10.7.

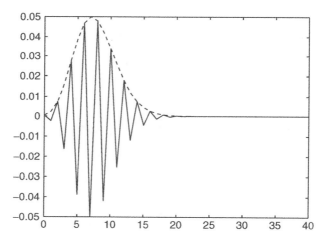

Figure 10.6 Terms in (10.66) for $\mu = 3$, $k = 10$, and $\theta = 5$, calculated from (10.66) using an approximation to ${}_1F_1$. The dashed curve plots the absolute value of the terms

While various expressions for the c.d.f. of the doubly noncentral T, denoted $F_T(t; k, \mu, \theta)$, are available (see the references in Johnson, Kotz and Balakrishnan, 1994, pp. 536–537), we present the one derived in Kocherlakota and Kocherlakota (1991), which is convenient for numerical computation. They show that

$$F_T(t; k, \mu, \theta) = \sum_{i=0}^{\infty} \omega_{i,\theta} F_{t'}\left(t\left(\frac{k+2i}{k}\right)^{1/2}; k+2i, \mu\right),$$

(10.69)

```
function y=doubnoncentterm(j,t,k,mu,theta)
a=t*mu*sqrt(2/k);
if a==0
   if j==0, term1=0; %%% a^j = 0^0 = 1 and log(1)=0
   else, y=0; return, end
else
   term1=j*log(a);
end
kon = (-(theta+mu^2)/2) - log(pi*k)/2; z=(k+j+1)/2; w=1+t^2/k;
term=kon+term1+gammaln(z)-gammaln(j+1)-gammaln(k/2)-z*log(w);
f = real(f11(z,k/2,theta/(2*w)) ); y=real(exp(term + log(f)));
```

Program Listing 10.13 Computes terms (10.67) for calculating (10.66), using the approximation to $_1F_1$ given in Listing 5.6. Roundoff error sometimes induces tiny imaginary components, so just use the `real` component

```
function f=doubnoncent(tvec,k,mu,theta)
f=zeros(length(tvec),1);
for loop=1:length(tvec)
   t=tvec(loop);   kum=0;
   for j=0:300,   kum=kum+doubnoncentterm(j,t,mu,k,theta);   end
   f(loop)=kum;
end
```

Program Listing 10.14 Uses Program 10.13 to compute (10.66). Values for j have been arbitrarily chosen as 0 to 300, which were adequate for the example shown. A smarter algorithm (such as used in Listings 10.12 and 10.15) will be required to compute the density for any parameter values

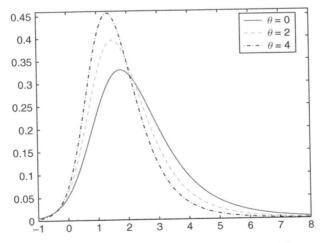

Figure 10.7 Doubly noncentral t density for $\mu = 2$, $k = 5$, and three values of θ

i.e., an infinite Poisson weighted sum of *singly* noncentral t c.d.f. values (see Problem 10.13 for derivation). This is implemented in a program given in Listing 10.15.

```
function cdf = tcdfdoub(tvec,k,mu,theta)
if theta==0; cdf=nctcdf(tvec,k,mu); return, end
cdf=zeros(length(tvec),1);
for tloop=1:length(tvec)
  t=tvec(tloop);
  partsum=0; inc=50; %    calculate the sum in chunks of inc to
  lo=0; hi=inc; done=0; %  capitalize on Matlab's vector computing
  while ~done
    ivec=lo:hi;
    lnw = -theta/2 + ivec*log(theta/2) - gammaln(ivec+1);
    k2i=k+2*ivec;
    logcdfpart = log(nctcdf(t*sqrt(k2i/k),k2i,mu));
    newpartvec = exp(lnw + logcdfpart);
    newpartsum = sum(newpartvec);
    partsum = partsum + newpartsum;
    lo=hi+1; hi=hi+inc;
    if ( newpartvec(1) >= newpartvec(end) ) & ( newpartsum<1e-12 )
      done=1;
    end
  end
  cdf(tloop) = real(partsum); % roundoff error produces imaginary comp
end                           % of the order of 1e-28. Just delete it.
```

Program Listing 10.15 Computes 10.69 using Matlab's program for computing the singly non-central t c.d.f., though the simple program in Listing 10.12 could also be used

10.4.2 Saddlepoint approximation

The s.p.a. for the singly and doubly noncentral t is a direct extension of Example 5.10, which applied the Daniels and Young (1991) marginal s.p.a. to the central Student's t. As in Broda and Paolella (2007), let $X_1 \sim N(\mu, 1)$, independent of $X_2 \sim \chi^2(k, \theta)$. Let $g : \mathbb{R} \times \mathbb{R}_{>0} \to \mathbb{R} \times \mathbb{R}_{>0}$, $(y_1, y_2) \mapsto (y_1 y_2, k y_2^2)$. Then g is bijective and smooth, with a smooth inverse. Define $\mathbf{Y} = (Y_1, Y_2) = g^{-1}(X_1, X_2) = \left(X_1/\sqrt{X_2/k}, \sqrt{X_2/k}\right)'$, so that $\mathbf{X} = (X_1, X_2) = g(\mathbf{Y}) = (Y_1 Y_2, k Y_2^2)$. Then $Y_1 \sim t''(k, \mu, \theta)$.

The joint cumulant generating function of $\mathbf{X} = (X_1, X_2)$ at $\mathbf{t} = (t_1, t_2)$ is, from independence,

$$\mathbb{K}_{\mathbf{X}}(\mathbf{t}) = \mathbb{K}_{X_1}(t_1) + \mathbb{K}_{X_2}(t_2) = t_1 \mu + \frac{1}{2} t_1^2 - \frac{k}{2} \log(1 - 2t_2) + \frac{t_2 \theta}{1 - 2t_2},$$

where $\mathbb{K}_{X_1}$ and $\mathbb{K}_{X_2}$ are the c.g.f.s of X_1 and X_2, respectively. The saddlepoint $(\hat{\mathbf{t}}, \hat{y}_2) = (\hat{t}_1, \hat{t}_2, \hat{y}_2)$, $\hat{t}_2 < 1/2$, $\hat{y}_2 > 0$, solves the system of equations

$$\mu + t_1 = y_1 y_2$$

$$\frac{k}{1 - 2t_2} + \frac{\theta}{(1 - 2t_2)^2} = k y_2^2$$

$$t_1 y_1 + 2k t_2 y_2 = 0.$$

Straightforward calculation reveals that

$$\hat{t}_1 = -\mu + y_1 \hat{y}_2, \quad \hat{t}_2 = -\frac{y_1 \hat{t}_1}{2k \hat{y}_2}, \tag{10.70}$$

and $\hat{y}_2$ solves the cubic $s(y_2) := a_3 y_2^3 + a_2 y_2^2 + a_1 y_2 + a_0 = 0$, where

$$a_3 = y_1^4 + 2k y_1^2 + k^2, \quad a_2 = -2y_1^3 \mu - 2y_1 k \mu, \quad a_1 = y_1^2 \mu^2 - k y_1^2 - k^2 - \theta k,$$

and $a_0 = y_1 k \mu$. Upon defining

$$c_2 = \frac{a_2}{a_3}, \quad c_1 = \frac{a_1}{a_3}, \quad c_0 = \frac{a_0}{a_3}, \quad q = \frac{1}{3} c_1 - \frac{1}{9} c_2^2,$$

$$r = \frac{1}{6}(c_1 c_2 - 3 c_0) - \frac{1}{27} c_2^3, \quad m = q^3 + r^2, \quad s_{1,2} = (r \pm \sqrt{m})^{1/3},$$

the roots of the cubic are given by

$$z_1 = (s_1 + s_2) - \frac{c_2}{3} \quad \text{and} \quad z_{2,3} = -\frac{1}{2}(s_1 + s_2) - \frac{c_2}{3} \pm \frac{i\sqrt{3}}{2}(s_1 - s_2).$$

The saddlepoint solution is always z_1, as proved in Broda and Paolella (2007). With

$$\nabla_{y_1} g(\mathbf{y}) = (y_2, 0)', \quad \nabla_{y_2} g(\mathbf{y}) = (y_1, 2k y_2)',$$

$$\nabla_{y_2}^2 g(\mathbf{y}) = (0, 2k)', \quad K''(\mathbf{t}) = \text{diag}\left(1, 2k(1 - 2t_2)^{-2} + 4\theta(1 - 2t_2)^{-3}\right),$$

$$\det\left[\partial g / \partial \mathbf{y}\right] = \det\left[\nabla_{y_1} g(\mathbf{y}), \nabla_{y_2} g(\mathbf{y})\right] = 2k y_2^2,$$

and after some simplification, the quantities entering approximations (5.23) and (5.24) take the simple form

$$d = (\hat{t}_1 \hat{y}_2)^{-1}, \quad u = \sqrt{(y_1^2 + 2k \hat{t}_2)(2k v^2 + 4\theta v^3) + 4k^2 \hat{y}_2^2} \Big/ \left(2k \hat{y}_2^2\right),$$

$$w = \sqrt{-\mu \hat{t}_1 - k \log v - 2\theta v \hat{t}_2} \, \text{sgn}\left(y_1 - \mu / \sqrt{1 + \theta / k}\right),$$

where $v = (1 - 2\hat{t}_2)^{-1}$, $\hat{\mathbf{t}} = (\hat{t}_1, \hat{t}_2)'$ is given by (10.70), and

$$\hat{y}_2 = \sqrt{-4q} \cos\left(\cos^{-1}\left(r / \sqrt{-q^3}\right)/3\right) - \frac{c_2}{3}. \tag{10.71}$$

With these quantities, (5.23) and (5.24) can be computed. The program in Listing 10.16 implements the approximation.

```
function [pdf,cdf] = nctspa(x,k,mu,theta,renorm,rec);
% pass renorm=1 to renormalize the SPA to the pdf.
% There is a last argument called rec. DO NOT PASS it!

if nargin<3, mu=0; end;      if nargin<4, theta=0; end;
if nargin<5, renorm=0; end;  if nargin<6, rec=0; end

n=k; a=x; alpha=mu/sqrt((1+theta/n)); normconst=1;
if renorm==1 & rec==0
   term1 = quadl(@nctspa,alpha-1e+10,alpha,[],[],n,mu,theta);
   term2 = quadl(@nctspa,alpha,alpha+1e+10,[],[],n,mu,theta);
   normconst=1/(term1+term2);
end
cdf=zeros(size(a));pdf=cdf;
c3=n^2+2*n.*a.^2+a.^4;c2=(-2*mu*(a.^3+n.*a))./c3;
c1=(-n^2-n*a.^2-n*theta+a.^2*mu^2)./c3;c0=(n*a*mu)./c3;
q=c1/3-(c2.^2)/9;r=1/6*(c1.*c2-3.*c0)-1/27.*c2.^3;
b0=sqrt(-4*q).*cos(acos(r./sqrt(-q.^3))/3)-c2/3;
t1=-mu+a.*b0;t2=-a.*t1./b0/n/2;nu=1./(1-2*t2);
w=sqrt(-mu*t1-n*log(nu)-2*theta*nu.*t2).*sign(a-alpha);
u=sqrt((a.^2+2*n.*t2).*(2*n*nu.^2+4*theta*nu.^3) ...
    +4*n^2*b0.^2)./(2*n*b0.^2);
pdf=normconst*normpdf(w)./u;
if nargout==2
   nz=(abs(t1.*b0)>=1e-10);iz=(abs(t1.*b0)<=1e-10);
   if any(nz)
      d(nz)=1./(t1(nz).*b0(nz));
      cdf(nz)=normcdf(w(nz))+normpdf(w(nz)).*(1./w(nz)-d(nz)./u(nz));
   end
   if any(iz)
      n=sum(iz==1);
      [garb cdfiz]=nctspa([a(iz)-1e-4 a(iz)+1e-4],n,mu,theta,0,rec+1);
      if rec>5
         cdf(iz)=0.5; warning('Too many recursions');
      else
         cdf(iz)=.5*(cdfiz(1:n)+cdfiz(n+1:end));
      end
   end
end
end
```

Program Listing 10.16 Computes the s.p.a. to the doubly noncentral t at value x

Remark: In the singly noncentral case with $\theta = 0$, u and w reduce to

$$u = \sqrt{(\mu y_1 \hat{y}_2 + 2k)/(2k)} \; /\hat{y}_2 \quad \text{and} \quad w = \sqrt{-\mu \hat{t}_1 - 2k \log(\hat{y}_2)} \; \text{sgn}(y_1 - \mu),$$

where d, $\hat{t}_1$ and $\hat{t}_2$ are as before, and

$$\hat{y}_2 = \frac{\mu y_1 + \sqrt{4k(y_1^2 + k) + \mu^2 y_1^2}}{2(y_1^2 + k)},$$

as given in DiCiccio and Martin (1991, p. 897).

10.4.3 Moments

First consider the singly noncentral case. Let $Z = X - \mu \sim N(0, 1)$, independent of $Y \sim \chi_k^2$. Then, for integer s, recalling (I.7.38) and that $\mathbb{E}[Z^i] = 0$ for $i = 1, 3, 5, \ldots,$

$$\mathbb{E}[X^s] = \mathbb{E}[(Z + \mu)^s] = \sum_{i=0}^{s} \binom{s}{i} \mathbb{E}[Z^i] \mu^{s-i} = \sum_{i=0}^{\lfloor s/2 \rfloor} \binom{s}{2i} \mathbb{E}[Z^{2i}] \mu^{s-2i}$$

$$= \sum_{i=0}^{\lfloor s/2 \rfloor} \binom{s}{2i} \frac{1}{\sqrt{\pi}} 2^i \Gamma\left(\frac{1}{2} + i\right) \mu^{s-2i} = \sum_{i=0}^{\lfloor s/2 \rfloor} \binom{s}{2i} \frac{(2i)!}{2^i i!} \mu^{s-2i},$$

where the last equality follows from (10.2). Thus, using (I.7.42), the raw moments of the singly noncentral t r.v. $T = X/\sqrt{Y/k}$ are, for $k > s$ and $s \in \mathbb{N}$,

$$\mathbb{E}[T^s] = k^{s/2} \mathbb{E}[Y^{-s/2}] \mathbb{E}[X^s] = \left(\frac{k}{2}\right)^{s/2} \frac{\Gamma(k/2 - s/2)}{\Gamma(k/2)} \sum_{i=0}^{\lfloor s/2 \rfloor} \binom{s}{2i} \frac{(2i)!}{2^i i!} \mu^{s-2i}.$$

In particular, with $T \sim t'(k, \mu)$, for $k > 1$ and $k > 2$ respectively,

$$\mathbb{E}[T] = \left(\frac{k}{2}\right)^{1/2} \frac{\Gamma(k/2 - 1/2)}{\Gamma(k/2)} \mu, \qquad \mathbb{E}[T^2] = \frac{k}{k-2}\left(1 + \mu^2\right), \qquad (10.72)$$

$$E[X^s] = \frac{2^s}{e^{\theta/2}} \frac{\Gamma(n/2 + s)}{\Gamma(n/2)} {}_1F_1(n/2 + s, n/2; \theta/2).$$

Now consider the doubly noncentral case, for which $T = X/\sqrt{Y/k}$ with $Z = X - \mu \sim N(0, 1)$, independent of $Y \sim \chi^2(k, \theta)$. Then $\mathbb{E}[T^s] = k^{s/2} \mathbb{E}[Y^{-s/2}] \mathbb{E}[X^s]$, and the moments of Y are given in (10.9); in particular,

$$\mathbb{E}[Y^{-r}] = \frac{e^{-\theta/2}}{2^r} \frac{\Gamma(k/2 - r)}{\Gamma(k/2)} {}_1F_1(k/2 - r, k/2, \theta/2)$$

$$= \frac{\Gamma(k/2 - r)}{2^r \Gamma(k/2)} {}_1F_1(r, k/2, -\theta/2), \qquad r < k/2, \qquad (10.73)$$

where the second equality follows from using the Kummer transformation (5.29), i.e., ${}_1F_1(a, b, x) = e^x {}_1F_1(b - a, b, -x)$. Thus, for $k > s$ and $s \in \mathbb{N}$,

$$\mathbb{E}[T^s] = \left(\frac{k}{2}\right)^{s/2} \frac{\Gamma(k/2 - s/2)}{\Gamma(k/2)} {}_1F_1(s/2, k/2, -\theta/2) \sum_{i=0}^{\lfloor s/2 \rfloor} \binom{s}{2i} \frac{(2i)!}{2^i i!} \mu^{s-2i},$$

as was derived in Krishnan (1967). For $s = 1$,

$$\mathbb{E}[T] = \mu \left(\frac{k}{2}\right)^{1/2} \frac{\Gamma(k/2 - 1/2)}{\Gamma(k/2)} {}_1F_1(1/2, n/2, -\theta/2), \qquad k > 1,$$

while for $s = 2$,[12]

$$\mathbb{E}\left[T^2\right] = \left(1 + \mu^2\right) \frac{k}{k-2} \, {}_1F_1\left(1, k/2, -\theta/2\right) \quad k > 2.$$

The absolute moments are also straightforward to derive. First, let $Z \sim N(\mu, 1)$ so that, from (10.9) with $n = 1$ and $s = m/2$, and applying the Kummer transformation,

$$\mathbb{E}\left[|Z|^m\right] = \frac{2^{m/2}\Gamma\left((1+m)/2\right)}{\exp(\mu^2/2)\sqrt{\pi}} \, {}_1F_1\left(\frac{1+m}{2}, \frac{1}{2}; \frac{\mu^2}{2}\right)$$

$$= \frac{2^{m/2}\Gamma\left((1+m)/2\right)}{\sqrt{\pi}} \, {}_1F_1\left(-\frac{m}{2}, \frac{1}{2}, -\frac{\mu^2}{2}\right).$$

Using this in conjunction with (10.73) yields, after a bit of simplifying, that

$$\mathbb{E}\left[|T|^m\right] = k^{m/2}\mathbb{E}\left[Y^{-m/2}\right]\mathbb{E}\left[|Z|^m\right],$$

$$= k^{m/2} \frac{\Gamma\left((k-m)/2\right)\Gamma\left((1+m)/2\right)}{\Gamma(k/2)\sqrt{\pi}} \, {}_1F_1\left(\frac{m}{2}, \frac{k}{2}, -\frac{\theta}{2}\right) {}_1F_1\left(-\frac{m}{2}, \frac{1}{2}, \frac{\mu^2}{2}\right),$$

for $0 < m < k$.

10.5   Saddlepoint uniqueness for the doubly noncentral F

For computing the s.p.a. to the doubly noncentral F distribution, the saddlepoint equation is a cubic polynomial with roots z_1, z_2, and z_3, given in (10.43), but the correct root for use in the approximation is always z_3, in (10.44). This was proved in Butler and Paolella (2002) and is reproduced here. The method of proof is of value, because it is applicable to similar situations, such as the s.p.a. to the normal–Laplace sum considered in Section 5.1.3 and the doubly noncentral t distribution in Section 10.4.2.

Recall that $F \sim F(n_1, n_2, \theta_1, \theta_2)$, for $n_1, n_2, \theta_1, \theta_2 \in \mathbb{R}_{>0}$, and $x > 0$ denotes the value at which the c.d.f. of F is to be evaluated (the case $x = 0$ is excluded because it is not in the interior of the support of F). Also, recall from (10.35) that the c.d.f. of F can be expressed as $\Pr(F \leq x) = \Pr(Y_x \leq 0)$, where Y_x is defined in (10.35).

The proof hinges on the important general fact mentioned in Section 5.1.1 that the unique root of the saddlepoint equation exists in $U_x := \{s : -(2x)^{-1} < s < n_1/(2n_2)\}$, the open convergence strip of the m.g.f. of Y_x, so that, in this setting, one may take any real z_i that is in U_x. Proving that z_3 is always the correct root (i.e., the unique saddlepoint solution) is done in three steps. First, we show that all three roots are real. The second step demonstrates that, if the roots are ordered along the real line, then $\mathbb{K}''_{Y_x}(s)$ is positive for the middle root and negative at the other two:

---

[12] These expressions, along with those for $s = 3$ and $s = 4$, are also given in Krishnan (1967). Kocherlakota and Kocherlakota (1991) and Johnson, Kotz and Balakrishnan (1995, p. 534) give them too, but, in both of these sources, the expressions for $\mathbb{E}\left[T^2\right]$ are not correct.

recall from Section 5.1.1 that, if $s \in U_x$, then $\mathbb{K}''_{Y_x}(s)$ is a variance of the tilted random variable T_s in (5.1), so it must be nonnegative, but this is not true for $\mathbb{K}''_{Y_x}$ when viewed as a function with domain $\mathbb{R}$ instead of U_x, and evaluated at values of $s \notin U_x$. The third and last step shows that $z_2 < z_3 \leq z_1$. Combining these results demonstrates that z_1 and z_2 cannot be the saddlepoints, i.e., z_3 is always the correct root.

Step 1. Showing that all three roots in (10.43) are real is equivalent to showing that m in (10.42) is negative (Abramowitz and Stegun, 1972, p. 17). Using Maple, m can be factored as

$$ m = -\frac{1}{6912} (n_1 x + n_2)^2 \times \frac{n_1}{x^6 n_2^5 (n_1 + n_2)^4} \times Q, $$

where Q consists of the sum of 47 terms in n_1, n_2, θ_1, θ_2 and x, all of which are positive, except for one, given by $-8\theta_2^2 n_2 x^2 \theta_1^2 n_1$. By combining this term with two other (positive) terms $4n_2^2 \theta_2^3 \theta_1 x$ and $4\theta_2 n_1^2 \theta_1^3 x^3$ in Q (found by trial and error), we have

$$ -8\theta_2^2 n_2 x^2 \theta_1^2 n_1 + 4n_2^2 \theta_2^3 \theta_1 x + 4\theta_2 n_1^2 \theta_1^3 x^3 = 4\theta_2 x \theta_1 \, (n_2\theta_2 - n_1\theta_1 x)^2 > 0, $$

showing that $Q > 0$, so that $m < 0$, and all three roots are real.

Step 2. Define

$$ L(s) = \left(1 - 2s\frac{n_2}{n_1}\right)^2 (1 + 2sx)^2 \, \mathbb{K}'_{Y_x}(s) $$

and note that a root of $\mathbb{K}'_{Y_x}$ is also a root for L. Differentiating gives

$$ L'(s) = \frac{\partial}{\partial s}\left\{\left(1 - 2s\frac{n_2}{n_1}\right)^2 (1 + 2sx)^2\right\} \mathbb{K}'_{Y_x}(s) + \left(1 - 2s\frac{n_2}{n_1}\right)^2 (1 + 2sx)^2 \, \mathbb{K}''_{Y_x}(s) $$

and, when evaluated at roots z_i, $i = 1, 2, 3$,

$$ L'(z_i) = \left(1 - 2z_i\frac{n_2}{n_1}\right)^2 (1 + 2z_i x)^2 \, \mathbb{K}''_{Y_x}(z_i), $$

so that

$$ \text{sgn}\left\{L'(z_i)\right\} = \text{sgn}\left\{\mathbb{K}''_{Y_x}(z_i)\right\}. \tag{10.74} $$

Now $L(s)$ is a cubic polynomial whose leading term is $-8n_2^2 x^2 \, (n_1 + n_2) \, n_1^{-2} s^3$ with a negative coefficient; thus $L'(s) > 0$ when s is the middle root and negative at the first and third ordered roots. From (10.74), the same holds for $\mathbb{K}''_{Y_x}(s)$ so the middle root must be the saddlepoint.

Step 3. To prove that $z_2 < z_3 \leq z_1$, first note that r in (10.42) is real and that $m < 0$, implying that $q < 0$. Then

$$r + \sqrt{m} = r + i\sqrt{-(q^3 + r^2)} = \sqrt{-q^3} \exp(i\lambda),$$

where $\lambda = \arg(r + \sqrt{m}) \in (0, \pi)$ and, more specifically, if $r < 0$, then $\lambda \in (\pi/2, \pi)$. Then

$$s_{1,2} = (r \pm \sqrt{m})^{1/3} = \left[\sqrt{-q^3} \exp(\pm i\lambda)\right]^{1/3} = \sqrt{-q} \exp(\pm i\lambda/3)$$

and, thus,

$$s_1 + s_2 = \sqrt{-q}\left[\exp(i\lambda/3) + \exp(-i\lambda/3)\right] = 2\sqrt{-q}\cos(\lambda/3) \tag{10.75}$$

is real and positive, as $0 \leq \lambda/3 < \pi/3 = 60°$. Similarly,

$$s_1 - s_2 = \sqrt{-q}\left[\exp(i\lambda/3) - \exp(-i\lambda/3)\right] = 2i\sqrt{-q}\sin(\lambda/3)$$

so that

$$i\frac{\sqrt{3}}{2}(s_1 - s_2) = -\sqrt{-3q}\sin(\lambda/3) < 0. \tag{10.76}$$

From (10.76) it follows directly that $z_2 < z_3$. Now, comparing z_1 and z_3,

$$z_3 = -\sqrt{-q}\cos(\lambda/3) - \frac{a_2}{3} + \sqrt{3}\sqrt{-q}\sin(\lambda/3) \overset{?}{\leq} 2\sqrt{-q}\cos(\lambda/3) - \frac{a_2}{3} = z_1$$

or $\sqrt{3}\sqrt{-q}\sin(\lambda/3) \overset{?}{\leq} 3\sqrt{-q}\cos(\lambda/3)$ or $\tan(\lambda/3) \overset{?}{\leq} \sqrt{3}$.   But $\tan(\lambda/3) \leq \tan(\pi/3) = \sqrt{3}$, so that $z_3 \leq z_1$ and $z_2 < z_3 \leq z_1$.   ∎

10.6   Problems

Every day[13] you may make progress. Every step may be fruitful. Yet there will stretch out before you an ever-lengthening, ever-ascending, ever-improving path. You know you will never get to the end of the journey. But this, so far from discouraging, only adds to the joy and glory of the climb.

(Sir Winston Churchill)

Personally I'm always ready to learn, although I do not always like being taught.

(Sir Winston Churchill)

A pessimist sees the difficulty in every opportunity; an optimist sees the opportunity in every difficulty.

(Sir Winston Churchill)

---

[13] Reproduced with permission from (From "Painting as a Pastime" by Winston Churchill. Reproduced with permission of Curtis Brown Ltd, London on behalf of The Estate of Winston Churchill Copyright Winston S. Churchill)

10.1. ★ Relate the incomplete gamma function to the 'Imhof method' developed in Section 10.1.5.

10.2. ★ Use the fact that the multivariate normal density function integrates to one to derive $\mathbb{M}_S(s)$, the moment generating function of $S = \sum_{i=1}^{n} \lambda_i Q_i$, where λ_i are known, nonzero weights and $Q_i \overset{\text{i.i.d.}}{\sim} \chi^2(1)$, i.e., i.i.d. central χ^2 r.v.s, each with one degree of freedom.

10.3. ★ Generalizing the previous question, let $\mathbf{X} \sim \mathrm{N}_n\left(\mathbf{0}, \sigma^2 \mathbf{I}\right)$ and $\mathbf{A}$ be a real, symmetric $n \times n$ matrix. Show that the m.g.f. of $S = \mathbf{X}'\mathbf{A}\mathbf{X}/\sigma^2$ is

$$\mathbb{M}_S(s) = \prod_{i=1}^{n} (1 - 2s\lambda_i)^{-1/2}, \tag{10.77}$$

where $\{\lambda_i\}$ are the eigenvalues of $\mathbf{A}$ and $|s|$ is sufficiently small.

10.4. ★ ★ It is easy to *confirm* that the matrix $\mathbf{B}_3$ in (10.1) is orthogonal with first row $\boldsymbol{\mu}'\theta^{-1/2}$, but it is not as simple to *derive* such a matrix. Using, say, Maple, derive a matrix $\mathbf{B}_3$ such as the one in (10.1).

10.5. ★ Confirm that setting $s = 1$ in (10.9) simplifies to $\mathbb{E}[X] = n + \theta$.

10.6. ★ Compute the m.g.f. of $X = \sum_{i=1}^{n} X_i^2$, $X_i \overset{\text{ind}}{\sim} \mathrm{N}(\mu_i, 1)$, $n = 1, \dots, n$, in two ways: first, by simplifying $\mathbb{M}_X(t) = \mathbb{E}\left[e^{t(\mathbf{X}'\mathbf{X})}\right]$, where $\mathbf{X} = (X_1, \dots X_n)'$; and second, by simplifying $\mathbb{M}_Y(t) = \mathbb{E}\left[e^{tY}\right]$, where r.v. Y has p.d.f. (10.3).

Hint: For the former, use the following general result. Let $A(\mathbf{x}) = \mathbf{x}'\mathbf{A}\mathbf{x} + \mathbf{x}'\mathbf{a} + a$ and $B(\mathbf{x}) = \mathbf{x}'\mathbf{B}\mathbf{x} + \mathbf{x}'\mathbf{b} + b$ be functions of $\mathbf{x}$, where $a, b \in \mathbb{R}$, $\mathbf{x}$, $\mathbf{a}$, and $\mathbf{b}$ are $n \times 1$ real vectors, and $\mathbf{A}$ and $\mathbf{B}$ are symmetric $n \times n$ matrices with $\mathbf{B}$ positive definite. Then

$$\int_{-\infty}^{\infty} \cdots \int_{-\infty}^{\infty} A(\mathbf{x})e^{-B(\mathbf{x})} d\mathbf{x} = \frac{1}{2}\pi^{n/2} |\mathbf{B}|^{-1/2} \exp\left\{\frac{1}{4}\left(\mathbf{b}'\mathbf{B}^{-1}\mathbf{b}\right) - b\right\}$$

$$\times \left[\operatorname{tr}\left(\mathbf{A}\mathbf{B}^{-1}\right) - \mathbf{b}'\mathbf{B}^{-1}\mathbf{a} + \frac{1}{2}\mathbf{b}'\mathbf{B}^{-1}\mathbf{A}\mathbf{B}^{-1}\mathbf{b} + 2a\right]. \tag{10.78}$$

For proof of (10.78), see Graybill (1976, p. 48).

10.7. ★ Let $X_i \overset{\text{ind}}{\sim} \mathrm{N}\left(\mu_i, \sigma_i^2\right)$, $i = 1, 2$. Compute the density of $R = (X_1/X_2)^2$. Then, for $\sigma_1^2 = \sigma_2^2 = 1$ and $\mu_2 = 0$, show that

$$f_R(x) = \frac{e^{-\mu_1/2}}{\pi} \frac{1}{\sqrt{x}(1+x)} {}_1F_1\left(1, \frac{1}{2}, \frac{\mu_1}{2} \frac{x}{1+x}\right).$$

10.8. ★ Derive (10.23) and (10.25).

10.9. Prove that expression (10.54) for $F_T(t)$ can also be written as

$$F_T(t) = \frac{2^{-k/2+1}}{\Gamma(k/2)} \int_0^\infty w^{k-1} e^{-w^2/2} \, \Phi\left(tk^{-1/2}z - \mu\right) dw,$$

where $\Phi(q)$ is the standard normal c.d.f.

10.10. Derive (10.58) from (10.57).

10.11. ★ ★ Derive (10.60) from (10.59) and (10.61) from (10.60).

10.12. ★ ★ Derive (10.65) and (10.66).

10.13. ★ Derive (10.69).

10.14. ★ ★ Recall from (1.100) and (1.101) that, for continuous r.v. X with c.g.f. $\mathbb{K}_X(s)$ converging on (c_1, c_2),

$$\overline{F}_X(x) = \frac{1}{2\pi i} \int_{c-i\infty}^{c+i\infty} \exp\{\mathbb{K}_X(s) - sx\} \, \frac{ds}{s}, \quad 0 < c < c_2,$$

and

$$F_X(x) = \frac{1}{2\pi i} \int_{c-i\infty}^{c+i\infty} \exp\{\mathbb{K}_X(s) - sx\} \, \frac{ds}{-s}, \quad c_1 < c < 0,$$

with $c_1 < 0 < c_2$.   (Contributed by Simon Broda)

(a) Denote the integrand in the above equations by

$$\exp\{I(s)\} = \exp\{\mathbb{K}_X(s) - xs - \ln(\operatorname{sgn}(c)s)\}. \tag{10.79}$$

Show that, with $s = a + ib$ and $\bar{s} = a - ib$,

$$\exp\{I(\bar{s})\} = \overline{\exp\{I(s)\}},$$

so that the integrals can be written

$$F_X(x) = H(c) - \frac{1}{2\pi i} \int_{c-i\infty}^{c+i\infty} \exp\{\mathbb{K}_X(s) - sx\} \frac{ds}{s}$$

$$= H(c) - \frac{1}{2\pi} \int_{-\infty}^{\infty} \exp\{\mathbb{K}_X(s) - sx\} \frac{dt}{s}$$

$$= H(c) - \frac{1}{\pi} \int_0^\infty \operatorname{Re}\left[\frac{\exp\{\mathbb{K}_X(s) - sx\}}{s}\right] dt, \tag{10.80}$$

where $s = c + it$, and $H(\cdot)$ is the Heaviside step function, i.e.,

$$H(c) = \begin{cases} 0, & c < 0, \\ \frac{1}{2}, & c = 0, \\ 1, & c > 0. \end{cases}$$

Hint: First show that, if the m.g.f. of X exists, then

$$\mathbb{M}_X(\overline{s}) = \overline{\mathbb{M}_X(s)}. \tag{10.81}$$

(b) Using (10.80), compute the c.d.f. of $Q = \sum_{j=1}^{N} \lambda_j \chi_j^2(v_j, \theta_j)$, i.e., a weighted sum of noncentral chi-square r.v.s with degrees of freedom v_j and noncentrality parameter θ_j. Recall from (10.19) and (10.20) that the m.g.f. of Q is

$$\mathbb{M}_Q(s) = \prod_{j=1}^{n} (1 - 2\lambda_j s)^{-v_j/2} \exp\left\{ \frac{\lambda_j \theta_j s}{1 - 2\lambda_j s} \right\}, \quad 1 - 2\lambda_i s > 0.$$

Choose $c = 0.25$, and plot the integrand for the c.d.f. of $X = X_1 - X_2$, where X_1, X_2 are central chi-square with one degree of freedom, at $x = 4$, over the range $[0, 5]$.

(c) In (1.100) and (1.101), integration is carried out along the so-called *Bromwich contour* $(c - i\infty; c + i\infty)$. Helstrom (1996) suggests deforming the contour in such a fashion that it closely resembles the path of steepest descent of the integrand. This path crosses the real axis at a point s_0 in the convergence strip (c_1, c_2) of $\mathbb{K}$, called a saddlepoint, where the integrand attains its minimum on the real axis; it is given by

$$\frac{dI(s)}{ds}\bigg|_{s_0} = 0, \tag{10.82}$$

where $I(s)$ is as in (10.79). There are two saddlepoints, one to the left and one to the right of the origin, each between the origin and the nearest singularity of the m.g.f. If all the singularities are positive (negative), it is therefore advantageous to use the positive (negative) saddlepoint, in order to have a bracketing interval for the required root search. If there are singularities on either side of the origin, Helstrom suggests using $s_0 > 0$ if $x > \mathbb{E}[X]$, and $s_0 < 0$ otherwise.

For a particular quadratic form (the least squares estimator in an AR(1) model), Helstrom finds that the path of steepest descent has roughly the form of the parabola given by

$$s = s_0 + \frac{1}{2} C y^2 + iy, \quad -\infty < y < \infty, \quad C = \frac{I^{(3)}(s_0)}{3 I^{(2)}(s_0)}, \tag{10.83}$$

superscripts in parentheses denoting derivatives. Using (10.81), the integral in (10.80) becomes

$$F_X(x) = H(c) - \frac{1}{\pi} \int_0^\infty \mathrm{Re}\left[e^{J(s)}(1 - iCy)\right] dy, \quad s = s_0 + \frac{1}{2}Cy^2 + iy.$$

(10.84)

Make a program that evaluates the c.d.f. of Q from part (**b**) using the above method. Plot the integrand for $x = 4$, and compare.

Notation and distribution tables

Table A.1 Abbreviations

	Description
c.d.f.	cumulative distribution function
c.f.	characteristic function
c.g.f.	cumulant generating function
CLT	central limit theorem
i.i.d.	independently and identically distributed
l.h.s. (r.h.s.)	left- (right-)hand side
m.g.f.	moment generating function
p.d.f. (p.m.f.)	probability density (mass) function
p.g.f.	probability generating function
RPE	relative percentage error
r.v.	random variable
SLLN	strong law of large numbers
s.p.a.	saddlepoint approximation
WLLN	weak law of large numbers
$\exists$	there exists
$\forall$	for each
iff, $\Leftrightarrow$	if and only if
$:=$	$a := b$ or $b =: a$ defines a to be b (e.g., $0! := 1$)
$\mathbb{C}, \mathbb{N}, \mathbb{Q}, \mathbb{R}, \mathbb{R}_{>0}, \mathbb{Z}$	sets of numbers (see Section I.A.1)
$\operatorname{sgn}(x)$	signum function; $\operatorname{sgn}(x) = \begin{cases} -1, & \text{if } x < 0, \\ 0, & \text{if } x = 0, \\ 1, & \text{if } x > 0. \end{cases}$

Intermediate Probability: A Computational Approach M. Paolella
© 2007 John Wiley & Sons, Ltd

Table A.2 Special functions

Name	Notation	Definition	Alternative formula
factorial	$n!$	$n(n-1)(n-2)\cdots 1$	$\Gamma(n+1)$
gamma function	$\Gamma(a)$	$\lim_{k\to\infty} k!\,k^{x-1}/x^{[k]}$	$\int_0^\infty t^{a-1}e^{-t}\,dt$
incomplete gamma function	$\Gamma_x(a)$	$\int_0^x t^{a-1}e^{-t}\,dt$	$_1F_1(a,a+1;-x)$
digamma function	$\psi(s)$	$\dfrac{d}{ds}\ln\Gamma(s) = \dfrac{\Gamma'(s)}{\Gamma(s)}$	$\int_0^\infty \left[\dfrac{e^{-t}}{t} - \dfrac{e^{-zt}}{1-e^{-t}}\right]dt$
beta function	$B(p,q)$	$\int_0^1 t^{p-1}(1-t)^{q-1}\,dt$	$\Gamma(p)\Gamma(q)/\Gamma(p+q)$; and for $p,q\in\mathbb{N}$, $B(p,q) = \dfrac{(p-1)!(q-1)!}{(q+p-1)!}$
confluent hypergeometric function	$_1F_1(a,b;z)$	$\sum_{n=0}^\infty \dfrac{a^{[n]}}{b^{[n]}}\dfrac{z^n}{n!}$	$[B(a,b-a)]^{-1}\times$ $\int_0^1 y^{a-1}(1-y)^{b-a-1}e^{zy}\,dy$ for $a>0,\quad b-a>0$
hypergeometric function	$_2F_1(a,b;c;z)$	$\sum_{n=0}^\infty \dfrac{a^{[n]}b^{[n]}}{c^{[n]}}\dfrac{z^n}{n!}$	$[B(a,c-a)]^{-1}\{\int_0^1 y^{a-1}\times$ $(1-y)^{c-a-1}(1-zy)^{-b}\,dy\}$ for $a>0,\ c-a>0,\ z<1$
modified Bessel function of the third kind	$K_v(z)$	$\dfrac{\Gamma(v+1/2)(2z)^v}{\sqrt{\pi}}\displaystyle\int_0^\infty \dfrac{\cos t}{(t^2+z^2)^{v+1/2}}\,dt$ (see Section 9.2)	

Table A.3 General notation I

Description	Notation	Examples
sets (measurable events, r.v. support, information sets, m.g.f. convergence strip, etc.)	calligraphic capital letters and Roman capital letters	$\mathcal{I}_{t-1} \subset \mathcal{I}_t \qquad A \subset \mathbb{R}^2$
Borel σ-field	$\mathcal{B}$	$\{\mathbb{R}, \mathcal{B}, \Pr(\cdot)\}$
events	capital Roman letters	$A \in \mathcal{A}, \quad A_1 \supset A_2 \supset A_3 \cdots$
event complement	A^c	$\left(\bigcup_{i=1}^n A_i\right)^c = \bigcap_{i=1}^n A_i^c$
random variables (r.v.s)	capital Roman letters	A, G, O, N, Y
support of r.v. X	S_X or, in context, just S	$\int_{S_X \cup \mathbb{R}_{>0}} x^2 dF_X(x)$
independence of two r.v.s	$X \perp Y$	$X \sim N(0,1) \perp Y \sim \chi_k^2$
value or set of values assumed by a r.v.	lower-case Roman letter (usually matching)	$\{X < a\} \cup \{X > b\} \quad \Pr(X > x)$
'unified' sum or integral (Riemann–Stieltjes)	$\int_{-\infty}^{\infty} g(x)\,dF_X(x) = \begin{cases} \int_S g(x) f_X(x)\,dx, & X \text{ continuous},\\[4pt] \sum_{i \in S} g(x_i) f_X(x_i), & X \text{ discrete}, \end{cases}$	
Landau's order symbol ('big oh', 'little oh')	$f(x) = O(g(x)) \Leftrightarrow \exists C > 0 \text{ s.t. } \lvert f(x) \rvert \le Cg(x)\rvert$ as $x \to \infty$, $f(x) = o(g(x)) \Leftrightarrow f(x)/g(x) \to 0$ as $x \to \infty$.	
proportional to	$f(x;\theta) \propto g(x;\theta) \Leftrightarrow f(x;\theta) = Cg(x;\theta)$, where C is a constant	$f_N(z; 0,1) \propto \exp\left\{-\tfrac{1}{2}z^2\right\}$
vectors (matrices)	bold lower (upper) case letters	$\mathbf{x}'\boldsymbol{\Sigma}\mathbf{x} > 0$
matrix transpose	$\mathbf{A}'$	$\mathbf{A} = \begin{bmatrix} a & b \\ c & d \end{bmatrix}, \mathbf{A}' = \begin{bmatrix} a & c \\ b & d \end{bmatrix}$
determinant of square $\mathbf{A}$	$\det(\mathbf{A})$, $\det \mathbf{A}$ or $\lvert \mathbf{A} \rvert$	$\begin{vmatrix} a & b \\ c & d \end{vmatrix} = ad - bc$

Table A.4 General notation II

Description	Notation	Example or definition
real and imaginary part	$\mathrm{Re}(Z), \ \mathrm{Im}(Z)$	$\mathrm{Im}(x + iy) = y$
eigenvalues of square $\mathbf{A}$	$\{\lambda_i\} = \mathrm{Eig}(\mathbf{A})$	$\det(\mathbf{A}) = \prod_i \mathrm{Eig}(\mathbf{A})$
diagonal of square $\mathbf{A}$	$\mathrm{diag}(\mathbf{A})$	$\mathrm{diag}\left[\begin{smallmatrix} a & b \\ c & d \end{smallmatrix}\right] = [a, d]'$
trace of square $\mathbf{A}$	$\mathrm{tr}(\mathbf{A}) = \sum \mathrm{diag}(\mathbf{A})$	$\mathrm{tr}(\mathbf{A}) = \sum_i \mathrm{Eig}(\mathbf{A})$
rank of $n \times m$ matrix $\mathbf{A}$	$0 \le \mathrm{rank}(\mathbf{A})$ $\le \min(n, m)$	$\mathrm{rank}(\mathbf{A}) = \mathrm{rank}(\mathbf{AA}')$ $= \mathrm{rank}(\mathbf{A}'\mathbf{A})$
approximate distribution of a random variable	$\overset{\mathrm{app}}{\sim} D(\theta)$	$\chi_n^2 \overset{\mathrm{app}}{\sim} N(n, 2n)$
k-norm of r.v. X	$\|X\|_k = \left(\mathbb{E}[|X|^k]\right)^{1/k},$ $k \ge 1$	$\|X\|_r \le \|X\|_s,$ $1 \le r \le s$
limit superior for sets	$A^* = \lim \sup_{i \to \infty} A_i$ $= \bigcap_{k=1}^{\infty} \bigcup_{n=k}^{\infty} A_n$	$A_n \to A$ iff $A = A_* = A^*$
limit inferior for sets	$A_* = \lim \inf_{i \to \infty} A_i$ $= \bigcup_{k=1}^{\infty} \bigcap_{n=k}^{\infty} A_n$	$(A^*)^c = \bigcup_{k=1}^{\infty} \bigcap_{n=k}^{\infty} A_n^c$
convergence in probability	$X_n \overset{p}{\to} X$	for $A_n = |X_n - X| > \epsilon,$ $\lim_{n \to \infty} \mathrm{Pr}(A_n) = 0$
almost sure convergence	$X_n \overset{a.s.}{\to} X$	for $A_n = |X_n - X| > \epsilon,$ $\lim_{m \to \infty} \mathrm{Pr}\left(\bigcup_{n=m}^{\infty} A_n\right) = 0$
convergence in r-mean	$X_n \overset{L_r}{\to} X$	$\lim_{n \to \infty} \mathbb{E}\left[|X_n - X|^r\right] = 0$
convergence in distribution	$X_n \overset{d}{\to} X$	$\lim_{n \to \infty} F_{X_n}(x) = F_X(x)$ $\forall x \in C(F_X)$

Table A.5 Generating functions and inversion formulae

Description	Notation, definition	Examples
probability generating function, for lattice r.v.s	$\mathbb{P}_X(t) = \mathbb{E}[t^X]$	$\mu'_{[g]}(X) = \mathbb{P}_X^{(g)}(1)$
moment generating function	$\mathbb{M}_X(t) = \mathbb{E}[\exp\{Xt\}]$	$\mathbb{E}[X^r] = \mathbb{M}_X^{(r)}(0)$
cumulant generating function	$\mathbb{K}_X(t) = \log \mathbb{M}_X(t)$	$\kappa_1 = \mu, \quad \kappa_2 = \mu_2,$
	$=: \sum_{r=0}^\infty \kappa_r t^r / r!$	$\kappa_3 = \mu_3, \; \kappa_4 = \mu_4 - 3\mu_2^2$
characteristic function	$\varphi_X(t) = \mathbb{E}\left[\exp[itX]\right]$	$\varphi_X(t) = \mathbb{M}_X(it)$
p.m.f. inversion	$f_X(x_j) = \lim_{T\to\infty} \frac{1}{2T} \int_{-T}^T e^{-itx_j} \varphi_X(t)\, dt$	
	$f_X(x_j) = \frac{1}{2\pi} \int_{-\pi}^\pi e^{-itx_j} \varphi_X(t)\, dt = \frac{1}{\pi} \int_0^\pi \mathrm{Re}\left[e^{-itx_j} \varphi_X(t)\right] dt$	
p.d.f. inversion	$f_X(x) = \frac{1}{2\pi i} \int_{-i\infty}^{i\infty} \exp\{\mathbb{K}_X(s) - sx\}\, ds$	
	$f_X(x) = \frac{1}{2\pi} \int_{-\infty}^\infty e^{-itx} \varphi_X(t)\, dt = \frac{1}{\pi} \int_0^\infty \mathrm{Re}\left[e^{-itx} \varphi_X(t)\right] dt$	
continuous c.d.f. inversion	$F_X(x) = \frac{1}{2\pi i} \int_{c-i\infty}^{c+i\infty} \exp\left(\mathbb{K}_X(t) - tx\right) \frac{dt}{t}$	
	$F_X(x) = \frac{1}{2} - \frac{1}{\pi} \int_0^\infty \frac{\mathrm{Im}\left[e^{-itx} \varphi_X(t)\right]}{t}\, dt$	

Table A.6 Conventions for distribution names

Description	Examples
A 'simple' or fundamental distribution name starts with a capital letter	Cau, Exp, Gam, Gum, Lap, Log, N
Preface the name with capital G to denote a generalization	GBeta, GGam, GHyp
Alternatively, use Gn if the generalization has n parameters	G3B, with 3 parameters, not including location and scale
Begin the name with capital I to indicate an inverse relation to another distribution	IGam (inverse gamma), IG (inverse Gaussian)
Add A for 'asymmetric'	GAt (generalized asymmetric t), ADWeib (asymmetric double Weibull)
Add D for 'discrete' or 'double'	DUnif (discrete uniform), DWeib (double Weibull)
Add H for 'hyper'	HGeo (hypergeometric), IHGeo (inverse hypergeometric)
Add L for 'log'	LN (log normal)
Add N for 'negative'	NBin (negative binomial)
Add N or Norm for 'normal'	NormLap (normal–Laplace), NIG (normal inverse Gaussian)
Some distributions have classical names which do not fit in this scheme	χ^2, t, F, GED, SN (skew normal), VG (variance–gamma)

Table A.7 Some distributional subsets[a]

Beta	⊆	G3B	Exp	⊆	Gam	Log	⊂	GLog
Beta	⊆	G4B	Exp	⊆	Weib	N	⊆	SN
Ber	⊆	Bin	F	⊆	G3F	N	⊆	GED
Ber	⊆	HGeo	Gam	⊆	GGam	N	⊂	Hyp
Geo	⊆	NBin	GED	⊆	FS–GED	N	⊆	SαS
Geo	⊆	Consec	Lap	⊆	GED	N	⊂	t
Cau	⊆	SαS	Lap	⊆	Hyp	Par II	⊆	GPar II
Cau	⊆	t	Lap	⊆	DWeib	SαS	⊆	$S_{\alpha,\beta}$
χ^2	⊆	Gam	Levy	⊂	IG	Unif	⊆	Beta
DWeib	⊆	ADWeib	Levy	⊆	$S_{\alpha,\beta}$	Weib	⊆	GGam

[a]Use of ⊂ instead of ⊆ means that the special case is attained only asymptotically. See Table A.8 for further subsets involving the Student's t distribution and Chapter 9 for tables and diagrams giving the subsets of the GIG and GHyp distributions.

Table A.8 Student's t generalizations[a]

McDonald and Newey	GT $\supseteq t$	(Symmetric)

$$f_{GT}(z; d, \nu) = K\left(1 + \frac{|z|^d}{\nu}\right)^{-(\nu+1/d)},$$

$$K^{-1} = 2d^{-1}\nu^{1/d}B\left(d^{-1}, \nu\right), \qquad d, \nu \in \mathbb{R}_{>0}$$

(generalized) Lye and Martin	LYMd $\supseteq$ GT	(Asymmetric)

$$f_{LYMd}(z; d, \nu, \theta) \propto \exp\left(\theta \arctan\left(\frac{z}{\nu^{1/d}}\right) - \left(\nu + \frac{1}{d}\right)\log\left(1 + \frac{|z|^d}{\nu}\right)\right)$$

$$d, \nu \in \mathbb{R}_{>0}, \ \theta \in \mathbb{R}$$

Fernández and Steel extension	FSt $\supseteq t$	(Asymmetric)

$$f_{FSt}(z; \nu, \theta) = \frac{2}{\theta + 1/\theta}\left\{f_T\left(\frac{z}{\theta}\right)\mathbb{I}_{[0,\infty)}(z) + f_T(z\theta)\mathbb{I}_{(-\infty,0)}(z)\right\}$$

$$\text{where } T \sim t(\nu), \qquad \nu, \theta \in \mathbb{R}_{>0}.$$

Generalized asymmetric t	GAT $\supseteq$ {FSt, GT, GED}	(Asymmetric)

$$f_{GAT}(z; d, \nu, \theta) = K\left\{\left(1 + \frac{(-z\cdot\theta)^d}{\nu}\right)^{-\left(\nu+\frac{1}{d}\right)}\mathbb{I}_{(-\infty,0)}(z)\right.$$

$$\left. + \left(1 + \frac{(z/\theta)^d}{\nu}\right)^{-\left(\nu+\frac{1}{d}\right)}\mathbb{I}_{[0,\infty)}(z)\right\}$$

$$K^{-1} = \left(\theta^{-1} + \theta\right)d^{-1}\nu^{1/d}B\left(\frac{1}{d}, \nu\right), \qquad d, \nu, \theta \in \mathbb{R}_{>0}$$

Jones and Faddy	JoF $\supseteq t$	(Asymmetric)

$$f_{JoF}(t; a, b) = C\left(1 + \frac{t}{(a+b+t^2)^{1/2}}\right)^{a+1/2}\left(1 - \frac{t}{(a+b+t^2)^{1/2}}\right)^{b+1/2},$$

$$C^{-1} = B(a, b)(a+b)^{1/2}2^{a+b-1}, \qquad a, b > 0$$

Hyperbolic asymmetric t	HA$t \supset t$	(Asymmetric)

$$f_{HAt}(x; n, \beta, \mu, \delta) = \frac{2^{\frac{-n+1}{2}}\delta^n}{\sqrt{\pi}\Gamma(n/2)}\left(\frac{y_x}{|\beta|}\right)^{-\frac{n+1}{2}}K_{-\frac{n+1}{2}}(|\beta|y_x)e^{\beta(x-\mu)},$$

$$y_x = \sqrt{\delta^2 + (x-\mu)^2}, \qquad n > 0, \beta, \mu \in \mathbb{R}, \beta \neq 0, \delta > 0$$

Singly noncentral t	$t'(k, \mu) \supseteq t$	(Asymmetric)

See Table A.10

Doubly noncentral t	$t''(k, \mu, \theta) \supseteq t'(k, \mu)$	(Asymmetric)

See Table A.10

---

[a] See Chapter 7 (and its Problems and Solutions) for details on the first five entries. The hyperbolic asymmetric t is discussed in Chapter 9. The noncentral t distributions are discussed in Section 10.4; see also Tables A.9 and A.10 below.

Table A.9 Noncentral distributions I[a]

$$\text{Let } \omega_{i,\theta} := e^{-\theta/2} (\theta/2)^i / i!$$

$$\text{noncentral } \chi^2, \quad X \sim \chi^2(n, \theta)$$

$$(n \text{ degrees of freedom, noncentrality } \theta)$$

$$X = \sum_{i=1}^{n} X_i^2, \text{ where } \theta = \boldsymbol{\mu}'\boldsymbol{\mu} \text{ and } \mathbf{X} \sim N_n(\boldsymbol{\mu}, \mathbf{I})$$

$$f_X(x; n, \theta) = \sum_{i=0}^{\infty} \omega_i f_X(x; n + 2i, 0), \quad F_X(x; n, \theta) = \sum_{i=0}^{\infty} \omega_{i,\theta} F_X(x; n + 2i, 0)$$

$$\mathbb{E}[X] = n + \theta, \quad \mathbb{V}(X) = 2n + 4\theta, \quad M_X(t) = (1 - 2t)^{-n/2} \exp\left\{\frac{t\theta}{1 - 2t}\right\}, \ t < \tfrac{1}{2}$$

$$\kappa_i = \mathbb{K}_X^{(i)}(0) = 2^{i-1}(i - 1)!(n + i\theta)$$

$$\text{doubly noncentral } F, \quad F \sim F(n_1, n_2, \theta_1, \theta_2)$$

$$(n_1, n_2 \text{ degrees of freedom, noncentrality } \theta_1, \theta_2)$$

$$F = \frac{X_1/n_1}{X_2/n_2}, \text{ where } X_i \overset{\text{ind}}{\sim} \chi^2(n_i, \theta_i)$$

$$f_F(x) = \sum_{i=0}^{\infty} \sum_{j=0}^{\infty} \omega_{i,\theta_1} \omega_{j,\theta_2} \frac{n_1^{n_1/2+i}}{n_2^{-n_2/2-j}} \frac{x^{n_1/2+i-1}(xn_1 + n_2)^{-(n_1+n_2)/2-i-j}}{B(i + n_1/2, j + n_2/2)},$$

$$F_F(x; n_1, n_2, \theta_1, \theta_2) = \sum_{i=0}^{\infty} \sum_{j=0}^{\infty} \omega_{i,\theta_1} \omega_{j,\theta_2} \overline{B}_y(i + n_1/2, j + n_2/2), \quad y = \frac{n_1 x}{n_1 x + n_2}$$

$$\mathbb{E}[X^r] =: {_2}\mu_r' = {_1}\mu_{r,1}' F_1(r, n_2/2, -\theta_2/2), \quad n_2 > 2r,$$

where ${_1}\mu_r'$ are the raw moments of the singly noncentral F

$$\text{singly noncentral } F, \quad F \sim F(n_1, n_2, \theta_1)$$

$$(n_1, n_2 \text{ degrees of freedom, noncentrality } \theta_1)$$

$$f_F(x) = \sum_{i=0}^{\infty} \omega_{i,\theta_1} \frac{n_1^{n_1/2+i}}{n_2^{-n_2/2}} \frac{x^{n_1/2+i-1}(xn_1 + n_2)^{-(n_1+n_2)/2-i}}{B(i + n_1/2, n_2/2)}$$

$$F_F(x) = \sum_{i=0}^{\infty} \omega_{i,\theta_1} \overline{B}_y(i + n_1/2, n_2/2), \quad y = \frac{n_1 x}{n_1 x + n_2}$$

$$\mathbb{E}[X] = \frac{n_2}{n_1} \frac{n_1 + \theta_1}{n_2 - 2}, \quad n_2 > 2$$

$$\mathbb{V}(X) = 2\frac{n_2^2}{n_1^2} \frac{(n_1 + \theta_1)^2 + (n_1 + 2\theta_1)(n_2 - 2)}{(n_2 - 2)^2(n_2 - 4)}, \quad n_2 > 4$$

$$\mathbb{E}[X^r] =: {_1}\mu_r' = \left(\frac{n_2}{n_1}\right)^r \frac{\Gamma(n_2/2 - r)}{\Gamma(n_2/2)} \sum_{i=0}^{\infty} \omega_{i,\theta_1} \frac{\Gamma(n_1/2 + i + r)}{\Gamma(n_1/2 + i)}, \quad n_2 > 2r$$

[a]Continued in Table A.10. See Chapter 10 (and its Problems and Solutions) for more detail.

Table A.10 Noncentral distributions II[a]

$$\text{Let } \omega_{i,\theta} := e^{-\theta/2} (\theta/2)^i / i!$$

doubly noncentral beta, $\quad B \sim \text{Beta}(n_1, n_2, \theta_1, \theta_2)$

$(n_1, n_2$ degrees of freedom, noncentrality $\theta_1, \theta_2)$

$$B = (nX)/(1 + nX), \quad \text{where } X \sim F(2n_1, 2n_2, \theta_1, \theta_2)$$

$$f_B(b; n_1, n_2, \theta_1, \theta_2) = \sum_{i=0}^{\infty} \sum_{j=0}^{\infty} \omega_{i,\theta_1} \omega_{j,\theta_2} f_{\text{Beta}}(b; n_1 + i, n_2 + j)$$

$$= \frac{n_2}{n_1(1-b)^2} f_X\left(\frac{b}{n(1-b)}; 2n_1, 2n_2, \theta_1, \theta_2\right)$$

$$F_B(b; n_1, n_2, \theta_1, \theta_2) = F_X\left(\frac{b}{n(1-b)}; 2n_1, 2n_2, \theta_1, \theta_2\right)$$

doubly noncentral t, $\quad T \sim t''(k, \mu, \theta)$

$$T = X/\sqrt{Y/k}, \quad \text{where } X \sim N(\mu, 1) \perp Y \sim \chi^2(k, \theta)$$

$$f_T = \frac{e^{-(\theta+\mu^2)/2}}{\sqrt{\pi k}} \sum_{j=0}^{\infty} \frac{1}{j!} \frac{\left(t\mu\sqrt{2/k}\right)^j}{\left(1 + t^2/k\right)^{(k+j+1)/2}}$$

$$\times \frac{\Gamma\left(\frac{k+j+1}{2}\right)}{\Gamma(k/2)} \, {}_1F_1\left(\frac{k+j+1}{2}, \frac{k}{2}; \frac{\theta}{2\left(1 + t^2/k\right)}\right)$$

$$F_T(t; k, \mu, \theta) = \sum_{i=0}^{\infty} \omega_{i,\theta} F_T\left(t \left(\frac{k + 2i}{k}\right)^{1/2}; k + 2i, \mu, 0\right)$$

$$\mathbb{E}[T] = \mu \left(\frac{k}{2}\right)^{1/2} \frac{\Gamma(k/2 - 1/2)}{\Gamma(k/2)} \, {}_1F_1(1/2, n/2, -\theta/2), \quad k > 1$$

$$\mathbb{E}[T^2] = (1 + \mu^2) \frac{k}{(k-2)} \, {}_1F_1(1, k/2, -\theta/2), \quad k > 2$$

$$\mathbb{E}[T^s] = \left(\frac{k}{2}\right)^{s/2} \frac{\Gamma(k/2 - s/2)}{\Gamma(k/2)} \, {}_1F_1(s/2, k/2, -\theta/2) \sum_{i=0}^{\lfloor s/2 \rfloor} \binom{s}{2i} \frac{(2i)!}{2^i i!} \mu^{s-2i}$$

singly noncentral t, $\quad T \sim t'(k, \mu)$

$$f_T(t; \mu, k) = \frac{2^{-k/2+1} k^{k/2}}{\Gamma(k/2)\sqrt{2\pi}} \int_0^{\infty} z^k \exp\left\{-\frac{1}{2}\left[(tz - \mu)^2 + kz^2\right]\right\} dt$$

$$f_T = e^{-\mu^2/2} \frac{\Gamma((k+1)/2) k^{k/2}}{\sqrt{\pi}\,\Gamma(k/2)} \left(\frac{1}{k + t^2}\right)^{\frac{k+1}{2}} \left(\sum_{i=0}^{\infty} \frac{(t\mu)^i}{i!} \left(\frac{2}{t^2 + k}\right)^{i/2} \frac{\Gamma((k + i + 1)/2)}{\Gamma((k+1)/2)}\right)$$

$$F_T(t; k, \mu) = \frac{2^{-k/2+1} k^{k/2}}{\Gamma(k/2)} \int_0^{\infty} \Phi(tz; \mu, 1) z^{k-1} \exp\left\{-\frac{1}{2}kz^2\right\} dt$$

$$\mathbb{E}[T] = \left(\frac{k}{2}\right)^{1/2} \frac{\Gamma(k/2 - 1/2)}{\Gamma(k/2)} \mu, \quad \mathbb{E}[T^2] = \frac{k}{k-2}(1 + \mu^2)$$

$$\mathbb{E}[T^s] = \left(\frac{k}{2}\right)^{s/2} \frac{\Gamma(k/2 - s/2)}{\Gamma(k/2)} \sum_{i=0}^{\lfloor s/2 \rfloor} \binom{s}{2i} \frac{(2i)!}{2^i i!} \mu^{s-2i}$$

[a] Continued from Table A.9. See Chapter 10 (and its Problems and Solutions) for more detail.

Table A.11   Some relationships among major distributions[a]

Input	Function	Output	Parameters
$B_i \overset{\text{ind}}{\sim} \text{Bin}(n_i, p)$	$S = \sum B_i$	$\text{Bin}(n, p)$	$n = \sum n_i$
$X_i \overset{\text{ind}}{\sim} \text{NBin}(r_i, p)$	$S = \sum X_i$	$\text{NBin}(r, p)$	$r = \sum r_i$
$P_i \overset{\text{ind}}{\sim} \text{Poi}(\lambda_i)$	$S = \sum P_i$	$\text{Poi}(\lambda)$	$\lambda = \sum \lambda_i$
$X_i \overset{\text{iid}}{\sim} \text{Bin}(n, p)$	$X_1 \mid (X_1 + X_2 = s)$	$\text{HGeo}(n, n, s)$	
$X_i \overset{\text{iid}}{\sim} \text{Geo}(p)$	$X_1 \mid (X_1 + X_2 = s)$	$\text{DUnif}(0, s)$	
$X_i \overset{\text{ind}}{\sim} \text{Poi}(\lambda_i)$	$X_1 \mid (X_1 + X_2 = s)$	$\text{Bin}(s, p)$	$p = \lambda_1/(\lambda_1 + \lambda_2)$
$X, Y \overset{\text{iid}}{\sim} \text{Gum}(0, 1)$	$D = X - Y$	$\text{Log}(0, 1)$	
$X, Y \overset{\text{iid}}{\sim} \text{Exp}(\lambda)$	$D = X - Y$	$\text{Lap}(0, \lambda)$	
$X, Y \overset{\text{iid}}{\sim} \text{Gam}(a, b)$	$D = X - Y$	$\mathbb{M}_D(t) = \left(1 - t^2/b^2\right)^{-a}$	
$X_i \overset{\text{iid}}{\sim} N(0, 1)$	X_1/X_2	$\text{Cau}(0, 1)$	
$X_i \overset{\text{ind}}{\sim} N(\mu_i, \sigma_i^2)$	$S = \sum X_i$	$N(\mu, \sigma^2)$	$\mu = \sum \mu_i,$ $\sigma^2 = \sum \sigma_i^2$
$X_i \overset{\text{iid}}{\sim} N(0, 1)$	$S = \sum_{i=1}^{n} X_i^2$	$\chi^2(n)$	
$X_i \overset{\text{ind}}{\sim} N(\mu_i, \sigma_i^2)$	$S = \sum_{i=1}^{n} Z_i^2,$ $Z_i = \frac{X_i - \mu_i}{\sigma_i}$	$\chi^2(n)$	
$Z \sim N(0, 1) \perp C \sim \chi_n^2$	$T = Z/\sqrt{C/n}$	t_n	
$C_i \overset{\text{ind}}{\sim} \chi^2(\nu_i)$	$\sum_i^n C_i$	$\chi^2(\nu)$	$\nu = \sum_i^n \nu_i$
$C_i \overset{\text{ind}}{\sim} \chi_{n_i}^2$	$F = (C_1/n_1) / (C_2/n_2)$	$F(n_1, n_2)$	
$X_i \overset{\text{iid}}{\sim} \text{Exp}(\lambda)$	$S = \sum_i^n X_i$	$\text{Gam}(n, \lambda)$	
$G_i \overset{\text{ind}}{\sim} \text{Gam}(a_i, c)$	$S = \sum G_i$	$\text{Gam}(a, c)$	$a = \sum a_i$
$G_i \overset{\text{ind}}{\sim} \text{Gam}(a_i, c)$	$B = G_1/(G_1 + G_2)$	$\text{Beta}(a_1, a_2)$	
$B \sim \text{Beta}(n_1, n_2)$	$X = \dfrac{B}{n(1-B)},$ $n = n_1/n_2$	$F(2n_1, 2n_2)$	
$X \sim F(n_1, n_2)$	$B = \dfrac{nX}{1 + nX},$ $n = n_1/n_2$	$\text{Beta}(p, q)$	$p = n_1/2,$ $q = n_2/2$
$G_i \overset{\text{ind}}{\sim} \text{Gam}(n_i, s_i)$	$R = G_1/(G_1 + G_2)$	$\text{G3B}(n_1, n_2, s)$	$s = s_1/s_2$
$G_i \overset{\text{ind}}{\sim} \text{Gam}(n_i, s_i)$	$W = G_1/G_2$	$\text{G3F}(n_1, n_2, s)$	$s = s_1/s_2$

[a] In the case of $X, Y \overset{\text{iid}}{\sim} \text{Gam}(a, b)$ and $D = X - Y$, the m.g.f. of D is given, and an expression for the density is available; see case 7 and the footnote in Table A.12 below.

Table A.12 Some mixture relationships[a]

Case	Input	Distribution of X	Parameters
1	$(X \mid R = r) \sim \text{Poi}(r),$ $R \sim \text{Gam}(a, b)$	$\text{NBin}(a, p)$	$p = b/(b+1)$
2	$(X \mid \Theta = \theta) \sim \text{Exp}(\theta),$ $\Theta \sim \text{Gam}(b, c)$	$\text{Par II}(b, c)$	
3	$(X \mid \Theta = \theta) \sim \text{Gam}(a, \theta),$ $\Theta \sim \text{Gam}(b, c)$	$\text{GPar II}(a, b, c)$	
4	$(X \mid V = v) \sim N\left(0, v^{-1}\right),$ $kV \sim \chi_k^2$	Student's t_k	
5	$(X \mid V = v) \sim N\left(\mu v^{-1/2}, v^{-1}\right),$ $kV \sim \chi_k^2$	$t'(k, \mu)$	
6	$(X \mid V = v) \sim N(0, v),$ $V \sim \text{Exp}(\lambda)$	$\text{Lap}(0, k)$	$k = (2\lambda)^{-1/2}$
7	$(X \mid \Theta = \theta) \sim N(0, \theta)$ $\Theta \sim \text{Gam}(b, c)$	$\mathbb{M}_X(t) = \left(1 - t^2/(2c)\right)^{-b}$	$\lvert t \rvert < \sqrt{2c}$
8	$(X \mid \Theta = \theta) \sim N(m\theta, \theta)$ $\Theta \sim \text{Gam}(b, c)$	$\mathbb{M}_X(t) = \left(1 - \dfrac{mt}{c} - \dfrac{t^2}{2c}\right)^{-b}$	for t such that $mt/c + t^2/(2c) < 1$
9	$(X \mid P = p) \sim \text{Bin}(n, p),$ $P \sim \text{Beta}(a, b)$	$\text{BetaBin}(n, a, b)$	

[a]See Section 7.3.2 for derivation of these results. Expressions for the p.d.f. in cases 7 and 8 are derived in Examples 7.19 and 7.20, respectively. The beta-binomial distribution in case 9 is discussed in Problem 7.9. More results on normal mean–variance mixtures can be found in Table 9.3.

References

Aas, K. and Haff, I. H. (2005). NIG and skew Student's t: Two special cases of the generalised hyperbolic distribution, Note SAMBA/01/05, Norwegian Computing Center, Oslo.

Aas, K. and Haff, I. H. (2006). 'The generalized hyperbolic skew Student's t-distribution', *Journal of Financial Econometrics*, **4(2)**:275–309.

Abadir, K. M. (1999). 'An introduction to hypergeometric functions for economists', *Econometric Reviews*, **18(3)**:287–330.

Abadir, K. M. and Magnus, J. R. (2003). 'Problem 03.4.1. Normal's deconvolution and the independence of sample mean and variance', *Econometric Theory*, **19**:691.

Abadir, K. M. and Magnus, J. R. (2005). *Matrix Algebra*, Cambridge University Press, Cambridge.

Abramowitz, M. and Stegun, I. A. (1972). *Handbook of Mathematical Functions*, Dover, New York.

Adler, R. J., Feldman, R. E., and Taqqu, M. S. (eds.) (1998). *A Practical Guide to Heavy Tails*, Birkhäuser, Boston.

Ahuja, J. C. and Nash, S. W. (1967). 'The generalized Gompertz–Verhulst family of distributions', *Sankhyā, Series A*, **29**:141–156.

Andrews, G. E., Askey, R., and Roy, R. (1999). *Special Functions*, Cambridge University Press, Cambridge.

Ayebo, A. and Kozubowski, T. J. (2003). 'An asymmetric generalization of Gaussian and Laplace laws', *Journal of Probability and Statistical Science*, **1(2)**:187–210.

Azzalini, A. (1985). 'A class of distributions which includes the normal ones', *Scandinavian Journal of Statistics*, **12**:171–178.

Azzalini, A. (1986). 'Further results on a class of distributions which includes the normal ones', *Statistica*, **46(2)**:199–208, Errata: http://azzalini.stat.unipd.it/SN/errata86.pdf.

Azzalini, A. and Capitanio, A. (2003). 'Distributions generated by perturbation of symmetry with emphasis on a multivariate skew t-distribution', *Journal of the Royal Statistical Society, Series B*, **65**:367–389.

Azzalini, A. and Dalla Valle, A. (1996). 'The multivariate skew-normal distribution', *Biometrika*, **83(4)**:715–726.

Bachman, G., Narici, L., and Beckenstein, E. (2000). *Fourier and Wavelet Analysis*, Springer-Verlag, New York.

Balakrishnan, N. and Kocherlakota, S. (1985). 'On the double Weibull distribution: order statistics and estimation', *Sankhyā, Series B*, **47**:161–178.

Barndorff-Nielsen, O. E. (1977). 'Exponentially decreasing distributions for the logarithm of particle size', *Proceedings of the Royal Society of London, Series A*, **353**:401–419.

Barndorff-Nielsen, O. E. (1998). 'Processes of normal inverse Gaussian type', *Finance and Stochastics*, **2**:41–68.

Barndorff-Nielsen, O. E., Blæsild, P., Jensen, J. L., and Sørensen, M. (1985). The fascination of sand. In Atkinson, A. C. and Fienberg, D. E. (eds.), *A Celebration of Statistics*, pp. 57–87. Springer-Verlag, New York.

Barndorff-Nielsen, O. E. and Cox, D. R. (1989). *Asymptotic Techniques for Use in Statistics*, Chapman & Hall, London.

Barndorff-Nielsen, O. E. and Halgreen, C. (1977). 'Infinite divisibility of the hyperbolic and generalized inverse Gaussian distributions', *Zeitschrift für Wahrscheinlichkeitstheorie und Verwandte Gebiete*, **38(4)**:309–311.

Barndorff-Nielsen, O. E., Kent, J., and Sørensen, M. (1982). 'Normal variance–mean mixtures and z distributions', *International Statistical Review*, **50**:145–159.

Bean, M. A. (2001). *Probability: The Science of Uncertainty, with Applications to Investments, Insurance, and Engineering*, Brooks/Cole, Pacific Grove, CA.

Beerends, R. J., ter Morsche, H. G., van den Berg, J. C., and van de Vrie, E. M. (2003). *Fourier and Laplace Transforms*, Cambridge University Press, Cambridge.

Beirlant, J., Goegebeur, Y., Segers, J., and Teugels, J. (2004). *Statistics of Extremes: Theory and Applications*, John Wiley & Sons, Inc., Hoboken, NJ.

Beirlant, J., Teugels, J. L., and Vynckier, P. (1996). *Practical Analysis of Extreme Values*, Leuven University Press, Leuven, Belgium.

Berger, M. A. (1993). *An Introduction to Probability and Stochastic Processes*, Springer-Verlag, New York.

Bernardo, J. M. and Smith, A. F. M. (1994). *Bayesian Theory*, John Wiley & Sons, Ltd, Chichester.

Bhatti, M. I. (1995). 'Optimal testing for equicorrelated linear regression models', *Statistical Papers*, **36**:299–312.

Bibby, B. M. and Sørensen, M. (2003). Hyperbolic processes in finance. In Rachev, S. T. (ed.), *Handbook of Heavy Tailed Distributions in Finance*, pp. 211–248. Elsevier Science, Amsterdam.

Bierens, H. J. (2004). *Introduction to the Mathematical and Statistical Foundations of Econometrics*, Cambridge University Press, Cambridge.

Billingsley, P. (1995). *Probability and Measure*, third edn., John Wiley & Sons, Inc., New York.

Bollerslev, T., Engle, R. F., and Nelson, D. B. (1994). ARCH models. In Engle, R. and McFadden, D. (eds.), *Handbook of Econometrics*, Volume 4, Chapter 49. Elsevier Science, Amsterdam.

Bowman, K. O., Shenton, L. R., and Gailey, P. C. (1998). 'Distribution of the ratio of gamma variates', *Communications in Statistics – Simulation and Computation*, **27(1)**:1–19.

Box, G. E. P. and Tiao, G. C. (1973, 1992), *Bayesian Inference in Statistical Analysis*, John Wiley & Sons, Inc., New York.

Boys, R. J. (1989). 'Algorithm AS R80 – a remark on Algorithm AS76: An integral useful in calculating noncentral-t and bivariate normal probabilities', *Applied Statistics*, **38**:580–582.

Broda, S. and Paolella, M. S. (2007). 'Saddlepoint approximations for the doubly noncentral t distribution', *Computational Statistics and Data Analysis*, **51**:2907–2918.

Brooks, C., Clare, A. D., Dalle-Molle, J. W., and Persand, G. (2005). 'A comparison of extreme value theory approaches for determining value at risk', *Journal of Empirical Finance*, **12**:339–352.

Bryc, W. (1995). *The Normal Distribution: Characterizations with Applications*, Springer-Verlag, New York, See also: http://math.uc.edu/~brycw/probab/books/.

Butler, R. J., McDonald, J. B., Nelson, R. D., and White, S. B. (1990). 'Robust and partially adaptive estimation of regression models', *Review of Economics and Statistics*, **72**:321–327.

Butler, R. W. (2007). *An Introduction to Saddlepoint Methods*, Cambridge University Press, Cambridge.

Butler, R. W., Huzurbazar, S., and Booth, J. G. (1992). 'Saddlepoint approximations for the generalized variance and Wilk's statistic', *Biometrika*, **79(1)**:157–169.

Butler, R. W. and Paolella, M. S. (1999). Calculating the density and distribution function for the singly and doubly noncentral F, Working Paper 120/99, Institute of Statistics and Econometrics, Christian Albrechts University, Kiel, Germany.

Butler, R. W. and Paolella, M. S. (2002). 'Calculating the density and distribution function for the singly and doubly noncentral F', *Statistics and Computing*, **12**:9–16.

Butler, R. W. and Wood, A. T. A. (2002). 'Laplace approximations for hypergeometric functions with matrix arguments', *Annals of Statistics*, **30**:1155–1177.

Cain, M. (1994). 'The moment-generating function of the minimum of bivariate normal random variables', *American Statistician*, **48(2)**:124–125.

Carr, P. and Madan, D. B. (1999). 'Option valuation using the fast Fourier transform', *Journal of Computational Finance*, **2(4)**:61–73.

Casella, G. and Berger, R. L. (1990). *Statistical Inference*, Wadsworth & Brooks/Cole, Pacific Grove, CA.

Castillo, E. and Galambos, J. (1989). 'Conditional distributions and the bivariate normal distribution', *Metrika*, **36**:209–214.

Černý, A. (2004). *Mathematical Techniques in Finance*, Princeton University Press, Princeton, NJ.

Chambers, J. M., Mallows, C. L., and Stuck, B. W. (1976). 'A method for simulating stable random variables', *Journal of the American Statistical Association*, **71**:340–344.

Chen, G. and Adatia, A. (1997). 'Independence and t distribution', *American Statistician*, **51(2)**: 176–177.

Choi, P., Nam, K., and Arize, A. C. (2007). Flexible Multivariate GARCH Modeling with an Asymmetric and Leptokurtic Distribution, *To Appear in: Studies in Nonlinear Dynamics & Econometrics* Technical report.

Chow, S. C. and Shao, J. (1990). 'On the difference between the classical and inverse methods of calibration', *Applied Statistics*, **39**:219–228.

Christiansen, C. and Hartmann, D. (1991). The hyperbolic distribution. In Syvitski, J. P. M. (ed.), *Principles, Methods and Application of Particle Size Analysis*. Cambridge University Press, Cambridge.

Coelho, C. A. (1998). 'The generalized integer gamma distribution–a basis for distributions in multivariate statistics', *Journal of Multivariate Analysis*, **64**:86–102, Addendum, 1999, **69**:281–5.

Coles, S. (2001). *An Introduction to Statistical Modeling of Extreme Values*, Springer-Verlag, London.

Cressie, N., Davis, A. S., Folks, J. L., and Policello II, G. E. (1981). 'The moment-generating function and negative integer moments', *American Statistician*, **35**:148–150.

Dalla Valle, A. (2004). The skew-normal distribution. In Genton, M. G. (ed.), *Skew-Elliptical Distributions and their Applications: A Journey beyond Normality*, Chapter 1. Chapman & Hall/CRC, Boca Raton, FL.

Daniels, H. E. (1954). 'Saddlepoint approximation in statistics', *Annals of Mathematical Statistics*, **25**:631–650.

Daniels, H. E. (1987). 'Tail Probability Approximation', *International Statistical Review*, **55**: 37–48.

Daniels, H. E. and Young, G. A. (1991). 'Saddlepoint approximation for the Studentized mean, with an application to the bootstrap', *Biometrika*, **78**:169–179.

David, H. A. (1981). *Order Statistics*, second edn., John Wiley & Sons, Inc., New York.

Davies, R. B. (1973). 'Numerical inversion of a characteristic function', *Biometrika*, **60(2)**: 415–417.

Davies, R. B. (1980). 'Algorithm AS 155. The distribution of a linear combination of χ^2 random variables', *Applied Statistics*, **29**:323–333.

De Vany, A. S. and Walls, W. D. (2004). 'Motion picture profit, the stable Paretian hypothesis, and the curse of the superstar', *Journal of Economic Dynamics and Control*, **28(6)**:1035–1057.

DeBenedictis, L. F. and Giles, D. E. A. (1998). Diagnostic testing in econometrics, variable addition, RESET and Fourier approximations. In Ullah, A. and Giles, D. E. A. (eds.), *Handbook of Applied Economic Statistics*, pp. 383–417. Marcel Dekker, New York.

Dennis III, S. Y. (1994). 'On the distribution of products of independent beta variates', *Communications in Statistics – Theory and Methods*, **23(7)**:1895–1913.

DiCiccio, T. J. and Martin, M. A. (1991). 'Approximations of marginal tail probabilities for a class of smooth functions with applications to Bayesian and conditional inference', *Biometrika*, **78**:891–902.

Dubey, S. D. (1968). 'A compound Weibull distribution', *Naval Research Logistics Quarterly*, **15**:179–188.

Durbin, J. and Watson, G. S. (1971). 'Testing for serial correlation in least squares regression. III', *Biometrika*, **58**:1–19.

Dyer, D. (1982). 'The convolution of generalized F distributions', *Journal of the American Statistical Association*, **77**:184–189.

Dyke, P. P. G. (2001). *An Introduction to Laplace Transforms and Fourier Series*, Springer-Verlag, New York.

Eberlein, E. and von Hammerstein, E. A. (2004). Generalized hyperbolic and inverse Gaussian distributions: limiting cases and approximation of processes. In Dalang, R. C., Dozzi, M., and Russo, F. (eds.), *Seminar on Stochastic Analysis, Random Fields and Applications IV, Progress in Probability*, Volume 58, pp. 221–264. Birkhäuser Verlag, Boston.

Embrechts, P., Klüppelberg, C., and Mikosch, T. (2000). *Modelling Extremal Events for Insurance and Finance*, Springer-Verlag, New York.

Everitt, B. S. and Hand, D. J. (1981). *Finite Mixture Distributions*, Chapman & Hall, London.

Faddy, M. J. (2002). 'Review of "Stochastic Processes: An Introduction", by P. W. Jones and P. Smith', *Biometrics*, **58**:1043.

Fan, D. Y. (1991). 'The distribution of the product of independent beta variables', *Communications in Statistics – Theory and Methods*, **20**:4043–4052.

Fang, K.-T., Kotz, S., and Ng, K.-W. (1990). *Symmetric Multivariate and Related Distributions*, Chapman & Hall, London.

Farebrother, R. W. (1980a), 'Algorithm AS 153: Pan's procedure for the tail probabilities of the Durbin–Watson statistic', *Applied Statistics*, **29(2)**:224–227.

Farebrother, R. W. (1980b), 'The Durbin–Watson test for serial correlation when there is no intercept in the regression', *Econometrica*, **48(6)**:1553–1555, Correction: 1981, **49**:277.

Farebrother, R. W. (1990). 'The distribution of a quadratic form in normal variables', *Applied Statistics*, **39**:294–309.

Feller, W. (1968). *An Introduction to Probability Theory and Its Applications*, third edn., Volume I, John Wiley & Sons, Inc., New York.

Feller, W. (1971). *An Introduction to Probability Theory and Its Applications*, second edn., Volume II, John Wiley & Sons, Inc., New York.

Fernández, C., Osiewalski, J., and Steel, M. F. (1995). 'Modelling and inference with V-spherical distributions', *Journal of the American Statistical Association*, **90**:1331–1340.

Fernández, C. and Steel, M. F. J. (1998). 'On Bayesian modelling of fat tails and skewness', *Journal of the American Statistical Association*, **93**:359–371.

Field, C. and Ronchetti, E. (1990). *Small Sample Asymptotics*, Institute of Mathematical Statistics, Hayward, CA.

Finkenstädt, B. and Rootzén, H. (eds.) (2004). *Extreme Values in Finance, Telecommunications, and the Environment*, Chapman & Hall/CRC, Boca Raton, FL.

Fisher, R. A. (1921). 'On the 'probable error' of a coefficient of correlation deduced from a small sample', *Metron*, **1**:3–32, Reprinted in *Collected Papers of R. A. Fisher*, University of

Adelaide, 1971. The collected papers of R. A. Fisher (relating to statistical and mathematical theory and applications) can also be viewed on line from the University of Adelaide Library, http://www.library.adelaide.edu.au/digitised/fisher/stat_math.html.

Garvin, J. B. (1997). Particle size distribution analysis of Mars Lander sites, 28th Lunar and Planetary Science Conference, LPI Contribution No. 1090.

Gassmann, H. I., Deák, I., and Szántai, T. (2002). 'Computing multivariate normal probabilities: a new look', *Journal of Computational and Graphical Statistics*, **11**:920–949.

Geary, R. C. (1944). 'Extension of a theorem by Harald Cramér on the frequency distribution of the quotient of two variables', *Journal of the Royal Statistical Society*, **17**:56–57.

Geary, R. C. (1947). 'Testing for normality', *Biometrika*, **34**:209–242.

Genton, M. G. (ed.) (2004). *Multivariate t Distributions and Their Applications*, CRC Press, Boca Raton, FL.

Genz, A. (1992). 'Numerical computation of the multivariate normal probabilities', *Journal of Computational and Graphical Statistics*, **1**:141–150.

Genz, A. (2004). 'Numerical computation of rectangular bivariate and trivariate normal and t probabilities', *Statistics and Computing*, **14**:251–260.

Gil-Peleaz, J. (1951). 'Note on the inversion theorem', *Biometrika*, **38**:481–482.

Giot, P. and Laurent, S. (2004). 'Modelling daily value-at-risk using realized volatility and ARCH type models', *Journal of Empirical Finance*, **11**:379–398.

Grad, A. and Solomon, H. (1955). 'Distribution of quadratic forms and some applications', *Annals of Mathematical Statistics*, **26**:464–77.

Gradshteyn, L. S. and Ryzhik, I. M. (2007). *Tables of Integrals, Series and Products*, seventh edn., Academic Press, New York.

Graybill, F. A. (1976). *Theory and Application of the Linear Model*, Duxbury Press, North Scituate, MA.

Graybill, F. A. (1983). *Matrices with Applications in Statistics*, Wadsworth, Pacific Grove, CA.

Greenberg, E. and Webster, Jr., C. E. (1983). *Advanced Econometrics: A Bridge to the Literature*, John Wiley & Sons, Inc., New York.

Greene, W. H. (1990). 'A gamma-distributed stochastic frontier model', *Journal of Econometrics*, **46**:141–163.

Gupta, R. C. and Ong, S. H. (2004). 'A new generalization of the negative binomial distribution', *Computational Statistics and Data Analysis*, **45**:287–300.

Gupta, S. S. (1963). 'Probability integrals of multivariate normal and multivariate t', *Annals of Mathematical Statistics*, **34(3)**:792–828.

Gut, A. (1995). *An Intermediate Course in Probability*, Springer-Verlag, New York.

Gut, A. (2005). *Probability: A Graduate Course*, Springer-Verlag, New York.

Ha, D. M. (2006). *Functional Analysis, Volume I: A Gentle Introduction*, Matrix Editions, Ithaca, NY.

Haas, M., Mittnik, S., and Paolella, M. S. (2004a). Mixed normal conditional heteroskedasticity. *Journal of Financial Econometrics*, **2(2)**:211–250.

Haas, M., Mittnik, S., and Paolella, M. S. (2004b). A new approach to Markov-switching GARCH models. *Journal of Financial Econometrics*, **2(4)**:493–530.

Haas, M., Mittnik, S., and Paolella, M. S. (2006). 'Modeling and predicting market risk with Laplace-Gaussian mixture distributions', *Applied Financial Economics*, **16**:1145–1162.

Hajivassiliou, V., McFadden, D., and Ruud, P. (1996). 'Simulation of multivariate normal orthant probabilities: methods and programs', *Journal of Econometrics*, **72**:85–134.

Hamouda, O. and Rowley, R. (1996). *Probability in Economics*, Routledge, London.

Hansen, B. E. (1994). 'Autoregressive conditional density estimation', *International Statistical Review*, **35(3)**:705–730.

Havil, J. (2003). *Gamma: Exploring Euler's Constant*, Princeton University Press, Princeton, NJ.

Hawkins, D. M. (1975). 'From the noncentral t to the normal integral', *American Statistician*, **29**:42–43.

Heijmans, R. (1999). 'When does the expectation of a ratio equal the ratio of expectations?', *Statistical Papers*, **40**:107–115.

Helstrom, C. W. (1996). 'Calculating the distribution of the serial correlation estimator by saddle-point integration', *Econometric Theory*, **12**:458–80.

Helstrom, C. W. and Ritcey, J. A. (1985). 'Evaluation of the noncentral F-distribution by numerical contour integration', *SIAM Journal on Scientific and Statistical Computing*, **6(3)**:505–514.

Henze, N. (1986). 'A probabilistic representation of the 'skew-normal' distribution', *Scandinavian Journal of Statistics*, **13**:271–275.

Hijab, O. (1997). *Introduction to Calculus and Classical Analysis*, Springer-Verlag, New York.

Hinkley, D. V. (1969). 'On the ratio of two correlated random variables', *Biometrika*, **56(3)**: 635–639.

Hoffmann-Jørgensen, J. (1994). *Probability with a View towards Statistics, Volume I*, Chapman & Hall, New York.

Imhof, J. P. (1961). 'Computing the distribution of quadratic forms in normal variables', *Biometrika*, **48**:419–26.

Jacod, J. and Protter, P. (2000). *Probability Essentials*, Springer-Verlag, Berlin.

Janicki, A. and Weron, A. (1994). *Simulation and Chaotic Behavior of α-Stable Stochastic Processes*, Marcel Dekker, New York.

Jensen, J. L. (1995). *Saddlepoint Approximations*, Oxford University Press, Oxford.

Jørgensen, B. (1982). *Statistical Properties of the Generalized Inverse Gaussian Distribution*, Lecture Notes in Statistics 9. Springer-Verlag, New York.

Johnson, N. L. (1949). 'Systems of frequency curves generated by method of translation', *Biometrika*, **36**:149–176.

Johnson, N. L. and Kotz, S. (1977). *Urn Models and Their Applications*, John Wiley & Sons, Inc., New York.

Johnson, N. L., Kotz, S., and Balakrishnan, N. (1994, 1995), *Continuous Univariate Distributions, Volumes 1 and 2*, second edn., John Wiley & Sons, Inc., New York.

Jones, F. (2001). *Lebesgue Integration on Euclidean Space*, revised edn., Jones & Bartlett, Boston.

Jones, M. C. and Balakrishnan, N. (2002). 'How are moments and moments of spacings related to distribution functions?', *Journal of Statistical Planning and Inference*, **103**:377–90.

Jones, M. C. and Faddy, M. J. (2003). 'A skew extension of the t distribution with applications', *Journal of the Royal Statistical Society, Series B*, **65**:159–174.

Jones, P. N. and McLachlan, G. J. (1990). 'Laplace-normal mixtures fitted to wind shear data', *Journal of Applied Statistics*, **17**:271–276.

Jones, P. W. and Smith, P. (2001). *Stochastic Processes: An Introduction*, Oxford University Press, Oxford.

Kafadar, K. (2001). 'In memoriam: John Wilder Tukey, June 16, 1915–July 26, 2000', *Technometrics*, **43(3)**:251–255.

Kamara, A. and Siegel, A. F. (1987). 'Optimal hedging in futures markets with multiple delivery specifications', *Journal of Finance*, **42(4)**:1007–1021.

Kanji, G. K. (1985). 'A mixture model for wind shear data', *Journal of Applied Statistics*, **12**: 49–58.

Kao, E. P. C. (1996). *An Introduction to Stochastic Processes*, Duxbury Press, Wadsworth, Belmont, CA.

Karian, Z. A. and Dudewicz, E. J. (2000). *Fitting Statistical Distributions – The Generalized Lambda Distribution and Generalized Bootstrap Methods*, CRC Press, Boca Raton, FL.

Karr, A. F. (1993). *Probability*, Springer-Verlag, New York.

Kass, R. E. (1988). 'Comment on: Saddlepoint Methods and Statistical Inference', *Statistical Science*, **3**:234–236.

Kleiber, C. and Kotz, S. (2003). *Statistical Size Distributions in Economics and Actuarial Sciences*, John Wiley & Sons, Inc., Hoboken, NJ.

Knautz, H. and Trenkler, G. (1993). 'A note on the correlation between S^2 and the least squares estimator in the linear regression model', *Statistical Papers*, **34**:237–246.

Knüsel, L. (1995). 'On the accuracy of the statistical distributions in GAUSS', *Computational Statistics and Data Analysis: Statistical Software Newsletter*, **20(6)**:699–702.

Knüsel, L. and Bablok, B. (1996). 'Computation of the noncentral gamma function', *SIAM Journal of Scientific Computing*, **17**:1224–1231.

Kocherlakota, K. and Kocherlakota, S. (1991). 'On the doubly noncentral t distribution', *Communications in Statistics –Simulation and Computation*, **20(1)**:23–32.

Koerts, J. and Abrahamse, A. P. J. (1969). *On the Theory and Application of the General Linear Model*, University Press, Rotterdam.

Kolassa, J. E. (1997). *Series Approximation Methods in Statistics*, second edn., Springer-Verlag, New York.

Kolassa, J. E. (2003). 'Multivariate saddlepoint tail probability approximations', *Annals of Statistics*, **31(1)**:274–286.

Komunjer, I. (2006). Asymmetric power distribution: theory and applications to risk measurement, Mimeo, University of California, San Diego.

Kotz, S., Balakrishnan, N., and Johnson, N. L. (2000). *Continuous Multivariate Distributions, Volume 1, Models and Applications*, second edn., John Wiley & Sons, Inc., New York.

Kotz, S. and Nadarajah, S. (2004). *Multivariate t Distributions and Their Applications*, Cambridge University Press, Cambridge.

Kotz, S., Podgorski, K., and Kozubowski, T. (2001). *The Laplace Distribution and Generalizations: A Revisit with Application to Communication, Economics, Engineering and Finance*, Birkhäuser, Boston.

Kowalski, C. (1973). 'Non-normal bivariate distributions with normal marginals', *American Statistician*, **27**:103–106.

Krishnan, M. (1967). 'The moments of a doubly noncentral t distribution', *Journal of the American Statistical Association*, **62**:278–287.

Krzanowski, W. J. and Marriott, F. H. C. (1994). *Multivariate Analysis, Part 1: Distributions, Ordination and Inference*, Edward Arnold, London.

Kuester, K., Mittik, S., and Paolella, M. S. (2006). 'Value-at-risk prediction: a comparison of alternative strategies', *Journal of Financial Econometrics*, **4(1)**:53–89.

Kuonen, D. (2000). 'A saddlepoint approximation for the collector's problem', *American Statistician*, **54(3)**:165–169.

Kurumai, H. and Ohtani, K. (1998). 'The exact distribution and density functions of a pre-test estimator of the error variance in a linear regression model with proxy variables', *Statistical Papers*, **39**:163–177.

Kuruoglu, E. E. (2001). 'Density parameter estimation of skewed α-stable distributions', *IEEE Transactions on Signal Processing*, **49(10)**:2192–2201.

Lange, K. (2003). *Applied Probability*, Springer-Verlag, New York.

Le Cam, L. (1990). 'Maximum likelihood: an introduction', *International Statistical Review*, **58(2)**: 153–171.

Lebedev, N. N. (1972). *Special Functions and Their Applications*, Dover, Mineola, NY.

Lee, J. C., Lee, C. F., and Wei, K. C. J. (1991). 'Binomial option pricing with stochastic parameters: a beta distribution approach', *Review of Quantitative Finance and Accounting*, **1**:435–448.

Lenth, R. V. (1987). 'Computing noncentral beta probabilities', *Applied Statistics*, **36**:241–244.

Lévy, P. (1925). *Calcul des probabilités*, Gauthier-Villars, Paris.

Light, W. A. (1990). *An Introduction to Abstract Analysis*, Chapman & Hall, London.

Liu, J. S. (1994). 'Siegel's formula via Stein's identities', *Statistics and Probability Letters*, **21**:247–251.

Lugannani, R. and Rice, S. O. (1980). 'Saddlepoint approximations for the distribution of sums of independent random variables', *Advances in Applied Probability*, **12**:475–490.

Lukacs, E. (1970). *Characteristic Functions*, second edn., Charles Griffin, London.

Lye, J. N. and Martin, V. L. (1993). 'Robust estimation, nonnormalities, and generalized exponential distributions', *Journal of the American Statistical Association*, **88(421)**:261–267.

Madan, D. B. and Seneta, E. (1990). 'The variance gamma (V.G.) model for share market returns', *Journal of Business*, **63**:511–524.

Mandelbrot, B. (1960). 'The Pareto–Levy law and the distribution of income', *International Economic Review*, **1**:79–106.

Mann, H. B. and Wald, A. (1943). 'On stochastic limit and order relationships', *Annals of Mathematical Statistics*, **14**:217–226.

Marsh, P. W. N. (1998). 'Saddlepoint approximations and non-central quadratic forms', *Econometric Theory*, **14**:539–559.

McCullough, B. D. (2000). 'The accuracy of Mathematica 4 as a statistical package', *Computational Statistics*, **15**:279–299.

McDonald, J. B. (1991). 'Parametric models for partially adaptive estimation with skewed and leptokurtic residuals', *Economics Letters*, **37**:273–278.

McDonald, J. B. (1997). Probability distributions for financial models. In Maddala, G. S. and Rao, C. R. (eds.), *Handbook of Statistics*, Volume 14. Elsevier Science, Amsterdam.

McDonald, J. B. and Newey, W. K. (1988). 'Partially adaptive estimation of regression models via the generalized t distribution', *Econometric Theory*, **4**:428–457.

McLachlan, G. J. and Peel, D. (2000). *Finite Mixture Models*, John Wiley & Sons, Inc., New York.

Mehta, J. S. and Swamy, P. A. V. B. (1978). 'The existence of moments of some simple Bayes estimators of coefficients in a simultaneous equation model', *Journal of Econometrics*, **7**:1–13.

Melnick, E. L. and Tenenbein, A. (1982). 'Misspecifications of the normal distribution', *American Statistician*, **36**:372–373.

Mencía, F. J. and Sentana, E. (2004). Estimation and testing of dynamic models with generalised hyperbolic innovations, CEMFI Working Paper 0411, Madrid.

Meng, X.-L. (2005). 'From unit root to Stein's estimator to Fisher's k statistics: if you have a moment, I can tell you more', *Statistical Science*, **20(2)**:141–162.

Menn, C. and Rachev, S. T. (2006). 'Calibrated FFT-based density approximations for α-stable distributions', *Computational Statistics & Data Analysis*, **50**:1891–1904.

Mittnik, S., Doganoglu, T., and Chenyao, D. (1999). 'Computing the probability density function of the stable Paretian distribution', *Mathematical and Computer Modelling*, **29**:235–240.

Mittnik, S. and Paolella, M. S. (2000). 'Conditional density and value-at-risk prediction of Asian currency exchange rates', *Journal of Forecasting*, **19**:313–333.

Mittnik, S., Paolella, M. S., and Rachev, S. T. (1998). 'Unconditional and conditional distributional models for the Nikkei index', *Asia-Pacific Financial Markets*, **5(2)**:99–128.

Mittnik, S., Paolella, M. S., and Rachev, S. T. (2002). 'Stationarity of stable power-GARCH processes', *Journal of Econometrics*, **106**:97–107.

Mudholkar, G. S., Freimer, M., and Hutson, A. D. (1997). 'On the efficiencies of some common quick estimators', *Communications in Statistics – Theory and Methods*, **26(7)**:1623–1647.

Mullen, K. (1967). 'A note on the ratio of two independent random variables', *American Statistician*, **21**:30–31.

Nikias, C. L. and Shao, M. (1995). *Signal Processing with Alpha-Stable Distributions and Applications*, Wiley-Interscience, New York.

Nolan, J. P. (2007). *Stable Distributions – Models for Heavy Tailed Data*, Birkhäuser, Boston, To appear. For a preview, see http://academic2.american.edu/~jpnolan/stable/stable.html.

O'Hagan, A. and Leonard, T. (1976). 'Bayes estimation subject to uncertainly about parameter constraints', *Biometrika*, **63**:201–203.

Pace, L. and Salvan, A. (1997). *Principles of Statistical Inference from a Neo-Fisherian Perspective*, World Scientific, Singapore.

Palka, B. P. (1991). *An Introduction to Complex Function Theory*, Springer-Verlag, New York.

Pan, J.-J. (1968). 'Distribution of the noncircular serial correlation coefficients', *IMS & AMS Selected Translations in Mathematical Statistics and Probability*, 7:281–292.

Paolella, M. S. (2004). 'Modeling higher frequency macroeconomic data: an application to German monthly money demand', *Applied Economics Quarterly*, 50(2):113–138.

Papageorgiou, H. (1997). Multivariate discrete distributions. In Kotz, S., Read, C. B., and Banks, D. L. (eds.), *Encyclopedia of Statistical Sciences, Update Volume 1*. John Wiley & Sons, Inc., New York.

Pham-Gia, T. and Turkkan, N. (2002). 'The product and quotient of general beta distributions', *Statistical Papers*, 43:537–550.

Pierce, D. A. and Dykstra, R. L. (1969). 'Independence and the normal distribution', *American Statistician*, 23:39.

Prause, K. (1999). *The Generalized Hyperbolic Model: Estimation, Financial Derivatives, and Risk Measures*, PhD thesis, Albert-Ludwigs-Universität Freiburg.

Prentice, R. L. (1975). 'Discrimination among some parametric models', *Biometrika*, 62:607–14.

Price, R. (1964). 'Some non-central *F*-distributions expressed in closed form', *Biometrika*, 51(1–2):107–122.

Rachev, S. T. (ed.) (2003). *Handbook of Heavy Tailed Distributions in Finance*, Elsevier Science, Amsterdam.

Rachev, S. T. and Mittnik, S. (2000). *Stable Paretian Models in Finance*, John Wiley & Sons, Inc., New York.

Ramberg, J. and Schmeiser, B. (1974). 'An approximate method for generating asymmetric random variables', *Communications of the ACM*, 17:78–82.

Ramsey, J. B. (1969). 'Tests for specification errors in classical linear least-squares regression analysis', *Journal of the Royal Statistical Society, Series B*, 31:350–371.

Randall, J. H. (1994). 'Calculating central and non-central *F* probabilities', *South African Journal of Statistics*, 28:67–72.

Rao, C. R. (1973). *Linear Statistical Inference and Its Applications*, second edn., John Wiley & Sons, Inc., New York.

Reid, N. (1988). 'Saddlepoint methods and statistical inference (with discussion)', *Statistical Science*, 3:213–238.

Reid, N. (1996). 'Likelihood and higher-order approximations to tail areas: a review and annotated bibliography', *Canadian Journal of Statistics*, 24(2):141–166.

Reid, N. (1997). Asymptotic expansions. In Kotz, S., Read, C. B., and Banks, D. L. (eds.), *Encyclopedia of Statistical Sciences, Update Volume 1*. John Wiley & Sons, Inc., New York.

Resnick, S. and Rootzén, H. (2000). 'Self-similar communication models and very heavy tails', *Annals of Applied Probability*, 10(3):753–778.

Resnick, S. I. (1992). *Adventures in Stochastic Processes*, Birkhäuser, Boston.

Resnick, S. I. (1999). *A Probability Path*, Birkhäuser, Boston.

Riordan, J. (1968). *Combinatorical Identities*, John Wiley & Sons, Inc., New York.

Rohatgi, V. K. (1984). *Statistical Inference*, John Wiley & Sons, Inc., New York.

Rohatgi, V. K. and Saleh, A. K. (2001). *An Introduction to Probability and Statistics*, second edn., John Wiley & Sons, Inc., New York.

Romanowski, M. (1979). *Random Errors in Observation and the Influence of Modulation on their Distribution*, Verlag Konrad Wittwer, Stuttgart.

Ross, S. (1988). *A First Course in Probability*, third edn., Macmillan, New York.

Ross, S. (1997). *Introduction to Probability Models*, sixth edn., Academic Press, San Diego, CA.

Ross, S. (2006). *A First Course in Probability*, seventh edn., Pearson Prentice Hall, Upper Saddle River, NJ.

Roussas, G. G. (1997). *A Course in Mathematical Statistics*, second edn., Academic Press, San Diego, CA.

Samorodnitsky, G. and Taqqu, M. S. (1994). *Stable Non-Gaussian Random Processes: Stochastic Models with Infinite Variance*, Chapman & Hall, London.

Sawa, T. (1972). 'Finite sample properties of the k-class estimator', *Econometrica*, **40(4)**:653–680.

Sawa, T. (1978). 'The exact moments of the least squares estimator for the autoregressive model', *Journal of Econometrics*, **8**:159–172.

Schader, M. and Schmid, F. (1986). 'Distribution function and percentage points for the central and noncentral F distribution', *Statistische Hefte*, **27**:67–74.

Scheffé, H. (1959). *The Analysis of Variance*, John Wiley & Sons, Inc., New York.

Schiff, J. L. (1999). *The Laplace Transform – Theory and Applications*, Springer-Verlag, New York.

Schilling, M. F., Watkins, A. E., and Watkins, W. (2002). 'Is human height bimodal?', *American Statistician*, **56(3)**:223–229.

Seneta, E. (1982). Bernstein, Sewrgei Natanovich. In Kotz, S. and Johnson, N. L. (eds.), *Encyclopedia of Statistical Sciences*, Volume 1. John Wiley & Sons, Inc., New York.

Seneta, E. (2004). 'Fitting the variance-gamma model to financial data', *Journal of Applied Probability*, **41A**:177–187.

Shephard, N. (1991). 'From characteristic function to distribution function: a simple framework for the theory', *Econometric Theory*, **7**:519–529.

Shiryaev, A. N. (1996). *Probability*, second edn., Springer-Verlag, New York.

Siegel, A. F. (1993). 'A surprising covariance involving the minimum of multivariate normal variables', *Journal of the American Statistical Association*, **88(421)**:77–80.

Skovgaard, I. M. (1987). 'Saddlepoint expansions for conditional distributions', *Journal of Applied Probability*, **24**:275–287.

Springer, M. D. (1978). *Algebra of Random Variables*, John Wiley & Sons, Inc., New York.

Srivastava, M. S. and Yao, W. K. (1989). 'Saddlepoint method for obtaining tail probability of Wilks' likelihood ratio test', *Journal of Multivariate Analysis*, **31**:117–126.

Steece, B. M. (1976). 'On the exact distribution for the product of two independent beta-distributed variables', *Metron*, **34**:187–190.

Steele, J. M. (2004). *The Cauchy–Schwarz Master Class: An Introduction to the Art of Mathematical Inequalities*, Cambridge University Press, Cambridge.

Stirzaker, D. (2003). *Elementary Probability*, second edn., Cambridge University Press, Cambridge.

Stoll, M. (2001). *Introduction to Real Analysis*, second edn., Addison-Wesley, Boston.

Stuart, A. and Ord, J. K. (1994). *Kendall's Advanced Theory of Statistics, Volume 1: Distributiuon Theory*, sixth edn., Edward Arnold, London.

Stuart, A., Ord, J. K., and Arnold, S. F. (1999). *Kendall's Advanced Theory of Statistics, Volume 2A: Classical Inference and the Linear Model*, sixth edn., Edward Arnold, London.

Subbotin, M. T. (1923). 'On the law of frequency of error', *Mathematicheskii Sbornik*, **31**:296–300.

Sungur, E. A. (1990). 'Dependence information in parameterized copulas', *Communications in Statistics – Simulation and Computation*, **19(4)**:1339–1360.

Tardiff, R. M. (1981). 'L'Hospital's rule and the central limit theorem', *American Statistician*, **35**:43.

Teichroew, D. (1957). 'The mixture of normal distributions with different variances', *Annals of Mathematical Statistics*, **28**:510–512.

Thompson, J. R. (2001). 'The age of Tukey', *Technometrics*, **43(3)**:256–265.

Tierney, L. (1988). 'Comment on: Saddlepoint Methods and Statistical Inference', *Statistical Science*, **3**:233–234.

Tiku, M. L. (1972). 'A note on the distribution of the doubly noncentral F distribution', *Australian Journal of Statistics*, **14(1)**:37–40.

Tiku, M. L. and Yip, D. Y. N. (1978). 'A four moment approximation based on the F distribution', *Australian Journal of Statistics*, **20(3)**:257–261.

Tippett, L. H. C. (1925). 'On the extreme individuals and the range of samples taken from a normal population', *Biometrika*, **17**:364–387.

Titterington, D. M., Smith, A. F. M., and Makov, U. E. (1985). *The Statistical Analysis of Finite Mixture Distributions*, John Wiley & Sons, Inc., New York.

Topp, C. W. and Leone, F. C. (1955). 'A family of *J*-shaped frequency functions', *Journal of the American Statistical Association*, **50**:209–219.

Tukey, J. W. (1962). 'The future of data analysis', *Annals of Mathematical Statistics*, **33**:1–67.

Uchaikin, V. V. and Zolotarev, V. M. (1999). *Chance and Stability, Stable Distributions and Their Applications*, VSP, Utrecht, Netherlands.

Vasicek, O. A. (1998). A series expansion for the bivariate normal integral, Working Paper 999-0000-043, KMV Corporation, San Francisco.

Vijverberg, W. P. M. (2000). 'Rectangular and wedge-shaped multivariate normal probabilities', *Economics Letters*, **68**:13–20.

Walls, W. D. (2005). 'Modelling heavy tails and skewness in film returns', *Applied Financial Economics*, **15(17)**:1181–1188.

Wang, S. (1990). 'Saddlepoint approximations for bivariate distributions', *Journal of Applied Probability*, **27**:586–597.

Wang, Y. H. (1993). 'On the number of successes in independent trials', *Statistica Sinica*, **3**: 295–312.

Weron, R. (1996). 'On the Chambers–Mallows–Stuck method for simulating skewed stable random variables', *Statistics & Probability Letters*, **28**:165–171.

Wilf, H. S. (1994). *generatingfunctionology*, Academic Press, San Diego, CA.

Wilks, S. S. (1963). *Mathematical Statistics (2nd Printing with Corrections)*, John Wiley & Sons, Inc., New York.

Young, M. S. and Graff, R. A. (1995). 'Real estate is not normal: a fresh look at real estate return distributions', *Journal of Real Estate Finance and Economics*, **10**:225–259.

Zielinski, R. (1999). 'A median-unbiased estimator of the AR(1) coefficient', *Journal of Time Series Analysis*, **20(4)**:477–481.

Zolotarev, V. M. (1986). *One-Dimensional Stable Distributions*, Translations of Mathematical Monographs 65. American Mathematical Society, Providence, RI, Translated from the original Russian verion (1983).

Index

Almost surely equal, 143

Banach's matchbox problem, 57, 160
Bayes' rule, 106
Berra, Yogi, 203
Bessel function
    modified, 302, 346
Bochner's theorem, 24
Borel–Cantelli lemmas, 141
Broda, Simon, 57, 62, 63, 388

Cantelli's inequality, 135
Cauchy–Schwarz inequality, 130
Central limit theorem, 158
Characteristic function, 17
Characterization, 11, 104
Chebyshev's inequality, 134
    one-sided, 135
    order, 135
Chernoff bound, 134
Chernoff's inequality, 134
Cholesky decomposition, 108, 124
Churchill, Sir Winston, 27, 124, 386
Closed under addition, 66
Confidence interval
    nonparametric, 209
Confluent hypergeometric function, 193, 347,
    362
Continuity correction, 167
Convergence
    almost surely, 146
    complete, 149
    in r-mean, 151
    in distribution, 154
    in probability, 143
Convolution, 60, 338
Correlation
    partial, 116
Covariance, 99
Cumulant generating function, 7

De Moivre–Jordan theorem, 230
Digamma function, 67, 95
Discrete Fourier transform, 40
Distribution
    beta
        generalized three-parameter, 272
    beta-binomial, 277
    binomial, 11, 66
        sum, 69
    bivariate
        with normal marginals, 104
    bivariate normal, 101, 102, 104, 108, 109
    Cauchy, 32, 58, 327
        asymmetric, 326
    consecutive, 9, 29
    Dirac, 317
    discrete uniform, 5
    exponential
        sum, 68
    Fernández–Steel
        generalized exponential, 247
        Student's t, 247
    G3B, 272
    G3F, 273
    G4B, 272

gamma
   difference, 72
   sum, 77
GAt, 275
generalized exponential
   GED, 240, 271
generalized gamma, 241
generalized lambda, 253
generalized logistic, 249, 274
generalized Student's t, 241
generalized (type II) Pareto, 241
Gumbel, 67
hyperbolic, 326
   asymmetric t, 323
   generalized, 317
   positive, 313
inverse gamma, 312
inverse Gaussian, 315
   generalized, 308
inverse hyperbolic sine, 252
Laplace, 34, 81, 306
   asymmetric, 325
Lévy, 26, 287, 314, 316
logistic, 9, 67, 274
   generalized, 274
mixture, 256, 305, 345
   continuous, 260
   countable, 258
   finite, 257
multinomial, 56, 98
multivariate normal, 97, 100
negative binomial, 66
noncentral
   beta, 371
   chi-square, 39, 343
   F, 359
   Student's t, 270, 372
normal
   bivariate, 101, 102, 104, 108, 109
   multivariate, 100
   ratio, 84
   skew, 246
normal inverse Gaussian, 327
normal–Laplace convolution, 81, 179, 271
Pareto
   Type II, 262
   Type III, 242
Poisson
   sum, 66

Pólya–Eggenberger, 240
stable Paretian, 281
Student's t, 79
   GAt, 275
   hyperbolic asymmetric, 323, 324
   Jones and Faddy, 248
   Lye–Martin, 247
   noncentral, 269
symmetric triangular, 93
Tukey lambda, 250
uniform
   sum, 71
variance–gamma, 267, 323
Weibull, 244, 261
   asymmetric double, 250
   double, 249

Einstein, Albert, 65, 256
Equal almost surely, 143
Equal in distribution, 11, 143
Equicorrelated, 120
Equivariance, 198
Euler formula, 18
Euler's constant, 67
Euler's reflection formula, 9
Exponential tilting, 171
Extreme value theory, 205
Extremes, 203

Fast Fourier transform, 40
Fourier transform, 61
Frontier function, 85

Geary's ratio result, 34
Generalized central limit theorem, 299
Generalized hypergeometric function, 193
Gramm–Schmidt process, 41

Haas, Markus, 57
Hölder's inequality, 131
Homoscedasticity, 115
Howler, 89
Hypergeometric function
   confluent, 193
   generalized, 193

Imhof's procedure, 356
Incomplete beta function, 207

Incomplete gamma function, 176
Inequality
Cantelli, 135
Cauchy–Schwarz, 130
Chebyshev, 134
one-sided, 135
order, 135
Chernoff, 134
Hölder, 131
Jensen's, 122, 130
Kolmogorov, 136
other, 136
Markov, 133
Minkowski, 131
triangle, 131
Infinite divisibility, 58
Infinite monkey theorem, 142
Infinitely divisible, 58, 338
Inverse Laplace transform, 22
Inversion formula, 27
Inversion formulae, 68

Jensen's inequality, 122, 130

Kolmogorov's inequality, 136
other, 136
Kummer transformation, 195, 362, 383

l'Hôpital's rule, 6, 166, 187, 287,
293
Laplace approximation, 171
Laplace transform, 22, 58, 92
inverse, 22
Leibniz' rule, 79, 81

MANOVA, 200
Markov's inequality, 133
Mean
approximation, 85
Mellin transform, 200
Midrange, 215
Minkowski's inequality, 131
Moment generating function, 3
Moments
multivariate, 97

Nonparametrics, 255
Normal variance mixture, 263
Null event, 143

Occupancy distributions, 69, 173
Order statistics, 203
Orthant probability, 109

Pan's procedure, 350
Paravicini, Walther, 62, 233, 301
Pólya, George, 158
Positive definite, 122
Probability integral transform, 254
Probability limit, 143

Quantile, 228

Range, 215
Regression function, 114
Runs, 9, 166

Saddlepoint
approximation, 171
c.d.f. approximation, 175
equation, 171
renormalized, 172
second-order approximations, 178
Sample midrange, 215
Sample range, 215
Sawa's ratio result, 15
Schur's decomposition theorem, 123
Semi-heavy tails, 332
Shape triangle, 334
Spectral decomposition, 123
Stirling's approximation, 13, 14, 162, 167,
172, 177, 369
Summability, 281

Tails
semi-heavy, 243
Time series, 125
Triangle inequality, 131
Tukey, John, 251

Uniqueness theorem, 24, 37

Variance
approximation, 85
of a sum, 99

Weak law of large numbers, 144

Zero–one law, 142